Next-Generation Wireless Networks Meet Advanced Machine Learning Applications

Ioan-Sorin Comşa
Brunel University London, UK

Ramona Trestian
Middlesex University, UK

A volume in the Advances in Wireless Technologies and Telecommunication (AWTT) Book Series

IGI Global
DISSEMINATOR OF KNOWLEDGE

Published in the United States of America by
IGI Global
Information Science Reference (an imprint of IGI Global)
701 E. Chocolate Avenue
Hershey PA, USA 17033
Tel: 717-533-8845
Fax: 717-533-8661
E-mail: cust@igi-global.com
Web site: http://www.igi-global.com

Library of Congress Cataloging-in-Publication Data

Names: Comsa, Ioan-Sorin, 1984- editor. | Trestian, Ramona, 1983- editor.
Title: Next-generation wireless networks meet advanced machine learning applications / Ioan-Sorin Comsa and Ramona Trestian, editors.
Description: Hershey, PA : Information Science Reference, an imprint of IGI Global, [2019] | Includes bibliographical references and index.
Identifiers: LCCN 2018028321| ISBN 9781522574583 (hardcover) | ISBN 9781522574590 (ebook)
Subjects: LCSH: Cognitive radio networks. | Wireless communication systems--Automatic control. | Self-organizing systems. | Machine learning.
Classification: LCC TK5103.4815 .N49 2019 | DDC 006.3/1--dc23 LC record available at https://lccn.loc.gov/2018028321

This book is published in the IGI Global book series Advances in Wireless Technologies and Telecommunication (AWTT) (ISSN: 2327-3305; eISSN: 2327-3313)

British Cataloguing in Publication Data
A Cataloguing in Publication record for this book is available from the British Library.

All work contributed to this book is new, previously-unpublished material. The views expressed in this book are those of the authors, but not necessarily of the publisher.

For electronic access to this publication, please contact: eresources@igi-global.com.

Advances in Wireless Technologies and Telecommunication (AWTT) Book Series

Xiaoge Xu
University of Nottingham Ningbo China, China

ISSN:2327-3305
EISSN:2327-3313

Mission

The wireless computing industry is constantly evolving, redesigning the ways in which individuals share information. Wireless technology and telecommunication remain one of the most important technologies in business organizations. The utilization of these technologies has enhanced business efficiency by enabling dynamic resources in all aspects of society.

The **Advances in Wireless Technologies and Telecommunication Book Series** aims to provide researchers and academic communities with quality research on the concepts and developments in the wireless technology fields. Developers, engineers, students, research strategists, and IT managers will find this series useful to gain insight into next generation wireless technologies and telecommunication.

Coverage

- Radio Communication
- Virtual Network Operations
- Telecommunications
- Cellular Networks
- Broadcasting
- Wireless Sensor Networks
- Wireless Broadband
- Grid Communications
- Mobile Communications
- Wireless Technologies

IGI Global is currently accepting manuscripts for publication within this series. To submit a proposal for a volume in this series, please contact our Acquisition Editors at Acquisitions@igi-global.com or visit: http://www.igi-global.com/publish/.

The Advances in Wireless Technologies and Telecommunication (AWTT) Book Series (ISSN 2327-3305) is published by IGI Global, 701 E. Chocolate Avenue, Hershey, PA 17033-1240, USA, www.igi-global.com. This series is composed of titles available for purchase individually; each title is edited to be contextually exclusive from any other title within the series. For pricing and ordering information please visit http://www.igi-global.com/book-series/advances-wireless-technologies-telecommunication/73684. Postmaster: Send all address changes to above address.

Titles in this Series

For a list of additional titles in this series, please visit: www.igi-global.com/book-series

Enabling Technologies and Architectures for Next-Generation Networking Capabilities
Mahmoud Elkhodr (Central Queensland University, Australia)
Information Science Reference • copyright 2019 • 384pp • H/C (ISBN: 9781522560234) • US $195.00 (our price)

Mobile Devices and Smart Gadgets in Human Rights
Sajid Umair (National University of Sciences and Technology (NUST), Pakistan) and Muhammad Yousaf Shah (Ministry of Human Rights, Pakistan)
Information Science Reference • copyright 2019 • 304pp • H/C (ISBN: 9781522569398) • US $195.00 (our price)

Emerging Automation Techniques for the Future Internet
Mohamed Boucadair (Orange, France) and Christian Jacquenet (Orange, France)
Information Science Reference • copyright 2019 • 351pp • H/C (ISBN: 9781522571469) • US $195.00 (our price)

Emerging Capabilities and Applications of Wireless Power Transfer
Alicia Triviño-Cabrera (University of Málaga, Spain) and José A. Aguado (University of Málaga, Spain)
Information Science Reference • copyright 2019 • 383pp • H/C (ISBN: 9781522558705) • US $195.00 (our price)

Sensing Techniques for Next Generation Cognitive Radio Networks
Ashish Bagwari (Uttrakhand Technical University, India) Jyotshana Bagwari (Uttrakhand Technical University, India) and Geetam Singh Tomar (THDC Institute of Hydropower Engineering and Technology, India)
Information Science Reference • copyright 2019 • 356pp • H/C (ISBN: 9781522553540) • US $215.00 (our price)

Mobile Applications and Solutions for Social Inclusion
Sara Paiva (Instituto Politécnico de Viana do Castelo, Portugal)
Information Science Reference • copyright 2018 • 354pp • H/C (ISBN: 9781522552703) • US $195.00 (our price)

Optimizing E-Participation Initiatives Through Social Media
Laura Alcaide-Muñoz (University of Granada, Spain) and Francisco José Alcaraz-Quiles (University of Granada, Spain)
Information Science Reference • copyright 2018 • 307pp • H/C (ISBN: 9781522553267) • US $195.00 (our price)

Centrality Metrics for Complex Network Analysis Emerging Research and Opportunities
Natarajan Meghanathan (Jackson State University, USA)
Information Science Reference • copyright 2018 • 183pp • H/C (ISBN: 9781522538028) • US $175.00 (our price)

IGI Global
DISSEMINATOR OF KNOWLEDGE

701 East Chocolate Avenue, Hershey, PA 17033, USA
Tel: 717-533-8845 x100 • Fax: 717-533-8661
E-Mail: cust@igi-global.com • www.igi-global.com

Table of Contents

Detailed Table of Contents

domain at each predefined transmission time intervals (TTIs), time windows used to trigger the user requests and to respond them accordingly. At each TTI, the scheduling procedure is conducted based on a scheduling rule that aims to focus only on particular scheduling objective such as fairness, delay, packet loss, or throughput requirements. The purpose of this chapter is to formulate and solve an aggregate optimization problem that selects at each TTI the most convenient scheduling rule in order to maximize the satisfaction of all scheduling objectives concomitantly TTI-by-TTI. The use of reinforcement learning is proposed to solve such complex multi-objective optimization problem and to ease the decision making on which scheduling rule should be applied at each TTI.

Fast advancements in equipment, programming, and correspondence advances have permitted the rise of internet-associated tangible gadgets that give perception and information estimation from the physical world. It is assessed that the aggregate number of internet-associated gadgets being utilized will be in the vicinity of 25 and 50 billion. As the numbers develop and advances turn out to be more develop, the volume of information distributed will increment. Web-associated gadgets innovation, alluded to as internet of things (IoT), keeps on broadening the present internet by giving network and cooperation between the physical and digital universes. Notwithstanding expanded volume, the IoT produces big data described by speed as far as time and area reliance, with an assortment of numerous modalities and changing information quality. Keen handling and investigation of this big data is the way to creating shrewd IoT applications. This chapter evaluates the distinctive machine learning techniques that deal with the difficulties in IoT information.

Disasters are happening due to drastic environmental destructions that may cause damage to wireless data transmission networks. There must be a system that monitors and takes necessary actions for reliable communication which can be provided by the wireless sensor network (WSN) that were organized as multiple nodes. In the heterogeneous environment these ubiquitous nodes are able to handle disasters like floods, drought, earthquake, and cyclone, or network fluctuations through fire accidents. Disasters can be monitored by augmenting a variety of sensors to sense and detect sudden changes in temperature, pressure, seismic wave, noises, etc. Large numbers of sensor nodes are distributed over a geographical area in WSN providing trustable data transfer with multi node sink. In this chapter, the authors review various WSN routing protocols for reliable data transfer in disaster management using multiple sink. The main objective of this chapter is to provide future research directions to enhance QoS in disaster management.

Cognitive radio technology (CRNs) will be the fundamental driving force behind the next generation (5G) mobile communication systems as it provides the optimal solution for the problem of spectrum scarcity via dynamic spectrum usage. The CRNs, however, pose several key challenges such as network management, spectrum allocation, and access, energy efficiency, interference, cost, spectrum sensing, security, and quality of service (QoS). In this chapter, the authors undertake a comprehensive analysis of 30 peer-reviewed scientific publications collated from 2017 to 2018 April that examine cognitive radio networks to identify practical solutions proposed to overcome critical challenges in this field. Nine distinct challenges were considered: network management, spectrum allocation, and access, energy efficiency, interference, cost, spectrum sensing, security, and QoS. The analysis demonstrates that the majority of research work related to CRN focuses on approaches to improve network management and, specifically, optimization of networks.

Machine learning is one of the most popular research areas, and it is commonly used in wireless communications and networks. Security and fast communication are among of the key requirements for next generation wireless networks. Machine learning techniques are getting more important day-by-day since the types, amount, and structure of data is continuously changing. Recent developments in smart phones and other devices like drones, wearable devices, machines with sensors need reliable communication within internet of things (IoT) systems. For this purpose, artificial intelligence can increase the security and reliability and manage the data that is generated by the wireless systems. In this chapter, the authors investigate several machine learning techniques for wireless communications including deep learning, which represents a branch of artificial neural networks.

As the circumstances are changing, mankind has turned out to be more inclined to snappy and speedier correspondence and access to information. The correspondence happens in numerous structures (e.g., presently, this correspondence is all the more a virtual substance than a physical one). So as to keep up fast correspondence, the coming age will depend on exceptionally tried and true, canny and self-learning/self-modifying correspondence organizers. In this context, this chapter reviews the most important machine learning techniques with the direct applicability in wireless ad-hoc systems. A guide of machine learning methods and their relevance is also provided. Different applications of ad-hoc wireless networks are discussed in terms of energy-aware communications, optimal node deployment and localization, resource allocation, and scheduling.

Section 2
Machine Learning and Artificial Intelligence Applications

The user experience constitutes an important quality metric when delivering high-definition video services in wireless networks. Failing to provide these services within requested data rates, the user perceived quality is strongly degraded. On the radio interface, the packet scheduler is the key entity designed to satisfy the users' data rates requirements. In this chapter, a novel scheduler is proposed to guarantee the bit rate requirements for different types of services. However, the existing scheduling schemes satisfy the user rate requirements only at some extent because of their inflexibility to adapt for a variety of traffic and network conditions. In this sense, the authors propose an innovative framework able to select each time the most appropriate scheduling scheme. This framework makes use of reinforcement learning and neural network approximations to learn over time the scheduler type to be applied on each momentary state. The simulation results show the effectiveness of the proposed techniques for a variety of data rates' requirements and network conditions.

The rapidly evolving 5G cellular system adapts cloud computing technology in its radio access networks (RAN), namely Cloud RAN or CRAN. CRAN enables better scalability, flexibility, and performance that allows 5G to provide connectivity for the vast volume of IoT devices. This chapter presents two major research results addressing the load balance (LB) problem in CRAN. First, the authors propose a generic online learning (GOL) system; GOL integrates reinforcement learning (RL) with deep learning method for an environment not fully visible, changing over time, while receiving feedbacks of fluctuating delays. Simulation results show that GOL successfully achieves the LB objectives of reducing both cache-misses and communication load. Next, they study eight practical LB based on real cellular network traffic characteristics provided by Nokia Research. Experiment results of these algorithms on queue-length analysis show that the simple, light-weight queue-based LB is almost as effectively as the much more complex waiting-time-based LB.

Video streaming services top the internet traffic surging forward a competitive environment to impart best quality of experience (QoE) to the users. The standard codecs utilized in video transmission systems eliminate the spatiotemporal redundancies in order to decrease the bandwidth requirement. This may adversely affect the perceptual quality of videos. To rate a video quality both subjective and objective parameters can be used. So, it is essential to construct frameworks which will measure integrity of video just like humans. This chapter focuses on application of machine learning to evaluate the QoE without requiring human efforts with higher accuracy of 86% and 91% employing the linear and support vector regression respectively. Machine learning model is developed to forecast the subjective quality of H.264 videos obtained after streaming through wireless networks from the subjective scores.

This chapter introduces statistical learning methods and findings of a group decision-making algorithm in internet of things (IoT) and edge computing environments. The discussed methodology locally detects outliers in an on-line and adaptive mode. It is driven by three perspectives—opinion, confidence, and independence—and exploits the incremental principal component analysis using the power method for eigenvector and eigenvalue estimation and Knuth and Welford's online algorithms for variance estimation. The methodology is implemented and evaluated over real contextual data in a wireless network of environmental sensors from where appropriate conclusions are drawn about the capabilities and limitations of the proposed solution in IoT environments.

Target localization and tracking has always been a hot topic in all eras of communication studies. Conventional system used radars for the purpose of locating and/or tracking an object using the classical methods of signal processing. Radars are generally classified as active and passive, where the former uses both transmitter and receivers simultaneously to perform the localization task. On the other hand, passive radars use existing illuminators of opportunity such as wi-fi or GSM signals to perform the aforementioned tasks. Although they perform detection using classical correlation methods and CFAR, recently machine learning has been used in various application of passive sensing to elevate the system performance. The latest developed models for intelligent RF passive sensing system for both outdoor and indoor scenarios are discussed in this chapter, which will give insight to the readers about their designing.

Preface

Next-generation wireless networks are expected to provide a multi-user multi-technology multi-application and multi-provider environment. Inevitable, this challenge comes with a new paradigm for radio technology that is expected to support high data rates, extremely low latencies and new types of applications with heterogeneous Quality of Service (QoS) requirements for end-users. While wireless technologies had a spectacular evolution over the past years, the present trend is to adopt a global network of shared standards which comes to meet user applications' requirements. To this end, Machine Learning (ML) is an artificial intelligence tool that is seen as a promising solution for next generation wireless networks that will enable intelligent radio adaptive learning and prediction or decision making so that the heterogeneous applications and users' requirements can be highly satisfied. The purpose of this book is to enhance the motivation of using powerful machine learning tools in order to determine optimal system configurations such that diverse requirements of next-generation wireless networks can be satisfied.

Machine learning is used with great success for a wide range of applications including finance, economics, statistics, project management, etc. Generally speaking, a machine learning tool is an agent that executes a task based on some input observations in order to optimize a given performance metric according to some historical experience. Iteratively, the optimization system aims to improve its goal metric while executing tasks by tasks while exploiting at the same time its own experience. The characteristics of the dataset that stores all input observations determine the type of learning. If the dataset is known and finite and the samples are labelled, the *supervised* learning is addressed. The term *supervised* stands from the fact that each input sample has its own target value or supervisory signal and the mapping function can be learnt based on this correspondence. Examples of supervised learning tool include regression models, Support Vector Machine (SVN), Artificial Neural Networks (ANN) and Bayesian learning. When the input samples are not labelled, the term of *unsupervised* learning is addressed since this model has to learn by itself the structure of the input samples. The clustering algorithms are the most know algorithms used for unsupervised learning. In most of the existing optimization problems in wireless communications, the input observations are not known in advance and they are provided as the system evolves according to the conducted tasks. In this case, *reinforcement* learning paradigm is addressed where the model is learnt according to the reward/cost values received after applying actions at every input state based on the given learner experience. This book aims to identify different applications in which supervised, unsupervised and reinforcement learning paradigms can be used to increase the satisfaction performance of various requirements for future wireless networks.

In wireless communications, machine learning is applied in various domains such as mobile data analysis, localization, Wireless Sensors Networks (WSN), Radio Resource Management (RRM), network security, signal processing, etc. As part of data analysis, supervised learning can be used to learn the functions that can map different QoS indicators into Quality of Experience (QoE) parameters in order to evaluate the user experience under different network conditions and applications. Also, to optimize the cloud Radio Access Networks (RAN), different learning tools can be used to forecast the mobile traffic under various networking conditions. For mobility and localization purposes, supervised and unsupervised learning techniques are used to predict the user location based on Channel State Information (CSI) samples received from each user. One goal would be to reduce the signalling overhead caused by large amount of CSI feedbacks at the radio interface by using prediction models. In WSNs, machine learning is used for node localization, failure detection and path optimization between sink and destination nodes in order to conserve the overall network energy. The RRM includes several functionalities such as: power control, interference management, scheduling and radio resource allocation. Machine learning is used in RRM with the scope of maximizing the system capacity while respecting the restrictive QoS requirements for new types of applications adopted in future wireless/cellular networks. Supervised and unsupervised learning can be used in cyber security to classify the malware and to deny certain services when the conditions will require. In signal processing, machine learning is used in encoding/decoding optimization problems, modulation recognition and classification, radio channel estimation and decoding.

CHALLENGES

One of the most important challenges when applying machine learning in wireless networks is given by the large dimensionality of input state space when different optimization problems are addressed. This aspect refers to the variety of network conditions such as: configurable parameters, high number of connected devices, large number of operating bands with high bandwidths and flexible sub-frames. Each connected device is characterized by different momentary vectors with large dimensions that may include, for instance, QoS indicators and requirements, applications' characteristics, channel conditions, etc. To this extent, the input data set used to train different decision-making tasks may take a very large and variable dimensionality since the number of connected devices changes over time. In many optimization problems, the input state space is infinite, and thus the state-action values cannot be enumerated exhaustively. In addition, most of the existing optimization tasks in wireless/cellular networks are stochastic which means that machine learning cannot fully control the system evolution from state to state. Thus, optimal solutions are very difficult to be found under these circumstances.

Another important challenge refers to the partial observability of the state space when the agent is located on both user and base station sides. In general, the agent state space consists of measurements taken by each user or by the overall network data collected at the base station level. For example, when machine learning optimizes the modulation recognition and classification, the CSI data received from each user may be delayed due to the network conditions. Thus, the decision must be taken based on only the CSI values that are already available at the base station level at each time instant. Also, by implementing new radio access technologies in future wireless/cellular networks such as mmWave communications will limit the number of controllable cells and thus, the agents' observability is serious diminished.

Implementing machine learning to make the radio systems more intelligent can be computational very expensive since powerful hardware and software are required to conduct the learning over very high dimensional state space. Thus, the learning process may take longer as the agents' state dimension and size increases. Also, the complexity for the entire machine learning engine is very important, since once learnt, the agent must perform its decisions within very strict time intervals. For instance, it is expected that in 5G radio access networks for very stringent requirements of latency-critical applications, the radio resource scheduling must be performed in less than 100 micro seconds. Then, the agent or the controller complexity must be reduced at minimum, while still providing acceptable accuracy when performing the decision-making at each time instant.

The learning quality depends on the learning system configuration and the ways how the explorations of different state spaces are performed. For instance, when using clustering algorithms for unsupervised classification, the number of clusters must be decided in advance. Higher number will provide better accuracy at the expense of higher computational complexity while a lower number will decrease the complexity by harming the accuracy. The supervised learning is performed based on two sets of input datasets, training and testing. Then, a strategy is required to decide which type of sample (training or testing) must be chosen each time to the agent. Also, the SVM tool used for supervised classification, needs to set-up considerable number of parameters before launching the learning process. Reinforcement leaning comes with some configuration parameters regarding the learning rate and the discount factor that must be set in order to perform the learning over a given time period. Above all of these aforementioned aspects, the learning performance strongly depends on the exploration duration of given input datasets or state spaces. For instance, if the learning process takes less than the optimal duration, the learnt policy of decisions may not generalize very well the entire state space. On the other side, too long explorations will degrade the learning quality since some observations may be visited more often than other ones. Therefore, a special care must be given to the way how input state space is explored and the learning stage duration.

OPPORTUNITIES

The dimensionality problem of the state space can be addressed by using additional pre-processing tools designed to reduce to more convenient dimensions designed for the machine learning purposes. The pre-processing stage itself may use some unsupervised and/or supervised learning tools to perform the classification of the original inputs samples. To solve the variability in the state space dimension, some other tools may be used such as Principal Component Analysis (PCA) in order to extract the most important features from the state space given its variable dimension. In many optimization problems, the state space cannot be reduced to a finite size and one-dimensional format due to the stochastic nature of the wireless networks conditions. Therefore, we can get only approximations of the best solutions. For example, in classification problems with infinite training set, semi-supervised learning tools can be used to obtain good performance. Radial Basis Function neural networks can be used as a semi-supervised learning tool to get good classification approximation for each sample. In reinforcement learning, the decision-making is approximated by using ANNs, convolutional neural networks, or different representations of non-linear functions. However, the purpose of these approaches is to find solutions as close as possible to the optimal ones.

To cope with the partial observability of the agent states for certain wireless networks applications, extensive radio measurements can be conducted in order to have a more precise positioning based on uplink communications and large antenna arrays at the base station. If additional measurements are not implementable, then predictions methods can be used instead. For instance, users with delayed CSIs can predict current values based on some historical information. The accuracy of these predictions will play a crucial role for the entire learning system.

Given the powerful hardware, more data storage capacity and advanced machine learning tools have motivated network operators to rethink most of the existing algorithms and to adopt optimization solutions that are able to better conserve the network resources. Graphical Processing Units and multi-core central process units can be used to perform the learning tasks when the agent state space has considerable dimension and size. Alongside these technologies, the online learning can be performed and the decision-making can be optimized over time in real-time systems. Distributed computing can be also used in multi-agent systems where the learning process is accomplished on different nodes and the state space cannot be reduced. In this case, the state space for each agent is much reduced and the decisions are performed separately. In scheduling problems, each user is characterised by one agent that performs its own action in fully concurrence with other users in order to get as many resources as possible. In device-to-device communications, agents are cooperating by exchanging different types of information from the state spaces in order to achieve the same goal. Coordination between intelligent agents is needed when each agent performs its own task and their results are centralized. Although the multi-agent learning systems can represent a viable solution to reduce the overall system complexity, these solutions may suffer from the difficulty in finding global accepted policies.

The system configuration for different machine learning algorithms must be decided before launching the exploration/learning processes based on some quality metrics. For instance when neural networks are used to approximate solutions in supervised and reinforcement learning tasks, the errors between the provided output values and given patterns or target values must be monitored. When setting large number of layers and nodes, the complexity is higher, the accuracy is higher but the input data can be over-fitted and the error can increase drastically. The same symptom in error evolution may show that the artificial neural network under-fits the input data by setting insufficient number of layers or nodes. Based on the complexity, accuracy, over-fitting and under-fitting symptoms, the configuration of the neural networks must be a priori decided. Also, the learning termination condition is one key element that may affect the quality of decision-making. Depending on the addressed optimization problem, some functions based on the temporal errors can be implemented as a stopping criterion for the learning stage.

BOOK ORGANIZATION

The objective of the proposed book is to introduce the reader to the latest trends in various technologies of next generation wireless networks. In particular, the book looks into the use of Machine Learning techniques and Artificial Intelligence to model various technical problems of next-generation systems that will enable the dynamic optimisation of system configuration.

This book is organized in two main sections. The first section titled "Machine Learning Evolution for Wireless Communications" provides an overview of the state-of-the-art of machine learning technologies applied in different technology sectors which could be considered as part of next-generation wireless networks. It includes important chapters written by researchers from prestigious laboratories from United Kingdom, Brazil, Switzerland, Australia, Turkey and India with the main focus on wireless and cellular networks, Self-Organizing Networks (SON), radio resource management, Internet of Things, wireless sensor networks and cognitive radio. A brief description of each of the chapters in this section is given below:

Chapter 1 presents an overview of the role and applications of different machine learning algorithms applied in the context of SON for future wireless and cellular networks. In the context of SON the machine learning tools can be used with great success since it analyses and uses historical data in order to improve and adapt themselves without human control to the changeable network conditions. To this extent, intelligent algorithms must be used in order to extract different patterns from input data and learn over time from previous actions the best adaptability strategies. The chapter comprises the introduction in ML and SON followed by the analysis of the state of the art solutions that involve the machine learning tools for self-configuration, self-optimization and self-healing. Also, the guidelines on the utilization of intelligent algorithms in SON and future research trends in the area are highlighted.

Chapter 2 addresses the radio resource scheduling in Orthogonal Frequency Division Multiple Access (OFDMA) downlink radio access scheme. The radio resource allocation is conducted in the frequency domain at each predefined Time Transmission Interval (TTI) by a given scheduling rule that is decided by the network operator. By using a single scheduling rule across the entire scheduling session, the overall QoS satisfaction may be degraded since each particular rule is designed to deal only with particular QoS objectives. Therefore, the purpose of this chapter is to formulate and solve an aggregate optimization problem that is able to select at each TTI the most convenient scheduling rule according to the network conditions such that the overall QoS satisfaction is maximized in terms of system throughput maximization, Packet Loss Rate (PLR), Guaranteed Bit Rate (GBR), delay and user fairness. A complex machine learning solution based on supervised, un-supervised and reinforcement learning technologies is proposed to find solutions for such complex optimization problem. Also, the system architecture based on machine learning is proposed for OFDMA and access schemes and beyond.

Chapter 3 evaluates different machine learning techniques used in Internet of Things (IoT). Big Data information is utilized to construct IoT structures which incorporate things-driven, information-driven, benefit-driven engineering, or cloud-based IoT. With the advancements in IoT, big data can be collected covering information from sensors, radio resources utilization, low power and vitality information. By managing the big data collected from these large scale sensor systems and IoT could improve the resource administration, security and protection, and increase the overall system efficiency. To adequately blend the Big Data information and convey among IoT devices, machine learning strategies are utilized. To this extent, an overview of the fundamental uses of machine learning techniques in IoT is provided.

Chapter 4 reviews various WSN routing protocols for reliable data transfer in disaster management using multiple sink nodes. The main objective of this chapter is to provide future research directions to enhance QoS satisfaction in disaster management. Different applications of WSN are analysed in terms of tsunami detection, earthquake sensing, forest fire, flood and water monitoring system. Various routing protocols are analysed in terms of delay, load balancing, energy consumption, robustness and network life-time. The role of machine learning in the routing algorithms is evaluated in order to address the multi-objective aim that encompasses the aforementioned targets.

Chapter 5 aims to examine the challenges for cognitive radio networks in terms of network management, spectrum allocation, and access, energy efficiency, interference, cost, spectrum sensing, security and QoS satisfaction. The analysis of this chapter demonstrates that the majority of research work related to cognitive radio networks focuses on approaches to improve network management and specifically, optimization of networks. The second most-researched area is spectrum sharing and allocation techniques in cognitive radio networks. The majority of authors have based their work on various approaches to providing access for both primary users and secondary users to the licensed spectrum. QoS satisfaction is another highly addressed issue in the context of CRN. Our findings suggest some techniques for enhancing network throughput in CRNs. One of the emerging trends in the field is to employ machine learning methods to enhance the access and spectrum sharing and to assist cooperative spectrum handovers and assure the security by deploying intelligent detection and mitigation of jamming attacks.

Chapter 6 analyses machine learning and big data processing approaches in wireless networks. According to the authors, the most popular machine learning techniques are: PCA, k-Nearest Neighbour, Support Vector Machines (SVN), Artificial Neural Networks (ANN), Extreme Learning Machines (ELM) and Deep Learning. These techniques are getting more important day-by-day since the types, amount and structure of data is continuously changing. Recent developments in smart phones and other devices like drones, wearable devices, machines with sensors need reliable communication within Internet of Things (IoT) system. For this purpose, artificial intelligence can provide secure systems and manage the data that is generated by the wireless system. In the next-generation of wireless and cellular networks, ANN will be more popular for the IoT systems, since more and more devices will be connected wirelessly and it has dynamic structure which can be managed intelligently.

Chapter 7 reviews the most important machine learning techniques with the direct applicability in wireless ad-hoc systems. A guide of machine learning methods and their relevance is also provided. Different applications of ad-hoc wireless networks are discussed in terms of energy-aware communications, optimal node deployment and localization, resource allocation and scheduling. Basically, machine learning technology can be used for both static and dynamic wireless ad-hoc networks. For static networks, machine learning can be used with the limitation that the obtained policies are able to provide good solutions only for the considered scenarios. On the other side, the network dynamicity will increase the generality of learnt policies due to more various conditions for the wireless channels. In some related work, improper machine learning technologies are used since the obtained solutions indicate very high system complexity and low learning quality.

The second section titled "Machine Learning and Artificial Intelligence Applications" presents various solutions based on unsupervised, supervised and reinforcement learning technologies in wireless networks' applications such as: tracking and positioning, wireless sensor networks, quality estimation in video streaming, 5G Cloud-RAN optimization and radio resource allocation and scheduling. This section includes important chapters written by researchers from prestigious laboratories from United Kingdom, Switzerland, United States of America, India and Pakistan and a brief description for each of them is given bellow:

Chapter 8 considers the optimization problems in radio resource scheduling for OFDMA downlink transmissions. The scheduler in OFDMA downlink systems works with simple scheduling rules that aim to satisfy certain scheduling objectives. In this chapter, an innovative reinforcement learning framework is propose in order to improve the user bit rates' satisfaction for different traffic types such as: full buffer, constant bit rate and variable bit rate. These traffic characteristics were chosen in order to cover a wide range of applications with various arrival rates and GBR requirements (e.g. video, VoIP, FTP, web browsing). The proposed frameworks aims to apply at each time instant based on the momentary scheduler state the most suitable scheduling rule in order to maximize the time when the bit rates requirements are satisfied for each mobile user. Different reinforcement learning algorithms are evaluated by taking into account the dynamic channel conditions and traffic loads. To reduce the complexity for the proposed framework being given by the high dimensionality of the scheduler state space, the decisions on which scheduling rule should be applied on each state is approximated by using artificial neural networks. The learnt policies are evaluated also in terms of different settings of the time windows that are used to compute the user average data rates. Simulation results indicate that the proposed RL framework is able to outperform other scheduling schemes when setting optimum time windows for each involved type of traffic.

Chapter 9 considers the concept of cloud computing technology in 5G radio access networks entitled Cloud Radio Access Networks (CRAN). CRAN enables better scalability, flexibility and performance that allow 5G networks to provide connectivity for the vast volume of IoT devices. This chapter presents two major results that address the problem of load balancing in CRAN. First, a Generic Online Learning (GOL) system is proposed to deal with load balancing in 5G CRAN. GOL integrates Reinforcement learning (RL) with deep learning method, for an environment not fully visible, changing over time, while receiving feedbacks of fluctuating delays. The problem domain has been mapped to the GOL architecture with a proper design of punishments and rewards for its learning algorithm. Simulation results show that GOL successfully achieves the load balancing objectives by reducing both cache-misses and communication load. Second, this chapter studies eight practical load balancer based on real cellular network traffic characteristics provided by Nokia Research. The results have demonstrated how well these eight algorithms are able to reduce latencies experienced by IoT communications in 5G system. Based on queue-length analysis and waiting time, the proposed GOL system achieves the shortest waiting time and shortest total service time when compared to other eight algorithms.

Chapter 10 presents a novel machine learning model that is used to evaluate the QoE of a video sequence from its objective quality metrics. The standard codecs utilized in video transmission systems eliminate the spatio-temporal redundancies in order to decrease the bandwidth requirement. This may adversely affect the perceptual quality of videos. To rate a video quality both subjective and objective parameters can be used. So, it is essential to construct frameworks which will measure integrity of video just like humans. This chapter focuses on application of supervised machine learning to evaluate the QoE without requiring human efforts with higher accuracy of 86% and 91% employing the linear and support vector regression, respectively. Machine learning model is developed to forecast the subjective quality of H.264 videos obtained after streaming through wireless networks from the subjective scores.

Chapter 11 introduces statistical learning methods and findings of a group decision-making algorithm in IoT and Edge Computing environments. The discussed methodology locally detects outliers in an on-line and adaptive mode. It is driven by three perspectives: opinion, confidence and independence; and exploits the incremental PCA unsupervised learning algorithm by using the Power Method for eigenvector and eigenvalue estimation and Knuth and Welford's online algorithms for variance estimation. The methodology is implemented and evaluated over real contextual data in a wireless network of environmental sensors from where appropriate conclusions are drawn about the capabilities and limitations of the proposed solution in IoT environments.

Chapter 12 provides detailed history, description, design metrics, design scheme and applications of passive sensing systems. The upcoming passive sensing systems are designed by using Deep learning/ Machine learning models instead of matched filters based on signal processing. They are efficient in computation and better in accuracy than the systems based on signal processing techniques. Both the indoor and outdoor applications of passive sensing system i.e. target positioning, activity recognition, through-the-wall monitoring, etc., are adopting the deep learning models for their detection phase. The novel approaches discussed in the chapter for designing of both indoor and outdoor passive sensing systems based on machine learning provided enough knowledge to develop and analyze any intelligent RF sensing system. This chapter also discusses the latest trends and methods for boosting performance of ML-based passive sensing system, which give diverse choices to research community for choosing their future research directions.

Chapter 13 explores the possibility of using Artificial Intelligence in creating bots that can cheer people up when they are sad, referred to as Cheerbots. The system aims at analysing the semantics of the sentences and analyzing the pitch of the person's voice to identify their emotional state, and then providing an n-way dialogue conversation, relevant within the existing context, instead of the mundane two-way dialogue conversation. In this context, this chapter examines the possibility of creating bots that can serve as an emotional support to humans by integrating Artificial Intelligence techniques.

The prospective audience of this book will be on one side undergraduate students, postgraduate students, and researchers who are interested in the emerging area of machine learning solutions for next generation wireless networks and on the other side industry people who are working in the mentioned areas. The potential use of this book could be: course supplement for upper-level courses; library reference for students' projects; resource for researchers in machine learning applications in mobile and wireless communication area as the book will give an overview of both major domains together with their current research progress, issues and ideas; reference for the industry field, as important research issues could be addressed in the real-world environment

The *Next-Generation Wireless Networks Meet Advanced Machine Learning Applications* book is published by IGI Global (formerly Idea Group Inc.), publisher of the "Information Science Reference" (formerly Idea Group Reference), "Medical Information Science Reference," "Business Science Reference," and "Engineering Science Reference" imprints. For additional information regarding the publisher, please visit www.igi-global.com.

The editors wish you a pleasant reading.

Ioan-Sorin Comşa
Brunel University London, UK

Ramona Trestian
Middlesex University, UK

Acknowledgment

The *Next-Generation Wireless Networks Meet Advanced Machine Learning Applications* book would not exist without the effort of many people whose names may not appear on the cover of the book. However, their hard work, cooperation, friendship and understanding were very important to the preparation and production of the book.

The editors would like to sincerely thank the entire team at IGI Global for their support and help in the publication of this book. In particular, we wish to thank Ms. Jordan Tepper the Assistant Development Editor at IGI Global. Her patience was amazing and appreciated, while her guidance was always timely, useful, and valued.

As the success of this book depends on the high-quality chapters submitted, the editors wish to express their deep appreciation to the authors of each chapter for their significant contribution. Without their insights and expertise this book would not be possible.

As in general the work associated with the chapter review is underestimated and forgotten, we would like to thank the team of reviewers for their generous commitment of time and effort they have put in the reviewing process and for providing their expertise to ensure a high-quality review process. We also thank the Editorial Advisory Board members for their implication in this project.

We would like to thank our families for being a continuous help along the way. In particular Ioan-Sorin would like to thank his dear parents Otilia and Iulian for their unconditional support and being an immense source of inspiration throughout his life. A special gratitude is for his brother Sergiu and sister-in-law Alina for their immense support during these years. His lovely nieces: Amalia and Patricia for making his life more beautiful. Ramona would like to thank her wonderful and loving parents Maria and Vasile, for their unconditional love and care that represents a source of inspiration in everything she is doing. A special gratitude goes to her dear brother Ionut and his lovely wife Andreea. A special thank you goes towards her husband Kumar, for his love, patience and continued support that were essential to this project as well as towards her little bundle of joy Noah Anthony for bringing light into her life.

Ioan-Sorin Comşa
Brunel University London, UK

Ramona Trestian
Middlesex University, UK

Section 1

Machine Learning Evolution for Wireless Communications

Chapter 1
The Role and Applications of Machine Learning in Future Self-Organizing Cellular Networks

Paulo Valente Klaine
University of Glasgow, UK

Oluwakayode Onireti
University of Glasgow, UK

Richard Demo Souza
Federal University of Santa Catarina (UFSC), Brazil

Muhammad Ali Imran
University of Glasgow, UK

ABSTRACT

In this chapter, a brief overview of the role and applications of machine learning (ML) algorithms in future wireless cellular networks is presented, more specifically, in the context of self-organizing networks (SONs). SON is a promising and innovative concept, in which future networks are expected to analyze and use historical data in order to improve and adapt themselves to the network conditions. For this to be possible, however, algorithms that are capable of extracting patterns from data and learn from previous actions are necessary. This chapter highlights the utilization and possible applications of ML algorithms in future cellular networks. A brief introduction of ML and SON is presented, followed by an analysis of current state of the art solutions involving ML in SON. Lastly, guidelines on the utilization of intelligent algorithms in SON and future research trends in the area are highlighted and conclusions are drawn.

DOI: 10.4018/978-1-5225-7458-3.ch001

INTRODUCTION

In the last couple of years, mobile wireless networks have become an essential part of our lives, due to a broad range of applications and services that are available. Nowadays, people are able to do business on the go, by performing teleconferences whenever and wherever needed, watch or listen to their favorite videos and music on the fly, talk to distant relatives, stream audio/video whenever a special event happens, instantly upload photos or videos about their daily lives in social media, and many more (Aliu, Imran, Imran, & Evans, 2013).

In the future, however, this demand is expected to be much larger, with the advents of new technologies being developed, such as ultra-high definition videos, Virtual Reality (VR) applications, the Internet of Things (IoT); Machine-to-Machine (M2M) communications; cloud computing, and various other services that are unimaginable today. In addition, not only new technologies will need to be addressed by the Next Generation of Mobile Networks (NGMN), but also the ever increasing demand of users in terms of both capacity and better services (Huawei, Technologies Co., 2016), (Samsung Electronics Co., 2015). Some of the requirements that are present in the current state-of-the-art literature of NGMNs are (P. Fettweis, 2012), (G. Andrews, et al., 2014) (Huawei, Technologies Co., 2016):

- Address the exponential growth required in both coverage and capacity;
- Provide better Quality of Service and Experience (QoS and QoE) to end users;
- Support the coexistence with other radio technologies;
- Provide peak data rates higher than 10Gbps;
- Support latency lower than one millisecond (enabling the concept of tactile internet);
- Support ultra-high reliability and network energy efficiency.

From these requirements, it is clear that NGMNs are under a heavy pressure in order to address current limitations of present mobile networks and to push their performance to a next level, allowing all these requirements and new technologies to become a reality in the near future. In order to address the expected exponential growth required in both coverage and capacity, one major shift that will probably occur in NGMNs is the network densification process, in which operators are expected to deploy a wide range of small cells in order to cover a relatively small area (G. Andrews, et al., 2014), (Alsedairy, Qi, Imran, Imran, & Evans, 2015). Although the densification process is probably going to solve some of the future cellular network problems, this process will require an even bigger change in paradigm on how future cellular networks are organized. The dense deployment of small cells will result in an exponential increase in terms of network complexity, while also increasing the total network CAPital and OPerational EXpenditures (CAPEX and OPEX), affecting the whole network in terms of configuration, optimization and healing (Valente Klaine, Imran, Onireti, & Demo Souza, 2017).

In order to address these issues created by the densification process, one possible solution is the addition of more intelligence in the network, by using algorithms that rely on data collected by operators and traffic patterns, such as Machine Learning (ML) algorithms. In addition, ML solutions would also address other network issues by simplifying the coordination, configuration and optimization procedures of the network, minimizing the network complexity, energy consumed and expenditures, while also enabling autonomous network healing (Imran, Zoha, & Abu-Dayya, 2014), (Aliu, Imran, Imran, & Evans, 2013). Furthermore, ML algorithms would also be a key enabler to new types of applications that today are unimaginable, ranging from the autonomous configuration of a new Base Station (BS) and network topology, optimization and allocation of backhaul resources, user mobility tracking and prediction, to fully enabling the concept of Self Organizing Networks (SONs) (Valente Klaine, Imran, Onireti, & Demo Souza, 2017).

Hence, the objective of this chapter is to provide the readers with a clear view of why ML is necessary for future networks, in particular, on how ML can enable the concept of SON. In addition, the authors also present a brief definition of the concept of SON, including its three main functions: self-configuration, self-optimization and self-healing. Furthermore, the authors also present the readers with some of the current state-of-the-art applications using ML applied in wireless cellular networks, in order to enable SON. Lastly, the authors will also comment on the suitability of each ML technique for each SON use case, while also providing the readers with some future research directions and a brief conclusion, highlighting the important role that ML will have in future wireless networks.

More specifically, the objectives of the chapter are:

- Provide the readers with a clear understanding of current cellular networks limitations.
- Provide the readers with a clear definition of Self Organizing Network (SON) and its subdivisions, mainly: self-configuration, self-optimization, and self-healing.
- Provide readers with a clear motivation of why ML is necessary for future networks and how it can enable future services.
- Provide current state-of-the-art examples of applications of intelligent solutions in cellular networks, highlighting which algorithm was utilized and which problem it aimed to tackle.
- Provide general guidelines on when to use some ML algorithms for each SON function.
- Provide future research trends and directions that are expected to be seen in the near future in the area of SON.

MOTIVATION

From the previous section, it is evident that the requirements that NGMNs must address are quite stringent. However, first, before trying to tackle these requirements, NGMNs should also be able to overcome current limitations of present cellular networks. Current methods today lack the adaptability and flexibility required in order to become feasible solutions for future cellular networks (Imran, Zoha, & Abu-Dayya, 2014). One simple example is the process of a BS configuration. Whenever a new BS is deployed in a network, hundreds of parameters need to be manually configured by expert engineers in order for it to become operable. These parameters can be basic operational parameters, such as IP address, default gateway address, BS identifier, the configuration of the neighbor cell list of the new BS, as well as its neighbors, and radio parameters, such as transmit power, handover (HO) parameters, like hysteresis and Time-To-Trigger (TTT), antenna azimuth and down-tilt angles, etc. (Valente Klaine, Imran, Onireti, & Demo Souza, 2017).

As it can be seen, just the simple task of adding a new BS in a network requires a lot of expertise and manual work, which is not the ideal approach. In addition, most of these parameters can also be sub-optimally configured, as the radio parameters of a BS can also affect the total network coverage or its overall topology. One can think of as an example, if the operator configures the power of the new BS too high, this will have an impact on nearby BSs, as their coverage area will be reduced and interference will increase. Thus, a simple miscalculation or change in network conditions can be disastrous.

This simple example also highlights that a static configuration of parameters is not necessarily the best solution. As the network constantly changes, it is also necessary to constantly optimize and adjust the parameters previously configured. However, as it is with the configuration process, the optimization process of a network, nowadays, also relies on manual intervention and the constant monitor of performance indicators of the network in order to determine which parameters of a cell need to be updated and by how much (Imran, Zoha, & Abu-Dayya, 2014).

Lastly, another example that points at the inefficiency of current methods is when a failure occurs in the network. Nowadays, current mobile networks are reactive (a failure happens and the mobile operators react to the failure by responding appropriately) (Farooq, Parwez, & Imran, 2015). This requires network operators to constantly monitor the network in terms of certain Key Performance Indicators (KPIs), which results in operators collecting huge amounts of data. This data then needs to be manually analyzed by expert engineers in order to determine the exact time and cause of the failure, and later the operator must then send experts to the failure location in order to fix it. This is extremely inefficient, as it can take a lot of time for the failure to be resolved, causing disruption in service, making customers unhappy and generating less revenue for the operator. Thus, from the three examples provided, it is clear that new paradigms are necessary in order to automate and make current mobile networks more efficient, and that is where ML algorithms can be extremely effective.

Machine Learning Definition and Categories

ML is the science of making computers take decisions without being explicitly programmed to do so (Simon, 2013). This is done by programming a set of algorithms that analyze a given set of data and try to make predictions about it. Depending on how learning is performed, these algorithms are classified differently, mainly into three broad categories: supervised learning, unsupervised learning and reinforcement learning.

Supervised Learning

In supervised learning, the algorithms require a data set that has information about both input and output data. As the name suggests, their process of learning is similar to having a supervisor, or a teacher, supervising the learning process. The teacher knows the answers (output) for every input data, and as the algorithm iteratively makes predictions during its training process, the teacher corrects it. More formally, supervised learning can be defined as algorithms that based on the input and output relationship of the data, learn a model that best represents the data and is able to make predictions for newly, unseen data examples (Kotsiantis, Zaharakis, & Pintelas, 2007), (Friedman, Hastie, & Tibshirani, 2001).

In addition, supervised learning algorithms are also split into two main categories, which varies according to the type of the output variable. If the output variable is a discrete variable, such as a class, for example: "Red", "Blue", or "Cat" and "Dog", the supervised learning problem is referred to as a *classification* problem. On the other hand, if the output variable is a real or a continuous value, such as the value of a house, or the birth rate of a certain region, then the supervised problem is considered as a *regression* problem (Friedman, Hastie, & Tibshirani, 2001).

Examples of algorithms from supervised learning range from very simple functions, such as linear regression, logistic regression, *k*-Nearest Neighbors, decision trees, Support Vector Machines (SVM), to more complex ones, such as neural networks, and its variations, like convolutional neural networks and deep neural networks (LeCun, Bengio, & & Hinton, 2015).

Unsupervised Learning

Unsupervised learning, on the other hand, does not have the luxury of having a supervisor in order to aid the learning process. In this case, a data set with only input labels is provided (there is no information about the correct labels for the data, so a teacher would be of no help). In this case, an algorithm is given an unlabeled data set and its goal is to correctly learn a model that best represents the dataset (Friedman, Hastie, & Tibshirani, 2001). Due to its nature of trying to estimate a model for data without labels, unsupervised learning algorithms consist mainly of clustering algorithms, such as *K*-Means, Self Organizing Maps (SOM), and anomaly detectors.

Reinforcement Learning

Differently than the previous two techniques, Reinforcement Learning (RL) is a ML technique based on a goal seeking approach (Sutton & Barto, 1998). In contrast to other ML techniques, such as supervised learning, in which the system learns by analyzing examples provided by a supervisor, in RL, the learner must discover which actions to take by trying them (Sutton & Barto, 1998), (Kaelbling, Littman, & Moore, 1996).

In RL, a system, called an agent, interacts with its surroundings, the environment. Then, based on the agent's current state, its desired goal, and the environment's conditions, the agent takes an action. After that, the agent receives a reward, if the action taken was good, or a penalty, if it was bad. The main goal of a RL system is to maximize its total reward. In order to achieve this, the agent must not only exploit the best actions currently known but also explore new actions, in order to determine if there are better possible actions to be taken. This trade-off is known as the exploration-exploitation trade-off (Sutton & Barto, 1998), (Kaelbling, Littman, & Moore, 1996).

RL can be divided, mainly, into three categories (Sutton & Barto, 1998):

1. **Dynamic Programming (DP):** In which the agent has a perfect model of the environment, given by a Markov Decision Process (MDP), and the goal is to learn the optimal policies in order to choose the best actions.
2. **Monte Carlo (MC) Methods:** In this case, it is not assumed that there is complete knowledge about the environment. Thus, the agent must learn either online, by experiencing the environment, or through simulated experiences, in which the environment is represented by a very simple model.
3. **Temporal-Difference Learning (TD Learning):** Which can be defined as a combination of MC methods and DP. Just like MC methods, TD Learning agents can learn directly from their experience with the environment, without the need of the complete environment's dynamics. Furthermore, similar to DP, TD algorithms update their estimates (either a policy or a value function) based on other learned estimates, or in other words, they are capable of bootstrapping.

Some examples of RL algorithms include armed bandit problems, Q-Learning, and SARSA (Sutton & Barto, 1998).

Figure 1 shows how ML is divided into the three main branches and its algorithms.

Figure 1. Block diagram showing the three main branches of ML and some of its algorithms

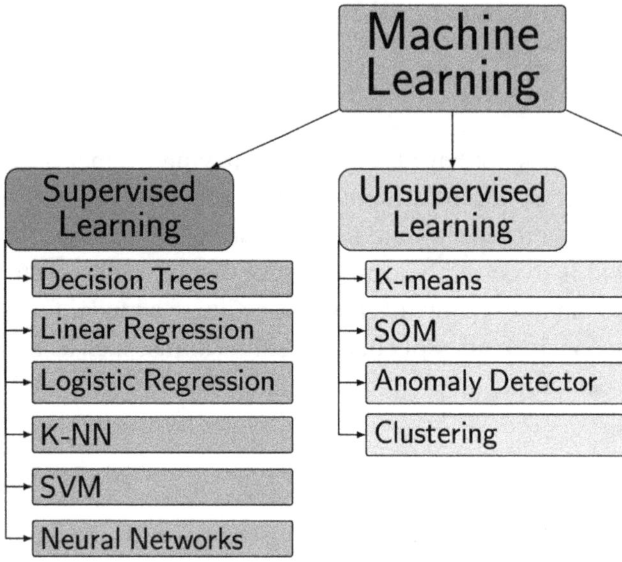

Machine Learning as an Enabler of SON

As it can be seen from the previous sections, there are lots of different algorithms that can be used, depending on the type of data available and application that is needed. In addition, since operators already collect a huge amount of data daily from their networks, it is only expected that in the near future such data can be leveraged to its full potential, moving away from manual solutions, which are both time and cost ineffective.

In addition, by deploying ML algorithms, new types of services and applications can be enabled. One example in the realm of SON can be the prediction of user mobility. By tracking each user of the network via intelligent algorithms, the network is able to predict whenever and wherever a user will be. This, in its turn, can be translated into gains in terms of control signals, in which the network is able to allocate resources in advance (before the user actually moves to the next cell) and reduce the signaling costs associated with HO, for example (Mohamed, et al., 2015).

Another interesting example can be the autonomous configuration of a mobile network from ground zero. This can be done with the usage of mobile BSs, such as truck BSs, or drones, for example, and intelligent solutions could then determine the best positions for the BSs to be, in order to maximize coverage area or minimize the number of users in an outage. This scenario can be extremely useful not only in self-configuration cases, as a planning strategy for network operators, but also in emergencies, in which a network has been destroyed and a new network must be deployed as quickly and as efficient as possible (Erdelj, Natalizio, Chowdhury, & Akyildiz, 2017). Thus, in this latter scenario, agility and adaptability are crucial, rendering intelligent solutions, depending on ML algorithms, essential in order to become a reality.

Another problem that can be solved by the analysis of data and the usage of ML algorithms can be to make current cellular networks more energy efficient. For example, by analyzing traffic patterns during different times of the day and different days of the week, operators could determine whenever it is best to turn on or off certain BSs in order to save energy, while also minimizing the number of unsatisfied users (Onireti, Mohamed, Pervaiz, & Imran, 2017). Since it is often not trivial for human experts to determine intrinsic relationships in data, especially whenever there is such a huge amount of data, as is the case of mobile networks, ML solutions are essential in order to explore these hidden correlations and determine which BS to turn on and off and for how long.

Lastly, another key concept that can be enabled through the use of ML is the detection and mitigation of faults in the network. As previously mentioned, current cellular networks are reactive, but in the future, by analyzing network patterns and historical data gathered from the network, ML algorithms will be able to predict, to a certain degree, faults that occur in the network (Farooq, Parwez, & Imran, 2015). This will enable a future network to become proactive instead of reactive and will provide a better and seamless experience to mobile users.

WHAT ARE SELF ORGANIZING NETWORKS (SON)?

The idea of self-organization is mainly inspired from biological systems and nature, in which certain species show a certain behavior in order to adapt themselves to the environment they are inserted in and achieve a particular objective. For example, a group of flocking birds is able to maintain organization without hitting one another when rapidly flying in different directions, or a group of ants is able to autonomously organize themselves in order to bring the food back to their hive (Aliu, Imran, Imran, & Evans, 2013).

Thus, based on this underlying principle of self-organization and intelligence, the idea of SON can be defined as an adaptive and autonomous network, that is not only able to maintain or achieve its desired objectives, but also that is scalable, stable and agile enough to do so (Aliu, Imran, Imran, & Evans, 2013) (Valente Klaine, Imran, Onireti, & Demo Souza, 2017). These networks are not only able to autonomously decide when and which actions to take, based on the analysis of historical data collected and their continuous interaction with the environment, but also able to improve their performance online, as the network runs.

Cellular SON can be divided into three main categories, and they are jointly referred to as *self-x* functions, they are: self-configuration, self-optimization, and self-healing (Aliu, Imran, Imran, & Evans, 2013), (Valente Klaine, Imran, Onireti, & Demo Souza, 2017).

Self-Configuration

The self-configuration function can be defined as the process of configuring all parameters necessary in order to make a network operable, such as the location of BSs in a network, software parameters of a BS, such as IP address and main gateway, network policies, topology and radio parameters, like antenna down-tilt and azimuth angles and transmission power.

The process of self-configuration can be done in several stages of a network. Most commonly, network operators would use it as a planning tool, in order to determine what the best parameters of a newly deployed network are. However, self-configuration can also be deployed when the network is already operable, by automatically configuring a newly deployed BS, for example. Additionally, self-configuration can also be utilized whenever a disaster happens and most of the network is disrupted. In this scenario, self-configuration algorithms could determine the best new positions of newly deployed BSs in order to restore service to the affected area (Erdelj, Natalizio, Chowdhury, & Akyildiz, 2017).

Self-Optimization

Another branch of SON is the self-optimization function. As the name implies, self-optimization is performed in order to continuously optimize and tune network parameters, while the network is operable. This procedure is performed due to the changes that can happen in the network, such as user mobility, high traffic demands at a specific location or time of the day, events taking place in a city, etc. In addition, by continuously monitoring the network, self-optimization algorithms are able to use reported measurements combined with historical data previously gathered and ensure that the network objectives are maintained and that its overall performance is near optimal (Aliu, Imran, Imran, & Evans, 2013).

Some examples of self-optimization functions can be in terms of caching content at BSs; backhaul congestion optimization; coverage and capacity; user mobility and HO parameters, and network energy efficiency, to name a few (Valente Klaine, Imran, Onireti, & Demo Souza, 2017).

Self-Healing

Finally yet importantly, is the self-healing function. This function, as the self-optimization one, is performed while the network is operational, as it must constantly monitor the network and its data in order to predict, detect, and classify faults as they occur. In addition, self-healing functions must also mitigate the effects that failures can have in mobile networks, not only by detecting and predicting when a failure is going to happen, but also by triggering the appropriate mechanisms in order to mitigate the effect that the fault can have in the network (Valente Klaine, Imran, Onireti, & Demo Souza, 2017).

Examples of self-healing functions can be the prediction of when a failure is likely to occur and pinpoint the BS that is most likely to fail. Another example can be the management of cell outages, in which whenever an outage happens, the appropriate self-healing function should be triggered so that the neighboring BSs can adjust their radio parameters in order to cover the out of coverage area, while the fault is not resolved.

Use Cases of Self-x Functions

Additionally, each *self-x* function can also be further divided into sub-functions, normally known as *use cases*. Figure 2 shows an outline of the most commonly found SON use cases in the recent literature.

As it can be seen, future networks are not only expected to address several issues that are commonly found in current cellular networks, but also act as an enabler for future technologies and domains that do not exist today.

MACHINE LEARNING APPLICATIONS IN SON

From the previous sections, it can be seen that intelligence in future networks is a promising concept. However, since each SON function and its respective use cases have different requirements and rely on

Figure 2. SON Use cases
adapted from (Valente Klaine, Imran, Onireti, & Demo Souza, 2017)

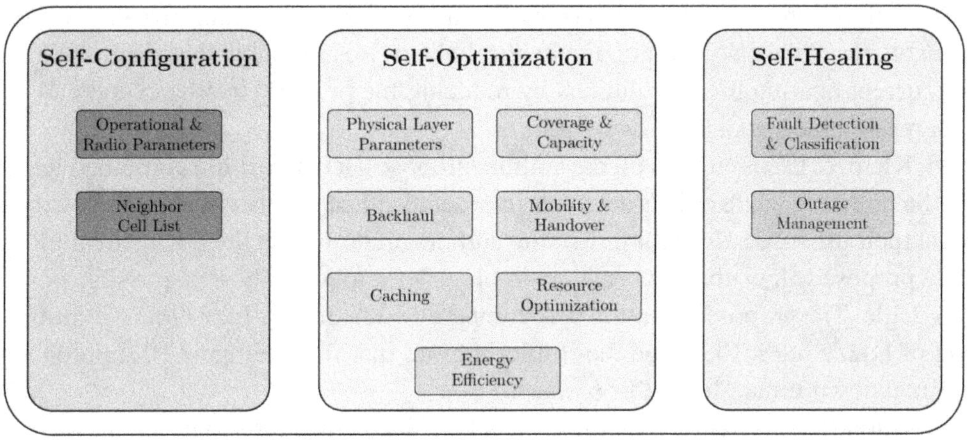

different types of data, demanding different types of ML solutions, some ML algorithms are more suitable for different *self-x* functions (Valente Klaine, Imran, Onireti, & Demo Souza, 2017).

In this section, first, some examples of applications of ML algorithms in each SON function will be analyzed and commented upon, highlighting which problem it aimed to tackle, how it was implemented and its results. Then, the authors would like to comment on the suitability of ML techniques for each SON use case, highlighting which techniques are more appropriate for which type of problem, while also giving some recommendations on which type of ML technique to use for every SON function.

Machine Learning in Self-Configuration

Although the concept of SON has been around for a while, the area of self-configuration is still underexplored. However, despite this fact, there are some interesting works covering the autonomous configuration of parameters of BSs.

Operational and Radio Parameters

The first step into configuring a new network should be the configuration of BSs basic parameters, such as IP address, main gateway, cell identifier and network topology. In addition, the configuration of BSs radio parameters is also important in order to allow users to connect to it. Examples of radio parameters are transmission power, in terms of both data and pilot symbols, antenna down-tilt and azimuth angles and HO parameters. Regarding the configuration of these parameters, a few examples can be found in the literature.

The work in (Binzer & Landstorfer, 2000), for example, proposes a solution based in Self Organizing Maps in order to configure and optimize network parameters of a Code Division Multiple Access (CDMA) network. The proposed solution aimed to determine not only planning parameters of a network, such as the number of BSs required in a certain area and their positions, but also radio parameters, such as the antenna's beam pattern and its transmit power.

Another work in this domain is the work in (Safdar, Hasbulah, & Rehan, 2015), in which the authors propose a RL approach to determine the best routing protocols for a cognitive radio network. The proposed intelligent algorithm, based on Q-Learning, has as its main goal to minimize the overall end-to-end delay of the network, while also maximizing its throughput. Results show that the intelligent algorithm outperforms current non-intelligent solutions by reducing the protocol overhead and end-to-end delay, while also increasing the package delivery ratio.

In (Razavi, Klein, & Claussen, 2010), the authors propose a RL algorithm combined with fuzzy logic to configure the down-tilt angle of BSs antennas in order to adjust the coverage and capacity of each cell. In order to test their proposed algorithm, first, the authors randomly set the down-tilt angles of each BS. After that, the proposed algorithm was enabled in order for it to learn the best possible configuration of the antenna's angle. The proposed solution was compared with another fuzzy logic solution (Evolutionary Learning of Fuzzy rules, ELF) and the results showed that the intelligent RL approach was able to learn the optimal down-tilt angle for this scenario.

Another solution that also combines the concepts of RL with fuzzy logic is the work in (Islam & Mitschele-Thiel, 2012). The proposed approach is very similar to the one previously mentioned, but in this case, despite the authors also trying to configure and learn the best down-tilt angle of antennas, they also consider different sources of noise in the data, such as thermal noise and receiver noise.

Neighbor Cell List (NCL)

Another important parameter to be configured whenever a new network must be deployed is the list of neighbors that a BS will communicate to. This is essential in order to establish HOs and for the network topology. For example, whenever a new BS is added to the network, this new BS should be capable of autonomously sensing and discovering its nearest neighbors, in order to add them to a list and be able to communicate to them. In addition, the other BSs that the new BS is now connected to must also add it to their list, otherwise, no communication will be possible.

However, despite its importance, this use-case of self-configuration has seen no usage of ML algorithms in recent years. Most solutions used still rely on the usage of feedback controllers and simple thresholds based on the distance between BSs or the perceived SINR (Valente Klaine, Imran, Onireti, & Demo Souza, 2017). Thus, the authors would like to emphasize that this area of SON is still an open issue and that further research should focus on it.

Table 1 shows a summary of the proposed ML solutions in the area of self-configuration.

Table 1. Application of ML algorithms in self-configuration

	Supervised Learning	Unsupervised Learning	Reinforcement Learning	Miscellaneous
Operational & Radio Parameters	✗	✓	✓	✗
Neighbor Cell List	✗	✗	✗	✗

Machine Learning in Self-Optimization

In contrast to self-configuration, which has received very few attention in the past couple years, self-optimization, on the other hand, is a very active research area, mainly due to a large number of parameters that can be adjusted and optimized. Since a network environment is not static with changes occurring in a network all the time, a wireless network might have to adjust its initially configured parameters in order to adapt itself and keep its objectives. Thus, self-optimization will play vital role in future cellular networks.

Physical Layer Parameters

One important aspect of future cellular networks is the optimization of physical layer parameters of communication. Nowadays, before the transmission of data begins, the transmitter must send to the receiver the communication parameters that it is going to use. Such parameters include the type of modulation and coding, type of waveform, FFT size and Cyclic Prefix (CP) length, to name a few (Su, 2013), (Lichtman, Headley, & Reed, 2014).

This is far from ideal, as any abrupt changes in the channel or network conditions may require a change in one of these parameters, requiring the transmitter to send them again to the receiver. Consequentially, this can waste a lot of bandwidth, while also increasing communication delay and signaling cost (Lichtman, Headley, & Reed, 2014). Thus, autonomous methods that attempt to dynamically learn or blindly estimate the transmitter's parameters at the receiver side are welcomed.

One example of such solution is the work in (Lichtman, Headley, & Reed, 2014), in which the authors attempt to determine at the receiver side which type of single carrier modulation was sent by the transmitter considering the effect of IQ imbalance. In this solution, the authors compare three different types of supervised learning methods, mainly *K*-NN, SVM and decision trees. Results show that the supervised learning methods are able to perform well in terms of classifying different types of modulation techniques, while also being able to compensate for the IQ imbalances introduced at the receiver side.

Another interesting work in the area of physical layer optimization is the work in (Dörner, Cammerer, Hoydis, & Brink, 2017), in which the authors design a whole communication system using nothing but neural networks and deep learning. The authors approach demonstrates that it is entirely possible to create a communication system in which the whole physical layer processing is performed by neural networks.

Backhaul Parameters

Another critical parameter that can be optimized is the backhaul connection of BSs, that is, the connection between the BSs and the core network. Some works that perform this optimization include the works in (Jaber M., Imran, Tafazolli, & Tukmanov, A Distributed SON-Based User-Centric Backhaul Provisioning Scheme, 2016), (Jaber M., Imran, Tafazolli, & Tukmanov, A Multiple Attribute User-Centric Backhaul Provisioning Scheme Using Distributed SON, 2016), (Jaber M., Imran, Tafazolli, & Tukmanov, 2015), in which a shift in current cellular network cell association paradigm is proposed.

Current cell association procedures look only at the Radio Access Network (RAN) parameters, such as the received signal power and Signal-to-Interference-plus-Noise Ratio (SINR). In the future, however, looking only at RAN parameters might not be enough in order to satisfy users, as the backhaul of the mobile network can become the bottleneck (Jaber M., Imran, Tafazolli, & Tukmanov, A Distributed

SON-Based User-Centric Backhaul Provisioning Scheme, 2016). Thus, solutions that look at this end-to-end connectivity, between the core network and user and not only the RAN parameters can be suitable.

In this sense, the authors propose to use RL in order to optimize the user cell association and improve overall user QoS in terms of throughput, latency and resiliency. The algorithm, based on Q-Learning, aims to tune the Cell Range Extension Offset (CREOs) of small cells so that users can connect to the small cell that has the backhaul that would better match a user needs (for example, if a user requires low latency, it would tend to connect to a backhaul that provides low latency, etc.). Results show that the proposed solution is capable of improving overall satisfaction, at the cost of a minor reduction in total cumulative throughput.

Caching

Due to the exponential growth in mobile devices and the rising popularity of social networks and multimedia services, video streaming will compose a huge part of the mobile network traffic. In 2016, it was already accounted that 60% of total mobile traffic came from mobile video, and this number is expected to be around 80% by 2021 (Index, 2017). Thus, in order to address this huge demand, caching must be provided in all levels of the network, but the questions of what, where and how to cache, remain.

In this context, some works that investigate the impact of caching in mobile networks is the work in (Bastug, Bennis, & Debbah, 2014), in which the authors propose two different solutions for the caching problem. The first one is based on collaborative filtering, a type of supervised learning technique that tries to build a model from a source's past behavior considering not only that source's data but also data from other similar sources. By using this algorithm, the authors were able to develop a caching solution that alleviated backhaul congestion.

The second solution proposed by them considered a scenario involving D2D communications, in which specific users would have contents cached into their devices so that it could be shared with other users. By using an unsupervised learning algorithm, *K*-means, the authors were able to cluster users into different groups and dynamically distribute content in the clusters.

Coverage and Capacity

In order to cope with the increasing traffic, it is essential that network coverage and capacity is constantly optimized in future networks. Based on that, several works propose intelligent solutions in order to tackle this problem.

One possible solution is the work proposed in (Debono & Buhagiar, 2005), in which the authors utilize the unsupervised learning technique, SOM, in order to achieve better network coverage. Their proposed algorithm would first try to optimize the number of BSs inside a cluster and then optimize each BSs antenna's parameters. The authors also propose two different scenarios, one in which only cluster sizes are changed, and another in which both cluster sizes and antenna parameters are optimized. Another approach to the coverage and capacity optimization problem is the work in (Ho, Ashraf, & Claussen, 2009), in which a Genetic Algorithm is proposed to minimize network coverage holes, perform load balancing and minimize pilot channel transmit power.

Mobility and Handover

The process of changing the communication channel in an on-going call is known as HO. HOs are of extreme importance in mobile networks due to the mobility of its users. Without this procedure, mobility could not be supported, as connections would not survive the process of changing cells. Thus, in order to optimize and improve the performance of future cellular networks, it is vital that NGMNs are able to predict user's movement in order to better manage network resources (Valente Klaine, Imran, Onireti, & Demo Souza, 2017). Figure 3 shows an example of how ML algorithms can be used in order to track user mobility.

One recent work in the area of mobility prediction is the work in (Mohamed, Onireti, Imran, Imran, & Tafazolli, 2017), in which the authors propose a discrete Markov chain in order to predict the next cell a user will be. The proposed solution is shown to be signaling efficient, being able to minimize overhead signals and HO latency.

Another solution in the realm of HO optimization is the optimization of HO parameters, such as the work in (Sinclair, Harle, Glover, Irvine, & Atkinson, 2013). In this work, the authors proposed a SOM in order to optimize two critical HO parameters, hysteresis and TTT, and achieve a balance between unnecessary HOs and call drop rates. Results show that the proposed solution is able to significantly reduce the number of unnecessary HOs and dropped calls by up to 70%.

Resource Allocation and Load Balancing

Another critical aspect of future mobile networks is the provisioning and allocation of network resources. Additionally, since traffic is usually unevenly distributed in mobile networks, solutions that are also able to intelligently manage and balance traffic load are also welcomed.

Figure 3. A user moving through the network from point A to point B. In this case, historical data from the user could be leveraged in order to predict for different days of the week, or different times of the day, the route the user is most likely to take.

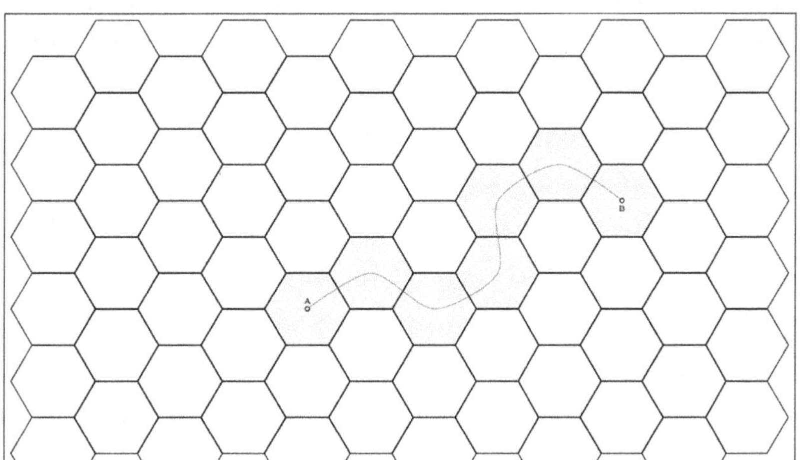

In this sense, solutions such as (Dirani & Altman, 2011), propose a framework for uplink transmission of mobile users in a LTE network in order to reduce blocking rates and file transfer times. The proposed solution relies on Q-Learning combined with Fuzzy Logic to dynamically optimize Radio Resource Management (RRM) functions and it is able to provide gains in terms of network capacity and user-perceived quality in data applications. Another work regarding the use case of load balancing is the work in (Bassoy, Jaber, Imran, & Xiao, 2016), in which the authors utilize a clustering algorithm, in a network with separated control and data planes in order to reduce high load on cells in hotspot areas and improve user satisfaction rates.

Energy Efficiency

In the future, with the increasing number of BSs being deployed by operators, the huge amount of devices connected to the network, as well as the development of cloud solutions and network function virtualization, the total energy consumption of wireless networks is expected to largely increase (Alsedairy, Qi, Imran, Imran, & Evans, 2015). Thus, intelligent solutions that are able to improve the network energy efficiency, providing a greener network in order not only to reduce operator's costs but most importantly, to benefit the natural environment are necessary.

In this context, a work that utilizes RL, more specifically Q-Learning, in order to determine which BSs to turn on or off and to improve the energy efficiency of mobile networks was proposed in (Miozzo, Giupponi, Rossi, & Dini, 2017). Another interesting approach to the energy optimization problem is the work in (Wang, Cheng, & Tsai, 2016). In this paper, the authors utilize a supervised learning algorithm, polynomial regression, combined with data collected by the network in order to improve the total cell throughput and energy efficiency of an ultra-dense small cell network. Results show that the proposed solution is able to improve throughput by 50% and energy efficiency by 135% when compared to other approaches.

Table 2 summarizes the ML methods utilized in state-of-the-art research in the area of self-optimization.

Machine Learning in Self-Healing

Another area that has received quite some attention in the last couple of years is the area of self-healing. Due to its critical nature, as whenever a fault happens in a network its service is disrupted, self-healing

Table 2. Application of ML algorithms in self-optimization

	Supervised Learning	Unsupervised Learning	Reinforcement Learning	Miscellaneous
Physical Layer	✓	✗	✗	✗
Backhaul	✗	✗	✓	✗
Caching	✓	✓	✗	✗
Coverage & Capacity	✗	✓	✗	✓
Mobility & Handover	✗	✓	✗	✓
Resource Allocation & Load Balancing	✗	✓	✓	✗
Energy Efficiency	✓	✗	✓	✗

functions must be precise, in order to detect or predict faults in advance and not generate false alarms, while also being agile in order to trigger the appropriate mechanisms and mitigate the effects of the failure (Onireti, et al., 2016).

Due to its critical nature, several research groups focus on the autonomous detection of faults, their classification and the management of two of the most common faults in a mobile network, cell outage and sleeping cell management. In the next couple of sub-sections, some examples of state-of-the-art solutions in self-healing will be presented.

Fault Detection and Classification

The first thing that a self-healing function must be able to do is to perform the detection of a fault. This includes determining where a fault has happened and when it has happened. In addition, in the case of proactive networks, instead of predicting when a fault has happened, the network can try and predict the likelihood of a fault happening in the next couple hours, for example (Valente Klaine, Imran, Onireti, & Demo Souza, 2017). Furthermore, not only the detection of a fault is important, but also its classification plays a crucial role in self-healing, as correctly classifying a fault, allows the correct mechanisms to be triggered in order to solve it.

In the context of proactive networks, one possible solution is the work proposed in (Farooq, Parwez, & Imran, 2015), in which the authors utilize a continuous time Markov chains combined with exponential distributions to model the reliability behavior of BSs. The proposed solution is evaluated in three different scenarios and in all of them the primary objective is to predict the occurrence of faults in the network based on historical data. Their results show that the proposed solution is able to learn to predict certain faults and reduce network recovery time, improving the overall network reliability.

Another approach to fault detection is the work in (Hashmi, Darbandi, & Imran, 2017). In this work, the authors compare five different unsupervised learning techniques in order to detect faults in mobile networks. Through the use of extensive data analysis, the proposed solutions are capable of extracting network trends and tackle similar faults in the future. Thus, the paper shows that the algorithms are capable of minimizing network recovery time and operational costs, paving the way for future self-healing approaches.

Other solutions, such as (Wang, Zhang, & Zhang, 2011) propose a framework based on transfer learning (which consists of applying a learning algorithm in a similar task and then using this already pre-trained model in a new task) in order to classify faults in femtocells. By leveraging the historical data of other femtocells, the authors are able to build a model in which two independent classifiers are trained and used as voters in the fault classification model. The proposed solution is compared with other approaches and their results show that their transfer learning approach is able to have higher accuracy than other methods.

Outage and Sleeping Cell Management

Lastly, one of the most common faults that can happen in a network is a cell being in an outage. Another particular type of problem in the outage realm is also the sleeping cell problem, in which an outage cell is not shown to the operator (users are not able to connect to that cell, but from the operator's perspective, the cell appears to be fully operable). Both these scenarios are equally important and have seen a great deal of attention recently, due to the increased scalability and complexity issues expected in

NGMNs (Valente Klaine, Imran, Onireti, & Demo Souza, 2017). Figure 4 shows an example on how ML algorithms will be able to detect that an outage has happened in the network and in which location and automatically trigger the correct functions, so that the outage effects can be mitigated.

One example of cell outage management includes the work in (Onireti, et al., 2016). In this work, the authors consider a framework with distinct control and data planes and propose an algorithm that is able to detect cell outages in both planes. The authors utilize both supervised and unsupervised techniques to perform outage detection and in order to perform outage compensation (mitigate the outage effects), the authors consider a RL approach in order to adjust antenna gains and transmit powers.

Regarding sleeping cell management, one example is the work in (Chernov, Chernogorov, Petrov, & Ristaniemi, 2014), in which a data mining framework focused on the detection of sleeping cells caused mainly by the failure of random access channel is presented. The proposed algorithm, based on supervised learning (K-Nearest Neighbors) is utilized, together with dimension reduction techniques in order to assign scores for possible cells in a sleeping state. According to each cell score, the algorithm is shown to be able to detect whenever a cell is in a sleeping state or not.

Table 3 presents the ML solutions found in the area of self-healing.

Guidelines on the Application of ML in SON

From the previous sections, it could be seen that certain algorithms are utilized more predominantly in one use case than the other, this is no coincidence, as depending on the type of data and the requirements

Figure 4. Use case of outage management, in which the central site of a network goes in outage (a). Based on that, the appropriate mechanisms are triggered and the neighbor BSs adjust their transmit powers in order to cover the outage area (b), adapted from (Valente Klaine, Imran, Onireti, & Demo Souza, 2017).

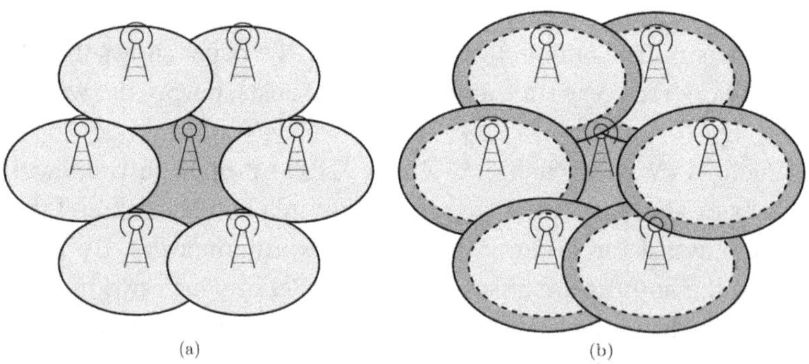

Table 3. Application of ML algorithms in self-healing

	Supervised Learning	Unsupervised Learning	Reinforcement Learning	Miscellaneous
Fault Detection & Classification	✓	✓	✗	✓
Outage & Sleeping Cell Management	✓	✓	✓	✗

needed, some ML algorithms are more suitable for the job than others. In (Valente Klaine, Imran, Onireti, & Demo Souza, 2017), the authors provide a detailed view and comparison of different ML solutions in terms of certain SON metrics, such as scalability, algorithm training, response and convergence time, the amount of training data needed and its accuracy.

For example, certain use cases of self-configuration, such as the planning of BSs location, or the configuration of network topology do not require a very stringent time to be computed. Thus, offline solutions that explore a wide range of possibilities and possible solutions, such as genetic algorithms, or more complex heuristics could be used. On the other hand, use cases such as mobility prediction or fault detection, which require almost instantaneous responses, would not benefit from these approaches, hence, solutions that have a lower response time would be required, such as supervised or unsupervised learning.

Based on these two simple examples, it is clear that the choice of ML algorithm should be done carefully, as it can have a big impact on the problem being tackled.

FUTURE TRENDS

As it could be seen, intelligence in future mobile networks is expected to play a huge role, from being able to solve current limitations of present cellular networks, to creating and enabling new types of services. However, despite the great amount of work being done in the past couple years in the area of SON, there are still some open issues and challenges that need to be addressed (Valente Klaine, Imran, Onireti, & Demo Souza, 2017).

Challenges in Self-Configuration

As it could be seen from the self-configuration section, this SON function is still underexplored, with few intelligent solutions being applied in order to solve its use cases. Hence, several challenges in this function are still present, such as how to autonomously configure a mobile network in scenarios with an ultra-dense deployment of BSs? In the future, BSs are expected to be deployed not only by the network operators but by regular users as well. Hence, some questions remain on how to autonomously and intelligently configure these nodes out of the control of the operator and how to efficiently integrate them into the network (Valente Klaine, Imran, Onireti, & Demo Souza, 2017).

In addition, a specific uncharted area in self-configuration is also the automatic configuration of NCLs, which plays a vital role in the configuration of the network and its topology. However, despite its importance, no work using ML algorithms in this use case was found and still simple solutions involving threshold-based or feedback controllers are still dominant.

Lastly, another area that is still not well covered in the realm of self-configuration is the reconfiguration of a network after natural disasters have occurred. In this context, the development of ML solutions that are able to rapidly and intelligently reconfigure a network from zero is essential in order to not only re-establish network service, but also to save human lives (Erdelj, Natalizio, Chowdhury, & Akyildiz, 2017).

Challenges in Self-Optimization

Despite self-optimization being a well-explored area, some issues still remain. In particular, use cases such as the optimization of backhaul resources are essential for future cellular networks, as the backhaul is expected to be the main bottleneck of future networks, instead of the RAN (Jaber M., Imran, Tafazolli, & Tukmanov, A Distributed SON-Based User-Centric Backhaul Provisioning Scheme, 2016). Hence, solutions that are able to optimize current backhaul links, without the need of upgrading all of them are welcomed.

Another area that can be further explored is the area of caching, in which the determination of which contents to cache and at which BSs still play a major role. In addition, in order to meet the stringent requirements in delay for NGMNs, caching contents at the edge of the network based on data collected from each cell in different times of the day can also be explored (Bastug, Bennis, & Debbah, 2014).

Finally, another use case that can play major role in future mobile networks is the optimization of physical layer parameters. Most of the current solutions attempt to blindly estimate the transmitter parameters at the receiver side considering only single carrier waveforms (Lichtman, Headley, & Reed, 2014). However, in the future, as multi-carrier waveforms, such as Orthogonal Frequency Division Multiplexing and its variants, are expected to play a major role, it is essential that the prediction of physical layer parameters can be done in multi-carrier systems.

Challenges in Self-Healing

In terms of self-healing functions, as it could be seen, most of them still rely on the concept of a reactive network, trying to learn from previous historical data in order to automatically detect and classify a fault in the network in order to trigger the appropriate mechanisms that will mitigate its effects. However, as already mentioned in this chapter, a major shift is required in future networks, in which networks are expected to become proactive instead of reactive. In this sense, the deployment of ML solutions in order to try and predict with a high level of certainty whenever faults are likely to occur in a mobile network is still needed.

CONCLUSION

In this chapter, a clear view of why ML is necessary for future networks was presented. In this context, the authors highlighted the importance of intelligent solutions in future networks as not only a solution in order to overcome current cellular networks obstacles but also as an enabler for future services and applications that today are unimaginable. With the advents of smart cities, vehicles, personal assistants and the ever increasing demand for better services, it is essential that NGMNs shift their paradigms, in order to keep on par with other state-of-the-art systems. Thus, artificial intelligence and ML algorithms are expected to play a huge role in the future of SONs.

Furthermore, for future networks to become a reality, a change in paradigm is also necessary for them to become just operable. As it is expected, in the future, thousands of small cells will be deployed in a mobile network, some without the control of the operator, thousands of cells and their parameters will need to be configured, optimized and monitored in order to detect and classify failures in a network. Thus, if all of these are done as the way it is done today, future networks are infeasible, as the amount of data generated and collected daily will be immense. Thus, in order to leverage the full potential of data analytics using all of this massive data collected by the network, intelligent solutions that rely on data analysis, such as ML algorithms are essential in order to enable SON.

ACKNOWLEDGMENT

This research was supported by the DARE project grant (No. EP/P028764/1) under the EPSRC's Global Challenges Research Fund (GCRF) allocation.

The authors would also like to thank CNPq (Brazil) for the support for this work.

REFERENCES

Aliu, O., Imran, A., Imran, M., & Evans, B. (2013). A survey of self organisation in future cellular networks. *IEEE Communications Surveys and Tutorials*, *15*(1), 336–361. doi:10.1109/SURV.2012.021312.00116

Alsedairy, T., Qi, Y., Imran, A., Imran, M., & Evans, B. (2015). Self organising cloud cells: A resource efficient network densification strategy. *Transactions on Emerging Telecommunications Technologies*, *26*(8), 1096–1107. doi:10.1002/ett.2824

Andrews, G., Buzzi, S., Choi, W., Hanly, S. V., Lozano, A., Soong, A. C. K., & Zhang, J. C. (2014). What will 5G be? *IEEE Journal on Selected Areas in Communications*, *32*(6), 1065–1082. doi:10.1109/JSAC.2014.2328098

Bassoy, S., Jaber, M., Imran, M. A., & Xiao, P. (2016). Load Aware Self-Organising User-Centric Dynamic CoMP Clustering for 5G Networks. *IEEE Access: Practical Innovations, Open Solutions*, *4*, 2895–2906. doi:10.1109/ACCESS.2016.2569824

Bastug, E., Bennis, M., & Debbah, M. (2014). Living on the edge: The role of proactive caching in 5G wireless networks. *IEEE Communications Magazine*, *52*(8), 82–89. doi:10.1109/MCOM.2014.6871674

Binzer, T., & Landstorfer, F. (2000). Radio network planning with neural networks. In *Vehicular Technology Conference Fall 2000. IEEE VTS Fall VTC2000. 52nd Vehicular Technology Conference* (pp. 811-817). Boston: IEEE. 10.1109/VETECF.2000.887116

Chernov, S., Chernogorov, F., Petrov, D., & Ristaniemi, T. (2014). Data mining framework for random access failure detection in LTE networks. In *2014 IEEE 25th Annual International Symposium on Personal, Indoor, and Mobile Radio Communication (PIMRC)* (pp. 1-6). Washington, DC: IEEE.

Debono, C., & Buhagiar, J. (2005). Cellular network coverage optimization through the application of self-organizing neural networks. In *IEEE 62nd Vehicular Technology Conference, 2005. VTC-2005-Fall* (pp. 1-5). Dallas, TX: IEEE.

Dirani, M., & Altman, Z. (2011). Self-organizing networks in next generation radio access networks: Application to fractional power control. *Computer Networks*, *55*(2), 431–438. doi:10.1016/j.comnet.2010.08.012

Dörner, S., Cammerer, S., Hoydis, J., & Brink, S. T. (2017). Deep Learning Based Communication Over the Air. *IEEE Journal of Selected Topics in Signal Processing*, 132–143.

Erdelj, M., Natalizio, E., Chowdhury, K. R., & Akyildiz, I. F. (2017). Help from the Sky: Leveraging UAVs for Disaster Management. *IEEE Pervasive Computing*, *16*(1), 24–32. doi:10.1109/MPRV.2017.11

Farooq, H., Parwez, M. S., & Imran, A. (2015). Continuous Time Markov Chain Based Reliability Analysis for Future Cellular Networks. In *2015 IEEE Global Communications Conference (GLOBECOM)* (pp. 1-5). San Diego, CA: IEEE. 10.1109/GLOCOM.2015.7417594

Fettweis, G. (2012). A 5G wireless communications vision. *Microwave Journal*, 24–36.

Friedman, J., Hastie, T., & Tibshirani, R. (2001). *The elements of statistical learning*. New York: Springer.

Hashmi, U. S., Darbandi, A., & Imran, A. (2017). Enabling proactive self-healing by data mining network failure logs. In *2017 International Conference on Computing, Networking and Communications (ICNC)* (pp. 1-7). Santa Clara, CA: IEEE. 10.1109/ICCNC.2017.7876181

Ho, L. T., Ashraf, I., & Claussen, H. (2009). Evolving femtocell coverage optimization algorithms using genetic programming. In *IEEE 20th International Symposium on Personal, Indoor and Mobile Radio Communications, 2009* (pp. 1-5). Tokyo, Japan: IEEE.

Huawei, Technologies Co. (2016). *5G opening up new business opportunities*. Shenzhen: Huawei Technologies Co.

Imran, A., Zoha, A., & Abu-Dayya, A. (2014). Challenges in 5G: How to empower SON with big data for enabling 5G. *IEEE Network*, *28*(6), 27–33. doi:10.1109/MNET.2014.6963801

Index, C. (2017). *Global Mobile Data Traffic Forecast Update*. San Jose, CA: Cisco White Paper.

Islam, M. N., & Mitschele-Thiel, A. (2012). Reinforcement learning strategies for self-organized coverage and capacity optimization. In *IEEE Wireless Communications and Networking Conference (WCNC)*, 2012 (pp. 1-6). Shanghai, China: IEEE. 10.1109/WCNC.2012.6214281

Jaber, M., Imran, M., Tafazolli, R., & Tukmanov, A. (2015). An adaptive backhaul-aware cell range extension approach. In *IEEE International Conference on Communication Workshop (ICCW)*, 2015 (pp. 1-6). London, UK: IEEE. 10.1109/ICCW.2015.7247158

Jaber, M., Imran, M. A., Tafazolli, R., & Tukmanov, A. (2016). A Distributed SON-Based User-Centric Backhaul Provisioning Scheme. *IEEE Access: Practical Innovations, Open Solutions*, *4*, 2314–1330. doi:10.1109/ACCESS.2016.2566958

Jaber, M., Imran, M. A., Tafazolli, R., & Tukmanov, A. (2016). A Multiple Attribute User-Centric Backhaul Provisioning Scheme Using Distributed SON. In *IEEE Global Communications Conference (GLOBECOM)*, 2016 (pp. 1-6). Washington, DC: IEEE. 10.1109/GLOCOM.2016.7841518

Kaelbling, L., Littman, M., & Moore, A. (1996). Reinforcement learning: A survey. *Journal of Artificial Intelligence Research*, *4*, 237–285. doi:10.1613/jair.301

Kotsiantis, S., Zaharakis, I., & Pintelas, P. (2007). Supervised machine learning: A review of classification techniques. In *Proceedings of the 2007 conference on Emerging Artificial Intelligence Applications in Computer Engineering*, (pp. 3-24). Amsterdam: Academic Press.

LeCun, Y., Bengio, Y., & Hinton, G. (2015). Deep learning. *Nature*, *521*(7553), 436–444. doi:10.1038/nature14539 PMID:26017442

Lichtman, M., Headley, W. C., & Reed, J. H. (2014). Automatic Modulation Classification under IQ Imbalance Using Supervised Learning. In *IEEE Military Communications Conference, MILCOM 2013* (pp. 1-6). San Diego, CA: IEEE.

Miozzo, M., Giupponi, L., Rossi, M., & Dini, P. (2017). *Switch-On/Off Policies for Energy Harvesting Small Cells through Distributed Q-Learning. In 2017 IEEE Wireless Communications and Networking Conference Workshops (WCNCW)* (pp. 1–6). San Francisco, CA: IEEE.

Mohamed, A., Onireti, O., Hoseinitabatabaei, S. A., Imran, M., Imran, A., & Tafazolli, R. (2015). Mobility prediction for handover management in cellular networks with control/data separation. In *2015 IEEE International Conference on Communications (ICC)* (pp. 1-6). London, UK: IEEE. 10.1109/ICC.2015.7248939

Mohamed, A., Onireti, O., Imran, M. A., Imran, A., & Tafazolli, R. (2017). Predictive and Core-Network Efficient RRC Signalling for Active State Handover in RANs With Control/Data Separation. *IEEE Transactions on Wireless Communications*, *16*(3), 1423–1436. doi:10.1109/TWC.2016.2644608

Onireti, O., Mohamed, A., Pervaiz, H., & Imran, M. (2017). Analytical approach to base station sleep mode power consumption and sleep depth. In *2017 IEEE 28th Annual International Symposium on Personal, Indoor, and Mobile Radio Communications (PIMRC)* (pp. 1-7). Montreal, Canada: IEEE.

Onireti, O., Zoha, A., Moysen, J., Imran, A., Giupponi, L., Imran, M. A., & Abu-Dayya, A. (2016). A Cell Outage Management Framework for Dense Heterogeneous Networks. *IEEE Transactions on Vehicular Technology*, *65*(4), 2097–2113. doi:10.1109/TVT.2015.2431371

Razavi, R., Klein, S., & Claussen, H. (2010). A Fuzzy reinforcement learning approach for self-optimization of coverage in LTE networks. *Bell Labs Technical Journal*, *15*(3), 153–175. doi:10.1002/bltj.20463

Safdar, T., Hasbulah, H., & Rehan, M. (2015). Effect of Reinforcement Learning on Routing of Cognitive Radio Ad-Hoc Networks. In *International Symposium on Mathematical Sciences and Computing Research (iSMSC)* (pp. 1-7). Ipon, Malaysia: IEEE. 10.1109/ISMSC.2015.7594025

Samsung Electronics Co. (2015). *5G Vision*. Samsung Electronics Co.

Simon, P. (2013). *Too Big to Ignore: the Business Case for Big Data*. John Wiley & Sons.

Sinclair, N., Harle, D., Glover, I. A., Irvine, J., & Atkinson, R. C. (2013). An Advanced SOM Algorithm Applied to Handover Management Within LTE. *IEEE Transactions on Vehicular Technology*, *62*(5), 1883–1894. doi:10.1109/TVT.2013.2251922

Su, W. (2013). Feature Space Analysis of Modulation Classification Using Very High-Order Statistics. *IEEE Communications Letters*, *17*(9), 1688–1691. doi:10.1109/LCOMM.2013.080613.130070

Sutton, R., & Barto, A. (1998). *Reinforcement Learning: An Introduction*. Cambridge, MA: MIT Press.

Valente Klaine, P., Imran, M., Onireti, O., & Demo Souza, R. (2017). A Survey of Machine Learning Techniques Applied to Self-Organizing Cellular Networks. *IEEE Communications Surveys and Tutorials*, *19*(4), 2392–2431. doi:10.1109/COMST.2017.2727878

Wang, L.-C., Cheng, S.-H., & Tsai, A.-H. (2016). Bi-SON: Big-Data Self Organizing Network for Energy Efficient Ultra-Dense Small Cells. In *2016 IEEE 84th Vehicular Technology Conference (VTC-Fall)* (pp. 1-5). Montreal, Canada: IEEE.

Wang, W., Zhang, J., & Zhang, Q. (2011). Transfer Learning Based Diagnosis for Configuration Troubleshooting in Self-Organizing Femtocell Networks. In *2011 IEEE Global Telecommunications Conference (GLOBECOM 2011)* (pp. 1-5). Kathmandu, Nepal: IEEE. 10.1109/GLOCOM.2011.6133802

ADDITIONAL READING

Fan, B., Leng, S., & Yang, K. (2016). A dynamic bandwidth allocation algorithm in mobile networks with big data of users and networks. *IEEE Network*, *30*(1), 6–10. doi:10.1109/MNET.2016.7389824

Farooq, H., & Imran, A. (2017). Spatiotemporal mobility prediction in proactive self-organizing cellular networks. *IEEE Communications Letters*, *21*(2), 370–373. doi:10.1109/LCOMM.2016.2623276

Gomez-Andrades, A., Barco, R., Munoz, P., & Serrano, I. (2017). Data analytics for diagnosing the RF condition in self-organizing networks. *IEEE Transactions on Mobile Computing*, *16*(6), 1587–1600. doi:10.1109/TMC.2016.2601919

Mehta, M., Rane, N., Karandikar, A., Imran, M. A., & Evans, B. G. (2016). A self-organized resource allocation scheme for heterogeneous macro-femto networks. *Wireless Communications and Mobile Computing*, *16*(3), 330–342. doi:10.1002/wcm.2518

Wang, W., Liao, Q., & Zhang, Q. (2014). COD: A cooperative cell outage detection architecture for self-organizing femtocell networks. *IEEE Transactions on Wireless Communications*, *13*(11), 6007–6014. doi:10.1109/TWC.2014.2360865

Wang, X., Chen, M., Taleb, T., Ksentini, A., & Leung, V. (2014). Cache in the air: Exploiting content caching and delivery techniques for 5G systems. *IEEE Communications Magazine*, *52*(2), 131–139. doi:10.1109/MCOM.2014.6736753

Zheng, K., Yang, Z., Zhang, K., Chatzimisios, P., Yang, K., & Xiang, W. (2016). Big data-driven optimization for mobile networks toward 5G. *IEEE Network*, *30*(1), 44–51. doi:10.1109/MNET.2016.7389830

Zoha, A., Saeed, A., Imran, A., Imran, M. A., & Abu-Dayya, A. (2016). A learning-based approach for autonomous outage detection and coverage optimization. *Transactions on Emerging Telecommunications Technologies*, *27*(3), 439–450. doi:10.1002/ett.2971

KEY TERMS AND DEFINITIONS

Machine Learning: Class of algorithms that analyze data and make predictions about it.

Mobile Networks: Radio communication network, in which the last hop of the network is wireless.

Reinforcement Learning: Algorithms that learn from their own experience with the environment.

Self-Configuration: Autonomous configuration of network parameters in order for a network to become operable.

Self-Healing: Autonomous detection, classification, and management of failures in a mobile network, so that the effects of a fault can be mitigated.

Self-Optimization: The process of autonomously adjusting network parameters so that the network performance can be back to near optimum.

Self-Organizing Cellular Networks: Mobile networks that are capable of autonomously adapting itself to changes in their environment, while maintaining its desired objectives.

Supervised Learning: Class of machine learning algorithms that rely on the knowledge of an external supervisor in order to learn.

Unsupervised Learning: A class of machine learning algorithms that, given a set of unlabeled data, is capable of learning a function that best represents the model.

Chapter 2
Machine Learning in Radio Resource Scheduling

Ioan-Sorin Comşa
Brunel University London, UK

Sijing Zhang
University of Bedfordshire, UK

Mehmet Emin Aydin
University of the West of England, UK

Pierre Kuonen
University of Applied Sciences of Western Switzerland, Switzerland

Ramona Trestian
https://orcid.org/0000-0003-3315-3081
Middlesex University London, UK

Gheorghiţă Ghinea
Brunel University London, UK

ABSTRACT

In access networks, the radio resource management is designed to deal with the system capacity maximization while the quality of service (QoS) requirements need be satisfied for different types of applications. In particular, the radio resource scheduling aims to allocate users' data packets in frequency domain at each predefined transmission time intervals (TTIs), time windows used to trigger the user requests and to respond them accordingly. At each TTI, the scheduling procedure is conducted based on a scheduling rule that aims to focus only on particular scheduling objective such as fairness, delay, packet loss, or throughput requirements. The purpose of this chapter is to formulate and solve an aggregate optimization problem that selects at each TTI the most convenient scheduling rule in order to maximize the satisfaction of all scheduling objectives concomitantly TTI-by-TTI. The use of reinforcement learning is proposed to solve such complex multi-objective optimization problem and to ease the decision making on which scheduling rule should be applied at each TTI.

DOI: 10.4018/978-1-5225-7458-3.ch002

INTRODUCTION

The continuous growth of mobile data usage and the increased interest for immersive video applications are pushing network operators to find suitable solutions to accommodate these services with very stringent Quality of Service (QoS) demands (Cisco, 2017). According to Trestian, Comsa, and Tuysuz (2018), the Quality of Experience provisioning will become the main differentiator between network operators, in which, the satisfaction of heterogeneous QoS requirements is playing a crucial role. In this context, the 5th Generation (5G) of mobile networks comes up with the promise of very low end-to-end latencies and much higher system capacity while implementing some important features such as (G. Andrews et al., 2014): new waveforms, densification of access networks, higher frequency bands, mass scale antennas and millimeter-wave communications. An issue to be addressed when delivering these bandwidth-hungry applications in multi-user scenario refers to the management of radio resources that can strongly affects the overall performance of QoS provisioning (Li et al., 2017). The responsible entity is the Radio Resource Management (RRM) that aims to ensure an efficient allocation of the disposable system bandwidth in order to maximize the QoS satisfaction while implementing advanced technologies as provided by Olwal, Djouani, and Kurien (2017) able to: save the energy, control the mobility and power allocation, mitigate interference, schedule users' packets in frequency domain at each TTI.

According to the performance of the scheduling process, the operator is struggling to provide the requested services while using the disposable radio infrastructure, regardless of the spatial/time positions of mobile terminals, user preferences, devices' types and application requirements (Comşa, 2014a). A major concern is to increase the system capacity or data rates of all active users while satisfying the application requirements. Users located in the proximity of base stations experience better channel quality and consequently can get higher data rates than those users with poorer channel quality located farer away from any available base station (Trestian, Muntean, & Ormond, 2009; Trestian, Ormond, & Muntean, 2012; Vien, Akinbote, Nguyen, Trestian, & Gemikonakli, 2015). By providing the disposable spectrum to those users with better channel conditions, other users are starved in receiving the requested data for longer time (Vien, Nguyen, Trestian, Shah, & Gemikonakli, 2016). Then, the fairness measure between different users with the same QoS profiles is impaired. Certain tradeoff measures between system throughput and user fairness can be adopted (Jain, Chiu, and Hawe, 1984; Comşa, 2014a). Together with these aspects, the requested services should be provided under some predefined QoS requirements as imposed by 3GPP (2012) in terms of Guaranteed Bit Rate (GBR), Head of Line (HoL) packet delay and Packet Loss Rate (PLR). The QoS requirements become more restrictive with the evolution of cellular standards, system architectures and applications. By encompassing the above discussed aspects, the packet scheduler is dealing with an optimization problem aiming to maximize the system throughput and being constrained by fairness and QoS requirements. If we further consider that the constraints' satisfaction refers to fairness and QoS objectives, then the Multi-Objective Optimization (MOO) is addressed as stated by Comşa, (2014a).

At each TTI, the scheduling procedure is conducted based on the scheduling rule that aims to prioritize active users in frequency domain (Toufik and Knopp, 2011). In literature, these rules are various, and each of them is targeting particular scheduling objective as surveyed by Capozzi, Piro, Grieco, Boggia, and Camarda (2013). For example, Zou, Trestian & Muntean propose a utility-based approach in order prioritize the multimedia streams within the radio resource scheduling process (2013). In general, the scheduling rule quantifies the benefit of allocating each user in the frequency domain from the perspective of the addressed objective. For this reason, almost all state-of-the-art scheduling rules impact differently when considering the multi-objective satisfaction criterion (Comşa, 2014a). For example, some scheduling rules are oriented more on the tradeoff between system throughput maximization and user fairness satisfaction while degrading the performance of QoS objectives (Proebster, Mueller, and Bakker, 2010). Other rules aim to focus more on particular QoS objective (i.e. delay) while harming the performance of other QoS objectives (i.e. GBR, PLR), system throughput and user fairness (Liu, Tian, and Xu, 2013). Therefore, the multi-objective performance strongly depends on the adopted scheduling rule.

Alongside the exploited scheduling rule, the multi-objective satisfaction depends also on the momentary scheduler states, observed at each TTI and being composed of: channel conditions, number of users, QoS parameters, queue loads, application types, etc. Actually, when the scheduler conditions permit (reduced number of users, very good channel conditions, low traffic load), the multi-objective target can be attained by almost all existing scheduling rules. However, under more generalized scheduler state space (variable traffic types, channel conditions, number of user, etc.) particular scheduling rules are unable to reach an acceptable level for the multi-objective satisfaction. Hence, it can be envisioned that a mixture of scheduling rules can be used instead of a single one adopted across the entire process, in which, the scheduling rule that gives the highest multi-objective outcome in each scheduler state must be found and applied in order to maximize the multi-objective satisfaction measure over time.

This chapter deals with the multi-objective optimization problem where the joint assignation of radio resources and scheduling rules is considered in downlink scheduling. The general idea is to find in each instantaneous scheduler state the most convenient scheduling rule to be applied in order to maximize the overall system throughput while keeping the QoS constraints satisfied as long as possible. Due to the increased complexity of the proposed aggregate optimization problem, we adopt the use of machine learning tools in order to find suitable solutions in each scheduler state. In this sense, the Reinforcement Learning (RL) framework is proposed to learn over time the most convenient scheduler rule for each momentary state as initially proposed by Comşa, Aydin, Zhang, Kuonen, and Wagen (2011; 2012). The implementation of RL framework for downlink Orthogonal Frequency Division Multiple Access (OFDMA) scheduling is also addressed. The choice of OFDMA is due to its simplicity, efficiency and its wide deployment, being one of the multiple access schemes to be considered in 5G networks (FANTASTIC-5G, 2016).

The remaining of this chapter is organized as follows. The Background section introduces the scheduling concept. In ODFMA Scheduling Process, the OFDMA scheduling elements are presented in terms of: scheduler state, resource allocation procedure, utility and objective functions. Single objective optimization problem presents the general form of single objective optimization problem when each scheduling objective is considered separately. Multiple Objective Optimization Problem formulates the proposed multi-objective aggregate problem based on each particular single-objective problems previously introduced. Multi-Objective Scheduling Based on Reinforcement Learning provides a sub-optimal solution to this complex problem based on the RL approach. Finally, future research directions and open issues are addressed in Future Research Directions and Open Issues.

BACKGROUND

In general, one scheduling rule is focused first on the main objective; once the main objective is satisfied, the static scheduling rule can optimize other objectives with the amendment that the first objective is always satisfied. In this case, the multi-objective optimization problem becomes Sequential MOO

(SMOO). As some studies show (Lundevall et al., 2004; Ning, Ying, and Ping, 2006; Zhang, Yuan, and Zhang, 2011), the SMOO problems are considering the GBR satisfaction as a primary objective, whereas the user fairness satisfaction or user throughput maximization is considered once the GBR objective is satisfied for all active users TTI-by-TTI. The same principles of SMOO are targeted by the proposed scheduling rules in (Rhee, Holtzman, and Kim; 2003; Sadiq, Madan, and Sampath, 2009; Bae, Choi, and Chung, 2011) where the delay is the first objective. Another scheduler (as proposed by Khan, Martini, Bharucha, and Auer, 2012) aims to focus first on the PLR satisfaction followed by other objectives such as fairness and throughput. The biggest disadvantage of these rules is the lack of consideration of all QoS objectives. Since these rules are static over the entire scheduling process, the optimization problem is entitled Static Scheduling Rule based SMOO (SSR-SMOO).

When a different scheduling rule is applied at each TTI, then the Dynamic Scheduling Rule (DSR) principle is addressed. Schwarz, Mehlfuhrer, and Rupp (2011) address the DSR-SMOO problem in the sense that the Generalized Proportional Fair (GPF) is parameterized in order to optimize the throughput-fairness trade-off in terms of Jain fairness index (first introduced by Jain et al., 1984). The RL framework is proposed by Comşa, Zhang, Aydin, Kuonen, and Wagen (2012) to achieve different tradeoff levels between system throughput and user fairness. But this quantitative fairness objective used in these approaches is static and does not depend on channel or network conditions. Instead, the qualitative fairness measures based on channel statistics can be used, such as Next Generation Mobile Networks (NGMN) fairness (Proebster et al., 2010). According to this metric, the GPF scheduling rule should be adapted at certain TTIs, such that, the Cumulative Distribution Function (CDF) of normalized user throughputs is adjusted to respect the requirement from the CDF domain, as studied by Proebster et al. (2010). However, when using actor-critic RL algorithm as proposed by Comşa et al. (2014b) to parameterize the GPF scheduling rule based on dynamic scheduler conditions in order to increase the time when the NGMN fairness requirement is satisfied, the obtained gain is more than 10% higher when compared with the method proposed by Schwarz et al. (2011), and higher than 20% when compared to the adaptation technique proposed by Proebster et al. (2010). But all these approaches use a simple parameterization of GPF scheduling rule (considering only α parameter adaptation) when matching against the NGMN fairness requirement. In the study conducted by Comşa et al. (2014c), the same actor-critic RL algorithm is used to perform the double parameterization of the GPF scheduling rule in terms of both α and β parameters. The results indicate a gain larger than 5% when compared to actor-critic RL framework that uses the simple parameterization. The advantages of these RL-based frameworks reveal a higher capacity to adapt to a more generalized scheduler state space and an increased level of satisfaction for the NGMN fairness requirement. The QoS objectives are not considered in these optimization models.

In other circumstances, one static scheduling rule can optimize multiple objectives at the same time when applied TTI-by-TTI. The optimization procedure is called Static Scheduling Rule based Concurrent MOO (SSR-CMOO). Another scheduling proposed by Khan et al. (2012) considers the joint optimization of PLR and delay objectives, while the fairness objective is considered once the primary multi-objective satisfaction is achieved. Other schedulers aim to split the scheduling process in two stages (Monghal, Laselva, Michaelsen, and Wigard, 2010; Chung, Chang, and Wang, 2012; Wang, Li, Ji, and Zhang, 2013; Avocanh, Abdennebi, and Ben-Othman, 2014): a) time domain where the group of users with more stringent QoS requirements is prioritized to be scheduled b) frequency domain, where the radio resources are allocated for the preselected users. In these cases, the QoS objectives are targeted in time domain, whereas the frequency domain deals more with the trade-off between system throughput maximization and user fairness provisioning. Although these rules aim to satisfy the entire set of objectives, there is not a clear evidence how these schedulers will perform under generalized RRM state space (channel types, variable number of users and traffic types, arrival rates in data queues, etc).

By combining the concurrent multi-objective optimization and the dynamic scheduling rule selection (DSR-CMOO), a mixture of different scheduling rules is used, where one rule is applied at each TTI in order to improve the satisfaction measure when multiple scheduling objectives are addressed. Different RL algorithms are implemented by Comşa et al. (2018) for Constant Bit Rate and Variable Bit Rate traffic types in order to maximize the number of TTIs when the PLR and delay constraints are satisfied. The proposed framework makes use of four scheduling rules that are focused on the delay objective only. Gains higher than 10% are obtained for the proposed RL framework when compared to other conventional static scheduling rules. In other studies (Comşa, De-Domenico, and Ktenas, 2017; Comşa, Trestian, and Ghinea, 2018), the actor-critic RL framework is used to optimize the scheduling rule selection in each state when PLR, GBR and delay objectives are considered. These proposals aim to minimize the disadvantages of each particular scheduling rule while maximizing their advantages by applying on each scheduler state the best scheduling rule such that the multi-objective satisfaction is maximized. The throughput maximization and user fairness objectives are not taken into account by these proposals.

This chapter considers the DSR-CMOO model that aims to combine the static scheduling rules with the main focus on the following objectives: throughput maximization, NGMN fairness requirement and QoS constraints satisfaction in terms of PLR, GBR and delay.

ODFMA SCHEDULING PROCESS

The scheduler process is conducted based on four main components: scheduler state space, scheduling procedure, scheduling rule, and MOO performance evaluation. The scheduler state space contains all the parameters necessary to conduct the scheduling procedure at each TTI. The scheduling procedure in OFDMA systems includes: the user selection for each radio resource according to the exploited scheduling rule; the Modulation and Coding Scheme selection (MCS) for each scheduled user; and the

Transport Block (TB) computation that actually gives the total number of bits to be transmitted to each scheduled user based on the amount of allocated radio resources and supportable MCS. A scheduling rule is associated with the resource allocation process in order to satisfy given objectives. The MOO performance evaluation gives a satisfaction measure for all scheduling objectives at each TTI according to the applied scheduling rule in the previous state. The OFDMA scheduling process is presented in Fig. 1.

In OFDMA scheduling, the available bandwidth is divided in B number of resource blocks every TTI. If we also consider I_t as the number of users that can change at each TTI t, then the most important task in OFDMA scheduling is to find the potential benefit of allocating each Resource Block (RB) $j \in \mathcal{B} = \{1,2,...,B\}$ to certain users $i \in \mathcal{I}_t = \{1,2,...,I_t\}$ when a given performance criterion is addressed based on the exploited scheduling rule. More precisely, being given a certain performance criterion, the scheduler should be aware about the exact price or cost value of allocating RB $j \in \mathcal{B}$ to UE $i \in \mathcal{I}_t$ for its targeted objective satisfaction. So, the scheduler is responsible for optimizing the obtained pricing structure problem every TTI.

Figure 1. OFDMA Scheduling Process

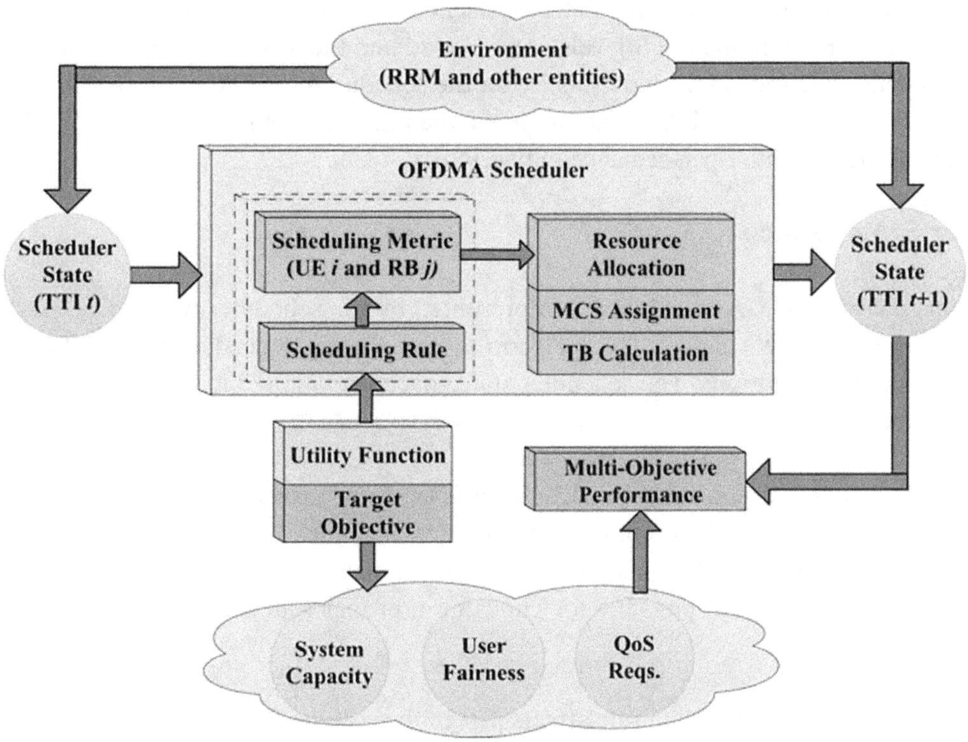

The potential benefit quantification of using some limited resources is inherited from the utility theory in economics which has been applied with great success in wireless networks in order to guarantee the QoS requirements and to exploit the multi-user diversity principle in opportunistic scheduling (Liu, Chong, and Shroff, 2001). In OFDMA networks, the proposed scheduling procedure maps the performance criteria in some utility metrics for each user $i \in \mathcal{I}_t$ and for each RB $j \in \mathcal{B}$. Then, the instantaneous optimization problems resume to the sum maximization of each user utility TTI-by-TTI.

Adopting the performance criteria in order to evaluate the performance of user centric objectives for different type of services represents a crucial task. By using classical scheduling procedures, it is difficult to satisfy multiple objectives simultaneously. Therefore, some priorities in satisfying certain objectives are given by adopting different performance criteria at once as denoted by SSR-SMOO problems. However, the performance criteria indicate basically the types of utility functions. For instance, if the utility function addresses the HoL packet delay performance, the scheduler is designed in such a way that the packet delay budget should be satisfied for certain requirements. If the optimality of the first condition is satisfied, then other objectives can be considered depending on the particularity of the utility function.

In Figure 1, based on the selected performance criterion (or scheduling objective to be satisfied) and the type of utility function, a scheduling rule is selected to conduct the scheduling procedure each TTI. At TTI t, the scheduler momentary state is observed and according to the decided scheduling rule, a number of $B \times I_t$ metrics is calculated by considering the metrics of all users $i \in \mathcal{I}_t$ for each RB $j \in \mathcal{B}$. The resource allocation performs for each RB from $\mathcal{B}$ the metric ordering of all users metrics calculated on that particular RB. Only users with the highest metric for each RB are selected to be scheduled at TTI t. This means actually the urgency for those users to be scheduled in order to maximize the given performance criterion. Then, the MCS levels are assigned and the TB determines the actual amount of data to be broadcasted to those selected users. After the scheduling procedure is completed, the system moves to the next state at TTI $t+1$. Only at this stage, the real multi-objective measure is determined in order to evaluate the scheduler performance in previous state, at TTI t.

Scheduler State Space

An important role in OFDMA scheduling is represented by the scheduler state space since both optimization approaches, SMOO and CMOO, perform the scheduling procedure at each TTI based on the momentary scheduler conditions. The scheduler state space is exploited in different ways based on the optimization type:

- In the SSR-SMOO optimization, the scheduler state space provides the necessary parameters for the utility function computation in the scheduling procedure;
- For the DSR-SMOO and DSR-CMOO problems, alongside the utility parameters provision, the scheduler state space is responsible for choosing a proper scheduling rule in order to meet the proposed objective.

Inevitably, the selected scheduling rule affects part of the scheduler state space evolution. The scheduler state space is divided into two disjoint sub-spaces:

- **Uncontrollable Scheduler State Space:** The CQI reports, HARQ retransmissions indicators, arrival bit rates, QoS requirements are included. We define the measurable uncontrollable state space $\mathcal{S}_U$ and $z[t] \in \mathcal{S}_U$ the momentary uncontrollable vector at TTI t.
- **Controllable Scheduler State Space:** The parameters responsible for the objective performance evaluation, such as HoL delay, average user rate, normalized user rate, packet loss rate, and queue size are included. Similarly, we define the measurable controllable state space $\mathcal{S}_C$ and $c[t] \in \mathcal{S}_C$ the momentary controllable vector at TTI t.

The overall measurable state space is defined as $\mathcal{S} = \mathcal{S}_U \cup \mathcal{S}_C$ and the momentary scheduler state at TTI t is $s[t] = [z, c] \in \mathcal{S}$. Each of these elements is detailed in the following sub-sections.

Uncontrollable Scheduler State

The uncontrollable scheduler state space represents the indices and parameters that reflect mainly the channel conditions, the service parameters from the upper layers, and the QoS requirements for each active data flow. Even if these parameters are modeled as random processes rather than the scheduling procedure results, the obtained subspace plays a crucial role in achieving satisfaction of different objectives. The uncontrollable state space $\mathcal{S}_U$ encompasses the following elements:

- **Channel Quality Indicator (CQI) Reports:** It is assumed that at each TTI t, each user $i \in \mathcal{I}_t$ reports the CQI value for each RB $j \in \mathcal{B}$ through the PUCCH control channel. Let us define $CQI_{i,j}[t]$ as the CQI report value of user $i \in \mathcal{I}_t$ and resource block $j \in \mathcal{B}$ at TTI t. Then, $CQI[t]$ is the momentary vector of all CQI reports for all active users defined as follows:

$$CQI[t] = \left[CQI_{i,j}[t]\right]_{\substack{j=1,..,B \\ i=1,...,I_t}} \tag{1}$$

- **Achievable User Rate:** Based on the $CQI_{i,j}[t]$ reports, the achievable user rate $r_{i,j}[t]$ is computed for the scheduling decision. For each $CQI_{i,j}[t]$, a MCS is determined in order to provide the number of bits that can be transmitted if RB $j \in \mathcal{B}$ would be allocated to user $i \in \mathcal{I}_t$. Similarly to CQI reports, the vector of instantaneous rates is obtained based on:

$$r[t] = \left[r_{i,j}[t]\right]_{\substack{j=1,..,B \\ i=1,...,I_t}} \tag{2}$$

- **Instantaneous Arrival Rate:** We define the arrival rate in data queue for user $i \in \mathcal{I}_t$ at TTI t as $\lambda_i[t]$. Then, the instantaneous vector of arrival rates is defined by $\lambda[t] = \left[\lambda_1, \lambda_2, ..., \lambda_{I_t}\right]$ with the amendment that for each user is considered only one data queue.

The momentary uncontrollable scheduler state is obtained by taking into account the indicators introduced above such as: $z[t] = \left[CQI, r, \lambda\right]$.

Controllable Scheduler State

The controllable scheduler subspace denotes the set of indices which are used for the multi-objective performance evaluation. Basically, when $\mathcal{S}_C$ is optimal, the entire scheduler state is considered optimal. Under these circumstances, other subspace $\mathcal{S}_U$ provides the necessary information to maintain the scheduler in the optimal region by satisfying all scheduling objectives. The controllable subspace being considered here comprises the following elements:

- **Instantaneous User Rate:** When performing the scheduling procedure, the instantaneous user rate $R_i[t]$ represents the total number of bits associated to the scheduled user $i \in \mathcal{I}_t$. The associated vector is: $R[t] = \left[R_1, R_2, ..., R_I\right]$.
- **Instantaneous User Throughput:** If the transmitted packets in the previous TTI were correctly decoded by each scheduled active user (HARQ report is zero), then the instantaneous user rate becomes the instantaneous user throughput $T_i[t]$. Consequently, the instantaneous throughput vector becomes $T[t] = \left[T_1, T_2, ..., T_I\right]$.

- **Average User Throughput:** It is used to improve the fairness among users. If $T_i[t]$ is used as the fairness satisfaction metric, then the scheduler should be fair at each TTI. This aspect is undesirable because it affects the spectral efficiency performance. Therefore, it is preferred to evaluate the fairness performance by using a time window or a predefined number of TTIs. So, the average user throughput $\bar{T}_i[t]$ is determined as follows:

$$\bar{T}_i[t] = (1-\beta)\cdot\bar{T}_i[t-1] + \beta\cdot T_i[t] \tag{3}$$

where β represents the forgetting factor which impacts in the scheduler performance. The lower values for parameter β implies in fact lower impacts of the current scheduling procedure of the optimization problem. The instantaneous vector is: $\bar{T}[t] = [\bar{T}_1, \bar{T}_2, \ldots, \bar{T}_I]$.

- **HoL Packet Delay:** We define by $d_i[t]$ the maximum waiting time for a given data packet in the MAC level data queue for user $i \in \mathcal{I}_t$. The momentary delay vector is $d[t] = [d_1, d_2, \ldots, d_I]$.
- **Packet Loss Rate:** We denote by $L_i[t]$ the packet loss rate at TTI *t* that indicates the number of lost packets over the total number of sent packets in a given time window of T_w. We define the momentary vector of PLRs as $L[t] = [L_1, L_2, \ldots, L_I]$.
- **Transmission Queues Size:** We define $q_i[t]$ the transmission queue size for user $i \in \mathcal{I}_t$ at TTI *t*. Consequently, the instantaneous queue vector is $q[t] = [q_1, q_2, \ldots, q_I]$.

Then, the momentary controllable scheduler state becomes $c[t] = [R, T, \bar{T}, L, d, q] \in \mathcal{S}_C$.

Radio Resource Allocation

The main focus for the OFDMA scheduler is to assign the available set of RBs to different active users in order to satisfy given scheduling objectives. The idea is to quantify the benefit (utility) of allocating each RB $j \in \mathcal{B}$ to user $i \in \mathcal{I}_t$ at each TTI *t*. In this sense, the utility function has to be defined.

For our scheduling purposes, the utility functions cannot be measured directly. The solution is to perform the instantaneous rate allocation based on the utility representation at each TTI *t* and to measure or to evaluate the allocation performance at each TTI *t*+1 by using the objective functions. We define by $r = [r_1, r_2, \ldots, r_I]^T$ the instantaneous achievable rate matrix being obtained based on the CQI reports, where $r_i = [r_{i,1}, r_{i,2}, \ldots, r_{i,B}]$ is the vector of instantaneous rates of each user $i \in \mathcal{I}_t$.

For each user $i \in \mathcal{I}_t$, let us consider $U_i(r_i)$ the utility function which is a benefit representation of allocating the vector of rates r_i to user $i \in \mathcal{I}_t$. In long term, the packet scheduler aims to maximize the aggregate user, such as:

$$\max_{t\to\infty, i\in\mathcal{I}_t} \sum_t \sum_i U_i\left(r_i\left[t\right]\right) \tag{4}$$

In multi-user scenario, the disposable set of RBs is allocated only to some users at each TTI according to some allocations variables $b\left[t\right]=\left\{b_{i,j}\left[t\right]\right\}, i=1,\ldots,I_t, j=1,\ldots,B$ that take the binary values as follows:

$$b_{i,j}=\begin{cases}1, & if\ RB\ j\in\mathcal{B}\ is\ allocated\ to\ UE\ i\in\mathcal{I}_t\\ 0, & otherwise\end{cases} \tag{5}$$

The instantaneous data rates $R_i\left[t\right]$ for each user are obtained after performing the scheduling decision under the allocation variables b at each TTI *t*. Let us define the instantaneous rate region constrained by policy b such as $\mathcal{R}_b$. Therefore, the definition domain for the utility function is $U_i:\mathcal{R}_b\to\mathbb{R}$ and the long-term optimization problem becomes (Song and Li Geoffrey, 2005):

$$\begin{aligned}&\max_b \sum_i U_i\left(\sum_j b_{i,j}\cdot r_{i,j}\right)\\ &s.t.\ \begin{array}{l}\sum_i b_{i,j}=1,\ \ j=1,\ldots,B\\ b_{i,j}\in\left\{0,1\right\},\ \ \forall i\in\mathcal{I},\forall j\in\mathcal{B}\end{array}\end{aligned} \tag{6}$$

According to Kelly (1997) the local maximum is also a global maximum in (6) if and only if the region $\mathcal{R}_b$ is a *convex set* and $U_i\left(R_i\right)$ is a *concave* function. However, the convexity problem of $\mathcal{R}_b$ in OFDM systems has been discussed intensively by Song (2005) and the authors came at the conclusion that the short-term optimization problem at each TTI *t* can be obtained by using the first order approximation of Taylor's expansion as expressed by (7) as stated by Song (2005):

$$\sum_i U_i\left(R_i\left[t\right]\right)-\sum_i U_i\left(R_i\left[t-1\right]\right)\approx\sum_i U_i^{'}\left(R_i\left[t-1\right]\right)\cdot\left(R_i\left[t\right]-R_i\left[t-1\right]\right) \tag{7}$$

where $U_i^{'}\left(R_i\left[t-1\right]\right)=\partial U_i\left(R_i\left[t-1\right]\right)/\partial R_i\left[t-1\right]$ is the marginal utility for user $i\in\mathcal{I}_t$. The instantaneous rate $R_i\left[t-1\right]$ at TTI *t*-1 is obtained after performing the scheduling procedure at TTI *t*-1 and this value is used in the optimization problem at TTI *t*. Therefore, the short-term optimization problem can be written under the following form:

$$\begin{aligned}&\max_{b[t]} \sum_i\sum_j b_{i,j}\left[t\right]\cdot U_i^{'}\left(R_i\left[t\right]\right)\cdot r_{i,j}\left[t\right]\\ &s.t.\ \begin{array}{l}\sum_i b_{i,j}\left[t\right]=1, j=1,\ldots,B\\ b_{i,j}\left[t\right]\in\left\{0,1\right\},\forall i\in\mathcal{I}_t,\forall j\in\mathcal{B}\end{array}\end{aligned} \tag{8}$$

The optimization model being exposed in (8) represents a *linear programming model* where the unknown variables are the resource assignment variables $b_{i,j}\left[t\right] \in \left\{0,1\right\}$ that need to be determined at each TTI *t*. Due to the reduced number of resources that has to be allocated at each TTI, the assignment is performed by using the following equation:

$$m_{j}\left[t\right] = \arg\max_{i \in \mathcal{I}_t} \left\{U_{i}^{'}\left(R_{i}\left[t\right]\right) \cdot r_{i,j}\left[t\right]\right\} \tag{9}$$

where, $m_{j}\left[t\right]$ indicates that RB $j \in \mathcal{B}$ is assigned or allocated to user $m \in \mathcal{I}_t \forall m \neq i$ at TTI *t*. Consequently, $b_{m,j}\left[t\right] = 1$ and $b_{i,j}\left[t\right] = 0, \forall i \in \mathcal{I}_t$ and $\forall i \neq m$. This way, the user assignment is performed for each RB for a given bandwidth. Once the resource allocation is finished, the transport block size is determined for each selected user.

As mentioned earlier, the instantaneous rate for each user $i \in \mathcal{I}_t$ and for each RB $j \in \mathcal{B}$ is determined based on the CQI reports. The marginal function is positive because the utility function $U_{i}\left(R_{i}\right)$ must be concave (the second derivative is negative) in order to assure the linearity of the considered optimization problem. When the utility function $U_{i}\left(R_{i}\right)$ takes the polynomial form, the role of its marginal utility is to schedule those users with the highest instantaneous rates by increasing at the same time the total system capacity if the radio channels are errorless. In the case of re-transmissions, the marginal utility as a function of instantaneous user throughput $U_{i}^{'}\left(T_{i}\right)$ should be used in order to provide more RBs to those users which require less retransmissions during the downlink scheduling session to avoid the waste of radio resources.

The linear programming model exposed in (8) is a typical SSR-SMOO problem being focused on the system throughput maximization without considering other objectives such as: user fairness, GBR, HoL delay, packet loss. The impact of the resource allocation problem in the scheduling objectives can be measured by using the objective functions. The objectives functions can be modeled by using the QoS constraints. When different objective(s) is (are) analyzed, the performance of the optimization problem from (8) can be improved if the marginal utility function considers the performance parameter(s) of the addressed objective(s). More details about this aspect are presented in the following section.

Utility and Objective Functions in OFDMA Systems

Utility functions are designed to quantify the benefit of allocating a given and finite number of RBs to active mobile users. The type of utility function can influence the optimization problem in the direction of different scheduling objectives. The classification of utility functions can be achieved by considering three perspectives: the argument function, the utility weight and the manufacturing methodologies. Based on the manufacturing methods, there are two methods that can be used to obtain utility functions (Song and Li Geoffrey, 2005; Song, 2005) exposed bellow together with the proposed methodology:

- **Application Based Utility Functions:** One way is to develop utility functions that characterize a specific type of application which can be obtained by using sophisticated subjective surveys. These utilities can suffer from the imperfection of the measurements, and different parameters are fixed to some objective values denoting the inflexibility for those situations which are not covered by the considered surveys.

- **Traffic Habits Based Utility Functions:** statistics about the percentage of different traffic types that can co-exist at different moments of time in different urban scenarios. The utility functions are designed based on these statistics of heterogeneous traffic types.
- **Scheduler State Based Aggregate Utility Function:** Based on the scheduler state $s \in \mathcal{S}$, different utility functions (being already proposed in the literature) are applied in order to maximize the long term aggregate utility function and to solve the DSR-SMOO/CMOO combinatorial problems. More precisely, it maximizes the sum of some existing utility functions subject to some objective functions' requirements.

The short-term optimization for the resource allocation is obtained when performing the first order of Taylor's expansion between two time consecutive utility functions. This way, the marginal utility function or the first derivative utility function is obtained. The term of marginal can refer also to a small change that can appear in the optimization problem between two consecutive momentary scheduler states. In fact, the marginal utility indicates high gains for those users with poor objective performance, whereas other users with much better multi-objective performance are getting much lower gains. In this sense it is encouraged to allocate higher amount of resources to those users with higher gains in the marginal utility.

In the optimization problem exposed in (8), when selecting any gain in the marginal utility function leads to the system capacity maximization without any consideration about other objectives. By designing the marginal utility with proper weights, different objectives can be addressed. So, the role of the marginal utility in the optimization problem is to reduce the impact of the instantaneous achievable rate $r_{i,j}\left[t\right]$ (or to annihilate any variation of the radio channel) and to focus the entire optimization problem on scheduling different users based on their performance of satisfying different objectives addressed by different marginal utility weights. To conclude, the multi-objective performance depends on the type of marginal utility which is used in the optimization problem.

To generalize the utility function representation, we define $x_i \in \mathcal{X}_i$ the argument of the utility function for user $i \in \mathcal{I}_t$ and $y_i \in \mathcal{Y}_i$ the argument for the utility weight, where $\mathcal{X} \cup \mathcal{Y} \subseteq \mathcal{S}_C$, $\mathcal{X} = \bigcup_i \mathcal{X}_i$ and $\mathcal{Y} = \bigcup_i \mathcal{Y}_i$. Therefore, the utility function for user $i \in \mathcal{I}_t$ can be decomposed as follows:

$$U_i\left(x_i\right) = F_i\left(x_i\right) \cdot W_i\left(y_i\right) \tag{10}$$

where function $F_i : \mathcal{X}_i \to \mathbb{R}$ is concave and differentiable, and the utility weight $W_i : \mathcal{Y}_i \to \mathbb{R}$ of user $i \in \mathcal{I}_t$ is a constant, but it is represented as a function in order to highlight the objective indicator $y = \left[y_1, y_2, \ldots, y_I\right] \in \mathcal{Y}$, where $y = \left\{R, T, \overline{T}, L, d\right\}$.

When one element for the controllable parameters set $y_i = \left\{R_i, T_i, \overline{T}_i, L_i, d_i\right\}$ respects the corresponding QoS constraints, then user $i \in \mathcal{I}_t$ is satisfied in terms of the addressed objective. The first derivative of the utility function is determined by using $U_i'\left(x_i\right) = F_i'\left(x_i\right) \cdot W_i\left(y_i\right)$, where $F_i'\left(x_i\right) = \partial F_i\left(x_i\right) / \partial x_i$. If the marginal utility function is developed in such a way that the radio channel variations are compensated at each TTI t for each user $i \in \mathcal{I}_t \left(r_{i,j}[t] \cdot F_i^{'}(x_i) \approx 1 \right)$, then the optimization problem is focused more on the scheduling objective evaluated by the weight argument $y \in \mathcal{Y}$.

SINGLE OBJECTIVE OPTIMIZATION PROBLEM

In this chapter, we study five objectives such as: throughput maximization, NGMN user fairness, guaranteed bit rate, packet delay satisfaction and packet loss minimization. We consider $\mathcal{O}$ the set of aforementioned objectives. For each objective $o \in \mathcal{O}$, let us define the pool of utilities $\mathcal{U}_o$, and then, the entire set of utilities for all objectives is defined as $\mathcal{U} = \bigcup_o \mathcal{U}_o$. As mentioned earlier, the type of marginal utility for objective $o \in \mathcal{O}$ is correlated with the utility weight. If $u_o \in \mathcal{U}_o$ is the weight index targeting the objective $o \in \mathcal{O}$, then the short-term optimization can be obtained for the general form of scheduling utilities by using the first order of Taylor's expansion and being similar to (7):

$$\sum_i U_{o,i}\left(x_i[t]\right) - \sum_i U_{o,i}\left(x_i[t-1]\right) \approx \sum_i W_{o,u,i}\left(y_{o,i}[t]\right) \cdot F_i^{'}\left(x_i[t]\right) \cdot \left(R_i[t] - R_i[t-1]\right) \tag{11}$$

where, $y_{o,i}[t] \in \left\{ R_i, T_i, \overline{T}_i, L_i, d_i \right\}$ is the controllable QoS parameter targeting objective $o \in \mathcal{O}$ and $W_{o,u,i}\left(y_{o,i}[t]\right)$ is the weight function $u_o \in \mathcal{U}_o$ that aims to maximize the resource allocation from the perspective of the same objective $o \in \mathcal{O}$. Then, the optimization problem when both $o \in \mathcal{O}$ and $u_o \in \mathcal{U}_o$ are static becomes as follows:

$$\begin{aligned} &\max_{b[t]} \sum_i \sum_j b_{i,j}[t] \cdot W_{o,u,i}\left(y_{o,i}[t]\right) \cdot F_i^{'}\left(x_i[t]\right) \cdot r_{i,j}[t] \\ s.t.\ &\sum_i b_{i,j}[t] = 1, \ j = 1, \ldots, B \\ &b_{i,j}[t] = \{0,1\}, \ \forall i \in \mathcal{I}_t, \forall j \in \mathcal{B} \end{aligned} \tag{12}$$

where the optimization problem is focusing on objective $o \in \mathcal{O}$ subject to the convex set of constraints. The optimization problem of (12) is a linear programming model. If $W_{o,u,i}[t] >> F_i^{'}[t] \cdot r_{i,j}[t]$, then the addressed objective $o \in \mathcal{O}$ is evaluated based on the weight argument $y_{o,i}[t]$ for each user $i \in \mathcal{I}_t$. At this extent, the weight matrix for objective $o \in \mathcal{O}$ is

$$\nabla U_o = \left\{ U_{o,u,i}^{'} = W_{o,u,i} \cdot F_i^{'}, u_o = 1, \ldots, U_o, i = 1, \ldots, I_t \right\}.$$

In SSR-SMOO problems, the weight matrix ∇U_o assigns the same type of marginal utility functions to each user $i \in \mathcal{I}_t$ for the entire scheduling session.

The optimal resource allocation when following the linear optimization problem from (12) for each RB $j \in \mathcal{B}$ and user $\forall i \in \mathcal{I}_t$ is given by the following metric calculation:

$$m_j[t] = \arg\max_{i \in \mathcal{I}_t} \left\{ W_{o,u,i}\left(y_{o,i}[t]\right) \cdot F_i'\left(x_i[t]\right) \cdot r_{i,j}[t] \right\} \quad (13)$$

where $m_j[t]$ indicates that RB $j \in \mathcal{B}$ is allocated to user $m \in \mathcal{I}_t, \forall m \neq i$ at TTI *t*. Then, $b_{m,j}[t] = 1$ and $b_{i,j}[t] = 0, \forall m \in \mathcal{I}_t, \forall i \in \mathcal{I}_t$ and $\forall i \neq m$. Therefore, the scheduling rule function can be defined in the following manner:

$$D_{o,u,i} : \mathcal{Y} \to \mathbb{R} D_{o,u,i}\left(y_{o,i}[t]\right) = W_{o,u,i}\left(y_{o,i}[t]\right) \cdot F_i'\left(x_i[t]\right) \quad (14)$$

If the objective $o \in \mathcal{O}$ and the utility weight $u_o \in \mathcal{U}_o$ stay fix during the entire scheduling session, then the linear optimization problem from (12) is a SSR-SMOO/CMOO problem. The CMOO refers to the fact that the argument of $W_{o,u,i}\left(y_i[t]\right)$ can be multi-dimensional and the optimization problem is focusing on two or more objectives, and the argument is a vector of $y_i = \left[y_{1,i}, y_{2,i}, \ldots, y_{O,i}\right]$.

The performance of the scheduling discipline $D_{o,u,i}$ can be evaluated by using the objective functions. Let us define the objective function $\phi_{o,i}\left(y_{o,i}[t]\right)$ for objective $o \in \mathcal{O}$ and user $i \in \mathcal{I}_t$, where the definition domain is $\phi_{o,i} : \mathbb{R} \to \mathbb{R}$. The objective condition for a particular SSR-SMOO problem for each user $i \in \mathcal{I}_t$ at each TTI is given by:

$$\phi_{o,i}\left(y_{o,i}[t]\right) \geq 0, \forall o \in \mathcal{O}, \forall i \in \mathcal{I}_t \quad (15)$$

When the condition is satisfied for each user $i \in \mathcal{I}_t$, then the scheduler becomes optimal at TTI *t* from the viewpoint of objective $o \in \mathcal{O}$. Therefore, the aggregate function for all users and objective $\forall o \in \mathcal{O}$ becomes: $\phi_o\left(y_o[t]\right) = \left(1/I_t\right) \cdot \sum_{i=1}^{I_t} \phi_{o,i}\left(y_{o,i}[t]\right)$.

In DSR-CMOO problems, the impact of each scheduling rule in each objective is strongly required. In this sense, the aggregate multi-objective function $\Phi_{o,u_o} : \mathbb{R}^{I \times O} \to \mathbb{R}$ when applying the scheduling rule $D_{o,u,i}$, can be represented as indicated in (16):

$$\Phi_{o,u_o}\left(y[t]\right) = \sum_{o_*}^{O} \delta_{o_*} \cdot \phi_{o_*,u}\left(y_{o_*}[t]\right) \quad (16)$$

where $y = \left[y_1, y_2, \ldots, y_O\right]$ and $\phi_{o_*,u_o} : \mathbb{R}^I \to \mathbb{R}$ is the aggregate function of objective $o_* \in \mathcal{O}$ when the scheduling rule $D_{o,u_o,i}$ is considered for each user $i \in \mathcal{I}_t$. Parameter δ_{o_*} is the weight for each particular objective function ϕ_{o_*,u_o}, $\forall o_* \in \mathcal{O}$. The necessary and not sufficient condition to be satisfied by the aggregate multi-objective function at each TTI *t* for the entire set of active users is:

$$\Phi_{o,w_o}\left(y\left[t\right]\right) \geq 0 \quad (17)$$

Equation (17) can be satisfied only if the condition (16) is met for each user $i \in \mathcal{I}_t$ and for each objective $o \in \mathcal{O}$. Precise details about each scheduling objective are provided in the following sub-sections.

Throughput Maximization

For throughput maximization ($o = 1$), the utility argument is $x_i\left[t\right] = R_i$ and the weight function is $W_{1,u_1,i}\left(y_i\left[t\right]\right) = 1$. The multi-objective evaluator grants the scheduling performance according to objective function $\phi_{1,i} : \mathbb{R} \rightarrow \mathbb{R}$, $\forall i \in \mathcal{I}_t$ where $\phi_{1,i}\left[t\right] = \sum_i^{I_t} T_i\left[t\right] - \sum_i^{I_t} T_i\left[t-1\right]$. The role of this particular optimization problem is to increase the total cell throughput at each TTI in such a way that $\phi_{1,i}\left[t\right] \geq 0$. If in (13) the product is $W_{o,u,i}\left(y_{o,i}\left[t\right]\right) \cdot F_i^{'}\left(x_i\left[t\right]\right) = 1$, then we obtain the Maximum Rate scheduling rule, aiming to maximize at each TTI the total cell spectral efficiency.

User Fairness

The user fairness should be guaranteed at each TTI meaning that it will impact the system capacity maximization. Then, a new scheme is required in order to give more flexibility to the system throughput improvement. The time window (a given number of TTIs) used in user throughput computation constrains or relaxes the fairness performance depending on its length. By adopting $x_i\left[t\right] = T_i$ as an argument for the marginal utility function, the resource allocation at TTI *t* depends on the allocation history in the previous TTIs. The averaging procedure can be achieved basically in two ways:

- **By Using the Exponential Moving Filter (EMF):** The forgetting factor β is used to control the system throughput and user fairness tradeoff as indicated in (3), where $\beta = 1_{TTI} / T_{Ew}$, and T_{Ew} is the time window length. This means that a higher average throughput implies a lower priority for that user to be selected on the considered RB. The only condition is to set β larger than the channel correlation time in order to exploit the time diversity principle as revealed by Song (2005). When the time window T_{Ew} is too large, the cell spectral efficiency is affected whereas when the time window is too small, the user fairness is not satisfied anymore.
- **By Using the Median Moving Filter (MMF):** The idea is to store the instantaneous user throughputs for a given time window T_{Mw} and to use the mean value of these observations at each TTI in order to balance the system throughput and user fairness tradeoff.

The NGMN fairness is one of many others criteria that can be used to compute the objective function According to this principle detailed by Proebster et al. (2010), the CDF calculated for a given set of average/instantaneous user throughputs should not exceed a given NGMN threshold. Based on some convergence studies presented by Song (2005), the local optimization considers $x_{2,i}\left[t\right]=\overline{T}_i$ as an argument for the utility function and the weight function is $W_{2,u,i}\left(y_{2,i}\left[t\right]\right)=1$ since the QoS requirements are not included at this stage. A particular type of marginal utility function is $U'_{2,i}\left(\overline{T}_i\left[t\right]\right)=1/\overline{T}_i\left[t\right]$ which implies the metric of $m_j\left[t\right]=\arg\max_{i\in\mathcal{I}}\left\{r_{i,j}\left[t\right]/\overline{T}_i\left[t\right]\right\}$, known in the literature as the Proportional Fair (PF) scheduling rule, proposed by Kelly (1997).

In particular, the NGMN fairness objective function can be determined according to the following formula: $\phi_{2,i}\left[t\right]=\psi_i^R\left(NT_i\right)-\psi_i\left(NT_i\right)$, where $NT_i=T_i/\sum T_m$ is the normalized user $i\in\mathcal{I}_t$ throughput, $\psi_i\left(NT_i\right)$ is the CDF function of user $i\in\mathcal{I}_t$ for a given distribution of input normalized observations and $\psi_i^R\left(NT_i\right)$ represents the NGMN fairness requirement.

Guaranteed Bit Rate

When the rate constraint satisfaction ($o=3$) is considered in the optimization problem, the utility function weight is composed of: $x_{3,i}\left[t\right]=\overline{T}_i$ and $y_{3,i}\left[t\right]=\overline{\overline{T}}_i$, where $\overline{\overline{T}}_i\left[t\right]$ represents the average user throughput calculated using the median moving filter. Thus, the first objective is to satisfy all users' GBR requirements and then to focus more on user fairness by scheduling those users with lower throughputs. In this way, we measure the performance of such optimization problem by developing the following objective function: $\phi_{3,i}\left[t\right]=\overline{\overline{T}}_i\left[t\right]-T_i^R\left[t\right]$, where T_i^R is the GBR requirement of user $i\in\mathcal{I}_t$. According to (15), the mean user throughput should be greater than the GBR requirement at each TTI t in order to meet its objective.

HoL Packet Delay

If the utility weight depends on the instantaneous HoL packet delay ($o=4$) such that $y_{4,i}\left[t\right]=d_i$ and $x_{4,i}\left[t\right]=\overline{T}_i$, then the optimization problem considers the HoL packet delay as the first priority in the satisfaction of the performance criterion. Then, the delay based objective function that has to be maximized for each active data queue at each TTI t becomes $\phi_{4,i}\left[t\right]=d_i^R\left[t\right]-d_i\left[t\right]$, where d_i^R is the HoL delay requirement.

Packet Loss Rate

The packet loss objective ($o=5$) represents an important performance target in OFDMA packet scheduling. The utility weight depends on $y_{5,i}\left[t\right]=L_i\left[t\right]$ and the utility argument keeps a similar form as $x_{5,i}\left[t\right]=\overline{T}_i$. The objective function used to measure the performance of radio resource allocation in terms of PLR performance becomes: $\phi_{5,i}[t] = L_i^R[t] - L_i[t]$, where L_i^R is the PLR requirement. The PLR rate $L_i[t]$ is computed as a ratio between the number of lost packets and the total number of transmitted packets. The same time moving window length used to compute the mean user throughput $\overline{\overline{T}}_i$ can be used in this case as well to calculate the instantaneous packet loss at each TTI.

MULTIPLE OBJECTIVE OPTIMIZATION PROBLEM

Optimization problems similar to (12) address only particular SMOO problems and each problem is linear guaranteeing at the same time the global optimal solution when selecting the decision variables for the radio resource assignment. By adopting different utility functions, the scheduling procedure impacts differently in multi-objective problems. We formulate the aggregate function by considering the utility functions introduced in the previous sub-sections, such as:

$$U(x) = 1/(U \cdot I_t) \cdot \sum_o^O \sum_{u_o}^{U_o} \sum_i^{I_t} U_{o,u_o,i}(x_{o,i}) \tag{18}$$

The proposed multi-objective problem aims to maximize the sum of utilities for each scheduling objective in the long term as indicated below:

$$\max_{\mathcal{X} \subset \mathcal{S}_C} \left[\sum_t U(x[t]) \right] \tag{19}$$

By adopting the first order approximation of Taylor's expansion for each of the utility functions, the instantaneous multi-objective optimization problem becomes:

$$\max_{r[t] \in \mathcal{R}_b} \left[\sum_o \sum_{u_o} \sum_i \sum_j W_{o,u_o,i}(y_{o,i}[t]) \cdot F'_{o,i}(x_{o,i}[t]) \cdot r_{i,j}[t] \right] \tag{20}$$

According to (20), we need a policy that selects different utility functions at each TTI from the pool of utilities $\mathcal{U}$. Similar to the decision matrix $b[t]$, instead of users, the new decision matrix will consider the number of objectives and instead of resource blocks, the policy takes into account the existing scheduling rules for each objective. Also, at each TTI, the selection of only one objective is required while multiple users can be selected within one TTI when following the decision matrix $b[t]$. Then, we define by $c[t] = \{c_{o,u_o}[t]\}$ the scheduling rule decision matrix, where $o = 1,...,O$, and $u_o = 1,...,U_o$. According to (20), for each active user $i \in \mathcal{I}_t$, the same marginal utility $u_o \in \mathcal{U}_o$ must be assigned at each TTI *t*. Also, we need an additional variable able to assign a marginal utility function to each active user. In this sense, the

$$\max_{c,w,b} \sum_{o=1}^{O} \sum_{u_o=1}^{U_o} \sum_{i=1}^{I_t} \sum_{j=1}^{B} c_{o,w_o}[t] \cdot w_{u_o,i}^{o}[t] \cdot b_{i,j}[t] \cdot W_{o,u_o,i}\left(y_{o,i}[t]\right) \cdot F'_{o,i}\left(x_{o,i}[t]\right) \cdot r_{i,j}[t]$$

$$s.t. \begin{array}{ll} \sum_{o}\sum_{u_o} c_{o,u_o}[t] = 1 & (a) \\ \sum_{o}\sum_{u_o} w_{u_o,i}^{o}[t] = 1,\ i = 1,\dots,I_t & (b) \\ \sum_{i} w_{u_o^*,i}^{o}[t] = U_t,\ u_o^* \in \mathcal{U}_o, o \in \mathcal{O} & (c) \\ \sum_{i} w_{u_o^{\otimes},i}^{o}[t] = 0,\ u_o^{\otimes} = 1,\dots,U_o, o = 1,\dots,O, \forall u_o^{\otimes} \neq u_o^* & (d) \\ \sum_{i} b_{i,j}[t] \leq 1,\ j = 1,\dots,B & (e) \\ c_{o,u_o}[t] \cdot \Phi_{o,u_o}[t+1] \geq 0,\ o = 1,\dots,O, u_o = 1,\dots,U_o & (f) \\ c_{o,u_o}[t] \in \{0,1\},\ \forall o \in \mathcal{O}, \forall u_o \in \mathcal{U}_o & \\ w_{u_o,i}^{o}[t] \in \{0,1\},\ \forall o \in \mathcal{O}, \forall u_o \in \mathcal{U}_o, \forall i \in \mathcal{I}_t & \\ b_{i,j}[t] \in \{0,1\},\ \forall i \in \mathcal{I}_t, \forall j \in \mathcal{B} & \end{array} \quad (21)$$

matrix

$$w[t] = \left\{w_{u_o,i}[t], u_o = 1,\dots,U_o, i = 1,\dots,I_t\right\}$$

assigns the scheduling rule for objective $\forall o \in \mathcal{O}$ to each user $i \in \mathcal{I}_t$ at TTI *t*. Also, this matrix differs from one TTI to another when addressing DSR-CMOO problems.

The multi-objective optimization problem in OFDMA scheduling is formulated in (21), where the first constraint denotes the necessary condition of selecting at each TTI *t* only one scheduling rule according to the addressed objective. Constraints (b) indicate that only one marginal utility function is selected for the entire set of active users at each TTI *t*. Constraints (c) and (d) indicate that the same marginal utility function is assigned to all users at each TTI. Constraints (e) are the well-known conditions of assigning resource blocks to different users. Finally, the set of constraints (f) considers the aggregate multi-objective condition from (16). This implies that for the selected rule ($c_{o,u_o}[t] = 1$), the sum of aggregate functions for each objective at TTI *t*+1 should be greater than zero since the evaluation of the scheduling procedure is performed in the next TTI. If the objective conditions from (21.f) are satisfied $\forall o \in \mathcal{O}$ and $\forall u_o \in \mathcal{U}$, then the scheduler is optimal when a given DSR-CMOO problem is addressed.

The idea is to find at each TTI *t* the optimal set of decision variables in order to maximize the optimization problem and to respect the set of constraints. Due to the product $c_{o,u_o}[t] \cdot w_{u_o,i}^{o}[t] \cdot b_{i,j}[t]$, the optimization problem becomes non-linear and thus, the optimal solution in not guaranteed. In (21), the MOO optimization can be divided into three categories based on the dynamicity of the rule selection:

- **SSR-SMOO:** Occurs when objective $\forall o \in \mathcal{O}$ and the marginal utility function $\forall u_o \in \mathcal{U}_o$ are static over the entire transmission session. In this case, the SSR-SMOO problem refers to the classical optimization problems from (12) and constraints (a-d) and (f) from (21) are not required;

- **DSR-SMOO:** If $\forall o \in \mathcal{O}$ is static over the entire downlink scheduling session and $u_o[t] \in \mathcal{U}_o$ is variable TTI-by-TTI. In this case, different marginal utility functions are used concurrently in order to achieve the same target or objective, and constraints (f) consider only the satisfaction of the particular aggregate objective function;
- **DSR-CMOO:** When both $o[t] \in \mathcal{O}$ and $u_o[t] \in \mathcal{U}_o$ are variable over time. Different utility functions with different objective targets may be applied in order to achieve the aggregate objective concomitantly. Only in this particular case, the aggregate multi-objective conditions or constraints (f) are fully taken into account.

By selecting the decision variable $c_{o,u_o}[t]$ at each TTI, the scheduler evolves from state $s[t] \in \mathcal{S}$ to the next one $s[t+1] = s' \in \mathcal{S}$. Due to the time dependency process, the optimization problem from (21) is *dynamic*. The newest state $s' \in \mathcal{S}$ contains the momentary uncontrollable subspace $z[t+1] \in \mathcal{S}_U$ which is not depended on decided variables $\left(c_{o,u_o}[t], w^o_{u_o,i}[t], b_{i,j}[t]\right)$ at TTI *t*. For these reasons, the problem from (22) is considered dynamic and stochastic. Solving these combinatorial problems is not trivial since to find the best decision variables requires consistent computational time and system complexity. Developing a policy of scheduling rules is one of the best ways to ease the decision making at each TTI. We define a policy of scheduling rules as

$$\pi = \left\{c_{o,u_o}[t]; o = 1,..,O, u_o = 1,..,U_o, t = 1,...,\infty\right\}$$

represented by a generic set of scheduling rules or marginal utilities that are applied dynamically TTI-by-TTI based on the momentary scheduler states. An example of such policy is indicated in (22):

$$\pi = \left\{c_{1,3}[t], c_{3,2}[t+1], c_{2,5}[t+2], c_{4,3}[t+3]...\right\} \tag{22}$$

The optimality of such policies (π^*) relies on the selection of the best decision variables at each TTI in such a way that the set of constraints from (21) is fully satisfied and the outcome of QoS satisfaction is maximized over time. The optimization and refinement of such sequences abovementioned are not trivial due to the stochastic nature of the process which requires an infinite state space for searching the optimal solution. Two main approaches can be proposed for the policy optimization:

1. **Evolutionary Methods:** e.g., expression and evolutionary programming;
2. **Dynamic Programming Methodologies:** e.g., real-time dynamic programming and temporal difference based learning algorithms such as reinforcement learning techniques.

Under the assumptions of constant power allocation and the sub-optimal MCS allocation, the complexity of DSR-CMOO problems is $\mathrm{O}(O \times U \times I_t \times B)$. Each algorithm above-mentioned requires a reasonable number of scenarios in order to fine tune the final policy for the real-time downlink scheduling.

MULTI-OBJECTIVE SCHEDULING BASED ON REINFORCEMENT LEARNING

Once the scheduling rule variable $c_{o,u_o}\left[t\right]$ is fixed, the entire aggregate problem is reduced to a simple resource allocation procedure. To solve such non-linear optimization problems, three approaches can be adopted, as follows:

- **Sequential Problem Linearization:** Converts the non-linear problem into its corresponding linear representation. Unfortunately, the computation complexity increases with the size of $\mathcal{U}$, and this approach becomes immediately unsuitable for real-time OFDMA scheduling.
- **Parallel Problem Linearization:** Divides the non-linear multi-objective problem into U linear sub-problems. Basically, this approach aims to run different schedulers in parallel by performing different scheduling rules. After the assignment of RBs is performed for each parallel process, the scheduling rule which maximizes the optimization problem and respects the constraint set is selected. This approach becomes unsuitable when optimizing the fairness objective and infinite number of utilities is considered due to the continuous parameterization of the PF rule.
- **Sequential Problem Linearization in Two Stages:** Divides the non-linear multi-objective problem in two different stages of linear optimization problems. The solution is sub-optimal, but with some optimization tools the optimal solution can be very well approximated.

Due to its reduced complexity, we adopt the last solution that actually is solving the multi-objective satisfaction maximization in the first instance, followed by the simple resource allocation problem in the second instance, being governed by the selected scheduling rule from the first optimization problem.

Sequential Linearization in Two Stages

The main task of the linear optimization problem in (21) is to determine the best decision variable $c_{o,u_o}\left[t\right]$ at each TTI t in order to maximize the problem and to respect the given set of constraints. But this procedure is not guaranteeing the satisfaction of constraints (f) which implicitly highlights the performance of the entire scheduling procedure when one scheduling rule has been applied. In order to tackle this problematic issue, these constraints must to be included in the optimization problem by using relaxation methods. In this sense, the Augmented Lagrangian function and the dual optimization problem are required (Nocedal and Wright, 2006). We define the augmented Lagrangian for our problem as:

$$\mathcal{L}_A : \mathbb{R}^{O\times U} \times \mathbb{R}^{U\times I} \times \mathbb{R}^{I\times B} \times \mathbb{R}^{O\times U} \rightarrow \mathbb{R}$$

$$\begin{aligned}\mathcal{L}_A\left(c, w, b, \Phi^A\right) = &\sum_{o=1}^{O}\sum_{u_o=1}^{U_o}\sum_{i=1}^{I_t}\sum_{j=1}^{B} c_{o,u_o}\left[t\right] \cdot w^{o}_{u_o,i}\left[t\right] \cdot b_{i,j}\left[t\right] \cdot U'_{o,u_o,i}\left(x_{o,i}\left[t\right]\right) \cdot r_{i,j}\left[t\right] + && (1)\\ &\sum_{o=1}^{O}\sum_{u_o=1}^{U_o} \Phi^{A}_{o,u_o}\left[t\right] \cdot c_{o,u_o}\left[t\right] \cdot \Phi_{o,u_o}\left[t+1\right] + && (2)\\ &\sum_{o=1}^{O}\sum_{u_o=1}^{U_o} \frac{\mu_{o,u_o}}{2} \cdot \left(c_{o,u_o}\left[t\right] \cdot \Phi_{o,u_o}\left[t+1\right]\right)^2 && (3)\end{aligned} \qquad (23)$$

where the first term is the optimization function to be maximized from (21), the second term represents the Lagrange relaxation function and finally, the third one is the penalty function as given by Nocedal and Wright (2006). Basically, the augmented Lagrangian is considered to be a combination of Lagrange relaxation and penalty methods in solving complex constrained optimization problems. In (23), μ_{o,u_o} is the penalty factor and $\Phi^A_{o,u_o}[t]$ is the accumulated Lagrange multiplier that has to be updated TTI-by-TTI. According to Nocedal and Wright (2006), the accumulated Lagrange multiplier is updated by using the following formula:

$$\Phi^A_{o,u_o}[t+1] = \Phi^A_{o,u_o}[t] + \mu_{o,u_o} \cdot c_{o,u_o}[t] \cdot \Phi_{o,u_o}[t+1] \tag{24}$$

and $\Phi^A[t] = \left\{\Phi^A_{o,u_o}[t]\right\}$ is the matrix of Lagrange multipliers at TTI t and $\mu = \left\{\mu_{o,u_o}\right\}$ is the penalty matrix for each objective $o \in \mathcal{O}$ and for each marginal utility function $u_o \in \mathcal{U}_o$. Then, let us define the concave Lagrange dual function $\mathcal{G}_A\left(\Phi^A\right)$ which is defined as shown in (25):

$$\mathcal{G}_A : \mathbb{R}^{O \times U} \to \mathbb{R}, \quad \mathcal{G}_A\left(\Phi^A\right) = \sup_{c,u,b} \mathcal{L}_A\left(c, w^o, b, \Phi^A\right) \tag{25}$$

The objective is to find the optimal Lagrange dual function $\mathcal{G}_A\left(\Phi^{A*}\right)$ at each TTI t in such a way that:

$$\mathcal{G}_A\left(\Phi^{A*}[t]\right) = \sup_{c,w,b}\left[\mathcal{L}_A\left(c[t], w^o[t], b[t], \Phi^{A*}[t]\right)\right] \geq \mathcal{L}_A\left(c^*[t], w^{o*}[t], b^*[t], \Phi^A[t]\right) \tag{26}$$

where $\left\{c^*[t], w^{o*}[t], b^*[t]\right\}$ are the optimal assignment matrices at TTI t and $\Phi^{A*}[t]$ is the optimal matrix of Lagrange multipliers being calculated online at each TTI t. The role of the Lagrange dual function is to learn the optimal Lagrange multipliers and to take the assignment decisions based on their optimized values at each TTI. When the learned matrix of Lagrange multipliers is optimal, then the scheduling decision variables are optimal. Based on these aspects, the dual optimization problem is presented in (28).

The multi-objective problem exposed in (27) is a non-linear programming problem where term (1) and (2) in the optimization problem aim to select the best scheduling decision matrix in order to maximize the accumulated Lagrange multiplier and the aggregate multi-objective function at TTI t+1, whereas the third term is the typical resource allocation procedure performed based on the selected marginal utility function

$$\max_{c,w,b}\left\{\begin{array}{lr}\sum_{o=1}^{O}\sum_{u_o=1}^{U_o}\Phi^A_{o,u_o}[t]\cdot c_{o,u_o}[t]\cdot\Phi_{o,u_o}[t+1] + & (1)\\ \sum_{o=1}^{O}\sum_{u_o=1}^{U_o}\frac{\mu_{o,u_o}}{2}\cdot\left(c_{o,u_o}[t]\cdot\Phi_{o,u_o}[t+1]\right)^2 + & (2)\\ \sum_{o=1}^{O}\sum_{u_o=1}^{U_o}\sum_{i=1}^{I_t}\sum_{j=1}^{B}c_{o,u_o}[t]\cdot w^o_{u_o,i}[t]\cdot b_{i,j}[t]\cdot U'_{o,u_o,i}\left(x_{o,i}[t]\right)\cdot r_{i,j}[t] & (3)\end{array}\right\} \tag{27}$$

$$
s.t.
\begin{array}{l}
\sum_o \sum_{u_o} c_{o,u_o}\left[t\right] = 1 \left(a\right) \\
\sum_o \sum_{u_o} w^o_{u_o,i}\left[t\right] = 1,\ i = 1,\ldots,I_t \left(b\right) \\
\sum_i w^o_{u_o^*,i}\left[t\right] = U_t, u_o^* \in \mathcal{U}_o, o \in \mathcal{O}\left(c\right) \\
\sum_i w^o_{u_o^{\otimes},i}\left[t\right] = 0, u_o^{\otimes} = 1,\ldots,U_o, o = 1,\ldots,O, \forall u_o^{\otimes} \neq u_o^* \left(d\right) \\
\sum_i b_{i,j}\left[t\right] \leq 1, j = 1,\ldots,B\left(e\right) \\
c_{o,u_o}\left[t\right] \in \left\{0,1\right\}, \forall o \in \mathcal{O}, \forall u_o \in \mathcal{U}_o \\
w^o_{u_o,i}\left[t\right] \in \left\{0,1\right\}, \forall o \in \mathcal{O}, \forall u_o \in \mathcal{U}_o, \forall i \in \mathcal{I}_t \\
b_{i,j}\left[t\right] \in \left\{0,1\right\}, \forall i \in \mathcal{I}_t, \forall j \in \mathcal{B}
\end{array}
$$

By selecting the optimal matrix $c^*\left[t\right]$ in the first term, the second term is also maximized. Thus, the proposed sequential linearization method aims to split the non-linear optimization problem into two sub-optimal linear sub-problems as follows:

- First stage, the scheduling rule that maximizes the product between the accumulated Lagrange multiplier at TTI t and the aggregate multi-objective function at TTI t+1 must be selected as indicated in the following equation:

$$
\begin{array}{ll}
\max\limits_{c} & \sum_{o=1}^{O} \sum_{u_o=1}^{U_o} \Phi^A_{o,u_o}\left[t\right] \cdot c_{o,u_o}\left[t\right] \cdot \Phi_{o,u_o}\left[t+1\right] \\
s.t. & \sum_{o=1}^{O} \sum_{u_o=1}^{U_o} c_{o,u_o}\left[t\right] = 1 \\
& c_{o,u_o}\left[t\right] \in \left\{0,1\right\},\ \forall o \in \mathcal{O}, \forall u_o \in \mathcal{U}_o
\end{array}
\quad (28)
$$

- Second stage, the allocation procedure of radio resources for the active users is performed based on the selected rule from the first stage and the optimization problem is similar to (12).

The linear optimization problem from (28) can be solved by selecting the decision variable $c_{o,u_o}\left[t\right]$ according to the accumulated Lagrange multiplier $\Phi^A_{o,u_o}\left[t\right]$ such that, the instantaneous aggregate multi-objective function $\Phi_{o,u_o}\left[t+1\right]$ would be maximized. There are two main problems in selecting the optimal decision variable $c^*_{o,u_o}\left[t\right]$ at each TTI: a) the scheduling policy π has to be optimized and the accumulated Lagrange multiplier must be updated by many TTI-to-TTI iterations; b) the optimization process of the scheduling policy is practically impossible since the scheduler state space is not considered when the Lagrange multiplier Φ^A_{o,u_o} and the aggregate multi-objective function Φ_{o,u_o} are computed.

Scheduler State Space in Multi-Objective Optimization

Let us define the set of neighbor states $\mathcal{N}\left(s\right) \subset \mathcal{S}$ being composed by those possible states to which the current state $s\left[t\right]$ could evolve based on different selections of $c_{o,u_o}\left[t\right]$, and then the next state is

$s' \in \mathcal{N}(s)$. By using the terminology from the machine learning domain as given by Sutton and Barto (2017), the action value $Q_{o,u_o}(s)$ and the reward $RW_{o,u_o}(s')$ functions are obtained based on the accumulated Lagrange multiplier and aggregate multi-objective functions, respectively, as follows:

$$\begin{aligned} &\Phi_{o,u_o}[t+1] \mapsto RW_{o,u_o}(s'), RW_{o,u_o} : \mathcal{O} \times \mathcal{U} \times \mathcal{N}(s) \to \mathbb{R} \\ &\Phi^{A}_{o,u_o}[t] \mapsto Q_{o,u_o}(s), Q_{o,u_o} : \mathcal{O} \times \mathcal{U} \times \mathcal{S} \to \mathbb{R} \end{aligned} \tag{29}$$

where the reward function $RW_{o,u_o}(s')$ measures the performance of applying the scheduling rule corresponding to the decision variable $c_{o,u_o}[t]$ when the current state is $s \in \mathcal{S}$; $Q_{o,u_o}(s)$ is the accumulated reward for the decision variable $c_{o,u_o}[t]$ being applied only in the scheduler state $s \in \mathcal{S}$ for an infinite number of visits. By using the above notations, the dual optimization problem is highlighted in the following equation:

$$\begin{aligned} &\max_{c,d} \sum_{o=1}^{O} \sum_{u_o=1}^{U_o} \sum_{s=1}^{O \times U} Q_{o,u_o}(s) \cdot c_{o,w_o}[t] \cdot d^{o}_{u_o,s}[t] \cdot RW_{o,u_o,s}(s') \\ &\text{s.t.} \quad \sum_{o=1}^{\mathcal{O}} \sum_{u_o=1}^{U_o} c_{o,w_o}[t] = 1 \\ &\qquad \sum_{s=1}^{O \times U} d^{o}_{u_o,s}[t] = 1, \; o = 1,\dots,O, u_o = 1,\dots,U_o \\ &\qquad c_{o,u_o}[t] \in \{0,1\}, \; \forall o \in \mathcal{O}, \forall u_o \in \mathcal{U}_o \\ &\qquad d^{o}_{u_o,s}[t] \in \{0,1\}, \; \forall o \in \mathcal{O}, \forall u_o \in \mathcal{U}_o, \forall s \in \mathcal{N}(s) \end{aligned} \tag{30}$$

where $d^{o}_{u_o,s}[t]$ is the variable that decides the next state $s' \in \mathcal{S}$ and the considered assignation matrix is

$d^{o}[t] = \{d^{o}_{u_o,s}[t], u_o = 1,\dots,U_o, s = 1,\dots,O \times U_o\}, \forall o \in \mathcal{O}$.

The optimization problem in (30) is non-linear due to the product between the scheduling rule and the next state variables such as: $c_{o,u_o} \cdot d^{o}_{u_o,s}$. An additional variable is needed for linearization, such that $e_{o,u_o,s}[t] = c_{o,u_o}[t] \cdot d^{o}_{u_o,s}[t]$, where the matrix which indicates the scheduler state evolution at TTI t+1 when the decision variable $c_{o,w_o}[t]$ has been applied in the previous state is

$e_o[t] = \{e_{o,u_o,s}[t], u_o = 1,\dots,U_o, s = 1,\dots,O \times U_o\}$.

Then, the obtained linearized optimization problem is exposed in (31), where the first set of constraints (a) acts as an AND gate, where the input variables $\{c_{o,u_o}; d^{o}_{u_o,s}\} \in \{0,1\}$ count four possible binary combinations and the parameter $e_{o,u_o,s}[t] \in \{0,1\}$ is the output variable, where $l \in \mathbb{R}_+$ is a large positive number. If $e_{o,u_o,s}[t] = 1$, then $\sum_o \sum_{u_o} \sum_s e_{o,u_o,s}[t] = 1$ which means that the scheduler is evolving to an unique state at TTI t+1 based on the selected objective $\forall o \in \mathcal{O}$ and utility function $\forall u_o \in \mathcal{U}_o$. Constraints (b) indicate that only one scheduling rule focused on a particular objective is selected. Constraints (c) associate only one scheduler state $s' \in \mathcal{N}(s)$ according to the selected scheduling rule in state $s \in \mathcal{S}$. Constraints (d)-(f) make the entire problem combinatorial.

$$
\begin{aligned}
&\max_{c,d} \sum_{o=1}^{O} \sum_{u_o=1}^{U_o} \sum_{s=1}^{N(s)} Q_{o,u_o}(s) \cdot e_{o,u_o,s}[t] \cdot RW_{o,u_o,s}(s') \\
&\qquad \left.\begin{aligned}
&e_{o,u_o,s}[t] \leq c_{o,u_o}[t] \\
&e_{o,u_o,s}[t] \leq d^o_{u_o,s}[t] \cdot l \\
&e_{o,u_o,s}[t] \geq c_{o,u_o}[t] - \left(1 - d^o_{u_o,s}[t]\right) \cdot l \\
&\quad \forall o \in \mathcal{O}, \forall u_o \in \mathcal{U}_o, \forall s \in \mathcal{N}(s)
\end{aligned}\right\} (a) \\
&s.t. \quad \sum_{o=1}^{O} \sum_{w_o=1}^{U_o} c_{o,u_o}[t] = 1 \qquad (b) \\
&\qquad \sum_{s=1}^{O\times U} d^o_{u_o,s}[t] = 1, \; o = 1,\ldots,O, u_o = 1,\ldots,U_o \qquad (c) \\
&\qquad c_{o,u_o}[t] \in \{0,1\}, \; \forall o \in \mathcal{O}, \forall u_o \in \mathcal{U}_o \qquad (d) \\
&\qquad d^o_{u_o,s}[t] \in \{0,1\}, \; \forall o \in \mathcal{O}, \forall u_o \in \mathcal{U}_o, \forall s \in \mathcal{N}(s) \qquad (e) \\
&\qquad f_{o,u_o,s}[t] \in \{0,1\}, \; \forall o \in \mathcal{O}, \forall u_o \in \mathcal{U}_o, \forall s \in \mathcal{N}(s) \qquad (f)
\end{aligned}
\tag{31}
$$

Temporal Difference Learning for Multi-Objective Optimization Problem

The idea is to select the best decision variable $c_{o,u_o}[t]$ that maximizes the accumulated reward $Q_{o,u_o}(s)$ and to assign the state $s' \in \mathcal{S}$ according to $d^o[t]$, such that, the RRM reward $RW_{o,u_o,s}(s')$ at TTI t+1 is maximized. In practice, when deciding the assignation variable $c_{o,u_o}[t]$, the future state $s' \in \mathcal{S}$ is automatically determined at TTI t+1 as a result of the scheduling procedure. For this reason, the linear optimization problem exposed in (31) is valid only when the scheduling policy is optimal. In this case, by selecting the action (scheduling rule) with the maximum action value in state $s \in \mathcal{S}$ the reward maximization in the next state is guaranteed (Sutton and Barto, 2017). On this extent, in each momentary state, the best scheduling decision is determined such that $c^*_{o,u_o}[t] = \arg\max Q^*_{o,u_o}(s)$, where $Q^*_{o,u_o}(s)$ is the optimal accumulated reward value for objective $\forall o \in \mathcal{O}$ and scheduling rule $\forall u_o \in \mathcal{U}_o$ (or the optimal action value for action (o, u_o) in state $s \in \mathcal{S}$). Then, the scheduler reward at TTI t+1 is maximized when the optimal decision is applied.

Similar to other control systems as described by Sutton and Barto (2017), the idea is to maximize the total expected return or the expected accumulated reward RW^a_π for a given policy of scheduling rules π starting from any initial state $s[t]=s$ until the optimal scheduler state $s[t_o]=s^*$ is reached, where t_o is the time needed to reach the optimal state $s^* \in \mathcal{S}$ when given a certain order of applied scheduling rules following π from any random initial state $s[t] \in \mathcal{S}$. The accumulated reward for the considered policy is discounted according to (32) (Sutton and Barto, 2017):

$$\begin{aligned} RW_\pi^a\left(s[t]\right) &= RW_\pi\left(s[t+1]\right) + \ldots + \gamma^{t_o} \cdot RW_\pi\left(s[t+t_o+1]\right) = \sum_t^{t_o} \gamma^{t_o} \cdot RW_\pi\left(s[t+t_o+1]\right) \\ &= RW_\pi\left(s[t+1]\right) + \gamma \cdot RW_\pi^a\left(s[t+1]\right) \end{aligned} \tag{32}$$

where $\gamma \in [0,1]$ is the discount factor that sets the importance of future rewards. Equation (32) is the case of *temporal difference learning,* where the reward value can be deducted by following the reasoning:

$$RW_\pi\left(s[t+1]\right) = RW_\pi^a\left(s[t]\right) - \gamma \cdot RW_\pi^a\left(s[t+1]\right) \tag{33}$$

We know that when the scheduling policy is optimal π^*, then the instantaneous scheduler reward $RW_\pi\left(s[t+1]\right)$ is equivalent with the difference between the accumulated rewards from two consecutive states as indicated in (33). Actually, the accumulated reward RW_π^a in state $s \in \mathcal{S}$ for a given policy π is similar to Q_{o,u_o} at that state. Under optimal conditions, the instantaneous reward from (33) can be re-written as follows:

$$RW_\pi\left(s[t+1]\right) = Q_{o,u_o}^*\left(s\right) - \gamma \cdot Q_{o',u_{o'}^{'}}^*\left(s'\right) \tag{34}$$

where $\forall o' \in \mathcal{O}$ and $\forall u_{o'}^{'} \in \mathcal{U}_o$ are the selected objective and scheduling rule in state $s' \in \mathcal{S}$.

The scheduling policy π needs to be improved by using many visits of state $s \in \mathcal{S}$ in order to learn the optimal objective $o^* \in \mathcal{O}$ and the utility function $u_o^* \in \mathcal{U}_o$ that maximize the accumulated reward value $Q_{o,u_o}^*\left(s\right)$. This stage is entitled the learning stage since all possible scheduling rules have to be tested for each given momentary scheduler states. Once this policy is refined and trained properly, the exploitation stage is performed. In this stage, the optimization problem exposed in (31) is satisfied since the scheduling rule that maximizes the accumulated reward $Q_{o,u_o}^*\left(s\right)$ is selected and maximizes at the same time the reward value in next state $s' \in \mathcal{S}$.

Reinforcement Learning in DSR-SMOO/CMOO Problems

The scheduler state space is continuous and multi-dimension and practically the size of the neighboring states $\left|\mathcal{N}\left(s\right)\right| \rightarrow \infty$ is infinite due to the stochastic nature of the momentary states. Also, when optimizing the fairness criterion, an infinite number of scheduling rules are considered due to the continuous parameterization of PF scheduling rule. For these reasons the state-action pairs cannot be exhaustively enumerated and hence, the simplest look-up table is not suitable to store the $Q_{o,u_o}(s)$ values for each momentary state and scheduling rule. Then, the proposed framework can only approximate the best decisions to be taken on each state. Function approximations are used such as neural networks, as proposed by Comşa et al. (2011) (2012).

The dimension of scheduler state is another important problem to be addressed. The main issue is the dependence on the number of active users and system bandwidth. For instance, the dimension of the controllable vector $c = \left[c_1, c_2, ..., c_{I_t}\right] \in \mathcal{S}_C$ at each TTI depends on the number of users I_t which is also variable given the fact that the amount of active users can change every TTI. Also, the CQI reports for each user depend on the number of resource blocks associated to a given bandwidth. Then, a procedure able to compact the momentary scheduler state is absolutely necessary in order to avoid such dependencies and to enhance the learning procedure.

The machine learning framework must approximate the best scheduling selection state-by-state. The Reinforcement Learning (RL) is a temporal-difference learning scheme able to provide very good solutions in various optimization problems (Sutton and Barto, 2017). The RL is considered a combination of Monte Carlo and Dynamic Programming methods. The Monte Carlo method provides the expected return (the sum of discounted rewards from state to state) only at the end of the learning stage and the tasks are non-episodic which means that the reward would eventually be maximized only at the end of the learning stage. On the other side, the dynamic programming method can approximate each action value function at each time and not at the end of the learning stage as is the case of Monte Carlo methods. Moreover, these action values can be updated according to the received reward in order to optimize a given policy. In DSR-SMOO/CMOO problems, the tasks are non-episodic for certain conditions (i.e. the satisfaction of all objectives may not be reached due to high number of users or/and poor channel conditions). However, the RL framework must approximate the best scheduling rule in each state such as the multi-objective satisfaction is maximized as much as possible.

Different RL algorithms can be used depending on the particularities of DSR-SMOO/CMOO problems, as stated in the research conducted by Comşa (2014a), Comşa et al. (2014b) (2014c), and Comşa et al. (2018). The role of the selected RL algorithm is to update function approximator for each selected action $\left(o, u_o\right)$ by reinforcing each time the error $e = \overline{Q}_{o,u_o}(s) - Q_{o,u_o}(s)$, where the $\overline{Q}_{o,u_o}(s)$ is the approximated action value given by the approximator and $Q_{o,u_o}(s)$ is calculated based on (34) and reloaded here:

$$Q_{o,u_o}(s) = RW_\pi(s') + \gamma \cdot \overline{Q}_{o',u_{o'}'}(s') \tag{35}$$

It is important to notice that by updating the previous learned action value based on the temporal-difference error, $\overline{Q}_{o,u_o}(s)$ is decreased or increased by providing lower or higher probabilities for the decision variable $c_{o,u_o}[t]$ to be selected in the future when transiting nearly the same states from s to s'.

From the architectural point of view, the RL based DSR-CMOO approach defines two modules: Marginal Utility State Informer (MUSI) and Marginal Utility Type Informer (MUTI). MUTI converts the action which is provided by the intelligent controller in the corresponding scheduling rule such that $a[t] \rightarrow c_{o,u_o}[t]$, where $a[t]$ is the controller action. Based on the MUTI decision, MUSI provides the necessary state parameters in order to compute the corresponding scheduling metrics for each user and for each resource block. Figure 2 presents the proposed RL–based DSR-CMOO architecture for the OFDMA downlink scheduling. The MCS assignment procedure for the transport block computation is executed in a separate stage once the radio resource allocation is performed. Studies conducted by Kwan, Leung, and Zhang (2009) indicate a degradation of the cell spectral efficiency of about 10% when the RB and MCS assignments are performed in separate stages, but at a much lower computational complexity when compared to the joint optimization approach.

As depicted in Fig. 2, in the first stage, the objective to be followed and the corresponding scheduling rule are selected and then, the scheduling procedure is performed based on the data provided by MUSI and MUTI entities. In the learning stage, at TTI t, the momentary scheduler state $s[t] \in \mathcal{S}$ is observed. The state aggregation module aims to reduce the dimension of the initial state to a more compact representation in order to decrease the complexity of the proposed framework and to learn faster. Based on the aggregate momentary state, the function approximator provides the action values $\overline{Q}_{o,u_o}(s)$ for each objective $\forall o \in \mathcal{O}$ and utility $\forall u_o \in \mathcal{U}_o$. According to some probabilities, the controller may decide to apply the scheduling rule corresponding to $(o^*, u^*_{o^*}) = \arg\max_{o,u_o} Q_{o,u_o}(s)$ or to select a different scheduling rule in order to explore more action-state possibilities and to increase the quality of

Figure 2. Reinforcement learning based solution for multi-objective optimization problem

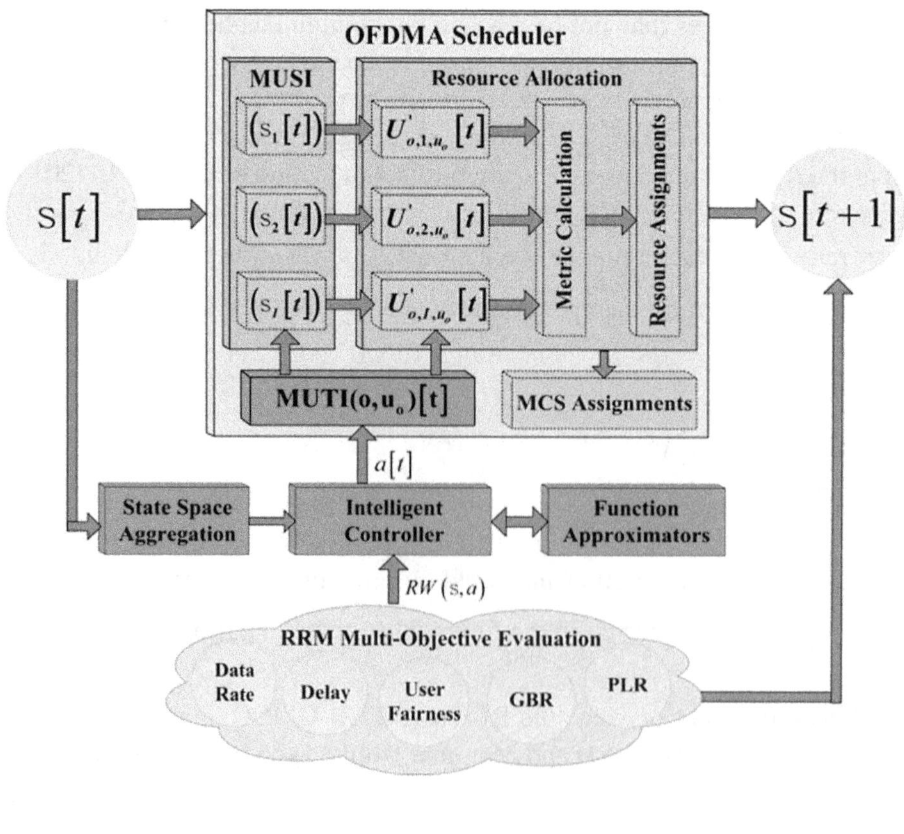

the learning stage. Once, the scheduling rule is decided, the resource allocation, MCS assignments and TB calculations are performed and the system evolves to the next state $s\left[t+1\right] = s' \in \mathcal{S}$. At TTI t+1, the RRM multi-objective evaluator determines the reward value, the error is computed according to (35) and reinforced in order to optimize the function approximator. The learning stage can continue for large number of state-to-state iterations until this error falls under a specified threshold. In the exploitation stage, the learnt function provides the scheduling rules to be applied in each state in order to maximize the multi-objective satisfaction.

FUTURE RESEARCH DIRECTIONS AND OPEN ISSUES

The relevant studies and related work based on SMOO and CMOO optimizations show that the scheduling rules have different behaviours under different circumstances such as the traffic load, scheduler states and different assumptions. Under these considerations, different scheduling rules may be applied for the DSR-SMOO/CMOO problems in order to find the optimal scheduler states much faster with a reduced system complexity.

The motivation of using the DSR-SMOO/CMOO concepts in next generation radio access networks is to increase the number of TTIs when all active bearers are satisfied in terms of different multi-objective criteria. Different scheduling rules work much better when compared with any other disciplines under given circumstances such as channel conditions, queue states, traffic load or QoS requirements. Hence, the principle of applying a mixture of scheduling rules becomes mandatory, and the most appropriate scheduling discipline is called when it can provide the highest scheduler reward.

Future possible research directions would be to include the present work in the presence of multiple traffic types. Basically, there are two ways of treating this innovative approach: to use the existing policies or to train other scheduling policies for each traffic type. The idea is to train different scheduling policies based on different traffic types by using slave controllers whereas the scheduling decision on the traffic classes is performed by a master controller.

For more realistic radio access network scheduling scenarios, the scheduling procedure is performed in the presence of multiple traffic types (i.e., heterogeneous traffic). The traffic heterogeneity implies different QoS requirements. In general, the reward functions for different objectives represent the sum of normalized sub-rewards from different traffic classes. Therefore, the traffic heterogeneity requires different reward functions for different priority classes. However, the controller state space dimension becomes much larger than the classical state space with homogeneous traffic.

The action and the state spaces of each slave agent are similar to the case of DSR-CMOO combinatorial problems with the homogeneous traffic type. The main problem is the state space dimensionality of the master agent which requires a high computational complexity for the function approximations and expensive exploration time. The current approach aims to avoid these drawbacks by partitioning the original state space into subspaces where each subspace represents the slave agent state space for a given traffic class. The slave agents can be trained separately from the master agent. When the QoS and fairness agents for each slave controller are trained enough, then these structures can be exploited when the master agent is trained.

CONCLUSION

This chapter investigates the radio resource management issues for next generation radio access networks and formulates an aggregate multi-objective problem that aims at selecting in each scheduler state the most suitable scheduling rule in order to maximize the system throughput while increasing the satisfaction of scheduling objectives in terms of: user fairness, packet delay, user rate and packet loss rate. The proposed framework aims to minimize the drawback of each individual scheduling rule and maximize their benefits when applying each one of them under the best matching scheduler conditions. Due to the complexity of this aggregate optimization problem, the proposed solution splits the entire scheduling framework in two parts where, the first one deals with the selection of the scheduling rule at each momentary scheduler state and the second one performs the classical resource allocation procedure. As part of the machine learning domain, the reinforcement learning is proposed to learn over time the most appropriated scheduling rule to be applied at each state. Due to the complexity and the dimensionality problems of the scheduler state space, additional methods imported from artificial intelligence domain (i.e. data mining to compress the scheduler state, neural networks to approximate the best scheduling rule under each state) must be adopted in conjunction with the proposed reinforcement learning framework in order to make the proposed solution suitable for real-time downlink schedulers.

REFERENCES

Andrews, J. G., Buzzi, S., Choi, W., Hanly, S. V., Lozano, A., Soong, A. C. K., & Zhang, J. C. (2014). What Will 5G Be? *IEEE Journal on Selected Areas in Communications*, *32*(6), 1065–1082. doi:10.1109/JSAC.2014.2328098

Avocanh, F. T. S., Abdennebi, M., & Ben-Othman, J. (2014). An Enhanced Two Level Scheduler to Increase Multimedia Services Performance in LTE Networks. In *Proceedings of IEEE International Conference on Communications (ICC)* (pp. 2351-2356). Sydney, NSW, Australia: IEEE. 10.1109/ICC.2014.6883674

Bae, S. J., Choi, B.-G., & Chung, M. (2011). Delay-Aware Packet Scheduling Algorithm for Multiple Traffic Classes in 3GPP LTE System. In *Proceedings of 17th Asia-Pacific Conference on Communications (APCC)* (pp. 33 – 37). Sabah, Malaysia: APCC. 10.1109/APCC.2011.6152755

Capozzi, F., Piro, G., Grieco, L. A., Boggia, G., & Camarda, P. (2013). Downlink Packet Scheduling in LTE Cellular Networks: Key Design Issues and a Survey. *IEEE Communications Surveys and Tutorials*, *15*(2), 678–700. doi:10.1109/SURV.2012.060912.00100

Chung, W. C., Chang, C. J., & Wang, L. C. (2012). An Intelligent Priority Resource Allocation Scheme for LTE-A Downlink Systems. *IEEE Wireless Communications Letters*, *1*(3), 241–244. doi:10.1109/WCL.2012.033012.120146

Cisco. (2017). *Cisco Visual Networking Index: Global Mobile Data Traffic Fore-cast Update, 2016-2021 White Paper* Retrieved from http://www.cisco.com/c/en/us/solutions/collateral/service-provider/visual-networking-index-vni/mobile-white-paper-c11-520862.html

Comşa, I.-S. (2014a). *Sustainable Scheduling Policies for Radio Access Networks Based on LTE Technology*. Bedfordshire, UK: University of Bedfordshire.

Comşa, I. S., Aydin, M., Zhang, S., Kuonen, P., & Wagen, J.-F. (2011). Reinforcement learning based radio resource scheduling in LTE-advanced. In *Proceedings of 17th International Conference on Automation and Computing* (pp. 219 - 224). Huddersfield, UK: Academic Press.

Comşa, I. S., Aydin, M., Zhang, S., Kuonen, P., & Wagen, J.-F. (2012). Multi Objective Resource Scheduling in LTE Networks Using Reinforcement Learning. *International Journal of Distributed Systems and Technologies*, *3*(2), 39–57. doi:10.4018/jdst.2012040103

Comşa, I.-S., Aydin, M., Zhang, S., Kuonen, P., Wagen, J.-F., & Lu, Y. (2014b). Scheduling Policies Based on Dynamic Throughput and Fairness Tradeoff Control in LTE-A Networks. In *Proceedings of IEEE Conference on Local Computer Networks (LCN)* (pp. 418-421). Edmonton, AB, Canada: IEEE. 10.1109/LCN.2014.6925806

Comşa, I.-S., De Domenico, A., & Ktenas, D. (2017). QoS-Driven Scheduling in 5G Radio Access Networks - A Reinforcement Learning Approach. In *Proceedings of IEEE Global Communications Conference (GLOBECOM)* (pp. 1-7). Singapore: IEEE. 10.1109/GLOCOM.2017.8254926

Comşa, I.-S., Trestian, R., & Ghinea, G. (2018). 360° Mulsemedia Experience over Next Generation Wireless Networks - A Reinforcement Learning Approach. In *Proceedings of Tenth International Conference on Quality of Multimedia Experience (QoMEX)* (pp. 1-6). Cagliari, Italy: Academic Press. 10.1109/QoMEX.2018.8463409

Comşa, I.-S., Zhang, S., Aydin, M., Chen, J., Kuonen, P., & Wagen, J.-F. (2014c). Adaptive Proportional Fair Parameterization Based LTE Scheduling Using Continuous Actor-Critic Reinforcement Learning. In *Proceedings of IEEE Global Communication Conference (GLOBECOM)* (pp. 4387 - 4393). Austin, TX: IEEE. 10.1109/GLOCOM.2014.7037498

Comşa, I.-S., Zhang, S., Aydin, M., Kuonen, P., Lu, Y., Trestian, R., & Ghinea, G. (2018). Towards 5G: A Reinforcement Learning-based Scheduling Solution for Data Traffic Management. *IEEE eTransactions on Network and Service Management*, 1–15.

Comşa, I.-S., Zhang, S., Aydin, M., Kuonen, P., & Wagen, J. (2012). A Novel Dynamic Q-Learning-Based Scheduler Technique for LTE-advanced Technologies Using Neural Networks. In *Proceedings of IEEE Conference on Local Computer Networks (LCN)* (pp. 332 – 335). Clearwater, FL: IEEE.

FANTASTIC-5G. (2016). *D4.1: Technical Results for Service Specific Multi-Node/MultiAntenna Solutions.* Retrieved from http://fantastic5g.com/wp-content/uploads/2016/06/FANTASTIC-5G_D4.1_ Final.pdf

3. GPP. (2012). *Technical Specification Group Services and System Aspects; Policy and Charging Control Architecture Release 12.* Retrieved from http://www.qtc.jp/3GPP/Specs/23203-a60.pdf

Jain, R., Chiu, D. M., & Hawe, W. A. (1984). Quantitative Measure of Fairness and Discrimination for Resource Allocation in Shared Computer System. In Research Report, DEC-TR-301, 1984.

Kelly, F. (1997). Charging and Rate Control for Elastic Traffic. *European Transactions of Telecommunications*, *8*(1), 33–37. doi:10.1002/ett.4460080106

Khan, N., Martini, M. G., Bharucha, Z., & Auer, G. (2012). Opportunistic Packet Loss Fair Scheduling for Delay-Sensitive Applications over LTE Systems. In *Proceedings of IEEE Wireless Communications and Networking Conference* (pp. 1456 – 1461). Shanghai, China: IEEE. 10.1109/WCNC.2012.6214010

Kwan, R., Leung, C., & Zhang, J. (2009). Resource Allocation in an LTE Cellular Communication System. In *Proceedings of IEEE International Conference on Communications*, (pp. 1-5). Dresden, Germany: IEEE. 10.1109/ICC.2009.5199412

Li, Y., Pateromichelakis, E., Vucic, N., Luo, J., Xu, W., & Caire, G. (2017). Radio Resource Management Considerations for 5G Millimeter Wave Backhaul and Access Networks. *IEEE Communications Magazine*, *55*(6), 86–92. doi:10.1109/MCOM.2017.1601118

Liu, B., Tian, H., & Xu, L. (2013). An Efficient Downlink Packet Scheduling Algorithm for Real Time Traffics in LTE Systems. In *Proceedings of IEEE Consumer Communications and Networking Conference (CCNC)* (vol. 1, pp. 364-369). Las Vegas, NV: IEEE.

Liu, X., Chong, E., & Shroff, N. (2001). Opportunistic Transmission Scheduling with Resource- Sharing Constraints in Wireless Networks. *IEEE Journal on Selected Areas in Communications*, *19*(10), 2053–2064. doi:10.1109/49.957318

Lundevall, M., Olin, B., Olsson, J., Wiberg, N., Wanstedt, S., Eriksson, J., & Eng, F. (2004). Streaming Applications over HSDPA in Mixed Service Scenarios. In *Proceedings of IEEE Vehicular Technology Conference (VTC-Fall)* (vol. 2, pp. 841 - 845). Los Angeles, CA: IEEE. 10.1109/VETECF.2004.1400139

Monghal, G., Laselva, D., Michaelsen, P.-H., & Wigard, J. (2010). Dynamic Packet Scheduling for Traffic Mixes of Best Effort and VoIP Users in E-UTRAN Downlink. In *Proceedings of IEEE Vehicular Technology Conference (VTC-Spring)* (pp. 1-5). Taipei, Taiwan: IEEE. 10.1109/VETECS.2010.5493737

Ning, X., Ting, Z., Ying, W., & Ping, Z. (2006). A MC-GMR Scheduler for Shared Data Channel in 3GPP LTE System. In *Proceedings of IEEE Vehicular Technology Conference (VTC-Fall)* (pp. 1-5). Montreal, Quebec, Canada: IEEE. 10.1109/VTCF.2006.279

Nocedal, J., & Wright, S. J. (2006). Penalty, Barrier, and Augmented Lagrangian Methods. In *Numerical Optimization* (pp. 488–524). Springer Series in Operations Research. doi:10.1007/978-0-387-40065-5_17

Olwal, T. O., Djouani, K., & Kurien, A. M. (2016). A Survey of Resource Management Toward 5G Radio Access Networks. *IEEE Communications Surveys and Tutorials*, *18*(3), 1656–1686. doi:10.1109/COMST.2016.2550765

Proebster, M., Mueller, C. M., & Bakker, H. (2010). Adaptive Fairness Control for a Proportional Fair LTE Scheduler. *Proceedings of IEEE International Symposium on Personal Indoor and Mobile Radio.* 10.1109/PIMRC.2010.5671970

Rhee, J. H., Holtzman, J. M., & Kim, D.-K. (2003). Scheduling of Real/Non-Real Time Services: Adaptive EXP/PF Algorithm. In *Proceedings of IEEE Semiannual Vehicular Technology Conference (VTC-Spring)* (vol. 1, pp. 462 – 466). Jeju, South Korea: IEEE.

Sadiq, B., Madan, R., & Sampath, A. (2009). Downlink Scheduling for Multiclass Traffic in LTE. *EURASIP Journal on Wireless Communications and Networking*, *2009*(1), 1–18. doi:10.1155/2009/510617

Schwarz, S., Mehlfuhrer, C., & Rupp, M. (2011). Throughput Maximizing Multiuser Scheduling with Adjustable Fairness. In *Proceedings of IEEE International Conference on Communications (ICC)* (pp.1-5). Kyoto, Japan: IEEE. 10.1109/icc.2011.5963489

Song, G. (2005). *Cross-Layer Resource Allocation and Scheduling in Wireless Multicarier Networks*. Atlanta, GA: Georgia Institute of Technology.

Sutton, R. S., & Barto, A. G. (2017). *Reinforcement Learning: An Introduction*. London: IMT Press. doi:10.1007/978-1-4899-7687-1_720

Toufik, I., & Knopp, R. (2011). Multi-User Scheduling and Interference Coordination. In LTE. The UMTS Long Term Evolution. From Theory To Practice (pp. 279-292). John Wiley & Sons Ltd. doi:10.1002/9780470978504.ch12

Trestian, R., Comşa, I.-S., & Tuysuz, M. F. (2018). Seamless Multimedia Delivery within a Heterogeneous Wireless Networks Environment: Are We There Yet? *IEEE Communications Surveys and Tutorials*, *20*(2), 945–977. doi:10.1109/COMST.2018.2789722

Trestian, R., Muntean, G., & Ormond, O. (2009). Signal Strength-based Adaptive Multimedia Delivery Mechanism. In *IEEE 34th Conference on Local Computer Networks, 2009. LCN 2009* (pp. 297–300). IEEE. 10.1109/LCN.2009.5355092

Trestian, R., Ormond, O., & Muntean, G.-M. (2012). On the Impact of Wireless Network Traffic Location and Access Technology on Mobile Device Energy Consumption. In *Proceedings of the 2012 IEEE 37th Conference on Local Computer Networks (LCN 2012)* (pp. 200–203). Washington, DC: IEEE. 10.1109/LCN.2012.6423608

Vien, Q. T., Akinbote, T., Nguyen, H. X., Trestian, R., & Gemikonakli, O. (2015). On the coverage and power allocation for downlink in heterogeneous wireless cellular networks. In *2015 IEEE International Conference on Communications (ICC)* (pp. 4641–4646). IEEE. 10.1109/ICC.2015.7249055

Vien, Q. T., Nguyen, H. X., Trestian, R., Shah, P., & Gemikonakli, O. (2016). A Hybrid Double-Threshold Based Cooperative Spectrum Sensing over Fading Channels. *IEEE Transactions on Wireless Communications*, *15*(3), 1821–1834. doi:10.1109/TWC.2015.2496598

Wang, K., Li, X., Ji, H., & Zhang, X. (2013). Heterogeneous Traffic Scheduling in Downlink High Speed Railway LTE Systems. In *Proceedings of IEEE Global Communications Conference (GLOBECOM)* (pp. 1452-1457). Atlanta, GA: IEEE. 10.1109/GLOCOM.2013.6831278

Zhang, J., Yuan, D., & Zhang, H. (2011). Joint Radio Resource Allocation and Scheduling in a Backhaul Constrained Multicell OFDMA Networks. In *Proceedings of 13th International Conference on Communication Technology (ICCT)* (pp. 47 – 51). Jinan, China: Academic Press.

Zou, L., Trestian, R., & Muntean, G.-M. (2013). A utility-based priority scheduling scheme for multimedia delivery over LTE networks. In *2013 IEEE International Symposium on Broadband Multimedia Systems and Broadcasting (BMSB)* (pp. 1–7). IEEE. 10.1109/BMSB.2013.6621745

ADDITIONAL READING

Moysen, J., & Giupponi, L. (2018). From 4G to 5G: Self-organized network management meets machine learning. *Computer Communications*, *129*, 248–268. doi:10.1016/j.comcom.2018.07.015

Polese, M., Jana, R., Kounev, V., Zhang, K., Deb, S., & Zorzi, M. (2018). Machine Learning at the Edge: A Data-Driven Architecture with Applications to 5G Cellular Networks. Retrieved from https://arxiv.org/abs/1808.07647

Vien, Q., Ogbonna, N., Nguyen, H. X., Trestian, R., & Shah, P. (2015). Non-Orthogonal Multiple Access for Wireless Downlink in Cloud Radio Access Networks. In Proceedings of European Wireless 2015; 21th European Wireless Conference (pp. 1–6).

KEY TERMS AND DEFINITIONS

Machine Learning: Category of algorithm that uses statistical techniques to analyze data and make predictions.

Multi-Objective Optimization: Mathematical optimization problems involving more than one objective function to be optimized simultaneously.

Radio Resource Management: Function of a mobile communication systems used for establishing, maintaining, and releasing of radio resources.

Reinforcement Learning: Training/learning method aiming to automatically determine the ideal behavior within a specific context based on rewarding desired behaviors and/or punishing undesired one.

Unsupervised Learning: Type of machine learning algorithm that is capable of learning a function that best represents a model from datasets consisting of unlabeled input data.

Chapter 3
Machine Learning for Internet of Things

Ramgopal Kashyap
https://orcid.org/0000-0002-5352-1286
Amity University Chhattisgarh, India

ABSTRACT

Fast advancements in equipment, programming, and correspondence advances have permitted the rise of internet-associated tangible gadgets that give perception and information estimation from the physical world. It is assessed that the aggregate number of internet-associated gadgets being utilized will be in the vicinity of 25 and 50 billion. As the numbers develop and advances turn out to be more develop, the volume of information distributed will increment. Web-associated gadgets innovation, alluded to as internet of things (IoT), keeps on broadening the present internet by giving network and cooperation between the physical and digital universes. Notwithstanding expanded volume, the IoT produces big data described by speed as far as time and area reliance, with an assortment of numerous modalities and changing information quality. Keen handling and investigation of this big data is the way to creating shrewd IoT applications. This chapter evaluates the distinctive machine learning techniques that deal with the difficulties in IoT information.

INTRODUCTION

The Internet of Things (IoT) worldview is developing through the far-reaching reception of detecting and catching smaller scale and nano-gadgets plunged in ordinary situations and interconnected in low-control, lossy systems. The sum and consistency of inescapable gadgets build day by day and after that the rate of crude information accessible for handling and investigation exponentially grows-up. Like never before, viable strategies are expected to treat information streams with the last objective to give a significant elucidation of recovered data (Puthal, 2018). The big data name was instituted to signify the innovative work of information mining procedures what's more, administration frameworks to manage "volume, speed, assortment and veracity" issues rising at the point when substantial amounts of data appear what's more, should be controlled. Consequently, Machine Learning (ML) is embraced to arrange

DOI: 10.4018/978-1-5225-7458-3.ch003

crude information and settle on expectations situated to choice help and computerization ("Special issue of Big Data Research Journal on "Big Data and Neural Networks"", 2018). Advance in ML calculations and improvement runs with advances of inescapable advances and Web-scale information administration designs, with the goal that certain advantages have been delivered from the information examination purpose of see. By some not insignificant shortcomings are as yet clear concerning the expanding multifaceted nature and heterogeneity of unavoidable figuring challenges. Especially, the absence of significant, machine-understandable portrayal of yields from established ML systems is a conspicuous cutoff for a conceivable abuse in completely autonomic application situations.

This chapter presents a general system pointing to upgrade traditional ML investigation on IoT information streams, partner semantic depictions to data recovered from the physical world, instead of inconsequential characterization names. The essential thought is to treat a commonplace ML order issue like an ontology-driven asset disclosure steps incorporate building a rationale based portrayal of measurable information dissemination and playing out fine-grained occasion recognition, misusing non-standard thinking administrations for matchmaking (Rathore, Paul, Ahmad & Jeon, 2017). The proposition grounds on the two thoughts and advances of disseminated information-based frameworks, whose people assertional information are physically fixing to objects spread in a given situation, without concentrated coordination. Every comment alludes to a metaphysics giving the conceptualization and vocabulary to the specific learning area besides, the proposed hypothetical show use a propelled matchmaking on metadata put away in detecting and catching gadgets plunged in a unique circumstance, lacking settled information bases. Induction assignments are circulated among gadgets which give negligible computational abilities. Stream thinking methods give the preparation to tackle the stream of semantically commented on refreshes gathered from low-level information, so as to empower versatile setting mindful practices. Along with this vision, inventive examination strategies connected to information extricated by modest off-the-rack sensor gadgets can give helpful outcomes in occasion acknowledgment without requiring extensive computational assets. Points of confinement of catching equipment could be offset novel programming side information elucidation approaches. The approach was tried and approved in a contextual investigation for street and movement checking on a genuine informational collection gathered for tests. Results were contrasted with great ML calculations keeping in mind the end goal to assess execution. The test crusade and early trials preparatory evaluate both possibility and maintainability of the various approaches. In this chapter machine learning for IoT, categorization of machine learning, model-based learning is discussed.

MOTIVATION

Principle inspiration for this chapter moves from the proof of genuine cutoff points in the IoT. Regardless of inescapability scaling down and availability interconnection ability reinforce physical frameworks, substantial information corpora appear without having truly the likelihood of dissecting them top to bottom locally. Usually received information mining strategies have two fundamental disadvantages: I) they essentially do close to a grouping errand and ii) their accuracy is expanded if connected on huge information sums so making unfeasible an on-line examination (Yildirim, Birant & Alpyildiz, 2017). These components forestall accepted the likelihood of acknowledging thinking things: the IoT is deciphered only as detected by the earth while is extremely hidden the likelihood of making choices and taking activities locally after the detecting arrange. It ought to be viewed as that in IoT situations, data is accumulated through miniaturized scale gadgets connected to regular things or sent in given conditions and interconnected remotely. Essentially, because of their little estimate, such questions have least handling capacities, a little stockpiling and low-throughput correspondence ability. They constantly deliver crude information whose volume makes important to be prepared by cutting-edge remote applications. An astute understanding of recovered data lessens measurements and potentially creates choices on what identified, regardless of whether at a remote stage. Traditional Machine Learning strategies have been to a great extent utilized however, their fundamental shortcoming is in the absence of a telling and noteworthy portrayal of uncovered occasions (Noel, 2016). IoT importance could be upgraded by explaining true questions, the information they assemble and the situations they are plunged in with brief, organized what's more, semantically rich depictions. The blend of the IoT with Semantic Web thoughts and innovations is achieving the alleged Semantic Web of Things (SWoT). This worldview plans to empower novel classes of clever applications and administrations grounded on Knowledge Representation (KR), misusing semantic-based programmed deductions to surmise certain data beginning from an unequivocal occasion also(Kashyap, 2018), setting discovery by partner an organized and machine-justifiable depiction in standard semantic web dialects, each grouping yield could expect a non-vague meaning (Jara et al., 2014). Besides, clarification capacities gave by the supporting semantic matchmaking permit to legitimize results, so expanding trust in framework reaction. On the off chance that inescapable small scale gadgets are fit for proficient on-board handling on the locally recovered information, they can depict themselves and the setting where they are situated toward outer gadgets what's more, applications. This would upgrade interoperability what's more, adaptability, encouraging unavoidable information based frameworks with high degrees of automaticity not yet permitted by normal IoT foundations and methods. Machine learning, internet on things and big data are dependent on each other while doing a great job as seen in figure 1.

Two imperative outcomes are actuated to start with of all likewise the human-computer association could be enhanced, by decreasing the client exertion required to profit from processing frameworks. In traditional IoT standards, a client unequivocally interfaces with one gadget at an opportunity to play out an errand (Urquhart & Rodden, 2016). Despite what might be expected, client specialists running on versatile figuring devices ought to have the capacity to cooperate at the same time with numerous inserted micro components, furnishing clients with setting mindful customized errand and choice help. Also, even on the off chance that machine learning systems, calculations, and instruments have empowered novel classes of investigations uncommonly valuable for Big Data Internet of Things viewpoint, the misuse of rationale based and rough revelation systems utilizing non-correct coordinating results remunerate conceivable blames in catching exercises, gadget unpredictability and the lack of quality of remote correspondences taking off versatile IoT foundations truly flexible for an across the board application ("Internet of Things & Creation of the Fifth V of Big Data", 2017).

Huge information investigation is a quickly extending research zone traversing the fields of software engineering, data administration, and has turned into a pervasive term in comprehension and taking care of complex issues in various disciplinary fields, for example, designing, connected arithmetic, drug, computational science, medicinal services, interpersonal organizations, fund, business, government, instruction, transportation and media communications. The utility of enormous information is discovered to a great extent in the region of the IoT. Enormous information is utilized to construct IoT structures which incorporate things-driven, information-driven, benefit-driven engineering, cloud-based IoT. Advancements empowering IoT to incorporate sensors, radio recurrence distinguishing proof, low power and vitality collecting, sensor systems and IoT benefits principally incorporate semantic administration,

Figure 1. Machine learning interoperability

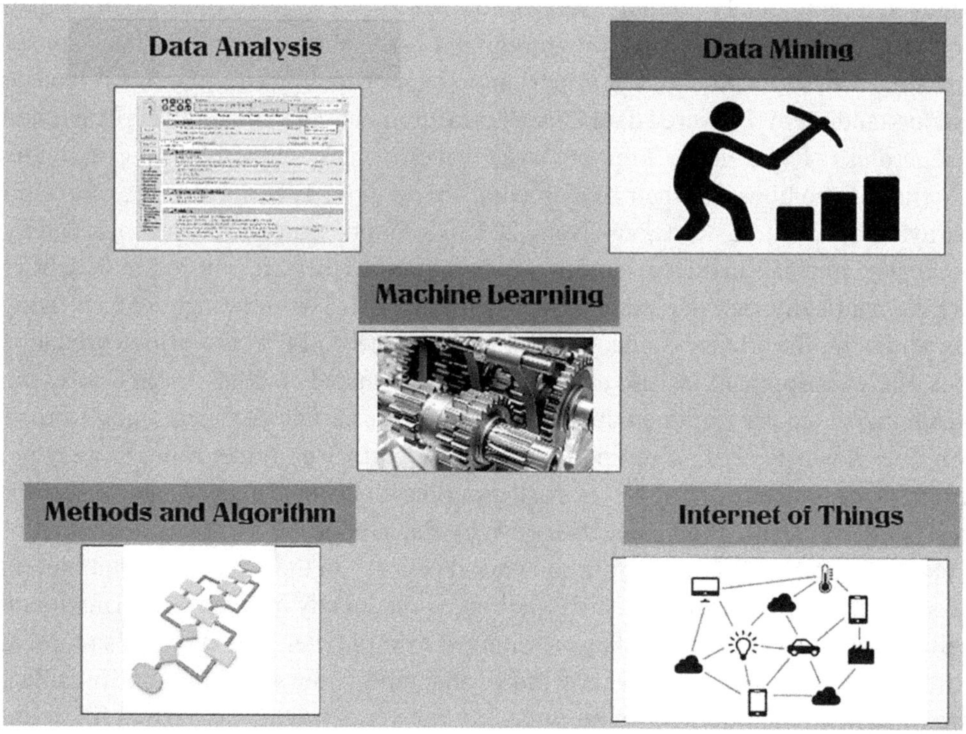

security and protection safeguarding conventions, plan cases of savvy administrations. To adequately blend enormous information and convey among gadgets utilizing IoT, machine learning strategies are utilized. Machine taking in separates importance from enormous information utilizing different systems which incorporate relapse investigation, bunching, Bayesian techniques, choice trees and irregular timberlands, bolster vector machines, support learning, group learning and profound learning.

FUNDAMENTAL USES OF MACHINE LEARNING IN IOT

Forecast of Information Patterns

Relapse calculations are the classes of calculations which help evaluate the pattern of the information focuses on by figuring a model in light of existing gathered information. The easiest relapse calculation is Linear Regression which is utilized to fit a straight line to information which is following a direct connection between the amount estimated by the sensor, and time. Frequently the information does not take after a straight relationship and more perplexing methodologies, for example, Polynomial Regression and more complex calculations must be utilized. The best in class Machine Learning calculations utilized for a forecast on account of IoT sensors i.e. time subordinate information incorporate Recurrent Neural Networks (RNNs) and Long Short-Term Memory (LSTM) Networks. Nonetheless, to begin with, understanding expectation, straight and polynomial relapse are a decent place to begin.

Bunching of Information

Another utilization case identified with IoT which can bridle the energy of ML calculations is that of an arrangement of information for binary arrangement (i.e. Hot or Not. with a temperature sensor, obviously), Logistic Regression (yes the name is intentionally befuddling) is the best place to begin. An application could be an examination of accelerometer, spinner and other Inertial Measurement Unit (IMU) information to distinguish practices, for example, strolling, standing, running or sitting.

Irregularity Detection

A third utilizes case which is firm material for IoT is Anomaly Detection, i.e. distinguishing anomalies in an informational collection. For example, if your temperature sensor itself turns sour, and begins giving a yield of 10 degrees Celsius out in the sun in a desert in summer, you would need to distinguish such an oddity. Figure 2 shows how a different machine learning algorithms give different result for that correct hypothesis is required. Measurable ways to deal with anomaly recognition include Gaussian disseminations and examining how far away the information point to be tried is from the verifiable information's mean.

When running profound learning investigation on cell phones, designers must adjust to a more asset obliged stage. Picture examination on asset obliged stages can expend critical process and memory assets. For instance, the SpotGarbage application utilizes convolutional neural systems to recognize junk in

Figure 2. Machine learning methods case-wise result comparison

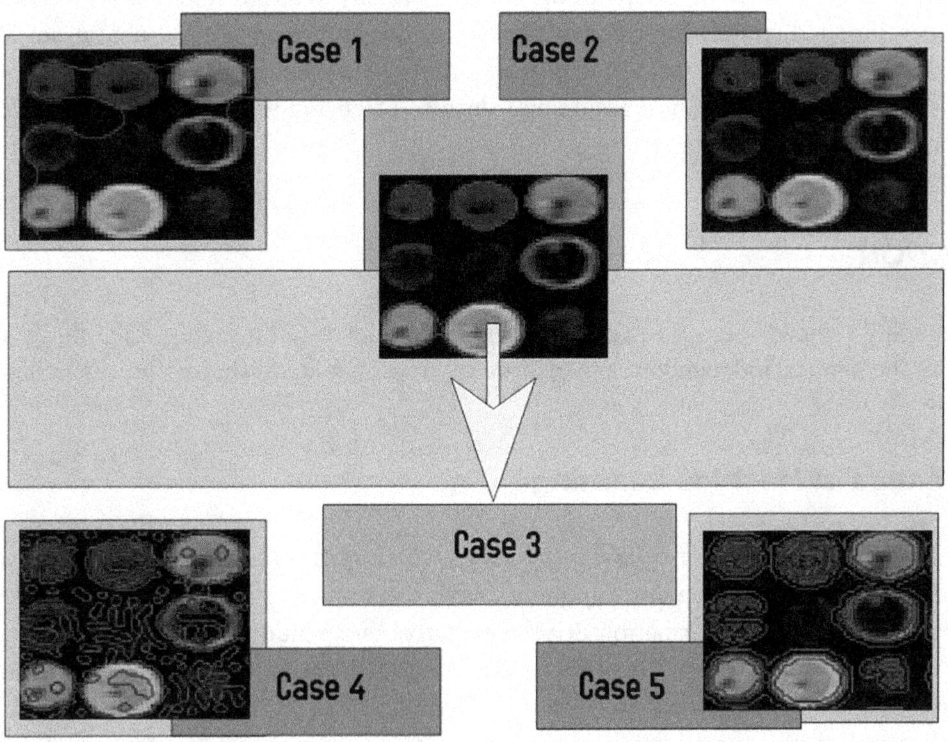

pictures, however, expends 83 percent of CPU and takes over five seconds to react. Luckily, late advances in arrange pressure, rough registering, and quickening agents are empowering profound learning on asset obliged IoT gadgets. These attributes request generous registering power, which can be a restricted asset on numerous IoT gadgets. Analysts have recognized three approaches to send profound figuring out how to asset compelled gadgets while lessening request on CPU, memory, and power.

System pressure is the way toward changing over a thickly associated neural system into an inadequately associated organizes. This system does not work with all profound learning systems but rather when appropriate, it can lessen both capacity and calculation stack. By wisely pruning excess associations, computational load can be lessened by a factor of 10 without antagonistically influencing precision. One examination found another essential favorable position of meager system models: these compacted models can keep running on generally utilized IoT stages, including those from Qualcomm, Intel, and NVidia. Another way to deal with decreasing computational load is inexact, instead of precisely process; the estimations of hubs with minimal impact on demonstrate exactness. For instance, in one examination on low power utilization outline, specialists produced rough outcomes by decreasing the number of bits per hub by 33 percent to 75 percent of bits in low-affect neurons. This change did not decrease exactness. While reasonable for some, applications, estimated processing isn't fitting for applications requiring high exactness.

A third way to deal with empowering profound learning on IoT gadgets is to convey specific quickening agent equipment. The Eyelids quickening agent, for instance, is intended to streamline for vitality utilization by convolutional neural system (CNN). The Dian Nao stage is intended for both convolutional and profound neural systems and upgrades for memory use and in addition vitality utilization. The drawback of this approach is that it requires particular equipment. Equipment based quickening agents are most appropriate for high-esteem applications that require low power utilization and quick calculation. Profound learning isn't out of the scope of IoT gadgets (Kashyap and Gautam, 2017). System pressure, inexact processing, and equipment quickening agents all empower profound neural net displaying on gadgets with restricted CPU, memory, and power. Likewise, with any particular approach, there are impediments and hindrances. System pressure and surmised figuring are appropriate to applications with resilience for an absence of accuracy.

FOUNDATION

This area quickly reviews ideas on machine learning and description logics, keeping in mind the end goal to make the chapter independent and effortlessly reasonable. At that point, it examines important related work.

Fundamentals of Machine Learning

Machine learning is a branch of artificial intelligence which means to manufacture frameworks competent in gaining from past involvement. Uniquely in contrast to master frameworks, ML calculations and methodologies are typically information-driven, inductive and general in nature; they are characterized and connected to settle on forecasts or choices in some broad class of undertakings, e.g., spam separating, penmanship acknowledgment or action location. Three noteworthy classifications of ML issues exist classification, comprising in the relationship of a perception test to one of an arrangement of conceivable classifications, e.g., an email message is spam or not as shown in figure 3. An order has in this manner a discrete n-ary yield. This is the fundamental issue considered in this chapter regression, characterized as the estimation of the relationship of a needy variable from at least one free factor, e.g., foreseeing the buy cost of house thinking about its size, age, location and different highlights (Chen & Liu, 2016). When all is said in done, both the sources of info and the yield in relapse can shift in constant esteem ranges. Clustering, i.e., separating an arrangement of perceptions into gatherings (bunches), which augment the comparability of tests inside each gathering and the contrast between gatherings. Numerous example acknowledgment issues depend on bunching.

The execution of a ML framework regularly incorporates two primary stages: preparing and testing. In the preparing stage, the embraced ML calculation constructs a model of the specific issue inductively from preparing information (Wang, McMahan & Gallagher, 2015). In light of the model, the calculation at that point produces yields no doubt coordinating the present data sources in the testing stage, which is executed amid framework approval keeping in mind the end goal to check wellness of the learning framework. Specifically, assessment of grouping execution depends on considering one of the yield classes as the positive class and characterizing (Figure 3).

Each accessible dataset to be characterized is isolated into a preparation set for display building and a test set for approval. There exist a few methodologies for choosing legitimately preparing and test segments. Among others, in k-overlay cross-approval, the dataset is divided into k subsets of equivalent size; one of them is utilized for testing and the rest of the k-1 for preparing. The procedure is rehashed k times, each time utilizing an alternate subset for testing. The less complex holdout technique, rather, isolates the dataset haphazardly, more often than not relegating a bigger extent of tests to the preparation set and a little one to the test set. ML techniques can be isolated into directed or unsupervised. Administered systems require a moderately expansive corpus of marked information to be worked for preparing, more often than not by hand. Besides, the subsequent models accomplish great exactness just for the

Figure 3. Machine learning applications

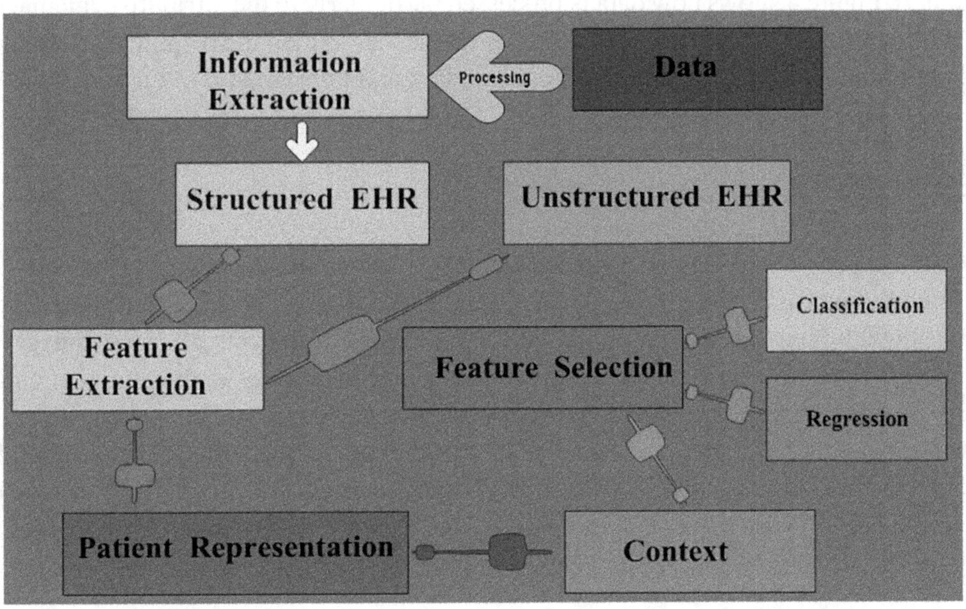

particular situations they are worked for. They are not reusable and versatile when conditions change. On the other hand, unsupervised strategies endeavor to assemble models specifically from unlabeled information and are common in bunching issues. The accessibility of huge information accumulations related to fractional human comment has as of late turned the thoughtfulness regarding semi-administered learning (Hussain & Cambria, 2018). By joining little scale master marked information and substantial scale unlabeled information in view of specific presumptions, semi-supervised techniques endeavor to locate the best tradeoff between framework precision and required human and computational exertion. As point by point in chapter arrangement execution of the proposed approach has been contrasted with the following famous ML approaches: Decision tree: it embraces an insatiable top down approach for building the characterization tree, beginning with the formation of the root hub. At each hub, the data pick up for each trait is computed and the character with the most astounding data pick up is chosen (Tofan, 2014). The pickup is characterized as a diminishment in entropy caused by part the occurrences in light of esteems taken by the trait.

- **Functional Tree:** A characterization tree with strategic relapse capacities at the internal hubs what's more, takes off. The calculation can manage twofold, multiclass target factors, numeric and ostensible characteristics and missing qualities. K-Nearest Neighbors, KNN, and an instance-based learning calculation. It finds the k closest examples to the information case and decides its class by distinguishing the absolute most regular class mark. It is considered not tolerant to clamor and missing qualities. In any case, KNN is profoundly precise, obtuse to exceptions and works well with both ostensible and numerical highlights (Yao, 2014). Random Tree: it consolidates two other ML calculations; demonstrates trees and arbitrary timberlands, in a request to accomplish both strength and adaptability. Demonstrate trees are decision trees where each leaf holds a direct model advanced for the neighborhood subspace of that leaf. Arbitrary backwoods are a group learning approach which constructs a few choice trees and picks the method of their yields. Including new constructors makes DL dialects more expressive. In any case, this prompts a development in the computational unpredictability of induction administrations. Henceforth a tradeoff is required. Figure 4 shows how data is processed particularly to the attributive language with unfit number limitations DL. It gives sufficient expressiveness to help the displaying designs while giving polynomial multifaceted nature to both standard and nonstandard induction administrations.

RELATED WORK

In IoT situations, keen interconnected articles assemble information tests, which can be utilized to recognize and foresee some true marvels displaying designs. Separating abnormal state data from the crude information caught by sensors interpreting in machine-reasonable dialects has a few intriguing applications. The work overviewed prerequisites, arrangements, and difficulties in the region of data reflection and presents an effective work process in view of the current cutting edge. Semantic Web inquires about tended to the undertaking of portraying sensor and information includes through ontologies. SSN-XG is the most important and broadly acknowledged proposition. It is sufficiently general to adjust to various applications and is good with the Open Geospatial Consortium (OGC) Sensor Web Enablement (SWE) benchmarks at the sensor and perception levels.

Figure 4. Big data processing model

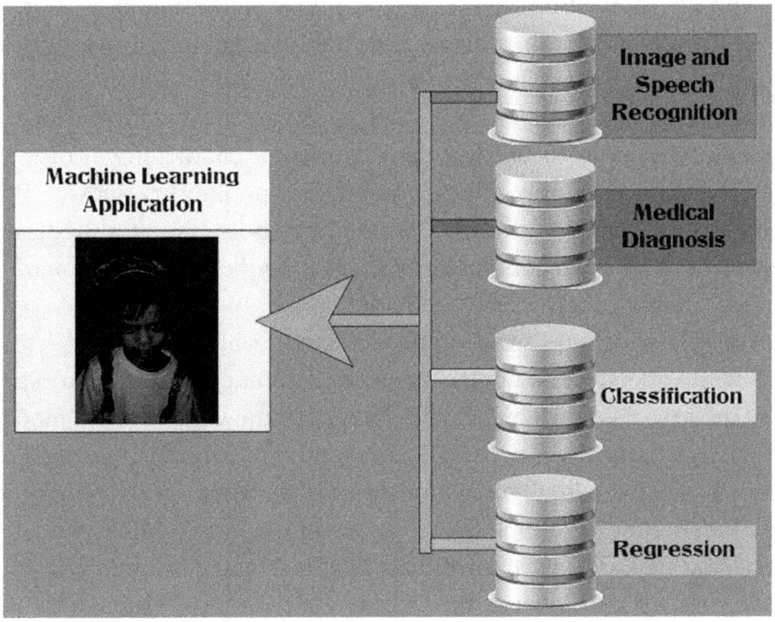

OGC SWE was utilized as a part of a few structures went for allowing access to sensor information as Restful administrations or on the other hand Linked Data. The issue of semantic information stream pressure in restricted asset spaces was confronted by building up an adaptable middleware stage to distribute semantically-commented on information streams on the Web through HTTP. Shockingly, the above arrangements just permitted rudimentary questions in SPARQL pieces on RDF explanations. More compelling strategies, for example, ontology-based Complex Event Processing (CEP) abused a common area of conceptualization to characterize occasions and activities to keep running on an occasion handling motor. Moreover, the ENVISION venture joined CEP with semantic advances to perform Semantic Event Processing from various sources: thinking was utilized to process the approaching actualities which populated a learning base (Dzbor, Stutt, Motta & Collins, 2007). Utilizing KR strategies on a lot of cases could be in actuality helpful to clarify crude information, deliver abnormal state depictions in a KB appropriate for cutting-edge thinking, expecting to enhance standard information mining and ML calculations. In a post-processing of ML tasks in light of cosmology consistency check meant to enhance consequences of affiliation control mining. Semantically conflicting affiliations were pruned and sifted through utilizing on rationale thinking for that. Augmentations to standard thinking calculations, supporting vulnerability and time connections, have been likewise demonstrated as viable in assignments such as action acknowledgment. The absolute most effective ML techniques, for example, Artificial Neural Networks (ANN) and profound learning methods, experience the ill effects of the haziness of models, which can't be deciphered by human specialists and along these lines can't clarify explanations behind the results they give (Kim, 2018). This is a difficult issue for ML reception in each one of those areas which require responsibility of choices and strength of yields against unplanned or intentional info control. Research endeavors to assemble understandable aftereffects of ML strategies and frameworks are in this way developing. An adroitly basic approach is to misuse outfit getting the hang of joining various low dimensional sub-models, where every individual sub model is sufficiently basic to be obvious by area specialists. Bayesian learning was utilized to create arrangements of principles in that point frame, which can give clear motivations to their expectations. The strategy was observed to be focused with condition of-the-art procedures in a stroke expectation assignment over an expansive dataset, in spite of the fact that preparation time shows up as rather long for IoT situations. The relapse device is capable to interpret naturally segments of the model into regular dialect portrayals of examples in the information. It depends on a compositional sentence structure characterized over a space of Gaussian relapse models, which is looked eagerly utilizing negligible probability and the Bayesian Information Criterion (BIC) (Neath & Cavanaugh, 2011). The approach bolsters variable dimensionality (number of parameters) in every relapse show, consequently permitting the determination of the coveted tradeoff amongst exactness and straightforwardness of elucidation.

Semantic-improved ML strategies can accomplish the same objectives through formal rationale based portrayals of models with a specific end goal to create logical functionalities, so expanding clients' trust. Besides, they can be incorporated into bigger intellectual frameworks, where models and expectations are utilized for computerized thinking. Especially, Semantic information mining alludes to information mining assignments which methodically consolidate area information into the procedure. The review included cosmology based control mining, grouping and bunching. Ontologies are helpful to connect the semantic hole between crude information and applications, and additionally to give information mining calculations earlier learning to manage the mining procedure or diminish the hunting space. They have been effectively utilized as a part of all means of the ordinary information mining work process. In Ontology-Based Information Extraction (OBIE) they are likewise abused to comment on the yield of information mining (Borisova, 2014). In the creators proposed an approach utilizing remote sensor systems and ontologies to speak to and surmise learning about movement conditions. Crude information was ordered through an ANN and mapped to philosophy classes for performing standard based thinking. The unsupervised model was utilized for characterizing Web Service data types in an expansive number of cosmology classes, by receiving an expanded ANN. Likewise for this situation; be that as it may, mining was abused just to outline to a solitary class. Promising semantic-based methodologies moreover incorporate fluffy DL learning, idea polynomial math, tensor systems in light of Real Logic. While their forecast execution seems great, computational productivity should, in any case, be assessed totally before thinking of them as reasonable for IoT situations. The system saves the established information mining and machine learning work process: information accumulation and purifying, show preparing, approval, framework utilization (Toivonen & Gross, 2015). In any case, as semantic-construct improvements grounded with respect to DLs change the way each progression is performed.

COSMOLOGY AND INFORMATION DEMONSTRATING

The work process begins with crude information accumulated e.g., by sensors dunked in a given situation for extricating a few distinct parameters, for the most part, named highlights. Keeping in mind the end goal to help semantic-based information explanation also, elucidation, a metaphysics T models the area conceptualization along appropriately indicated designs. T is accepted non-cyclic and communicated in the reasonably expressive ALN DL (Zhang, Xiang & Wang, 2010). This is required by the ensuing non-standard deductions for semantic matchmaking. For each estimating parameter, T must incorporate an idea progressive system as in framing a partonomy of the highest idea. As it were, every parameter will be spoken to by means of a class/subclass scientific categorization highlighting all noteworthy esteem reaches and setups it can have in the area of intrigue. The profundity of the chain of the importance and the broadness of each level will be picked by the learning modeler; they will regularly be relative to both determination and scope of detecting/catching hardware, and in addition to the required level of detail in information portrayal. For instance, the class chain of importance of the area cosmology displayed for the contextual investigation. Two distinctive displaying approaches were explored to speak to information ranges for the deliberate parameters inside the information base.

PREPARING

Like in traditional ML, the objective of this progression is to utilize preparing information to characterize the model to be utilized a short time later by the ML calculation to make expectations on test information (Punyavachira, 2013). In the proposed approach, the model comprises of a semantic explanation for every conceivable yield class, meaning they watched occasion/wonder as indicated by input information.

Costly Unplanned Downtime

In various segments, impromptu downtime coming about because of gear breakdown can cause substantial misfortunes. For example, as indicated by one investigation, such misfortunes normal $38 million every year for seaward oil and gas operators.14 Another source evaluated that for mechanical assembling altogether, impromptu downtime costs $50 billion every year, with gear disappointment being the reason for 42 percent of the outages. Prescient maintenance utilizing investigation to foresee gear disappointment early with a specific end goal to plan deliberate support strategies can relieve the harming financial matters of spontaneous downtime. In assembling, for example, Deloitte finds that prescient upkeep can diminish the time required to design support by 20– 50 percent, expansion hardware uptime and accessibility by 10– 20 percent, and lessen general upkeep costs by 5– 10 percent (Huby, Cockram & Fleming, 2013).

Since AI advancements especially machine learning can help recognize examples and abnormalities and make forecasts in light of vast arrangements of information, they are ended up being especially valuable in actualizing prescient support. Driving South Korean oil refiner SK Innovation, for instance, hopes to spare "billions of won" by utilizing machine figuring out how to foresee disappointment of associated compressors. Similarly, Italian prepare administrator Trenitalia hopes to keep away from impromptu downtime and spare 8– 10 percent on its €1.3 billion yearly upkeep costs. Meanwhile, French power utility EDF Group has effectively spared over $1 million with machine learning-driven early cautioning on hardware failure.

Operational Efficiency

AI-controlled IoT can accomplish more than help evades spontaneous downtime, it can likewise help enhance operational productivity. This is expected to some extent to the energy of machine figuring out how to produce quick and exact expectations and profound bits of knowledge and to AI advances' capacity to computerize a developing assortment of undertakings. For instance, for Hershey, dealing with the heaviness of their items amid the creation procedure is basic: Every 1 percent change in weight accuracy can mean more than $500,000 in reserve funds for a 14,000-gallon clump of the item, for example, Twizzlers. The organization utilized IoT and machine figuring out how to fundamentally decrease weight fluctuation amid generation. Information is caught and investigated by the second, and weight fluctuation can be anticipated by machine learning models, empowering 240 process changes for every day, contrasted with only 12 every day prior to the ML-controlled IoT arrangement was installed.

AI-based expectation is likewise helping Google cut 40 percent of server farm cooling costs. The arrangement, prepared on information from sensors in the office, predicts temperature and weight throughout the following hour to direct activities for restricting force consumption (Islam, 2018). Machine learning delivered bits of knowledge that induced one delivery armada administrator to make an outlandish move that spared them huge cash. Information gathered from transport board sensors were utilized to distinguish the relationship between's the measure of cash spent on cleaning the boats' structures and fuel proficiency. The examination demonstrated that by cleaning their boats frames two times every year as opposed to like clockwork and in this manner quadrupling their cleaning spending plan they would wind up sparing $400,000 because of more noteworthy fuel efficiency. The comments will be communicated in Concept Components as indicated by the accompanying recursive definition:

CONTEXTUAL INVESTIGATION: STREET AND ACTIVITY OBSERVING

Portability administrations are one of the fundamental IoT application territories. As a contextual investigation, a prototypical framework for street and activity observing was made e.g., to enhance the usefulness of route frameworks with continuous driver help. Valuable understanding on movement conditions is given both among adjacent vehicles in a shared manner through VANETs – Vehicular Ad-hoc Networks and on a substantial scale e.g., by refreshing a remote Geographical Information System with ongoing (Misra, Krishna & Abraham, 2010) and history data toward street approach creators).

The proposed information based framework misuses the semantic depictions of vehicles and setting explanations to:

1. Translate vehicle information separated by means of the obligatory
2. Coordinate ecological data;
3. Identify potential hazard factors.

Other than giving admonitions, the recognized information enables offering recommendations to the driver and assessing auto productivity and ecological effect continuously. The Java dialect was decided for the usage, keeping in mind the end goal to be good with both Java SE (Standard Edition) and Android stages. The model incorporated the Mini-ME lightweight go-between, which gives the expected derivations to the ALN DL under the supposition of non-cyclic TBoxes. The above ML structure was utilized to separate high-level signs beginning from an expansive number of low-level parameters procured by the auto through OBD-II and through the small-scale gadgets installed in the client cell phone, with the objective of precisely describing the general framework formed by driver, vehicle and condition. A committed dataset was gathered to promoting experiments. Crude information was recovered and put away utilizing the Torque Lite (OBD-II and Car) Android application on seven unique courses: rural, urban and blended ones, with medium and long separation. A normal of five follows per course were recorded, examining OBDII parameters and cell phone information at 1 Hz recurrence. Around 10,000 records were gathered by and large for each course; go up against various days, in different rush hour gridlock conditions and with three distinct autos (and drivers). The contextual investigation meant to recognize the driving style, the street attributes as far as consistency and activity conditions, by examining parameters accumulated by the auto and the client cell phone. It is deliberately basic to give only a prompt confirmation of idea, yet order can be to a great extent enhanced freely without changing hypothetical settings. In detail, the framework ought to identify the accompanying classes:

- Even, slightly uneven or uneven road;
- Low, medium or high traffic;
- Aggressive or even pace driving style.

Amid the dataset creation, every driver who gathered a following was requested to mark physically the records with the occasion trademark for every one of the above classifications. Assembled data speak to the crude information in the ML issue. Timestamp and GPS facilitate additionally taken through the cell phone (Abdolkarimi, Abaei & Mosavi, 2017) were added to each record. The examined information comprised of:

- Elevation change, figured more than 10 seconds;
- Speed: current esteem, normal and change in the most recent 60 seconds and change in speed for each second of discovery;
- Longitudinal and vertical quickening, estimated by the cell phone accelerometer and pre-prepared with a low-pass channel to erase high recurrence flag segments because of electrical commotion and outer powers;
- Motor load, communicated as a rate;
- Motor coolant temperatures;

Machine learning is firmly attached to computational measurements, it is likewise used to make expectations, reenactments and it is fixing to scientific advancement at the point when utilized in mechanical settings, machine learning strategies might be eluded as a prescient investigation or displaying. The quintessence of machine learning is to make smaller numerical models that speak to unique space ideas of profiles, tastes, connections fit well the current perceptions of the area can extrapolate well to new perceptions (Hasan & Mishra, 2012). These two targets characterize the two principle issues in the greater part of machine learning subfields which are:

1. **Under-Fitting:** Where the machine learning calculation was not ready to learn well from the present perceptions and has poor exactness and execution even on the officially notable examples (Kashyap and Gautam, 2017).
2. **Over-Fitting:** Where the machine learning calculation fits too well the preparation information with great execution and precision to a specific point that it loses its capacity, to sum up on new information (outside of the preparation set).

CATEGORIZATIONS OF MACHINE LEARNING

A few conceivable arrangements of machine learning procedures are conceivable, the systems as indicated by the kinds of issues they are endeavoring to tackle, or as per the sort of discovering that is utilized, or as per the recurrence of execution of the learning calculation. In this segment, we introduce these distinctive conceivable classifications.

Type of Issues

The principal arrangement is as indicated by the kind of issues they are endeavoring to understand.

1. **Speculation Based:** In these issues, there exists a best numerical capacity or speculation that connections the normal yield to the info. For instance, a straight relapse speculation, as on account of assessing the cost of the house from a few of its highlights: we can state that the cost is a straight blend of alternate qualities of the house. The learning comprises then on fitting the multiplicative weight parameters of the diverse ascribes to the extraordinary introduced preparing tests (Tango, Minin, Tesauri & Montanari, 2010).
2. **Thickness Estimation:** The issue here is to take in a likelihood of appropriation of contributions over a specific area space. This can make a test system that creates information with comparative measurable properties of the area space. For instance taking in the power control utilization profile by Gaussian portions, at that point we can think about that the scholarly circulation will be the same the following day (Mawdsley, Tyson, Peressotti, Jong & Yaffe, 2009).
3. **Grouping:** It is the issue of recognizing to which of an arrangement of classes (subpopulations) perception information has a place with, based on a preparation set of information containing perceptions with a formerly known classification enrollment.
4. **Bunching:** It is a comparable issue as the arrangement, yet the diverse classes or classes are not known from the earlier.
5. **Irregularity Discovery:** It is like the arrangement, however, these issues generally have just two classes (ordinary and strange) and there is a major predisposition in the appropriation of things in the classes (substantially more things in the ordinary class and few in the irregular).
6. **Dimensionality Lessening:** It is the issues that, from an arrangement of info traits, select the most fitting ones significant for a particular learning issue. Dimensionality Reduction strategies can be helpful to disentangle input information (by evacuating repetitive qualities) which can quicken a directed learning technique a short time later.
7. **Proposal Framework:** The point is to give from an arrangement of the user, product, rating, suggestions for clients of potential items they may like. Some suggestion frameworks are composite from linear relapse and grouping calculations. For case: bunch the clients or items to comparable client/items gatherings, then the suggestions are created from the weighted normal from the client/item neighbors.

Type of Learning

The second order is as per the sort of machine learning calculations used to take care of the beforehand characterized issues.

- **Supervised:** In administered taking in, the information has predefined and surely understood fields to fill in not surprisingly yield of the learning procedure, for example, a characterization classification (for ex.: spam, not spam) or to assess the estimation of an area particular field, for example, the cost of houses. In this sort of taking in, a scientific model is set up through a preparation procedure where it is given by input highlights and their comparing expected yields. The learning comprises of enhancing the numerical model, so they gave input highlights delineate to the grave expected yield (Sądel & Śnieżyński, 2017). The preparation process proceeds until the point when the numerical model accomplishes a coveted level of exactness on the preparation information between the data sources and yields. Some case issues of regulated learning are arrangement and connection issues. For example: Learning the cost of a house from a few of its highlights (GPS position, region, number of rooms, and year of construction...). The preparation set comprises of a few cases of houses characterized by their highlights and their costs. The ML makes numerical capacities, for illustration a straight relapse of the properties. The preparation procedure is to locate the best parameters of this relapse can fit well the information. In this sense, directed learning can be viewed as a capacity guess amongst yield and information properties.

- **Unsupervised:** For unsupervised taking in, the information isn't named and does not have a referred to field characterize as yield. The machine learning calculations attempt to reason structures introduce in the info information. Unsupervised learning can discover covered up designs in the info information. An illustration issue is grouping: a rundown of clients and their inclinations is given and the calculation needs to isolate them into gatherings or bunches of clients of comparative taste. In this sense, unsupervised learning can be viewed as a succinct smaller and compacted portrayal of the information.
- **Semi-Supervised:** For this situation, the info information is a blend of named and unlabeled illustrations. There is a coveted expectation issue expected yet in the meantime, the show must take in the structures to sort out the information and in addition to make expectations (Zhang, 2016).
- **Reinforcement Learning:** In this sort of taking in, the yield information is given a boost to the machine taking in demonstrate from the earth to which the ML show must react and respond by taking a few activities. The input is given in terms of disciplines and rewards from nature. The support learning calculations endeavor to discover an arrangement that maps conditions of the world to the best activities to be taken in those states with a specific end goal to amplify the normal reward from the earth. Fortification taking in varies from the managed learning issues in the way that adjust input/yield sets are never introduced expressly and straightforwardly. In this sense, support learning can be viewed as administered gaining however from postponed compensates work (Chen, Liu & Yang, 2016). In the meantime, the distinction with the unsupervised learning is that in fortification realizing there exists by one means or another a sort of reward work that aides the realizing, wherein the unsupervised adapting, no normal yield or criticism is displayed.
- **Meta-Learning:** It is the way toward taking in the parameters of a learning calculation itself to guarantee that both under-fitting and over-fitting issues are illuminated. Generally it includes par-

Table 1. Types of machine learning

S.N.	Supervised Learning	Unsupervised Learning	Semi-Supervised Learning
1.	Regular Target Variable	No Target Variable	Categorical Target Variable
2.	Regression and Classification	Clustering and Association	Clustering and Classification
3.	Prediction and Image Processing	Segmentation and Analysis	Classification and Segmentation

titioning the current accessible information into 2 sets (preparing and testing sets), at that point preparing the learning calculation on the primary set and testing on the second to get two execution measures on how well the calculation fits the information and how well it can sum up to new information it was not prepared on. The meta-learning streamlines the learning parameters of the learning calculations until getting great outcomes in both bearings.

FREQUENCY OF LEARNING

The third order is as indicated by how regularly and when we are executing the machine learning calculations.

1. **Full Online Learning:** For each new perception of Input or yield regulated or simply input (unsupervised), the learning calculation is executed, and its state is refreshed, at that point the preparation vector is disposed of. The upside of this sort of machine learning is that the learning is nuclear, incremental and continually advancing as indicated by new perceptions. Then again, the inconvenience is that it may have brought down precision than full-disconnected learning calculations for little preparing sets.
2. **Lethargic Learning:** Whenever there is a need to appraise a yield for a particular info vector, the learning calculation is prepared by a little cluster of perceptions close or like the one required (for instance the closest K neighbors like the asked input). This learning compose offers a case-based or setting based thinking to mean that the learning is custom fitted to the asked for particular info. In addition, it is on-request, implying that no preparing will be lost on preparing for every single conceivable setting or information sources, for whatever length of time that they are not asked for the forecast (Houeland & Aamodt, 2017). Anyway, the disservice of this type is that it is slower than the full on the web (because we have to determine, bring and prepare for K closest vectors previously having the capacity to foresee), in this manner it will take more defer to react. Besides, it requires a grouping capacity that gives the closest neighbors to gain from at whatever point info is exhibited.
3. **Cluster:** This is a moderate and tedious process, the learning is completely disconnected and works as a rule in general dataset of preparing cases. In any case, once prepared, the scholarly model is utilized to assess esteems till the following learning refresh stage where everything should be recreated and retrained without any preparation (Kashyap and Piersson, 2018). The benefit of this composes is that for the most part, it accomplishes the most noteworthy exactness, and anyway, this precision gets most exceedingly terrible with time till the following learning refresh stage (Schwab & Ray, 2017). The hazard here is that the educated model can be obsolete if the area develops immediately between two refresh stages. Another inconvenience is that generally, this kind of learning is exceptionally expending in term of computational assets up until now, we gathered that the setting rules and the characteristics to streamline per setting are physically sustained to the AmI framework either by framework architects or by the client themselves. In request to influence AmI frameworks to blur out of sight, these parameters should be scholarly consequently with minimal intercession of the client or framework fashioners. With a specific end goal to accomplish this, the last commitment of this proposal is to open the guide on the best way to coordinate machine learning systems.

The association between relevant thinking and machine learning is in two ways on one hand, machine learning can be utilized to distinguish setting rules, and the characteristics to improve for every unique situation, as appeared in figure 5. Then again, the unique circumstance data can be utilized to better prepare other machine learning calculations. For example, profiling client propensities or electrical utilizations will be enhanced when setting data is considered. For instance, knowing whether the present day is an occasion or a working day can influence a considerable measure electrical load forecast. In this section, we show the advantages of model-based setting mindful machine learning, as well as a demonstrating dialect for machine discovering that we execute in KMF. It exhibit and assess our approach just on the second course, where it utilize case from the brilliant network area. This utilization case, how to utilize displaying dialect in the machine learning space, and demonstrate the advantages of logical thinking for the machine learning.

The primary course, which is the utilization of machine learning for setting mindfulness, is as of now a continuous work. It isn't developing enough to be incorporated into this section; it will be talked about later on work of this theory. Be that as it may, this part displays the demonstrating thoughts and the foundations on the best way to set up a framework and the fundamental structure to empower machine getting the hang of thinking for setting mindfulness, in this manner the title of the proposition is Towards Ambient Intelligent Applications utilizing Machine Learning for Context-Awareness.

MODEL BASED MACHINE LEARNING

The focal thought of the model-based way to deal with machine learning is to make a custom show custom fitted particularly to each new application. At times, the model (together with a related surmising calculation) may compare to a conventional machine learning system, while much of the time it won't. Normally, show based machine learning will be executed utilizing a model particular dialect in which the model can be characterized utilizing reduced code, from which the product actualizing that model can be created consequently (Dehestani et al., 2011) as shown in figure 5 how model base and case base model are playing important role in machine learning.

Main Standards

The key objectives of a model-based approach incorporate the accompanying thoughts:

- Segregation of the learning concerns. If the application requires a blend of bunching and grouping with regards to time-arrangement information, it isn't important to squash together conventional calculations for every one of these components, yet rather a solitary, incorporated model catching the coveted conduct can be developed from a few free building pieces. The demonstrating dialect ought to have the capacity to course/chain the diverse learning models. In this manner, we want to reuse great auxiliary properties of models, for example, sluggish stacking and versatility, which can profit a ton the machine learning world too (Little, 2010).
- Segregation between the model conditions and the surmising calculation the model ought to indicate the space particular substances that the learning assignment relies upon, autonomously from the induction calculation itself. Along these lines, the surmising calculation can be changed effectively without the need to change the interface for the area application.

Figure 5. Model-based machine learning

- Segregation between display conditions and properties extraction the displaying ought to permit, for similar conditions, to extricate distinctive machine learning traits. For instance, profiling of power utilization relies upon brilliant meter and temperature sensor information (Kashyap and Tiwari, 2017). Be that as it may, numerous machine learning numerical theory models can be assembled utilizing these same conditions, for instance by utilizing the square estimation of the temperature, or the block estimation of the utilization as learning properties, or on the other hand attempting distinctive mixes of forces of these qualities. For the space application, this procedure ought to be straightforward (Sun & Betti, 2015).
- Re-convenience of the learning models a similar learning occurrence can be utilized as a part of a few errands without the need to copy it a few times. For instance, in the brilliant metering area, a power utilization profile of a client can be utilized as a part of a few errands: to anticipate the electrical load, to characterize the client as indicated by his profile and to recognize peculiar utilization conduct. On this point, demonstrating methods, for example, DSML, offer answers to characterizing a Meta structure that can be reused to characterize new models of machine learning. Straightforwardness of usefulness: the model is portrayed by conservative code inside a non-specific displaying dialect, thus the structure of the model is promptly evident. Such demonstrating code can without much of a stretch be shared and reached out inside a group of model manufacturers.
- Combining one taking in demonstrates from a few sub-models. This is normally known as outfit techniques. It is utilized to permit ml models to be made out of different weaker ml models that are freely prepared and whose outcomes are consolidated all together to improve the general learning. There are a few methods to separate the sub-models. A few systems part the preparation information, this is called stowing, different procedures split the highlights, and different strategies split the two information and highlights (for instance irregular timberlands).

- Suspicious utilization esteem identification in this segment we portray our approach for suspicious utilization esteem identification. To start with, we introduce a review of the approach and the included parts. Next, a clarification of the Gaussian blends model. At long last, how to take in various, context-dependent client profiles, and how we select and utilize them for suspicious utilization esteem location.

OVERVIEW: TOWARDS CONTEXTUAL LEARNING AND DETECTION

Shrewd meters consistently report their utilization esteems, in general interims, to utilities for checking and charging purposes. In Luxembourg, these interims are 15 minutes for power and an hour for gas. In request to recognize suspicious utilization esteems, this constant stream of savvy meter estimations is first broke down by a setting solver. For each new esteem, the setting solver chooses the most fitting profile for this client keeping in mind the end goal to choose what the ordinary is the scope of the currently estimated power utilization. For instance, for estimation esteem on a bright Monday evening working day in summer, the setting solver will choose a profile with amount setting parameters for this client. The specific situation, for instance, can incorporate highlights like: client write singular, family, industry, business, a fleeting setting period of the year, month, weekday, occasions, et cetera. The setting determination yields a rundown of positive and negative profiles. Positive profiles contain data to judge on the off chance that the deliberate esteem is known to be in the typical range, while negative profiles are known to be suspicious qualities from these profiles, a basic leadership step is executed, in light of a certainty rate and the likelihood appropriation gave by the profiles. In the event that the deliberate esteem is acknowledged, it is utilized to prepare the positive profiles. In the event that it isn't acknowledged, an alarm is made. What's more, an intelligent approval asks for is raised to physically approve if the esteem is for sure suspicious or not. At last, the negative evaluated esteems suspicious qualities are used to prepare the negative profiles. An underlying preparing period is required to bootstrap the profiles. The refresh of profiles is an exceptionally incremental advance and can subsequently be performed for each esteem. In any case, it would be additionally attainable to join a few qualities as opposed to checking each esteem. Especially, we utilize Gaussian blend models (GMM), which are known to be an intense instrument in approximating appropriations notwithstanding when their shape isn't near Gaussian (Coufal, 2016). A GMM is a probabilistic model that accepts that all information focuses are gotten from a blend of Gaussian circulations with obscure parameters. Blend models are essentially summing up the k-implies grouping.

Lawful Information Extraction

Given the assignment of breaking down characteristic dialect strategies, the inquiry turns out to be the way striking data can be separated from unordered approach writings. While most works in lawful data extraction identify with spaces other than security, regardless they give some direction. Different errands concentrated on the extraction of data from statutory and administrative laws, the recognition of lawful contentions or the ID of case law segments. There are a few works in the protection approach space, most remarkably, as a major aspect of the Usable Privacy approach project. They finished up the attainability of regular dialect examination in the security arrangement area all in all. Wilson et al. talked about the creation and investigation of a security arrangement corpus. When all is said in done, the talked about works affirm the appropriateness of lead and ML classifiers in the protection strategy space (Schumaker, 2013). Nonetheless, neither gives a thorough idea nor addresses, for instance, how to make utilization of crowdsourcing comes about. The programmed arrangement characterization all alone is characteristically constrained. None of the past works eases the client from really perusing the broke down approach. Interestingly, it is the objective of the work in this chapter to give clients a protection approach synopsis in lieu of the full approach (Kashyap and Piersson, 2018). I need to separate from a strategy fundamental arrangements, make it more understandable, give direction on the examined practices, and give a general assessment of its protection level.

Privacy Requirement Inconsistencies

Given the investigation into protection arrangement content, I trust, it is an advantageous errand to check the degree to which arrangements line up with genuine information rehearses. In such a manner, it especially smart to investigate whether versatile applications' practices are predictable with the exposures made in their strategies furthermore, chose prerequisites from different laws. The legitimate measurement is an essential one that gives importance to the application investigation comes about (Ott & Houdek, 2014). Machine learning challenges like privacy, technical challenges and interoperability are shown in figure 6. For instance, for applications that don't give area benefits the exchange of area information may seem intolerable. However, an exchange may be allowable assuming satisfactorily uncovered in a protection approach.

Just a couple of endeavors have endeavored to consolidate code examination of versatile applications with the examination of protection approaches to fill this void by recognizing security necessity irregularities associating the investigations of applications, protection arrangements, and security laws.

Figure 6. Challenges with machine learning

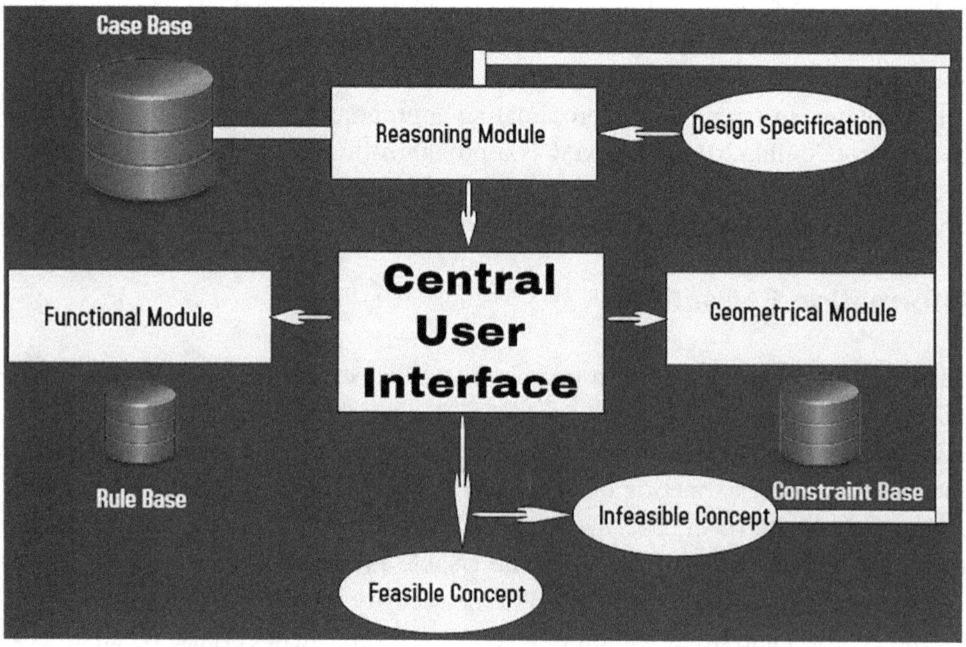

Cross-Gadget Tracking

ML strategies are assuming a focal part in cross-gadget following to investigate the space Drawbridge; a promotion arranges having some expertise in cross-gadget following facilitated Drawbridge Cross-Device Connections rivalry requesting that analysts use machine learning systems to connect gadgets to clients. Rivalry members were offered access to an anonym zed exclusive dataset to prepare and test their highlights and calculations. The opposition brought about eight short chapters by the absolute most effective members. Unique in relation to the dialog in this examination, these chapters took the point of view of a promotion arrange and concentrated solely on enhancing machine learning procedures and accomplishing a high F score. While these points are likewise part of my examination inspired by the protection of cross-gadget following. The primary spot arrangement in the Drawbridge rivalry which achieved the F-0.5 score of 0.9 is from numerous points of view delegate for the methods utilized as a part of the opposition. As other members' answers, it distinguished IP tends to that gadget of a similar client were associated with as the most imperative element. Hence, just depending on IP history would already be able to prompt F-0.5 score of 0.86. In any case, different examinations found that not all IP addresses are similarly important. Specifically, on the grounds that the same cell IP addresses happen for some gadgets of various clients they harbor less recognizing potential (Brookman, Rouge, Alva & Yeung, 2017).

While the Drawbridge rivalry was about the relationship of various client gadgets, it did not address the motivation behind the relationship: the distinguishing proof of socioeconomics, interests, and other monetizable data of the individual behind the gadgets. Different investigations exist on this subject, in any case, not with regards to cross-gadget following. For instance have demonstrated that to some degree identity can be anticipated from standard call detail records (CDRs), e.g., metadata about got and set telephone calls and instant messages (Ghotekar, 2016). Somewhat nearer to the exertion here, dissected the issue of anticipating Internet clients' sexual orientation and age in view of their perusing practices. They accomplished an F-1 score of 0.8 for anticipating sex and a score of 0.6 for classifying clients into five distinctive age gatherings. Keeping in mind the end goal to safeguard against these sorts of surmising assaults while as yet permitting customized notice, Blossom treats that encode a client's profile in a minimized and protection saving way. Perceiving the significance of IP addresses for recognizing clients they go for unlinkability of all inquiries from a similar IP. In such a manner, the investigate the impact of connecting gadgets through IP addresses on the exactness of learning.

Web Tracking

Much research was distributed on web following of individual gadgets. Nonetheless, to the best of my information, none of the current endeavors talks about following crosswise over gadgets. Such following is prominently not the same as conventional following that is centered on one gadget or program. To track web clients crosswise over gadgets organizations' need initially to recognize diverse program cases on the Internet. Ordinarily, HTTP treats are utilized for this reason. Since HTTP treats and other customary trackers keep upstate they are broadly used to track singular programs. Call attention to, if two sites are inserting a similar tracker an enemy can interface visits to those pages from a similar program case regardless of whether the client's IP address changes (Li, Jiang, Zhang, Pang & Huang, 2014). In their investigation they find that a foe with the capacity to inactively watch web movement on the Internet spine can reproduce up to 73% of a run of the mill client's perusing history. They got their outcomes utilizing their web estimation stage Open WPM, which they presented in conjunction with a substantial scale estimation of web following in light of a creep of a million sites. On the off chance that a program does not acknowledge HTTP or different treats, it can even now be followed by means of program fingerprinting, which was brought up and widely studied. Online gadget fingerprinting is the way toward gathering adequate data through the program to perform stateless gadget ID. Such fingerprinting is too used to re-recognize a program in the event that treats have been erased. Displaying machine learning procedures and coordinating them with information models and parts models, will enable us to make a keen model to react for the inconspicuousness challenge for AmI framework, and permits completely or semi-computerizing setting mindfulness. Besides, the coordination of machine learning in displaying methods has two directional advantages for the relevant thinking process. In the primary heading, these methods can help computerizing the disclosure of the setting principles and settings by mean of grouping or order or theory premise. In a second way, setting mindfulness can enhance the nature of the after effects of some machine learning methods. Crowd sourcing has the preferred standpoint that it joins the information of an expansive number of benefactors, which, on a basic level, can prompt a substantially more nuanced translation of vague approach arrangements than current classifiers could give. Notwithstanding, all crowd sourcing approaches experience the ill effects of an absence of cooperation and, thusly, don't scale well.

While the examination aftereffects of the most well-known sites might be accessible, those for some lesser known destinations are most certainly not. Likewise, a few storehouses just give the likelihood to gaze upward comes about on the web without offering helpful client access, for instance, by methods for a program augmentation or other programming. The utilization of directed ML methods, as utilized recommended here, requires ground-truth to help the advancement of these systems crowdsourcing has been proposed as a practical approach for social occasion rich explanations from unstructured security approaches. While crowd sourcing postures challenges because of the approaches' many-sided quality, doling out comment assignments to specialists what's more, setting stringent understanding limits and assessment criteria can in actuality prompt solid arrangement explanations. In any case, as it is a repeating issue that security arrangement comments catch with low between annotator understanding, I am presenting a measure for breaking down their unwavering quality in light of the thought that high annotator difference does not chiefly restrain the utilization of the comments for ML purposes as long as the contradiction isn't deliberate.

In an exceptionally intriguing commitment a huge rate of solidness in program history impressions, they set that it isn't easy to change one's perusing propensity. Their outcomes demonstrate that for 69% of clients the perusing history is one of a kind and that clients for whom they could identify no less than four went by sites were interestingly identifiable by their accounts in 97% of cases. They consider: if web perusing designs were exceptional for a given client, history examination could possibly recognize a similar client over different programs and gadgets. I need to address this inquiry. There are additionally different contrasts among clients that could single them out. Based on utilization follows from 255 clients of two diverse cell phone stages with 7-28 weeks of information per client found, for instance, that the mean number of associations every day for a client changes from 10 to 200, that the mean communication length fluctuates from 10 to 250 seconds, furthermore, that the quantity of uses utilized fluctuates from 10 to 90. This is particularly significant as found that the best outsider spaces crosswise over various classes of gadgets are significantly comparative. They discovered just a couple of versatile particular promotion systems prompting fundamentally the same as arrangements of the best work area and portable outsider trackers. While the web following examination researching how clients are followed crosswise over gadgets, there are different investigations on human-computer collaboration that give profitable insights on how it may work. The objective of these examinations is to enhance site route, program expectation of client goals, and query item significance for web crawlers.

CONCLUSION

Profound learning is notable for taking care of apparently obstinate issues in PC vision and characteristic dialect handling, yet it ordinarily does as such by utilizing gigantic CPU and GPU assets. Customary profound learning procedures aren't appropriate to tend to the difficulties of the Internet of Things (IoT) applications, in any case, since they can't make a difference a similar level of computational assets. An advance lessening in system or hub size may deliver inadmissible drops inexactness and accuracy. Profound learning is the best answer for various examination and identification issues. At the point when utilized with an eye to saving process, memory, and power assets, profound learning can convey insight to IoT gadgets. These frameworks involve a substantial number of conveyed gadgets consistently creating a high volume of information. Diverse sorts of gadgets create distinctive kinds of information, prompting heterogeneous informational collections, yet the information normally incorporates timestamps and area data. At long last, IoT applications must be intended to endure clamor in the information because of transmission and securing blunders. Counterfeit consciousness is assuming a developing part in IoT applications and deployments. Funding interests in IoT new businesses that are utilizing AI are up forcefully. Organizations have obtained many firms working at the convergence of AI and IoT over the most recent two years. Also, significant sellers of IoT stage programming are currently offering incorporated AI abilities, for example, machine learning-based examination. AI is assuming a featuring part in IoT in light of its capacity to rapidly wring bits of knowledge from information. Machine taking in, an AI innovation, brings the capacity to naturally distinguish designs and identify inconsistencies in the information that brilliant sensors and gadgets create data, for example, temperature, weight, moistness, air quality, vibration, and sound. In future machine learning can have noteworthy focal points over conventional business knowledge apparatuses for breaking down IoT information, including having the capacity to influence operational expectations up to 20 to times prior and with more prominent precision than edge based checking systems, for example, discourse acknowledgment and computer vision can help extricate understanding from information that used to require human audit. The effective mix of AI and IoT innovation is helping organizations keep away from spontaneous downtime, increment working productivity, empower new items and administrations, and improve hazard administration.

REFERENCES

Abdolkarimi, E., Abaei, G., & Mosavi, M. (2017). A wavelet-extreme learning machine for low-cost INS/GPS navigation system in high-speed applications. *GPS Solutions*, *22*(1), 15. doi:10.100710291-017-0682-x

Borisova, N. (2014). An Approach for Ontology Based Information Extraction. *Information Technology and Control*, *12*(1). doi:10.1515/itc-2015-0007

Brookman, J., Rouge, P., Alva, A., & Yeung, C. (2017). Cross-Device Tracking: Measurement and Disclosures. *Proceedings On Privacy Enhancing Technologies, 2017*(2). doi:10.1515/popets-2017-0020

Chen, W., Liu, T., & Yang, X. (2016). Reinforcement learning behaviors in sponsored search. *Applied Stochastic Models in Business and Industry*, *32*(3), 358–367. doi:10.1002/asmb.2157

Chen, Z., & Liu, B. (2016). Lifelong Machine Learning. *Synthesis Lectures On Artificial Intelligence And Machine Learning*, *10*(3), 1–145. doi:10.2200/S00737ED1V01Y201610AIM033

Coufal, D. (2016). On convergence of kernel density estimates in particle filtering. *Kybernetika*, 735–756. doi:10.14736/kyb-2016-5-0735

Dehestani, D., Eftekhari, F., Guo, Y., Ling, S., Su, S., & Nguyen, H. (2011). Online Support Vector Machine Applicationfor Model Based Fault Detection and Isolationof HVAC System. *International Journal Of Machine Learning And Computing*, 66-72. doi:10.7763/ijmlc.2011.v1.10

Dzbor, M., Stutt, A., Motta, E., & Collins, T. (2007). Representations for semantic learning webs: Semantic Web technology in learning support. *Journal of Computer Assisted Learning*, *23*(1), 69–82. doi:10.1111/j.1365-2729.2007.00202.x

Ghotekar, N. (2016). Analysis and Data Mining of Call Detail Records using Big Data Technology. *IJARCCE*, *5*(12), 280–283. doi:10.17148/IJARCCE.2016.51264

Hasan, M., & Mishra, P. (2012). Robust Gesture Recognition Using Gaussian Distribution for Features Fitting. *International Journal Of Machine Learning And Computing*, 266-273. doi:10.7763/ijmlc.2012.v2.128

Houeland, T., & Aamodt, A. (2017). A learning system based on lazy metareasoning. *Progress In Artificial Intelligence*, *7*(2), 129–146. doi:10.100713748-017-0138-0

Huby, G., Cockram, J., & Fleming, M. (2013). Through-life Data Exploitation to Reduce Downtime and Costs. *Procedia CIRP*, *11*, 50–55. doi:10.1016/j.procir.2013.07.070

Hussain, A., & Cambria, E. (2018). Semi-supervised learning for big social data analysis. *Neurocomputing*, *275*, 1662–1673. doi:10.1016/j.neucom.2017.10.010

Internet of Things & Creation of the Fifth V of Big Data. (2017). *International Journal Of Science And Research*, *6*(1), 1363–1366. doi:10.21275/art20164394

Islam, N. (2018). *Business Intelligence and Analytics for Operational Efficiency*. SSRN Electronic Journal. doi:10.2139srn.3163429

Jara, A., Olivieri, A., Bocchi, Y., Jung, M., Kastner, W., & Skarmeta, A. (2014). Semantic Web of Things: An analysis of the application semantics for the IoT moving towards the IoT convergence. *International Journal of Web and Grid Services*, *10*(2/3), 244. doi:10.1504/IJWGS.2014.060260

Kashyap, R. (2018). Object boundary detection through robust active contour based method with global information. *International Journal Of Image Mining*, *3*(1), 22. doi:10.1504/IJIM.2018.093008

Kashyap, R., & Gautam, P. (2017). *Fast Medical Image Segmentation Using Energy-Based Method*. Pattern and Data Analysis in Healthcare Settings, Medical Information Science Reference.

Kashyap, R., & Gautam, P. (2017). 'Fast Medical Image Segmentation Using Energy-Based Method', Biometrics. *Concepts, Methodologies, Tools, and Applications*, *3*(1), 1017–1042. doi:10.4018/978-1-5225-0983-7.ch040

Kashyap, R., Gautam, P., & Tiwari, V. (2018). Management and Monitoring Patterns and Future Scope. In Handbook of Research on Pattern Engineering System Development for Big Data Analytics. IGI Global. doi:10.4018/978-1-5225-3870-7.ch014

Kashyap, R., & Piersson, A. (2018). Big Data Challenges and Solutions in the Medical Industries. In Handbook of Research on Pattern Engineering System Development for Big Data Analytics. IGI Global. doi:10.4018/978-1-5225-3870-7.ch001

Kashyap, R., & Piersson, A. (2018). *Impact of Big Data on Security. In Handbook of Research on Network Forensics and Analysis Techniques* (pp. 283–299). IGI Global. doi:10.4018/978-1-5225-4100-4.ch015

Kashyap, R., & Tiwari, V. (2017). Energy-based active contour method for image segmentation. *International Journal of Electronic Healthcare*, *9*(2–3), 210–225. doi:10.1504/IJEH.2017.083165

Kashyap, R., & Tiwari, V. (2017). Energy-based active contour method for image segmentation. *International Journal of Electronic Healthcare*, *9*(2–3), 210–225. doi:10.1504/IJEH.2017.083165

Kim, L. (2018). DeepX: Deep Learning Accelerator for Restricted Boltzmann Machine Artificial Neural Networks. *IEEE Transactions on Neural Networks and Learning Systems*, *29*(5), 1441–1453. doi:10.1109/TNNLS.2017.2665555 PMID:28287986

Li, G., Jiang, S., Zhang, W., Pang, J., & Huang, Q. (2014). Online web video topic detection and tracking with semi-supervised learning. *Multimedia Systems*, *22*(1), 115–125. doi:10.100700530-014-0402-0

Little, B. (2010). Concerns with Learning-Management Systems and Virtual Learning Environments. *Elearn, 2010*(7), 2. doi:10.1145/1833513.1837142

Mawdsley, G., Tyson, A., Peressotti, C., Jong, R., & Yaffe, M. (2009). Accurate estimation of compressed breast thickness in mammography. *Medical Physics*, *36*(2), 577–586. doi:10.1118/1.3065068 PMID:19291997

Misra, S., Krishna, P., & Abraham, K. (2010). A stochastic learning automata-based solution for intrusion detection in vehicular ad hoc networks. *Security and Communication Networks*, *4*(6), 666–677. doi:10.1002ec.200

Neath, A., & Cavanaugh, J. (2011). The Bayesian information criterion: Background, derivation, and applications. *Wiley Interdisciplinary Reviews: Computational Statistics*, *4*(2), 199–203. doi:10.1002/wics.199

Noel, K. (2016). *Application of Machine Learning to Systematic Allocation Strategies*. SSRN Electronic Journal; doi:10.2139srn.2837664

Ott, D., & Houdek, F. (2014). Automatic Requirement Classification: Tackling Inconsistencies Between Requirements and Regulations. *International Journal of Semantic Computing*, *08*(01), 47–65. doi:10.1142/S1793351X14500020

Punyavachira, T. (2013). *Forecasting Stock Indices Movement Using Hybrid Model: A Comparison of Traditional and Machine Learning Approaches*. SSRN Electronic Journal. doi:10.2139srn.2416494

Puthal, D. (2018). Lattice-modelled Information Flow Control of Big Sensing Data Streams for Smart Health Application. *IEEE Internet of Things Journal*. doi:10.1109/jiot.2018.2805896

Rathore, M., Paul, A., Ahmad, A., & Jeon, G. (2017). IoT-Based Big Data. *International Journal on Semantic Web and Information Systems*, *13*(1), 28–47. doi:10.4018/IJSWIS.2017010103

Sądel, B., & Śnieżyński, B. (2017). Online Supervised Learning Approach for Machine Scheduling. *Schedae Informaticae*, *1/2016*. doi:10.4467/20838476SI.16.013.6194

Schumaker, R. (2013). Machine learning the harness track: Crowdsourcing and varying race history. *Decision Support Systems*, *54*(3), 1370–1379. doi:10.1016/j.dss.2012.12.013

Schwab, D., & Ray, S. (2017). Offline reinforcement learning with task hierarchies. *Machine Learning*, *106*(9-10), 1569–1598. doi:10.100710994-017-5650-8

Šumrada, R. (2011). Free OGC standardized web services for spatial data. *Geodetski Vestnik, 55*(1), 046-056. doi:10.15292/geodetski-vestnik.2011.01.046-056

Sun, H., & Betti, R. (2015). A Hybrid Optimization Algorithm with Bayesian Inference for Probabilistic Model Updating. *Computer-Aided Civil and Infrastructure Engineering*, *30*(8), 602–619. doi:10.1111/mice.12142

Tango, F., Minin, L., Tesauri, F., & Montanari, R. (2010). Field tests and machine learning approaches for refining algorithms and correlations of driver's model parameters. *Applied Ergonomics*, *41*(2), 211–224. doi:10.1016/j.apergo.2009.01.010 PMID:19286165

Tofan, C. A. (2014). Optimization Techniques of Decision Making - Decision Tree. *Advances In Social Sciences Research Journal*, *1*(5), 142–148. doi:10.14738/assrj.15.437

Toivonen, H., & Gross, O. (2015). Data mining and machine learning in computational creativity. *Wiley Interdisciplinary Reviews. Data Mining and Knowledge Discovery*, *5*(6), 265–275. doi:10.1002/widm.1170

Urquhart, L., & Rodden, T. (2016). *A Legal Turn in Human Computer Interaction? Towards Regulation by Designn for the Internet of Things*. SSRN Electronic Journal. doi:10.2139srn.2746467

Wang, D., McMahan, C., & Gallagher, C. (2015). A general regression framework for group testing data, which incorporates pool dilution effects. *Statistics in Medicine*, *34*(27), 3606–3621. doi:10.1002im.6578 PMID:26173957

Yao, M. (2014). Research on Learning Evidence Improvement for kNN Based Classification Algorithm. *International Journal Of Database Theory And Application*, *7*(1), 103–110. doi:10.14257/ijdta.2014.7.1.10

Yildirim, P., Birant, D., & Alpyildiz, T. (2017). Data mining and machine learning in textile industry. *Wiley Interdisciplinary Reviews. Data Mining and Knowledge Discovery*, *8*(1), e1228. doi:10.1002/widm.1228

Zhang, B., Xiang, Y., & Wang, J. (2010). Information Filtering Algorithm Based on Semantic Understanding. *Dianzi Yu Xinxi Xuebao*, *32*(10), 2324–2330. doi:10.3724/SP.J.1146.2009.01393

Zhang, N. (2016). Semi-supervised extreme learning machine with wavelet kernel. *International Journal Of Collaborative Intelligence*, *1*(4), 298. doi:10.1504/IJCI.2016.10004854

KEY TERMS AND DEFINITIONS

Artificial Neural Network: An artificial neural network (ANN) is a data handling worldview that is roused by the way organic sensory systems, for example, the cerebrum, process data. The key component of this worldview is the novel structure of the data handling framework. It is made out of an extensive number of exceptionally interconnected preparing components (neurones) working as one to take care of particular issues. ANNs, similar to individuals, learn by case. An ANN is designed for a particular application, for example, design acknowledgment or information grouping, through a learning procedure. Learning in natural frameworks includes changes in accordance with the synaptic associations that exist between the neurons.

Artificial Intelligence: Computerized reasoning is insight shown by machines, rather than the characteristic knowledge showed by people and different creatures. In software engineering, AI is characterized as the investigation of "keen specialists": any gadget that sees its condition and takes activities that amplify its risk of effectively accomplishing its goals. Colloquially, the expression "manmade brainpower" is connected when a machine impersonates "subjective" capacities that people connect with other human personalities, for example, "learning" and "critical thinking."

Chapter 4
A Survey on Routing Protocols of Wireless Sensor Networks:
A Reliable Data Transfer Using Multiple Sink for Disaster Management

Chandana Rani Kandru
VIT AP University, India

Ravi Sankar Sangam
VIT AP University, India

ABSTRACT

Disasters are happening due to drastic environmental destructions that may cause damage to wireless data transmission networks. There must be a system that monitors and takes necessary actions for reliable communication which can be provided by the wireless sensor network (WSN) that were organized as multiple nodes. In the heterogeneous environment these ubiquitous nodes are able to handle disasters like floods, drought, earthquake, and cyclone, or network fluctuations through fire accidents. Disasters can be monitored by augmenting a variety of sensors to sense and detect sudden changes in temperature, pressure, seismic wave, noises, etc. Large numbers of sensor nodes are distributed over a geographical area in WSN providing trustable data transfer with multi node sink. In this chapter, the authors review various WSN routing protocols for reliable data transfer in disaster management using multiple sink. The main objective of this chapter is to provide future research directions to enhance QoS in disaster management.

INTRODUCTION

Natural Disasters have been continuously increasing worldwide due to environmental destructions, happening due to huge human population. The situation is highly severe in developing countries as advanced detection technologies are either impractical or unaffordable. The environmental intensity of disaster makes it hard and, in some cases, entirely not possible for humans to react or face that critical

DOI: 10.4018/978-1-5225-7458-3.ch004

problem immediately (Erdelj. *et al.,* 2017). The first and foremost issue that needs to be addressed is to preserve human lives during and after disaster. At present efforts has been made in order to recognize and forecast the possibility that when a disaster happen, react as soon as possible and efficient manner in course of happening, quickly assess the damage, fix and restore to normal state.

Recent developments in wireless communication technologies, energy storage, computing power make a system composed of WSN, plays a critical role in natural disaster. Today the world, which is vastly developing in technologies, is much likely to find a way to control the occurring of disasters. Disasters can be identified by enlarging a variety of sensors such as Temperature, pressure, seismic wave detection, noises. Large numbers of sensor nodes are distributed over a geographical field in WSN providing trustable data transfer with multi node sink. By using WSN one can manage occurring disasters by implementing reliable data transfer between nodes. WSNs have been successfully make use for a number of purposes when dealing with disasters, like environment controlling, pre-disaster forecast and prevention, emergency preparedness, early warning system, post-disaster response (Chen *et al.,* 2013). In this chapter, our contributions are as follows:

- We present various state of art hierarchal routing algorithms under WSN, suitable for disaster management, like LEACH, PEGASIS, TEEN, APTEEN and HEED.
- We discuss the merits and demerits of these algorithms in the real time environment.
- Finally, we make a summary note on hierarchal routing algorithms, in a tabular form, based on the parameters that influences the lifetime of the network when it is applied in real-time environment.

An over view of WSN is discussed in the Section 1. Section 2 discuss different applications of WSNs. Section 3 and 4 presents data aggregation and reliable data transfer in WSN, based on single sink and multiple sink, respectively. In Section 5, we describe formation of Clusters in WSN using different algorithms. In Section 6, hierarchal protocols that are used in WSN are reviewed. We review Routing procedure in WSN in Section 7. In Section 8 the role of machine learning in WSN routing is discussed. Finally, we present conclusion and future scope in Section 9.

1. WIRELESS SENSOR NETWORKS

WSN is the outcome of sequence of wireless communication, internet and sensor technologies. It can be applied in various fields like military, health care monitoring environmental monitoring, natural disasters and many other fields. The WSN technology has been applied in controlling naturally occurring disasters for more than a decade. A WSN is an upcoming technology which is taken as a major consideration by the research community. WSN consist of various small, low cost devices that contain the motes, also referred as sensor nodes, placed in a network, which is actually a self-organizing ad-hoc system. The main functionality is to examine the physical environment, collect and transmits information (sensed data) to one or more sink nodes. The internal functionality of a sensor node is represented in Figure 1. The main operations performed in WSN are related to monitoring how the physical environment is, processing of sensed data and transfer these data to particular nodes. To perform these tasks, the sensor nodes are filled with powered batteries which contain limited energy (Yacoab & Sundaram, 2011). The important issue to be considered while dealing with wireless sensor is their energy consumption.

It is normally not possible or unusual to replace or recharge sensor batteries as the sensor nodes are typically deployed in remote areas. Therefore, the limited energy of sensor nodes should be managed efficiently to perform tasks for the available period of sensor lifetime. The efficiency of communication module is mainly depending on structure and size of each sensor node. These sensor nodes can easily be deployed into the error prone environment, where sensor nodes can easily transmit the error and disaster related sensed data from source to sink node. In small networks sensors can transmit data directly to the sink node, whereas in large network multi-hop communication is necessary, where sensors transmit and receive the data toward the sink.

Forwarding and receiving of data takes place which consume more energy depending upon the communication distance. Distance of communication can be reduced in many ways. One of the solutions is, to have a mobile sink which progress innermost to particular area and collects the data from the sensors through which it passes through. However, this arriving couldn't fit in the real-time applications as it delays the data transfer, where sink node needs to receive the information promptly. Another solution is to deploy multiple sinks; this makes to lessen the average number of hops when a message pass through before receiving and processed by a sink, as data will always send to the closest sink. But in this approach, we must have an idea about how to deploy and where to deploy multiple-sinks in particular field. This is the problem for "facility location": where there is given a set of sink nodes and a set of sensors to be served from these sink nodes where to deploy those sink nodes and which sink should serve which sensor node is to be considered to reduce the overall energy consumption.

The main function of sensor network is data gathering. In WSN, data aggregation is considered as a main primary distributed data processing procedure which reduces the medium access layer contention and saves energy.

2. WSN APPLICATIONS

There are many wide range of applications in WSN such as natural disaster forecasting (Erdelj et al., (2017)), agricultural environment monitoring (Kassim & Harun (2016)), State of art Industrial WSN(IWSN) (Ehrlich *et al.,* (2018)), forest wildlife monitoring application (Jan *et al.,* (2018)), Cloudburst (Goswami *et al.,* (2016)) etc. However, in this section, we discuss the various applications of WSN in natural disaster forecasting.

Figure 1. Representation of a sensor node

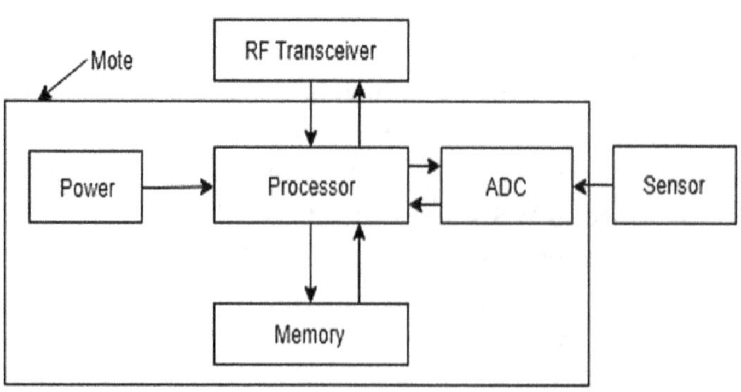

2.1 WSN for Flood and Water Monitoring System

Every year there's an enormous loss of thousands of lives and billions price of property in Asian nature owing to floods. Previous, attributable to floods in Bihar and Jharkhand states there's a loss of human lives and cattle, besides there is huge economical loss of institutions has been reportable. Indiscriminately, Ganga and Yamuna rivers break their boundaries and cause various losses. If protecting measures is taken before the disaster has smitten within the sort of floods, of these losses can't be place associate degree finish totally, however the losses to lives and property is reduced to somewhat minimum level. This is possible with the help of latest communication technology Wireless Sensor Networks. The sensors in WSN collectively form a cluster to communicate one another with the near Local Base Stations (LBS). These LBS will directly communicate with each other using wireless communications. The data transmitted from the sensors are combined in LBS to provide as inputs to the data processing centers.

2.2 Detection of Forest Fire

Forest house contains countless rare species of birds, animals and insects. Because of Forest Fires, it doesn't cause solely the loss of their shelters however additionally causes vast loss of flora and fauna. Occurring of forest fires are common in middle ranges of mountain chain region of Asian nation because of Lightening, excessive hear or carelessness on the part of local natives (Kaur *et al.,* 2012). There are several occasions once the fire out because of the negligence of the natives in Uttarakhand, Chhattisgarh, and north-eastern states. Large amount of motes is arranged into clusters in order that every node features a corresponding Cluster Head (CH). Motes can measure surroundings temperature and smoke. Motes also are assumed to understand their location data by equipment's like Global Positioning System. Here, each mote transfers data related to measurement information and location data, to the corresponding CH. The CH computes the weather information and transfers the information to the manager node via sink (Kaur *et al.,* 2012).

2.3 WSN for Earthquake Sensing

The most furious earthquake that happened in India was on twenty sixth Jan, 2001 that clatters not only in India but also in neighboring countries viz. Iran, Afghanistan and Pakistan, additionally couldn't get away from turbulence. Several precious lives will be saved by timely action. To handle the earthquake dynamics like extremely projectile magnitude and variable source location, every sensing element maintains separate applied statistical methods of frequency spectrum for distinction scales of seismic signal energy received by sensing element (Endo & Murray, 1991). Additionally, detection of earthquake occurrences, node level earthquake onset time is essential for localizing earthquake supply. Here, the base station identifies a personal earthquake and estimates onset time (Sleeman & Van Eck, 1999).

2.4 WSN for Tsunami Detection

It is calculable that tens of thousands of individuals died within the event occurred in 2004 Tsunami and lakhs in alternative countries of Asia. A system for tidal wave detection and mitigation employing a wireless ad-hoc sensing element network defines 3 styles of node; sensing element, commander and barrier (Kaur *et al.*, 2012). Comparatively huge amounts of sensing element of nodes collect underwater pressure reading across a costal field. In (Casey *et al.,* 2006), associate algorithmic approach is discussed based on General Regression Neural Network (GRNN) (Specht, 1991) to predict the trail of the wave. The GRNN analyzes the pressure information from sensing element nodes and predicts that barriers ought to fireplace to most effectively impede the tidal wave. In the next section we are going to discuss the aggregation of data issues based on single sink and multiple sink.

3. DATA AGGREGATION BASED ON DIFFERENT SINKS

3.1 Data Aggregation Related Issues Based on Single Sink

1. When gathering sensor data, many-to-one traffic pattern may lead to imbalance in energy expenditure to a higher extent in the whole network, causing early termination of the network lifetime. Here, the main challenge is to maintain the energy efficiency by maintaining the network lifetime which often results in less energy consumption (Wang & Liu, 2011).
2. Scalability is the problem that faces in single sink network architecture. The data that is collected at the sink may grow into more than its communication limit due to the large number of motes in the network. Further, the radio channel limit near to the sink may be overloaded if the common number of hops encoded by the source and the destination i.e. sink becomes excess (Cipollone *et al.,* 2007).
3. In single sink network architecture, if the sink is hugely loaded, then the data will not reach the destination, which results in transmission of failure.

3.2 Data Aggregation Related Issues Based on Multiple Sink

To overcome the problem of scalability, multiple sinks concept is introduced. Efficient data collecting trees are produced by the nodes, and then the finest sink is preferred for transferring that data in multiple sink networks. Decreasing the mean distance between sink and the nodes leads to improve the lifetime and to save energy in the multiple sink networks. The sink performs work as a gateway in multi sink network which forwards the collected data towards the storage system in the network. In the next section, the discussion is on how the reliable data is transferred using various sinks in WSN.

4. RELIABLE DATA TRANSFER USING SINGLE SINK AND MULTI-SINK

4.1 Reliable Data Transfer Using Single Sink in WSN

The routing path is initiated by the single sink node. In single sink sensor networks, refute of the sink node can cause the breakdown of full sensor network. The transmission of data goes on the routing path within the opposite way. The reliability of data transferred in WSN is categorized into Coverage, Exchange of secure data, Network Latency, Packet Delivery, and Availability of Network. The detail of data transmission is representation as shown in Figure 2.

Figure 2. Categorization of reliability in Wireless Sensor Nodes

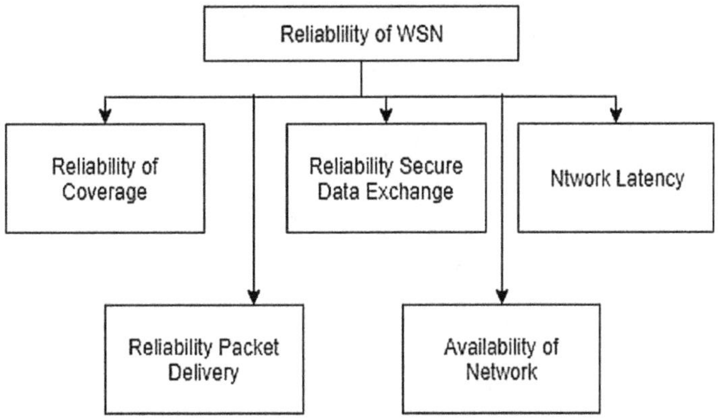

The nodes close to the sink consume a lot of energy and exhaust quickly. The important drawback to be solved among the sensor node is, the way to prolong the period of the network by equalization of the energy consumption.

Typically, the transmission of data in single-sink sensor networks considers the quantity of hops and total energy consumption from supply nodes to the sink. The routing path includes, nodes with the closest path which contains less quantity of energy and huge energy consumption for transmission of data. These routing ways cannot guarantee the utmost period of the network. The sink node has property of providing energy and is reusable. Meng *et al., (*2007) have proposed a multi-sink sensor networks and a routing formula Priority Based Routing algorithm for balancing of equalization of energy consumption of sensor nodes.

4.2 Reliable Data Transfer Using Multiple-Sink Sensor Networks

In-order to prolong the lifetime of networks, here multiple sink sensor networks are introduced. The topology and data collection depend on the system architecture of sensor networks. The energy model and Low Energy Adaptive Clustering Hierarchy (LEACH) proposed in (Heinzelman *et al.,* 2000) has been employed in determination of various numerous issues in WSNs.

In multi-sink networks, we can redistribute multiple sink nodes as it is needed. TCP/IP is a simple protocol where sensor nodes can associate with each other. As shown in the Figure 3, multi-sink sensor consists of sensor nodes, sink nodes, proxy and task manager.

The users send queries to task manager and it further send to a proxy by internet or satellite. The tasks of the proxy are speed in carrying out data and collecting data by broadcasting or P2P. The optimized sink node will receive the data transmitted by sensor nodes (Meng *et al.*, 2007). Thomas & Poonguzhali (2014) have assumed that the network is composed of *N* uniformly dispersed sensor nodes, denoted as $\left\{S_1, S_2, \ldots, S_N\right\}$. The randomly deployed *k* sink nodes is denoted as $\left\{BS_1, BS_2, \ldots, BS_N\right\}$. The energy model, shown in Equation (1) and (2), is adopted as in (Rappaport, 1996). Any sensor node consume E_{Tx} amount of energy to transmit *l* bits packet over *d* distance and E_{Rx} for reception, where the energy dissipated on circuit is E_{elec}, and E_{fs} and E_{mp} represent conditions of a channel (Thomas & Poonguzhali, 2014).

Figure 3. Diagrammatic representation of Multi-Sink

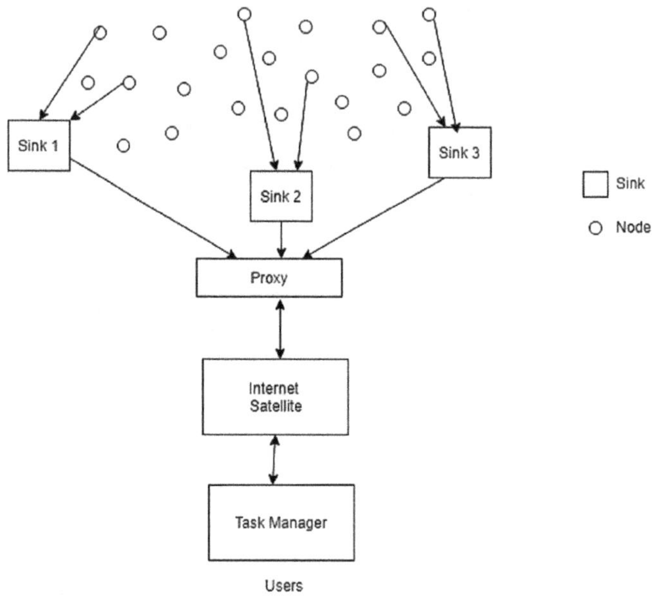

$$E_{Tx}\left(l,d\right)=\begin{cases} lE_{elec}+lE_{fs}d^{2}, d<d_{0} \\ lE_{elec}+lE_{fs}d^{4}, d\geq d_{0} \end{cases} \quad (1)$$

$$E_{Rx}\left(l\right)=lE_{elec} \quad (2)$$

5. FORMATION OF CLUSTER IN WSN USING VARIOUS ALGORITHMS

5.1 Cluster Formation

Clustering of motes allows gathering of sensed data, and transmission of data within the cluster by reducing the traffic. Cluster formation is done by sending alert signals to the neighboring sensor nodes and finding nearby sensor nodes. An energy-efficiency algorithm will prefer only few cluster heads in order to save energy, but if the cluster heads do not have good connectedness or if they are unstable, then the retransmission and the dropped packets lessen the network performance and the total wastage of energy will be higher. Bandyopadhyay & Coyle (2003) discussed two routing protocols namely, Energy-Efficient Multi-sink Clustering Algorithm (EMCA) and Mobile-sink Based Energy-Efficient Clustering Algorithm (MECA). These two routing protocols have following advantages.

1. The aggregation of data lessens the traffic capacity.
2. The CHs locate in a further uniform way when compared to probabilistic arrangement in LEACH, and it can also prolong the network lifetime (Gharavi & Ban, 2003).

When Cluster head need to transmit information among the motes in a cluster it can transmit precisely to their CH without any in-between sensor nodes, which leads to low energy consumption (As shown in Figure 4). The consumption of energy in the wireless is proportional to the square of the distance between sensor nodes. We need to consider, how much amount of data is transmitted in each cluster and between any clusters, the repetitiveness of transmission and utmost allowable latency, in order to conclude the optimal value of distance that minimize the energy consumption. Later, by equalizing the sensing field into various equal fields, the cluster head selection is done, by firstly taking the lowest ID node as cluster head. Secondly, it selects the sensor node with the highest degree (largest number among the neighbors). Lastly, the sensor node which has highest residual energy is chosen as the CH.

5.2 Energy Efficiency Multi-Sink Clustering Algorithm

In multi-sink sensor network, clusters are equalized into *n* equal clusters. There is a cluster head in each cluster where these cluster head perform data synthesis after collecting the data. Choosing of cluster head is done by, collecting the lowest ID node as cluster head, by choosing the sensor node with highest degree (largest among the neighbors), and sensor nodes that contain highest residual energy is selected as the CH. Then the CH transfers the aggregated information to the appropriate sink. In clustering algorithm, if the CHs located in ordered way, the energy consumption can be lower and maintained in proper way, which extends the network lifetime. This type of network work is highly suitable for huge deployment of networks which increase the network lifetime.

6. HIERARCHAL PROTOCOLS USED IN WSN

Well-architecture network protocols can lower energy consumption and extends network lifetime. Commonly, routing protocols is classified as Data Centric protocols, Hierarchical routing protocols,

Figure 4. Diagrammatic representation of clustering of sensor nodes

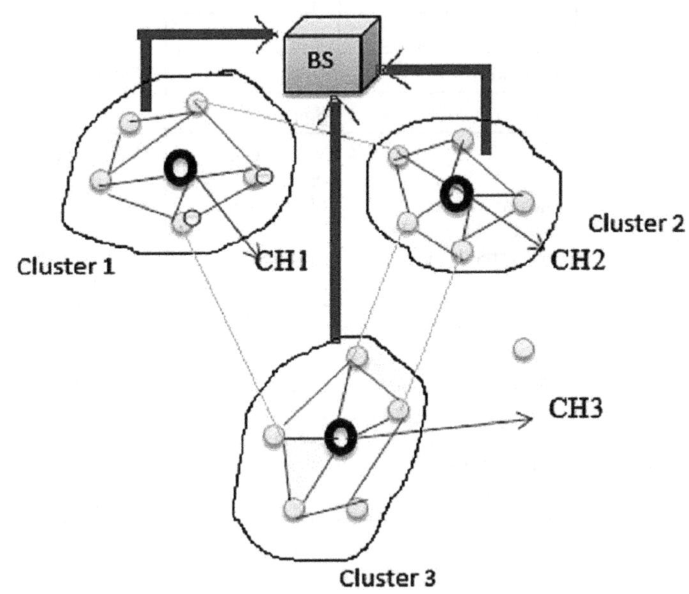

Location Based protocols, Network Flow and QoS Aware Protocol. Most of the common hierarchical routing protocols are considered in which CHs are selected in an energy efficient manner. Any CH is the representative for the respective cluster for collecting the news from other motes and transmits to the BS in the direct way or indirectly by using intermediate nodes. Hierarchical protocols are characterized to lower energy consumption by combining the data and by reducing the transportation of information to the BS. One among them is LEACH, PEGASIS, TEEN, APTEEN, and HEED. Besides, the precise merits and demerits of these algorithms along with the performance influencing parameters are summarized in Table 1.

6.1 LEACH

LEACH routing protocol is a hierarchal routing protocol which describes for Low Energy Adaptive Clustering Hierarchy protocol, LEACH protocol, sometimes it is said to be that there exist some problems like energy imbalance and highly energy consumption. For solving the above-mentioned problems, LEACH protocol is introduced to improve over analyzing the characteristic clustering routing protocol. This protocol is mainly for the function of the lessening power consumption, extending the network lifetime and accomplish energy balance. LEACH is self-adaptive and self-organized, which makes use of a cyclic process that include cluster setup phase and steady state phases. In order to shorten unnecessary energy cost, the steady state phase must be longer than setup phase. In the phase of Cluster formation, a

Table 1. Comparison of hierarchal routing protocols in WSN

Protocol Name	LEACH	PEGASIS	TEEN	APTEEN	HEED
Clustering	Random	Chain-based	Reactive-based	Random	Random
Network lifetime	Shortest	Longer	Long	Long	Long
Robust	Moderate	Good	Limited	Good	Good
Energy efficiency	Poor	Good	Moderate	Moderate	Very good
Scalability	Good	Good	Good	Good	Good
Load balancing	Medium	Medium	Good	Medium	Medium
Delay	Very small	Very large	Small	Small	Medium
Advantages	Low energy, ad-hoc, distributed protocol.	The transmitting distance for most of the node is reduced.	Works well in the conditions like sudden changes in the sensed attributes such as temperature	The threshold value for the energy consumption can be adjusted by changing the count time.	1.Nodes require only local information to form clusters. 2.Terminates in constant time, reduces energy load which leads to extension of network lifetime.
Disadvantages	Not applicable to networks deployed in large regions and the dynamic clustering brings extra overhead.	Doesn't consider about base station's location regarding energy of nodes when one of the nodes is chosen as the head node.	Consumes more energy and overhead in case of large network.	Additional overhead required to implement the threshold functions and the count time.	1.Random selection of cluster heads, may cause higher communication overhead. 2.Periodic cluster head rotation or election needs extra energy to rebuild clusters.

node picks a number between 0 and 1 randomly, which is distinguished to the threshold values. Provided, if the number is lower than the threshold value, the node has the right to be a CH in that cycle, if it is higher than the threshold value it becomes common node.

Steps for Algorithm LEACH:

1. Firstly, construct the nodes and frame the network.
2. After framing the network, select the CHs among the nodes.
3. After choosing cluster heads, separate the network into many clusters
4. Check the energy consumption of every node at the end of the first round. If the energy level is too short, then visit step2 or show the output through preferred CH.

In CH selection any node calculates the threshold value depending on the below formula

$$T(n) = \begin{cases} \dfrac{P}{1 - P * \left[r \bmod \left(\dfrac{1}{P} \right) \right]} \; if\, n \in G \\ 0.5 * \dfrac{P}{1 - P * \left[r \bmod \left(\dfrac{1}{P} \right) \right]} \; if\, n \in G \\ 0\; otherwise \end{cases} \tag{3}$$

Here

P: stands for desired percentage of CH
r: stands for present round in protocol operation.
G: defines the set of nodes that have been neither CHs nor been helper nodes in the last $1/P$ rounds.
H: defines the set of nodes that do not been CHs but took a role of helper node in the last $1/P$ rounds.

Each sensor assigns itself to be a CH by selecting a number between 0 and 1 randomly and comparing it to be low compared to threshold.

One of the main disadvantage in LEACH algorithm is when a cluster head dies with any of the reasons, the cluster will become ideal as the information that is collected by the cluster nodes can never reach the destination i.e. BS. It doesn't show any idea about the number of CHs in the network. Clusters are separated randomly, which cause uneven distribution of clusters. It does not supply any clearness about location of sensor nodes and the number of CHs in the network. Every CH directly makes communication with the BS it doesn't matter if the distance between Cluster Head and BS is far which consumes more amount of energy.

6.2 PEGASIS

PEGASIS is a hierarchal routing protocol which stands for Power Efficient Gathering in Sensor Information System. PEGASIS routing protocol is a chain-based routing protocol performs unlike leach. PEGASIS protocol builds chain instead of clusters so that work of dynamic clustering can be shortened. In PEGASIS protocol, as all the nodes communicate with their nearby neighbors and further continue their transmission until the collection of combined data reaches the BS. PEGASIS protocol develops the network lifetime, because it shortens the power consumption needed for particular round. The deviation from leach protocol is usage of multi-hop routing by building chains with choosing only a node to send the data to the BS instead by using multiple nodes to transmit the data. The main characteristics of PEGASIS are: In PEGASIS the BS is steady at a far distance from the sensor nodes. No movement of sensor nodes i.e. nodes are stationary.

PEGASIS protocol depends on two ideas: 1) chaining and 2) Data Fusion. In PEGASIS, any node can take a role of being a leader head of the chain; where for the constructing of the chain greedy algorithms can be used that are arranged by the sensor nodes. PEGASIS is not scalable as global knowledge of the network is not that much easy to gain.

6.3 TEEN

Akyildiz & Vuran (2010) have discussed TEEN is a hierarchal clustering protocol stands for Threshold-sensitive energy-efficient sensor network. TEEN protocol provides the event-based delivery in the network. Here, the multi-hop routes are automatically generated according to threshold value related to sensor data. TEEN protocol establishes and arrange the motes into multiple levels of CHs, which gather, combine the data and then forwards that information to a higher-level cluster head until it reaches to the BS. Based on structure of this hierarchy network structure, TEEN provides event-based communication over two thresholds: 1). Soft threshold $\left(S_T\right)$, 2). Hard Threshold $\left(H_T\right)$. Soft Threshold $\left(S_T\right)$ is used to decrease the redundancy in the transmission. At any time if the hard threshold exceeds, the sensor node refers the soft threshold for subsequent attention. If the difference between subsequent observations does not exceed the soft threshold, the information is not transmitted by the sensor nodes. It means, that it informs to the cluster head that there is a similar value observed in the data that is transmitted. Only new observations or data is transmitted if the soft threshold exceeds. Thus, TEEN depends on fixed threshold limits, which is not suitable for periodic reports required by some applications.

6.4 APTEEN

To provide periodic information improvement, the Adaptive Threshold-Sensitive Energy- Efficient Sensor Network (APTEEN) routing protocol has been introduced, as an Advanced Protocol of TEEN, in (Akyildiz & Vuran, 2010). APTEEN is a multi-hop clustering algorithm which provides a TDMA – based architecture for transmission of information in every cluster. APTEEN, is a hybrid protocol that modifies the values of periodicity or threshold used in TEEN, according to the user's specification and the type of application (Liu, X. (2012)). This algorithm depends upon query system which allows 3 types of questions viz. historical, on-time and persistent that can be used in a hybrid network. The CHs selection in APTEEN follows four parameters: (1) Attributes, (2) value of the hard and soft thresholds,

(3) Schedule, (4) Count Time. Subsequently, each node transmits the data systematically to the CHs. Thus, both monitoring and event-based applications can be served.

6.5 HEED

Younis & Fahmy (2004) introduced a Hybrid Energy-Efficient Distributed Clustering (HEED), follows multi-hop clustering. Cluster formation is performed mainly depending on the hybrid combination of 2 parameters called node's residual energy and intra-cluster communication cost. One of the main goals of HEED algorithm is to get an event-distributed CHs all over the network. The probability that a node can become a CH is based on the formula called

$$CHpr = Cpr\frac{Eresidual}{E\max} \tag{4}$$

Cpr, is set to assume that an optimal percentage cannot be computed a priori. $E\max$ is a reference maximum energy which is typically, $Eresidual$ is the estimated current energy of the node. Every node doubles its $CHpr$ value and moves to next iteration till its $CHpr$ reaches to 1. Thus, there are two types of status that a node could broadcast to the neighbors, a tentative status and final status (Liu, X. (2012). The node becomes a permanent CH if the $CHpr$ reaches to 1, otherwise the node becomes a tentative cluster.

7. ROUTING PROCEDURE IN WSN

Our main concept is how to lower the energy consumption and maintain the reliable data transfer between the nodes efficiently. The network is collection of *N* uniformly dispersed sensor nodes, represented as $\{S_1, S_2, \ldots, S_N\}$. There are randomly deployed k-sink nodes, represented as $\{BS_1, BS_2, \ldots, BS_N\}$. The representation of cluster head CH_n, energy consumption to sink BS_k is $E(CH_n, BS_k)$. The calculation of energy model is as follows.

$$E(CH_n, BS_k) = \begin{cases} lE_{elec} + lE_{fs}d(CH_n, BS_k)^2, d(CH_n, BS_k) < d_0 \\ lE_{elec} + lE_{fs}d(CH_n, BS_k)^4, d(CH_n, BS_k) \geq d_0 \end{cases} \tag{5}$$

Thomas & Poonguzhali (2014) have compared the distances from every CH to various sinks and select the minimum path. Any CH selects one optimal sink to transmit the aggregated information. For each CH_n, the energy consumption to sink BS_k is represented as

$$E(CH_n, BS_k) = \{lE_{elec} + lE_{fs}d(CH_n, BS_k)^2\} \tag{6}$$

In most of the clustering algorithms like LEACH, the sensor nodes in that particular cluster transmit the data directly to the CH. Some of the motes consume large amount of energy as long distance of transmission while they were placed in a various location. For this purpose, we use a multi-hop routing protocol. For every individual sensor node S_i in each cluster, find other sensor node S_j, send the data to its intermediate nodes to save energy by preventing direct communication with the CH. Now the energy consumption is represented as $E_2\left(S_i, S_j, CH_{Si}\right)$. As multiple sink is randomly deployed, few nodes may consume less energy by transmitting data directly to the sink than to the CH. These are the protocols and clustering algorithms that are used in reliable data transfer using multiple sink in WSN for the application of Disaster management.

8. THE ROLE OF MACHINE LEARNING IN WSN ROUTING

In present decade, the researchers have made some attempts to use machine learning algorithms in WSN routing to address some state of art issues like communication costs, limited energy, mobility and topology changes. In this section, we discuss the role and importance of machine learning in WSN applications. By using machine learning methods, one can discover important correlations in the sensor data and propose the improvement in sensor deployment for maximum data coverage to address the following major trending issues and/or challenges in WSN.

- WSN usually monitor dynamic environments that changes rapidly over the time. Ex: location of a node rapidly changes due to soil erosion. It is important to develop the sensor network to adopt and operate efficiently in such surroundings.
- WSNs might be used for gathering new information around unreachable and dangerous locations (like volcano eruption) in exploratory applications. Because of their unexpected behavior patterns, they may arise in those scenarios, system designers may progress solutions that initially may not operate as expected.
- Sensor networks are typically deployed in complicated environments where the researchers cannot figure accurate mathematical models to describe the behavior of the system.
- Sensor network designers frequently have access to large amount of data, but it might be unable to extract significant correlations in them.

However, there are some drawbacks that should be considered while using machine learning techniques in WSN.

- As a resource restricted framework, WSN always considers the percentage of energy budget to forecast the correct hypothesis and extract the promising relationship among data samples.
- In general, learning through examples needs a large dataset of samples to achieve the planned simplification capabilities and the algorithms designers will not have the full control on the information formulation process.

9. CONCLUSION AND FUTURE SCOPE

Since Multi-sink assured the researchers to make communication reliable between distributed sensor nodes. Based on the foregoing discussion, we have surveyed, and summarized research works which are recently focused mainly on energy efficient hierarchal clustering- based routing protocols for reliable data transfer based on multiple sink an application of disaster management using WSN. In addition, the process of data aggregation and fusion among the clusters is similarly an interesting problem to explore. In this chapter, we have come up with different clustering algorithms and routing protocols that effectively maintain a network i.e., either Terrestrial or underwater that is organized with multi hop WNS in view of Disaster Management through providing Fault Tolerance. Therefore, the application of appropriate routing protocols will increase the network lifetime and at the same time it will make sure the network connectivity and efficient data delivery.

Since past one decade many efforts have been taken in WSN to predict the disasters based on various sources of data. However, many challenging issues have not been addressed in many recent works. In future, the following issues in WSN routing can be addressed to enhance the accuracy of early prediction of disasters.

- Maintaining maximum coverage as well as node connectivity in a critical resource constrained situation, is a significant optimization problem.
- Adaptability and scalability of the clustering protocol across sensor nodes with various resource capability are to be examined.
- Routing protocol should achieve redundancy to improve energy and bandwidth utilization.
- Supplying energy and reducing energy consumption are not meeting in the real time scenario.
- There is a high possibility of node failure and change of topology which results to Fault Tolerance.
- There is a requisite to have most cost (time complexity) effective Machine learning based routing protocols to prolong the lifetime of WSN.

REFERENCES

Akyildiz, I. F., & Vuran, M. C. (2010). *Wireless sensor networks* (Vol. 4). John Wiley & Sons. doi:10.1002/9780470515181

Bandyopadhyay, S., & Coyle, E. J. (2003, April). An energy efficient hierarchical clustering algorithm for wireless sensor networks. In *INFOCOM 2003. Twenty-Second Annual Joint Conference of the IEEE Computer and Communications* (Vol. 3, pp. 1713-1723). IEEE.

Casey, K., Lim, A., & Dozier, G. (2006, July). Evolving general regression neural networks for tsunami detection and response. In *Evolutionary Computation. CEC 2006. IEEE Congress on* (pp. 2451-2458). IEEE.

Chen, D., Liu, Z., Wang, L., Dou, M., Chen, J., & Li, H. (2013). Natural disaster monitoring with wireless sensor networks: A case study of data-intensive applications upon low-cost scalable systems. *Mobile Networks and Applications*, *18*(5), 651–663. doi:10.100711036-013-0456-9

Ehrlich, M., Wisniewski, L., & Jasperneite, J. (2018). State of the art and future applications of industrial wireless sensor networks. In Kommunikation und Bildverarbeitung in der Automation (pp. 28-39). Springer.

Endo, E. T., & Murray, T. (1991). Real-time seismic amplitude measurement (RSAM): A volcano monitoring and prediction tool. *Bulletin of Volcanology*, *53*(7), 533–545. doi:10.1007/BF00298154

Erdelj, M., Król, M., & Natalizio, E. (2017). Wireless sensor networks and multi-UAV systems for natural disaster management. *Computer Networks*, *124*, 72–86. doi:10.1016/j.comnet.2017.05.021

Erdelj, M., Natalizio, E., Chowdhury, K. R., & Akyildiz, I. F. (2017). Help from the sky: Leveraging UAVs for disaster management. *IEEE Pervasive Computing*, (1), 24–32.

Gharavi, H., & Ban, K. (2003). Multihop sensor network design for wide-band communications. *Proceedings of the IEEE*, *91*(8), 1221–1234. doi:10.1109/JPROC.2003.814919

Goswami, S., Chakraborty, S., Ghosh, S., Chakrabarti, A., & Chakraborty, B. (2016). *A review on application of data mining techniques to combat natural disasters. Ain Shams Engineering Journal.*

Heinzelman, W. R., Chandrakasan, A., & Balakrishnan, H. (2000, January). Energy-efficient communication protocol for wireless microsensor networks. In *System sciences, Proceedings of the 33rd annual Hawaii international conference on* (pp. 10-pp). IEEE. 10.1109/HICSS.2000.926982

Jan, M. A., Nanda, P., He, X., & Liu, R. P. (2018). A Sybil attack detection scheme for a forest wildfire monitoring application. *Future Generation Computer Systems*, *80*, 613–626. doi:10.1016/j.future.2016.05.034

Kassim, M. R. M., & Harun, A. N. (2016, October). Applications of WSN in agricultural environment monitoring systems. In *Information and Communication Technology Convergence (ICTC), 2016 International Conference on* (pp. 344-349). IEEE.

Kaur, H., Sawhney, R. S., & Komal, N. (2012). Wireless sensor networks for disaster management. *International Journal of Advanced Research in Computer Engineering & Technology (IJARCET), 1*(5).

Liu, X. (2012). A survey on clustering routing protocols in wireless sensor networks. *Sensors, 12*(8), 11113-11153.

Meng, M., Xu, H., Wu, X., d'Auriol, B. J., Jeong, B. S., Lee, S., & Fan, X. (2007, June). PBR: Priority Based Routing in Multi-Sink Sensor Networks. In ICWN (pp. 380-386). Academic Press.

Rappaport, T. S. (1996). *Wireless communications: Principles and practice* (vol. 2). Prentice Hall PTR.

Sleeman, R., & Van Eck, T. (1999). Robust automatic P-phase picking: An on-line implementation in the analysis of broadband seismogram recordings. *Physics of the Earth and Planetary Interiors, 113*(1-4), 265–275. doi:10.1016/S0031-9201(99)00007-2

Specht, D. F. (1991). A general regression neural network. *IEEE Transactions on Neural Networks*, *2*(6), 568–576. doi:10.1109/72.97934 PMID:18282872

Thomas, R., & Poonguzhali, P. K. (2014). An energy efficient multi-sink clustering algorithm for wireless sensors networks. *International Journal of Advanced Engineering and Global Technology*, *2*(4).

Wang, F., & Liu, J. (2011). Networked wireless sensor data collection: Issues, challenges, and approaches. *IEEE Communications Surveys and Tutorials*, *13*(4), 673–687. doi:10.1109/SURV.2011.060710.00066

Yacoab, M. M., & Sundaram, V. (2011). Multiple sink based compressive data aggregation technique for wireless sensor network. *International Journal of Wireless & Mobile Networks*, *3*(2). doi:10.5121/ijwmn.2011.3215

Younis, O., & Fahmy, S. (2004). HEED: A hybrid, energy-efficient, distributed clustering approach for ad hoc sensor networks. *IEEE Transactions on Mobile Computing*, *3*(4), 366–379. doi:10.1109/TMC.2004.41

KEY TERMS AND DEFINITIONS

Base Station (BS): Base stations are one or more components of the wireless sensor networks with more energy, computational, and communication resources. Base station acts as a gateway between sensor nodes and end user where they forward information from the WSN to a server.

Cluster Heads (CH): Cluster head can be chosen randomly or based on some criteria in a cluster. Selecting of CH mainly effects on wireless sensor network lifetime. The optimal CH is the one which has highest residual energy, the maximum number of neighbor nodes, and minimum distance from BS.

Data Aggregation: Data aggregation is an actual way of avoiding redundant data transmission and it also minimizes the number of transmissions. In WSN, CH performs data aggregation on all the received data from cluster members and forwards the aggregated information to Base Station through single hop or multi-hop.

Load Balancing: Load balancing plays a vital role for prolonging the network lifetime. Uniform distribution of load among the CHs is important as to avoid hotspot problem. Uniform load distribution is assured in unequal clustering technique where, CHs consume approximately same amount of energy.

Multi-Sink: Maximum amount of energy is required to nodes in order to communicate data to the sink. For minimizing the energy consumption during transmission of data to the sink, multiple sinks are used.

Mobile-Sink: Mobile sink minimizes energy usage of sensor nodes and helps in minimizing data delivery delay for all the nodes. A mobile sink follows various kinds of mobility patterns in the sensor field, like predictable/fixed path mobility or controlled mobility, random mobility.

Wireless Sensor Network (WSN): WSN is a wireless network that contains spatially distributed self-governing devices using sensors to monitor environmental and physical conditions. The WSN system incorporates a gateway that supply wireless connectivity back to the wired world and distributed nodes.

Chapter 5
Review:
Effective Solutions for Challenges in Cognitive Radio Networks

Madhushi P. Ranasinghe
Charles Sturt University, Australia

Malka Halgamuge
The University of Melbourne, Australia

ABSTRACT

Cognitive radio technology (CRNs) will be the fundamental driving force behind the next generation (5G) mobile communication systems as it provides the optimal solution for the problem of spectrum scarcity via dynamic spectrum usage. The CRNs, however, pose several key challenges such as network management, spectrum allocation, and access, energy efficiency, interference, cost, spectrum sensing, security, and quality of service (QoS). In this chapter, the authors undertake a comprehensive analysis of 30 peer-reviewed scientific publications collated from 2017 to 2018 April that examine cognitive radio networks to identify practical solutions proposed to overcome critical challenges in this field. Nine distinct challenges were considered: network management, spectrum allocation, and access, energy efficiency, interference, cost, spectrum sensing, security, and QoS. The analysis demonstrates that the majority of research work related to CRN focuses on approaches to improve network management and, specifically, optimization of networks.

INTRODUCTION

The rapid growth in mobile communications, wireless technologies, big data, cloud computing and Internet of Things (IoT) has significantly raised the demand for both licensed and unlicensed spectrum bands. Therefore, increasing utilization of the limited radio frequency spectrum has become one of the biggest concerns in the communication world. The most prominent contribution towards fulfilling this objective is the development of Cognitive Radio Network (CRN). As its name implies, CRN is an intelligent radio network that can automatically detect available channels in a spectrum and then

DOI: 10.4018/978-1-5225-7458-3.ch005

dynamically change its reception and transmission parameters in order to allow more concurrent wireless communications in a given spectrum band (Sharma V., 2015). The main idea is to allow secondary users (SUs) who do not have access to the licensed spectrum, access to the licensed spectrum which is allocated for primary users (Pus). But it is important that this access does not cause any interference to the PUs' communication. Hence, it is apparent that minimizing interference with PUs is one of the key challenges in CRNs. Likewise, there are a number of challenging aspects that affect the performance of a CRN such as, accurate spectrum sensing, spectrum sharing and allocation, energy efficiency, scalability, mobility, cost, security, heterogeneity, scheduling, network management and quality of service (QoS). In order to get the maximum benefit out of this novel technology, it is important to realize the research gaps in this end and analyze the existing work to obtain an overall understanding of the research trends associated with CRNs and compare solutions provided in order to determine the most ideal solution to minimize the foregoing challenges.

Recent research focuses on these pressing challenges of CRNs and has introduced a wide range of mechanisms to reduce their impact and thus, improve the performance of CRNs. Some researchers have based their studies on specific applications. For example, Bräysy et al. (2017) discuss issues in CRN management in the context of military applications, Eze et al. (2017) focus on vehicular ad-hoc networks while Salameh et al. (2018) base their study on delay-sensitive IoT applications. Others discuss about the general concept of cognitive radio. Some other authors have paid attention to CRNs under various conditions such as security attacks (Salameh, Almajali, Ayyash, & Elgala, 2018), mobile networks (Bennaceur, Souihi, Idoudi, Saidane, & Mellouk, 2017) and multi-user networks (Wang, Ekin, & Serpedin, 2018). Very few researchers have performed survey or review articles on the concepts and issues of CRNs. For instance, Chen et al. (2018) have presented a comprehensive overview of the research contributions towards spectrum sharing process in CRNs that covers spectrum sensing, utilization, allocation and handoffs and related issues. Also Sumi and Ganesh (2017) analyzed techniques to enhance energy efficiency and network performance in their work.

An important concept is centralized spectrum sensing where a centralized server will estimate the spectrum access opportunity of a SU (Kim, Ko, Cha, & Kim, 2017). Other research has determined the effect of imperfect spectrum monitoring on the performance of a CRN (Thakur, Kumar, Pandit, Singh, & Satashia, 2017). Couturier et al. (2015) claim that cluster-based cooperative spectrum sensing can be applied to improve the performance of sensing process as clustering the SU network allows SUs to report sensing results to a common receiver with a spatial diversity. Furthermore, the study also defines what QoS is in the context of cognitive radio. QoS management in CRNs is similar to network management where appropriate network configuration such as routing and bandwidth allocation is allowed so that traffic with distinct QoS requirements can be controlled (Couturier, Krygier, Bentstuen, & Nir, 2015).

However, the fact that "to which extent these challenges are addressed" by the research and "what are the research trends" associated with CRNs seems to be under-researched. Further, the majority of authors so far have only discussed techniques to mitigate very few challenges of CRNs in their work, but no one discusses all challenging aspects in a single study.

Therefore, this chapter will play a significant role in overcoming the foregoing concerns by undertaking a comprehensive analysis of data from the 30 peer-reviewed scientific publications (2017-2018 April) that are published under CRNs. It will highlight the key challenges in cognitive radio networks that need to be addressed, determine the most suitable techniques to overcome those challenges by comparing solutions provided by each article and discover various research trends associated with challenges of CRNs.

The remainder of this chapter is as follows. In Section II we include raw data collection methodology, data inclusion criteria, and data analysis methods. Finally, the results of review on effective solutions for challenges in CRNs based on eight distinct challenges of CRNs including network management, spectrum allocation, and access, energy efficiency, interference, cost, spectrum sensing, security and QoS are presented in Section III, and in Section IV we discuss the results. Section V concludes the chapter. The raw data collected from 30 peer-reviewed publications used in this chapter are summarized in Table 3.

MATERIAL AND METHODS

This research is based on challenging areas of CRNs and effective solutions to overcome those challenges. This section describes the methodology that was used in this chapter.

Data Collection

The raw data for this chapter were collected from 30 peer-reviewed scientific publications including conference articles, journal articles and magazine articles published on various aspects of CRNs published from 2017 to mid-2018. These publications focused on various concepts and techniques to overcome the challenges in CRNs.

Data Inclusion Criteria

The data from the articles was compared using the following attributes: Article Reference Number, network management, spectrum allocation and access, energy efficiency, interference, cost, spectrum sensing, security and QoS. A detailed description of each attribute is presented in Table 1. These attributes were found to be the most significant contents in the published articles in regards to this topic.

Data Analysis

First, the comparison table was analyzed to assess the contribution of each publication towards the topic. There were a myriad of scientific publications that assess different challenging areas in CRNs. Studies that contributed only in less than 5 attributes and which were not rated as peer-reviewed publications were excluded. Then, the reviewed articles were clustered according to various criteria under each attribute (see Table 2).

A quantitative comparison for each challenging area of CRNs was made including the likelihood of each challenge being studied in research articles being illustrated graphically. This analysis revealed to what extent the challenges have been addressed, what were the most focused areas and what are the areas that need immediate attention and so on. This is important because only when more research and studies are devoted to this end, can we expect some breakthrough in this area.

Table 3 provides a summary of considerations, improvements and innovations related to challenging aspects of CRNs using data extracted from 30 peer-reviewed publications (2017-2018 April).

Table 1. Selected attribute description

Attribute Name	Description
Network Management	Efficient methods to manage performance, security, configuration, fault and administration in CRNs.
Spectrum Allocation and Access	Spectrum allocation and access techniques in CRNs. *Spectrum Allocation:* Spectrum is only allocated to selected set of SUs based on different criteria. For example, in auction game, the licensed spectrum is allocated by the PU to a certain SU after a bidding process. *Spectrum Access:* Explains how the spectrum is utilized by SUs with PUs at a given time. For example, in underlay mode, the SUs are permitted to utilize the licensed spectrum concurrently with PUs given that the SU's transmission power does not exceed the interference temperature which indicates the degree of interference that PU can tolerate caused by SUs during the spectrum access period (Chen, Hu, & Zhu, 2018).
Energy Efficiency	Protocols, algorithms or techniques to preserve energy in a CRN.
Interference	Different techniques to mitigate interference between PUs when accessing the licensed spectrum, interference caused by SUs to PUs when sharing the licensed spectrum, interference between SUs in multi-user access model and other forms of interferences such as self-interference in a SU due to close proximity between transmitting and receiving antenna and so on.
Cost	Methods to reduce cost of energy, bandwidth, computational overhead and other expenses like hardware costs.
Spectrum Sensing	Techniques used in spectrum sensing process. The spectrum sensing process can be explained in two stages as local spectrum sensing and sensing decision finalization. In local spectrum sensing, SU or sensor determine the state of the channel (idle or busy) in terms of PU activity. In spectrum decision finalization, SU confirms the state of channel either by itself or through the cooperation with other SUs or a Fusion Centre.
Security	The security threats in CRNs and methods to prevent or mitigate those.
QoS	Efficient techniques that enable CRNs to offer quality service to selected network traffic in terms of throughput, availability, latency, data loss, reliability and other aspects like traffic prioritization and reduced bit error rate.

RESULTS

The aim of this chapter was to review and evaluate the challenges in CRNs and the research contributions towards each aspect. The analysis of the data gathered from 30 peer-reviewed scientific publications during the 2017 to mid-2018 demonstrated eight distinct areas that can affect the performance of CRNs. These areas are network management, spectrum allocation and access, energy efficiency, interference, cost, spectrum sensing, security and QoS. An overview of the research contribution towards each area is illustrated in Figure 1. Furthermore, the research contributions (such as algorithms, techniques, protocols and important facts) in each of the foregoing areas were graphically represented in the Figures 2-9.

Analysis of Research Contribution Towards Each Key Challenging Attribute in CRNs

Figure 1 indicates that network management, spectrum sharing and allocation and QoS were the key areas of CRNs that have gained the attention of researchers. Out of the 30 research articles outlined in Table 1, network management was addressed in 97% of articles while spectrum sharing and allocation and QoS were addressed in 93% of studies. However, security and cost were only evaluated in 40% and 50% of publications respectively.

Table 2. Attribute and selected criteria

Attribute Name	Criteria
Network Management	1. Performance Management 2. Security Management 3. Fault Management 4. Optimization
Spectrum Allocation and Access	1. Spectrum Allocation Methods 2. Spectrum Access Methods
Energy Efficiency	1. Clustering 2. Energy Harvesting 3. Other Protocols
Interference	1. PU-PU Interference 2. SU-SU Interference 3. PU-SU Interference 4. Other
Cost	1. Energy Cost 2. Computational Cost 3. Bandwidth Cost 4. Other
Spectrum Sensing	1. Local Spectrum Sensing 2. Spectrum Decision Finalization
Security	1. Jamming Attacks 2. SSDF Attacks 3. Other
QoS	1. Network Throughput 2. Latency 3. Availability 4. Data Loss 5. Reliability 6. Other

Cognitive Radio Network Management

In Figure 2, it is clearly shown that 35% of the publications have suggested suitable solutions for optimizing CRNs, while 29% have provided effective measures to manage the performance of CRNs.

Spectrum Sharing and Allocation in CRNs

Figure 3 showcases that more than 70% of authors have discussed spectrum access methods in their work. On the other hand, only 29% of authors have paid attention to spectrum allocation methods.

Energy Efficiency in CRNs

Clustering and energy harvesting are two popular power allocation techniques used in current CRNs. However, as shown in Figure 3 most of the reported studies (73%) recommend a number of other different protocols to minimize energy consumption in CRNs and enhance the performance. Some of them are TTCM scheme, channel-aggregation diversity technology, relaying, CAODV protocol, primal-dual approach and joint power allocation techniques.

Table 3. Effective solutions to overcome challenges in Cognitive Radio Networks using data extracted from 30 peer-reviewed scientific publications (2017-2018 April)

Article	Network Management	Spectrum Allocation and Access	Energy Efficiency	Interference	Cost (Computational Cost, Energy Cost, Bandwidth Cost, Other Costs)	Spectrum Sensing	Security	QoS
1 Bräysy et al. (2017)	• Performance Management • Security Management • Fault Management (Utilize Cognitive Networks Engine and perform policy, fault, configuration, administration, performance and security management before, during and after the mission)		Other Protocols (Topology control)	SU-SU Interference (Topology control)	Computational Cost (Topology control)		Other (via Network wide end-to-end optimization)	• Latency • Reliability (Network wide end-to-end optimization)
2 Fathy et al. (2017)	Fault Management (Multi-route forwarding)	Spectrum Access Methods (SU is allowed to access any channel provided that PU is detected not to be using those)	Not Energy-efficient as more transmission energy is required			Sensing Decision Finalization (Cooperative distributed spectrum sensing)	SSDF Attacks (Multiple routes forwarding)	Reliability (Enhanced accuracy of sensing reports)
3 Thakur et al. (2017)	Fault Management (Traffic intensity and probability of monitoring error-based fault management to calculate data loss)	Spectrum Access Methods (Interweave approach)	Other Protocols (Parallel spectrum monitoring and data transmission)	PU-SU Interference (Parallel spectrum monitoring and data transmission)		Sensing Decision Finalization (Distributed spectrum sensing)		Data Loss (Modified frame structure)
4 S & Sudha (2017)	Performance Management (Double threshold censoring)		• Clustering (Sub clustering, special subset formation) • Other Protocols (Double threshold censoring, shortest distance-based CU (Cognitive User) and Sensor node joining algorithm)	PU-SU Interference (Increased minimum global detection probability)	Energy Cost (Energy efficiency optimization)	• Local Spectrum Sensing • Sensing Decision Finalization (Separate dedicated wireless sensor network with clusters, sub clusters and special subsets)		• Reliability (Double threshold censoring) • Other (Increased energy efficiency in spectrum sensing)

continued on following page

Table 3. Continued

Article	Network Management	Spectrum Allocation and Access	Energy Efficiency	Interference	Cost (Computational Cost, Energy Cost, Bandwidth Cost, Other Costs)	Spectrum Sensing	Security	QoS
5 Chen et al. (2018)	Performance Management (Queuing theory)	• Spectrum Allocation Methods (Game based spectrum allocation for low-latency applications and Career aggregation based spectrum allocation for massive-capacity applications, graph based spectrum allocation and auction based spectrum allocation) • Spectrum Access Methods (Queuing based spectrum access, overlay mode and hybrid mode)	• Energy Harvesting • Other Protocols (Spectrum database)	• PU-SU Interference (Power control, Interference Temperature Model in underlay mode, spectrum database) • Other (Non-orthogonal Multiple Access (NOMA))	Computational Cost (Utilization of Spectrum database minimizes the cost of analyzing mobile environmental parameters)	• Local Spectrum Sensing (Interference Temperature Model) • Sensing Decision Finalization (Cooperative sensing for low-latency applications and spectrum-database based sensing for wide-coverage applications, compressive sensing, full-duplex sensing and non-cooperative sensing)		Availability (Graph based spectrum allocation)
6 Liang et al. (2017)	• Performance Management (Cooperative communication is used to manage every task in the network) • Fault Management (Powerful maximum likelihood multiuser detection (MUD) to eliminate avalanche-like error propagation)	• Spectrum Allocation Methods (Game based spectrum allocation) • Spectrum Access Methods (Overlay spectrum access)	• Energy Harvesting • Other Protocols (Trellis Turbo Coded Modulation (TTCM) scheme)	• PU-PU Interference (Time Division Multiple Access (TDMA)) • Other (Invoke an advanced MUD technique at the relay node to cancel the self-interference)				Network Throughput (PUs lease their spectrum to a certain number of CUs and rely on CU's assistance)
7 Eze et al. (2017)	Performance Management (Traffic and network simulation software)	Spectrum Access Methods (Wireless Media Access Control (MAC) protocols based on opportunistic spectrum access do not guarantee fair and efficient spectrum sharing between competing SUs)		PU-SU Interference (Improved coordination between PUs and SUs)		Sensing Decision Finalization (Centralized, decentralized and hybrid spectrum sensing)	Jamming Attacks	Availability (Dynamic channel access scheme specific to cognitive radio assisted Vehicular Ad-Hoc Networks (VANETs))

continued on following page

Table 3. Continued

Article	Network Management	Spectrum Allocation and Access	Energy Efficiency	Interference	Cost (Computational Cost, Energy Cost, Bandwidth Cost, Other Costs)	Spectrum Sensing	Security	QoS
8 Wei & Hu (2018)	Performance Management (Fair-multi channel assignment scheme)	• Spectrum Allocation Methods (Fair-multi channel assignment scheme) • Spectrum Access Methods (Overlay spectrum access)	Other Protocols (Channel-aggregation diversity technology)	PU-SU Interference (Periodic spectrum sensing by the dedicated sensors)	• Energy Cost • Other (Channel-aggregation diversity technology)	Local Spectrum Sensing (New MAC framework and dedicated sensors)		
9 Sumi & Ganesh (2017)	Performance Management (Throughput and energy efficiency as performance measures)	Spectrum Access Methods (Underlay access for SUs far away from PU and Overlay access for SUs nearer to PU)	• Clustering • Other Protocols (Relaying, Double threshold censoring	PU-SU Interference (Power control)		Sensing Decision Finalization (Cooperative spectrum sensing)		• Network Throughput (Relaying) • Latency (Reduced sensing time)
10 Xie et al. (2018)	Security Management (Enhanced physical-layer security)	Spectrum Access Methods (Underlay spectrum access)		PU-SU Interference (Interference temperature model)			Other (Artificial noise transmission by PS where expectation signal is combined with artificial noise by superimposed coding to interfere eavesdropping)	Availability (New transmission protocol with artificial noise transmission improves outage performance for the secondary system)
11 Wang et al. (2018)	Performance Management (Prioritized Secondary User Access Control framework)	Spectrum Allocation Methods (Priority based spectrum allocation)	Other Protocols (Radiate jamming signal only when PU is idling)	SU-SU Interference (Oblique projection based jamming cancellation method)		Local Spectrum Sensing (Energy Detection method)	Jamming Attacks (Multi-tone jamming injection)	Other (Perfect/ Imperfect oblique projection to provide spectrum access priority levels among SUs)
12 Salameh et al. (2018)	Optimization (via minimized transmission invalidity ratio)	• Spectrum Allocation Methods (Probabilistic-based channel assignment) • Spectrum Access Methods (SUs opportunistically utilize same frequency bands owned by PR users)	• Other Protocols (Security solved at CRN provider so that no jamming detection stage or redundancy in IoT nodes required)	• SU-SU Interference (Multi-channel CSMA/CA (Carrier Sense Multiple Access/ Collision Avoidance)-based MAC protocol)		• Sensing Decision Finalization (cooperative spectrum sensing is performed with neighboring CR devices)	• Jamming Attacks (Probabilistic-based channel assignment, Maximum Likelihood Estimation (MLE)-based jamming detection)	• Network Throughput • Availability (Selection of channel with Minimum packet-invalidity ratio)

continued on following page

Table 3. Continued

Article	Network Management	Spectrum Allocation and Access	Energy Efficiency	Interference	Cost (Computational Cost, Energy Cost, Bandwidth Cost, Other Costs)	Spectrum Sensing	Security	QoS
13 Gottapu & Vallabhaneni (2018)		Spectrum Access Methods (SUs can only exploit a channel if the PU's current state is off)	Other Protocols (SU sense the assigned primary channels, CFC (Coordinator of Fusion Centers) and Div. N algorithm)	SU-SU Interference (CSMA/CA-based channel access)	Energy Cost (CFC and coordination algorithms)	• Local Spectrum Sensing (Energy detection) • Sensing Decision Finalization (Cooperative sensing-limited centralized topology)		Network Throughput (Div. Both algorithm)
14 Rajkumari & Marchang (2018)	Security Management (Secure Non-consensus based distributed spectrum sensing (NCSS) scheme with outlier detection)	Spectrum Access Methods (Interweave spectrum access)		PU-SU Interference (Improved coordination between PUs and SUs via cooperative spectrum sensing)	Computational Cost (Non-consensus scheme)	Sensing Decision Finalization (Non-centralized spectrum sensing)	SSDF Attacks (Secure NCSS scheme with outlier detection)	Availability (Secure NCSS scheme with outlier)
15 Youssef et al. (2017)	Performance Management (New power allocation mechanism for fair performance)	• Spectrum Access Methods (Underlay spectrum access) • Spectrum Allocation Methods (Centralized sub-carrier allocation scheme)	Other Protocols (New power allocation mechanism based on high fairness)	PU-SU Interference (Imposed IPC (Interference power constraint) on sub channels)		Sensing Decision Finalization (Centralized spectrum sensing)		Network Throughput (Centralized sub-carrier allocation scheme allocates the subcarrier to the SU with lowest interference channel gain)
16 Bharathi et al. (2018)	Security Management (EGT (Evolutionary Game Theory) -based anti-jamming defense module)	Spectrum Allocation Methods (Game based spectrum allocation)	Other Protocols (Selection of available low frequency bands for data transmission and not using distraction strategy for non-responsive jammers)	SU-SU Interference (FHSS (Frequency hopping spread spectrum) technique and Evolutionary game theory		Local Spectrum Sensing (Energy detection)	• Jamming Attacks (EGT-based anti-jamming defense module; Jammer distraction strategy for responsive jammer and mixed Nash Equilibrium for non-responsive jammer) • Other (Efficient encryption standards)	• Network Throughput • Other (EGT-based anti-jamming defense module)

continued on following page

Table 3. Continued

Article	Network Management	Spectrum Allocation and Access	Energy Efficiency	Interference	Cost (Computational Cost, Energy Cost, Bandwidth Cost, Other Costs)	Spectrum Sensing	Security	QoS
17 Fan et al. (2017)	• Performance Management (EE (energy efficiency) and HE (harvesting efficiency) as performance evaluators) • Optimization (Antenna diversity allows simultaneous spectrum sensing, data transmission and energy harvesting)	Spectrum Access Methods (Interweave spectrum access)	Energy Harvesting (RF-based)	Other (Passive self-interference suppression method, antenna polarization diversity)		Sensing Decision Finalization (Cooperative: Centralized spectrum sensing)		
18 Elnahas et al. (2018)	Optimization (Game theory framework, heuristic algorithm)	Spectrum Access Methods (each SU can use any of the M subands when the PU is absent)	• Energy Harvesting (the dedicated source RF energy harvesting) • Other Protocols (Energy causality constraint)	PU-SU Interference (Collision constraint)		• Local Spectrum Sensing (Energy detection) • Sensing Decision Finalization (Cooperative centralized spectrum sensing)		• Network Throughput (Evolutionary game model) • Reliability (Stackelberg game)
19 Zhang et al. (2018)	Optimization (Modified binary artificial bee colony algorithm)	• Spectrum Access Methods (Underlay spectrum access) • Spectrum Allocation Methods (Utilization-oriented spectrum allocation)		• PU-SU Interference (Resort to different channels other than the one used by PU) • SU-SU Interference (Suitable tuning of transmission power)	Computational cost (Ns=30 in MBABC (modified binary ABC) Algorithm)	Local Spectrum Sensing (Interference temperature model)		Availability (Utilization-oriented allocation model)
20 Bennaceur et al. (2017)	Optimization (Binary messaging between SU and DFC (Data fusion center))	Spectrum Access Methods (opportunistic spectrum accesses)	Other Protocols (DFC switch to the passive mode)	SU-SU Interference (GSS+ (an adaptation of game-based secure sensing) to prevent interference caused by SSDF attack)		Sensing Decision Finalization (cooperative SUs sending their local detections to a centralized DFC which makes a final decision about the PU availability)	SSDF Attacks (GSS+)	• Network Throughput • Other (Trust game-based model)
21 Sabbah& Ibnkahla (2017)	Optimization (CLH (cap-limited heuristic) algorithm)	Spectrum Access Methods (Hybrid: interweave/ underlay spectrum access)	Other Protocols (Primal-Dual approach)	PU-SU Interference (interference threshold-based power allocation)	Computational cost (CLH algorithm)			Network Throughput (Primal-Dual approach)

continued on following page

Table 3. Continued

Article	Network Management	Spectrum Allocation and Access	Energy Efficiency	Interference	Cost (Computational Cost, Energy Cost, Bandwidth Cost, Other Costs)	Spectrum Sensing	Security	QoS
22 Ye & Zhang (2017)	Security Management (Evidence-theory-based CSS (collaborative spectrum sensing) scheme with efficient creditability)	Spectrum Access Methods (opportunistic spectrum accesses)	Other Protocols (RMD (Rational mass distribution) method)		Bandwidth Cost (RMD method)	• Local Spectrum Sensing (Energy detection) • Sensing Decision Finalization (Centralized spectrum sensing)	SSDF Attacks (Evidence-theory-based CSS scheme with efficient creditability evaluation)	Reliability (Evidence-theory-based CSS scheme with efficient creditability evaluation)
23 Wang et al. (2018)	• Optimization (Optimal and suboptimal monotonic optimization techniques) • Fault Management (Sensing error constraint)	Spectrum Access Methods (Interweave spectrum access)		• SU-SU Interference (Single SU occupied channels) • PU-SU Interference (Interference threshold)	Computational cost (Suboptimal monotonic algorithm)	Sensing Decision Finalization (Multiband joint detection)		Network Throughput (Joint spectrum sensing and resource allocation)
24 Jaya et al. (2017)	Optimization (Joint optimization algorithm)	Spectrum Access Methods (SUs can data transmits to the base station during idle period of PUs)	Other Protocols (Joint optimization algorithm and cooperative decision-making)			• Local Spectrum Sensing (Energy detection) • Sensing Decision Finalization (Cooperative centralized spectrum sensing)		Availability (Joint optimization algorithm)
25 Abhipriya et al. (2017)	Fault Management (RERR (route request message))	Spectrum Access Methods (opportunistically allows users to access white spaces in frequency spectrum)	• Energy Harvesting (Radio identification-based operations) • Other Protocols (CAODV (Cognitive Ad hoc On Demand Distance Vector) protocol)	PU-SU Interference (CAODV protocol)	Other (CAODV protocol reduces packet drop)	Sensing Decision Finalization (Spectrum sensing technique in CAODV)		• Network Throughput • Latency • Data Loss (CAODV protocol) Other (PU will be prioritized to use single hop communication while CU use multi-hop communication using radio identification technology)
26 Sharma et al. (2017)	Optimization (Genetic algorithm)	Spectrum Allocation Methods (GA (Genetic algorithm)-based channel allocation)		PU-SU Interference (PMX (partial mapped crossover) and mutation)	Computational Cost	Sensing Decision Finalization (Distributed spectrum sensing)		Other (Genetic algorithm)
27 Sriharipriya & Sanju (2017)	Optimization (Levenberg-Maquardt ANN (artificial neural network) implementation method)	Spectrum Access Methods (The SU has to function the PU action for a little duration and accordingly appropriate the slot when it is sensed empty)	Other Protocols (Levenberg-Maquardt ANN implementation method)		Computational Cost (Incorporating SU info in fusion process)	Sensing Decision Finalization (Cooperative distributed spectrum sensing)		Reliability (Incorporating SU info in fusion process)

continued on following page

Table 3. Continued

Article	Network Management	Spectrum Allocation and Access	Energy Efficiency	Interference	Cost (Computational Cost, Energy Cost, Bandwidth Cost, Other Costs)	Spectrum Sensing	Security	QoS
28 Lan et al. (2018)	Optimization (Joint power allocation with I-CSI/S-CSI (instantaneous channel state information / statistical channel state information))	Spectrum Access Methods (Underlay spectrum access)	Other Protocols (Joint power allocation with I-CSI/S-CSI)	PU-SU Interference (Power control)				Other (Power control)
29 Jayasri & Elanagai (2017)	Optimization (EE-CSS (Energy efficient collaborative spectrum sensing))	Spectrum Access Methods (Interweave spectrum access)	• Clustering (including Sub clustering) • Other Protocols (Data fusion scheme)	SU-SU Interference (MAC protocol)	Bandwidth Cost (Hybrid routing algorithm)	• Local Spectrum Sensing (Energy detection) • Sensing Decision Finalization (Cooperative centralized spectrum sensing)	SSDF Attacks (TRMS (Trust and Reputation Management System))	Latency (Hybrid routing algorithm)
30 Kotobi & Bilén (2018)	Security Management (Blockchain verification protocol)	• Spectrum Allocation Methods (Auction mechanism-based spectrum allocation) • Spectrum Access Methods (Interweave spectrum access)	Other Protocols (Currency conversion mechanisms)	PU-SU Interference (Blockchain verification protocol)	• Computational Cost (Reset blockchain) • Other (Distributed topology eliminates the need of firewalls)	No need for spectrum sensing	Other (Blockchain verification protocol)	• Network Throughput • Availability (Blockchain verification protocol)

Figure 1. "Research Contribution" Overview of the percentage of scientific publications addressing each challenging aspect of CRNs
Pooling the data from the 30 peer-reviewed scientific articles published in 2017-2018 April

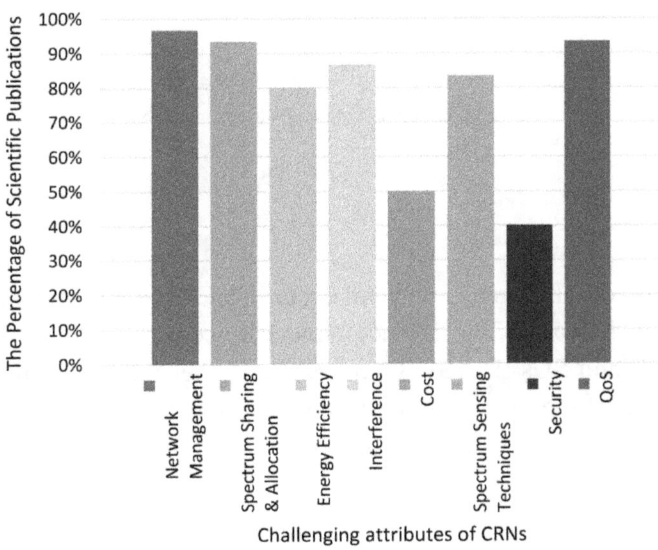

Figure 2. "Network Management" The research contribution under key management functions in CRNs
Pooling the data from the 30 peer-reviewed scientific articles published in 2017-2018 April.

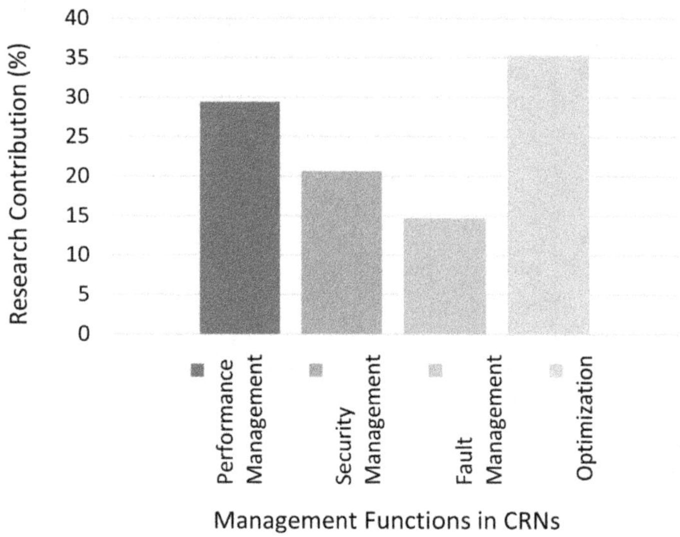

Figure 3. "Spectrum Sharing and Allocation" The comparison between research contribution towards spectrum access methods and spectrum allocation methods
Pooling the data from the 30 peer-reviewed scientific articles published in 2017-2018 April

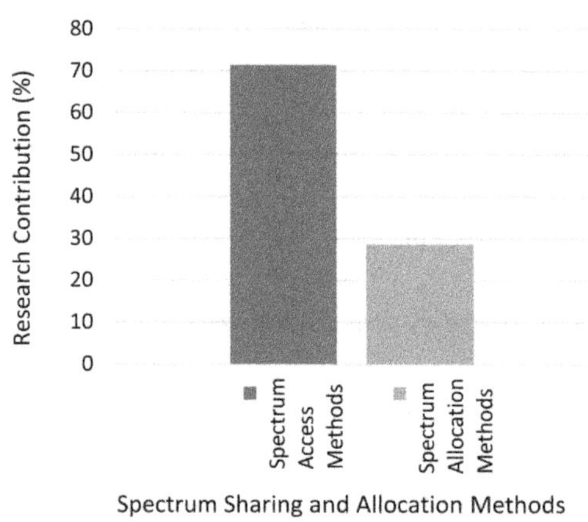

Interference in CRNs

There are three main types of interference that can occur in a CRN environment. They can be identified as interference between PUs (PU-PU Interference), interference between SUs (SU-SU Interference) and interference between PUs and SUs (PU-SU Interference). Other than these three, SUs can also suffer from self-interference due to the close proximity between their transmitting and receiving antennas. Significantly, the majority (57%) of studies focused on PU-SU interference mitigation techniques.

Figure 4. "Energy Efficiency" The analysis of energy-efficient power allocation techniques used in CRNs
Pooling the data from the 30 peer-reviewed scientific articles published in 2017-2018 April

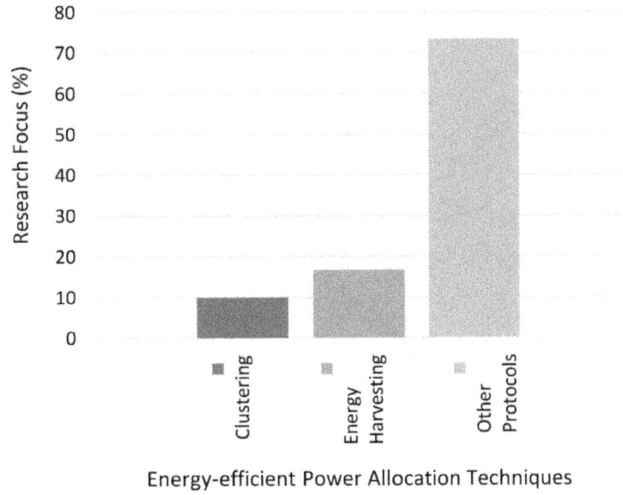

Figure 5. "Interference" The categorization of interference mitigation techniques
Pooling the data from the 30 peer-reviewed scientific articles published in 2017-2018 April

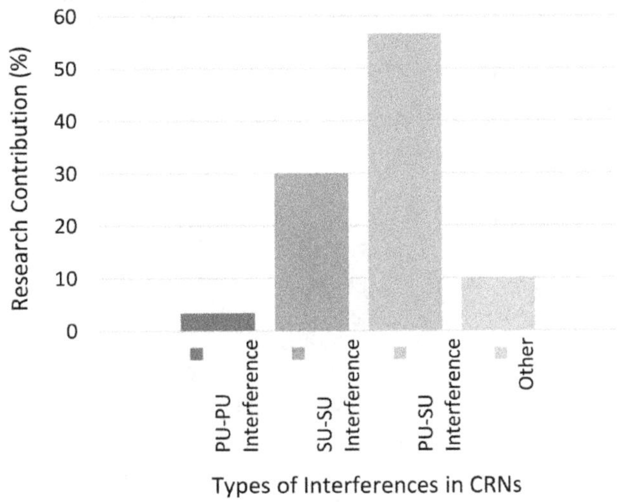

Costs Associated With CRNs

According to Figure 6, 53% of reported studies identified computational cost as one of the significant cost components in a CRN and discussed ways to minimize the computational overhead in CRNs. Specifically, 18% of authors discussed how to reduce other costs using such techniques as channel-aggregation diversity technology, CAODV protocol and distributed topology. The energy costs were discussed in 17% of the publications and 12% described how to mitigate bandwidth costs.

Figure 6. "Cost" The analysis of various types of costs associated with CRNs
Pooling the data from the 30 peer-reviewed scientific articles published in 2017-2018 April

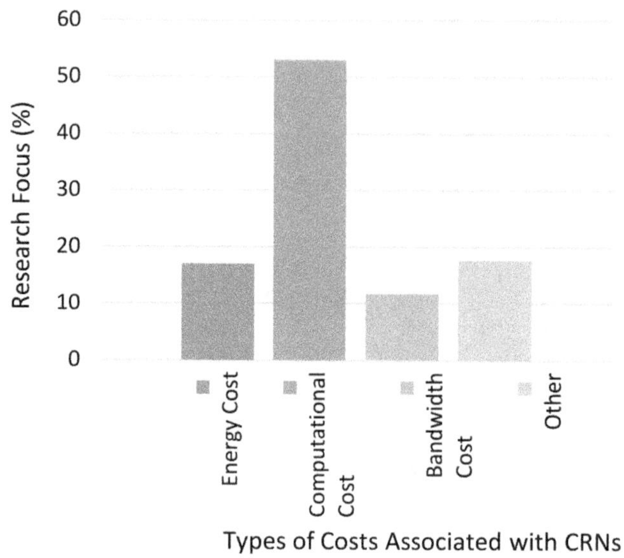

Figure 7. "Spectrum Sensing" The analysis of spectrum sensing process in cognitive radio networks
Pooling the data from the 30 peer-reviewed scientific articles published in 2017-2018 April

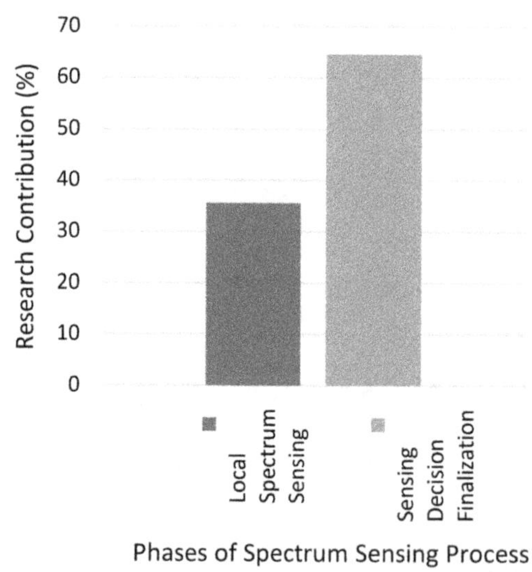

Spectrum Sensing Process in CRNs

The spectrum sensing process can be divided into two stages: local spectrum sensing and sensing decision finalization. Local spectrum sensing is a task performed by each SU to identify the spectrum holes in the licensed spectrum used by PUs. After performing local spectrum sensing, the SUs can either take the sensing decision by themselves or cooperate with a Fusion Centre or other SUs to take a more accurate decision. There are four primary methods for local spectrum sensing as energy detection, feature detection, matched filtering and coherent detection (Sharma V., 2015) and interference temperature model (used in underlay mode). The spectrum sensing decision is finalized either via non-cooperative sensing or cooperative sensing (centralized, distributed or limited centralized). Among the 30 reviewed publications, 65% of authors have explained how the spectrum sensing decision is finalized, while 35% have mentioned about how the local spectrum sensing is being performed. Moreover, Table 1 shows that the Energy detection method is a widely used technique for local spectrum sensing by secondary/ cognitive users and cooperative spectrum sensing is recommended for sensing decision finalization.

Security in CRNs

The most prominent security threats in CRNs are jamming attacks and SSDF attacks. Therefore 38% and 31% of authors have drawn their attention to jamming and SSDF attacks. Jamming attacks can interfere with the legitimate wireless access communications to adversely affect and degrade the overall QoS (Eze, Zhang, Liu, & Eze, 2017) of networks. In SSDF attacks, the attackers report falsified sensing information in order to corrupt the functioning of cooperative spectrum sensing process (Rajkumari & Marchang, 2018). Other than jamming and SSDF attacks, there are various security issues associated with CRNs such as confidentiality, authentication, spoofing attack, PU emulation attack and denial of service attack (DOS) (Salameh, Almajali, Ayyash, & Elgala, 2018). Some of the authors (31%) have explained how to prevent or mitigate these security issues in their studies.

Figure 8. "Security" The analysis of major security threats to CRNs and effective security measures
Pooling the data from the 30 peer-reviewed scientific articles published in 2017-2018 April

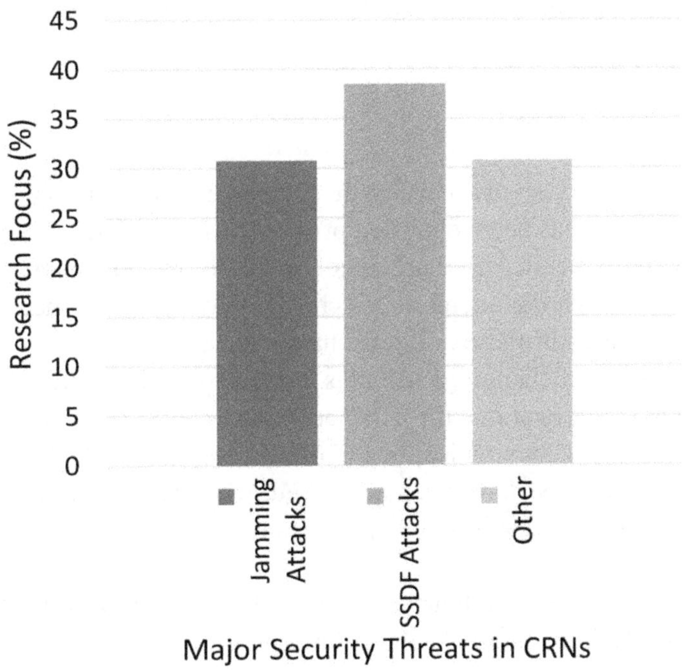

Figure 9. "QoS" The analysis of QoS considerations in CRNs and associated QoS-enhancing techniques
Pooling the data from the 30 peer-reviewed scientific articles published in 2017-2018 April

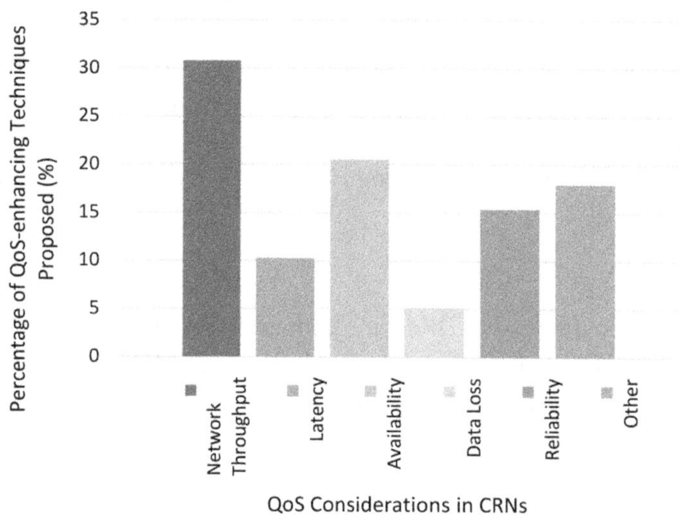

Achieve QoS in CRNs

Quality of service is one of the essential aspects in CRNs. QoS can be achieved in CRNs in terms of network throughput, availability, reliability and low latency and data loss. Other QoS considerations in a CRN can include traffic prioritization and protection of SU transmission at the PU presence. According to Figure 9, 31% of publications have introduced effective techniques to improve network throughput and similarly, 21% of publications have focused on improving availability in CRNs. Reliability enhancing methods are presented in 15% of the reviewed articles while latency mitigation methods are explained in 10%. However, only 5% of authors have addressed the issue of data loss.

DISCUSSION

CRNs are highly useful in the process of satisfying the critical requirements of the next generation mobile networks (5G). In particular, it has been identified primarily as a promising solution for the problem of radio spectrum scarcity. With the increased utilization of CRNs, the challenges that are associated with this technology have also drawn the attention of scholars. However, the mechanisms and algorithms through which CRNs can overcome these different challenges remain unclear as the majority of publications only focus on one or two challenging attributes in their work. Identification of effective solutions to meet all these challenges will facilitate the enhancement of cognitive radio technology and therefore pave the way for more advanced communications. Furthermore, the results indicate the extent to which each challenging attribute of CRNs have been addressed, what were the most focused areas and what areas need immediate attention so that it will help interested authors to identify potential research areas associated with CRNs.

In this chapter, data from 30 research articles was collected, analyzed in order to identify the key challenging attributes in CRNs and suitable algorithms, protocols, methods or techniques to overcome them. After a detailed analysis using a comparison table, it was found that the majority of reviewed articles presented novel techniques to improve network management process, spectrum sharing and allocation and QoS in CRNs. On the other hand, very few focus on finding solutions for cost and security related issues. This denotes that the research trend under the topic "challenges in CRNs" is more biased towards investigating network management, QoS and spectrum sharing and allocation issues in CRNs.

Our analysis from the reported studies provides an insight into each challenging attribute by further dividing them into subcategories. Among network management activities, optimization of networks is seen as the key concern in most of the existing work. More studies interrogated spectrum access methods which include overlay, underlay, interweave and hybrid spectrum access over spectrum allocation techniques. Many authors discussed improving network throughput as a means of enhancing QoS. The studies that explained interference in CRNs paid more attention on mitigating interference between PU and SU. The cost in CRNs was also identified as energy consumption, computational complexity, bandwidth, and hardware. However, the most prominent type of cost was computational cost which covers most of the expenses in CRNs as spectrum sensing, sharing, optimizing and security measures require a lot of computational power. The primary security issues in CRNs can be listed as jamming attacks, SSDF attacks, privacy and confidentiality issues and investigating techniques to prevent SSDF attacks was the highlight in many research articles. Spectrum sensing is one of the primary tasks in a CRN and it appears that many studies focused on approaches for successful spectrum sensing decision finalization over spectrum sensing process at individual nodes.

Our findings also suggest the use of customized protocols for efficacious energy consumption over clustering and energy harvesting methods. For instance, RMD approach is proposed by an author to reduce the amount of data transmitted between SUs and the fusion center by half and save the data transmission energy (Ye & Zhang, 2017). Another study suggested the use of spectrum database to save energy as the database can eliminate the requirement for every SU to detect the spectrum continuously (Chen, Hu, & Zhu, 2018). Our analysis from the reported studies highlights the research gaps in cost and security analysis of CRNs. According to Salameh et al. (2018), when security-unaware channel assignment schemes are employed in a CRN that are vulnerable to jamming attacks, the number of forced-terminated cognitive radio packets will increase due to jamming attacks, resulting in severe degradation in network availability, poor network throughput and wasted spectrum opportunities. Hence, it is necessary to focus on methods to prevent or mitigate security issues and reduce various costs in CRNs while promoting other aspects. Moreover, there were only two articles among 30 reviewed articles that discussed in their studies about all the eight challenging attributes that this chapter is based on. Hence, it is evident that the majority only addressed few attributes. For an instance, Fathy et al. (Fathy, Tammam, & Saafan, 2017) consider only about security and reducing errors and claims that their approach is not energy efficient. Nevertheless, this chapter gathers all solutions in a single document so that readers could identify the optimal solution based on its effectiveness in addressing all attributes. Because, in order to get the maximum benefit out of this novel technology, it is important to minimize all its existing issues and find the most ideal solutions for them.

Although results do not provide a clear perspective of the impact of various types of interferences, security attacks and costs on CRNs, they enable readers to anticipate it by observing the amount of research contribution under each parameter.

To summarize, this research work highlights the key challenges in CRNs, presents novel approaches provided by literature to address them and analyze to which extent those aspects have been addressed by the literature. Furthermore, our analysis discover research trends associated with CRNs while highlighting gaps in the respective research area. Nevertheless, this reinforces the need for discovering more potential approaches to overcome all the key challenges in cognitive radio technology without any tradeoffs.

FUTURE RESEARCH DIRECTIONS

Cognitive radio networking has gained a lot of attention of academic and industry researchers in the field of wireless communication. One of the emerging trends in the field is to employ machine learning methods to enhance spectrum sharing, access and cognitive communications. Sriharipriya & Sanju [30] have proposed the implementation of a neural network to maximize the energy efficiency and performance in CRNs. Clancy et al. (2007) have described the application of machine learning to tackle capacity maximization and dynamic spectrum access issues. Anandakumar and Umamaheswari (2017) explain how supervised machine learning techniques could assist cooperative spectrum handovers. However, another important area that machine learning applications could benefit CRNs is security. Our findings show that the majority of research work focus on detection and mitigation of jamming attacks, SSDF attacks, privacy and other security issues in CRNs via artificial noise injecion or multiple route forwarding mechanisms. But we could corperate machine learning, specifically, reinforcement learning to identify and prevent eavesdroppers, jammers and other types of malicious attacks to CRNs and balance the trade-offs between energy, cost and security. Furthermore, cloud based cognitnive radio networking and crowdsourcing based cognitnve radio spectrum sensing and access could also be considered as trending resaerch aspects in the field.

Our findings highlight the key challenges in CRNs. However, it does not cover other remaining challenges of CRNs such as scalability, mobility, signal processing, fragmentation, and scheduling. Moreover, cognitive radio networking involves some open research issues due to heterogeneity and unpredictable nature of availability of channels, incompatibility with conventional MAC layer protocols and complexity in route optimization. Investigating these challenges in CRNs could be an interesting avenue to explore in the future.

CONCLUSION

In this chapter, we performed a quantitative analysis of the data obtained from 30 peer-reviewed scientific publications (2017 to mid-2018) describing effective solutions that can be taken to mitigate key challenges in CRNs. The results highlighted the research gaps under the project title, trends associated with CRNs and comparison of the solutions provided by each study to face challenges of CRNs more effectively in order to determine the optimal solutions. Our observations showed significant contribution of research towards network management (97%), quality of service (93%) and spectrum sharing and allocation (93%) in CRNs; however, on the other hand, there was a considerable gap in research in terms of security and costs of CRNs. The chapter also found the novel techniques utilized in CRNs for network management, energy saving and enhancing quality of service. This chapter also provides an insight into how various types of interferences: interference between primary and SUs appears to be a major concern, costs, especially computational costs and security threats, mainly Spectrum sensing data falsification attacks could impact the operation of CRNs. Furthermore, our data collection presents a number of spectrum sharing and allocation techniques and spectrum sensing approaches. The comprehensive analysis presented in this chapter may be used by researchers to determine the optimal solution and develop advanced CRN design concepts as it discusses all challenging aspects in a single document, unlike the majority of authors so far who have only discussed one or two aspects of CRNs in their work. However, there are still some issues with the use of CRNs such as scalability, mobility, scheduling, signal processing and fragmentation that need to be addressed, but have not yet drawn the authors' attention. This research can be further extended to examine those challenges.

REFERENCES

Abhipriya, R., Aysha, S., Gayathri, K., & Kathiravan, K. (2017). 3S: A Radio Identification based Continuous Spectrum Sensing Protocol for Safety of Women in Cognitive Radio Networks. In *2017 International Conference on Communication and Signal Processing (ICCSP)* (pp. 2042 - 2046). Chennai: IEEE. 10.1109/ICCSP.2017.8286762

Anandakumar, H., & Umamaheswari, K. (2017). Supervised machine learning techniques in cognitive radio networks during cooperative spectrum handovers. *Cluster Computing*, *20*(2), 1505–1515. doi:10.100710586-017-0798-3

Bennaceur, J., Souihi, S., Idoudi, H., Saidane, L. A., & Mellouk, A. (2017). Game-based Secure Sensing for the Mobile Cognitive Radio Network. In *2017 IEEE 28th Annual International Symposium on Personal, Indoor, and Mobile Radio Communications (PIMRC)* (pp. 1 - 7). Montreal: IEEE.

Bharathi, S., Kumar, D., & Ram, D. (2018). Defence against responsive and non-responsive jamming attack in cognitive radio networks: an evolutionary game theoretical approach. *The Journal of Engineering, 2018*(2), 68 - 75.

Bräysy, T., Couturier, S., Smit, N., Le Nir, V., Tuukkanen, T., Verheul, E., ... Krygier, J. (2017). Network management issues in military cognitive radio networks. *International Conference on Military Communications and Information Systems (ICMCIS)*. 10.1109/ICMCIS.2017.7956487

Chen, B., Hu, F., & Zhu, K. (2018). Full Spectrum Sharing in Cognitive Radio Networks toward 5G: A Survey. *IEEE Access: Practical Innovations, Open Solutions*, (99), 1–1.

Clancy, C., Hecker, J., Stuntebeck, E., & O'Shea, T. (2007). Applications of Machine Learning to Cognitive Radio Networks. *IEEE Wireless Communications*, *14*(4), 47–52. doi:10.1109/MWC.2007.4300983

Couturier, S., Krygier, J., Bentstuen, O. I., & Nir, V. L. (2015). Challenges for Network Aspects of Cognitive Radio. *2015 International Conference on Military Communications and Information Systems (ICMCIS)*. 10.1109/ICMCIS.2015.7158680

Elnahas, O., Elsabrouty, M., Muta, O., & Furukawa, H. (2018). Game Theoretic Approaches for Cooperative Spectrum Sensing in Energy-Harvesting Cognitive Radio Networks. *IEEE Access: Practical Innovations, Open Solutions*, *6*, 11086–11100. doi:10.1109/ACCESS.2018.2810107

Eze, J., Zhang, S., Liu, E., & Eze, E. (2017). Cognitive Radio Technology assisted Vehicular Ad-Hoc Networks (VANETs): Current Status,Challenges, and Research Trends. In *23rd International Conference on Automation and Computing (ICAC)* (pp. 1 - 6). Huddersfield: IEEE. 10.23919/IConAC.2017.8082035

Fan, M., Zhang, H., Hu, H., Li, J., & Wang, Q. (2017). Cooperative Spectrum Sensing in RF-powered Full-Duplex Cognitive Radio Networks. In *2017 First International Conference on Electronics Instrumentation & Information Systems (EIIS)* (pp. 1 - 6). Harbin: IEEE. 10.1109/EIIS.2017.8298712

Fathy, M., Tammam, A., & Saafan, A. (2017). Mitigating the Impact of Malicious Behavior via Utilizing Multiple Routes in a Cooperative Sensing Cognitive Radio Network. *IEEE 15th Student Conference on Research and Development (SCOReD).*

Gottapu, S. K., & Vallabhaneni, P. (2018). Coordination between Multiple Collocated Cognitive Radio Networks. In *2018 Conference on Signal Processing And Communication Engineering Systems (SPACES)* (pp. 33 - 36). Vijayawada: IEEE. 10.1109/SPACES.2018.8316310

Jaya, T., Raaja, V. K., Sharanya, C., & Rajendran, V. (2017). An Efficient Power Allocation and Joint Optimal Sensing of Spectrum in CDMA-based Cognitive Radio Networks. In *2017 International Conference on Communication and Signal Processing* (pp. 1769 - 1773). Chennai: IEEE. 10.1109/ICCSP.2017.8286697

Jayasri, C., & Elanagai, G. (2017). Implementation of network security based data hauling by collaborative spectrum sensing in cognitive radio network. In *2017 International Conference on Innovations in Information, Embedded and Communication Systems (ICIIECS)* (pp. 1 - 5). Coimbatore: IEEE.

Kim, J., Ko, S.-W., Cha, H., & Kim, S.-L. (2017). Testbed Verification of Spectrum Access Opportunity Detection in Cognitive Radio Networks. *23rd Asia-Pacific Conference on Communications (APCC).* 10.23919/APCC.2017.8304086

Kotobi, K., & Bilén, S. G. (2018). Secure Blockchains for DYNAMIC Spectrum Access. *IEEE Vehicular Technology Magazine*, *13*(1), 32–39. doi:10.1109/MVT.2017.2740458

Lan, P., Zhai, C., Chen, L., Gao, B., & Sun, F. (2018). Optimal power allocation for bi-directional full duplex underlay cognitive radio networks. *IET Communications*, *12*(2), 220–227. doi:10.1049/iet-com.2016.1384

Liang, W., Ng, S. X., & Hanzo, L. (2017). Cooperative Overlay Spectrum Access in Cognitive Radio Networks. *IEEE Communications Surveys and Tutorials*, *19*(3), 1924–1944. doi:10.1109/COMST.2017.2690866

Rajkumari, R., & Marchang, N. (2018). Secure Non-consensus Based Spectrum Sensing in Non-Centralized Cognitive Radio Networks. *IEEE Sensors Journal*, *18*(9), 3883–3890. doi:10.1109/JSEN.2018.2816595

S, J. P., & Sudha, T. (2017). Energy Efficient Wireless Sensor Network Assisted Spectrum Sensing for Cognitive Radio Network. *IEEE International Conference on Intelligent Techniques in Control, Optimization and Signal Processing (INCOS).*

Sabbah, A., & Ibnkahla, M. (2017). Optimizing Power Allocation in Mission Critical Cognitive Radio Networks. In *2017 IEEE 28th Annual International Symposium on Personal, Indoor, and Mobile Radio Communications (PIMRC)* (pp. 1-7). Montreal: IEEE.

Salameh, H. B., Almajali, S., Ayyash, M., & Elgala, H. (2018). Spectrum Assignment in Cognitive Radio Networks for Internet-of-Things Delay-sensitive Applications under Jamming Attacks. *IEEE Internet of Things Journal.*

Sharma, D. K., Singh, A., Khanna, A., & Jain, A. (2017). Evaluation of Parameters and Techniques for Genetic Algorithm based Channel Allocation in Cognitive Radio Networks. In *2017 Tenth International Conference on Contemporary Computing (IC3)* (pp. 1 - 6). Noida: IEEE. 10.1109/IC3.2017.8284341

Sharma, V. (2015, May 21). *Cognitive radio networks: Technology Survey & Research Challenges.* Retrieved April 15, 2018, from https://www.slideshare.net/vatsalasharma180/cognitive-radio-networks-48446892

Sriharipriya, K. C., & Sanju, R. (2017). Artifical Neural Network Based Multi Dimensional Spectrum Sensing In Full Duplex Cognitive Radio Networks. In *2017 International Conference on Computing Methodologies and Communication (ICCMC)* (pp. 307 - 312). Erode: IEEE. 10.1109/ICCMC.2017.8282697

Sumi, M. S., & Ganesh, R. S. (2017). Performance Enhancing Techniques in Cognitive Radio Networks. In *IEEE International Conference on Circuits and Systems (ICCS 2017)* (pp. 172 - 178). Kerala: IEEE. 10.1109/ICCS1.2017.8325984

Thakur, P., Kumar, A., Pandit, S., Singh, G., & Satashia, S. N. (2017). Effect of Imperfect Spectrum Monitoring on Cognitive Radio Network Performance. *Fourth International Conference on Image Information Processing (ICIIP).* 10.1109/ICIIP.2017.8313716

Wang, H., Yao, Y.-D., & Peng, S. (2018). Prioritized Secondary User Access Control in Cognitive Radio Networks. *IEEE Access: Practical Innovations, Open Solutions*, *6*, 11007–11016. doi:10.1109/ACCESS.2018.2811341

Wang, X., Ekin, S., & Serpedin, E. (2018). Joint Spectrum Sensing and Resource Allocation in Multi-Band-Multi-User Cognitive Radio Networks. *IEEE Transactions on Communications*, 1–1.

Wei, Z.-H., & Hu, B.-J. (2018). A Fair Multi-Channel Assignment Algorithm With Practical Implementation in Distributed Cognitive Radio Networks. *IEEE Access: Practical Innovations, Open Solutions*, *6*, 14255–14267. doi:10.1109/ACCESS.2018.2808479

Xie, P., Zhang, M., Zhang, G., Zheng, R., Xing, L., & Wu, Q. (2018). On physical-layer security for primary system in underlay cognitive radio networks. *IET Networks*, *7*(2), 68–73. doi:10.1049/iet-net.2017.0138

Ye, F., & Zhang, X. (2017). Evidence-theory-based Collaborative Spectrum Sensing with Efficient Creditability Evaluation in Cognitive Radio Networks. In *Progress in Electromagnetics Research Symposium - Fall (PIERS - FALL), 2017* (pp. 46 - 54). Singapore: IEEE.

Youssef, K., Messiha, N., & Abd-Elnaby, M. (2017). Throughput and fairness enhancement based resource allocation scheme for underlay cognitive radio networks. In *2017 Japan-Africa Conference on Electronics, Communications and Computers (JAC-ECC)* (pp. 86 - 90). Alexandria: IEEE. 10.1109/JEC-ECC.2017.8305780

Zhang, X., Zhang, X., Han, L., & Xing, R. (2018). Utilization-Oriented Spectrum Allocation in an Underlay Cognitive Radio Network. *IEEE Access: Practical Innovations, Open Solutions*, *6*, 12905–12912. doi:10.1109/ACCESS.2018.2808473

KEY TERMS AND DEFINITIONS

Cooperative Spectrum Sensing: The process used by secondary user (SU) to confirm the state of channel (idle or busy) through the cooperation with other SUs or a fusion center.

Energy Detection: A method of spectrum sensing used by SU to identify the state of channel by itself. The knowledge of signal and noise power is used to identify the presence or absence of a PU.

Game-Based Spectrum Allocation: A method of allocating spectrum among PUs and SUs based on the game theory which includes players, strategies, and pay-offs.

Hybrid Spectrum Access: The method that allows SU to switch between overlay, underlay and interweave spectrum access methods based on the state of PUs.

Interweave Spectrum Access: A method that allows SU access to the licensed spectrum only when PU is not utilizing the channel.

Overlay Spectrum Access: A method of enabling SU and PU concurrent access to the spectrum by offsetting the effect of interference caused by SU using a part of SU's power for relaying PU's information.

Underlay Spectrum Access: A method of enabling SU and primary user PU concurrent access to the spectrum based on the degree of interference that PU can tolerate during the SU's spectrum access period.

Chapter 6
Overview of Machine Learning Approaches for Wireless Communication

Tolga Ensari
Istanbul University, Turkey

Melike Günay
Istanbul Kultur University, Turkey

Yağız Nalçakan
Altinbas University, Turkey

Eyyüp Yildiz
Erzincan University, Turkey

ABSTRACT

Machine learning is one of the most popular research areas, and it is commonly used in wireless communications and networks. Security and fast communication are among of the key requirements for next generation wireless networks. Machine learning techniques are getting more important day-by-day since the types, amount, and structure of data is continuously changing. Recent developments in smart phones and other devices like drones, wearable devices, machines with sensors need reliable communication within internet of things (IoT) systems. For this purpose, artificial intelligence can increase the security and reliability and manage the data that is generated by the wireless systems. In this chapter, the authors investigate several machine learning techniques for wireless communications including deep learning, which represents a branch of artificial neural networks.

INTRODUCTION

Convolutional neural networks and deep learning is very popular not only for wireless communication but also other areas. Deep learning is a type of machine learning that researchers have shown interested in recent years because of its recognition performance. Nowadays, it is mostly preferable by the sci-

DOI: 10.4018/978-1-5225-7458-3.ch006

entists for improving the performance of learning techniques and building intelligent systems to help people. Intelligent systems on speech recognition, pattern recognition, image identification, robotics, bioinformatics, and others are used for different areas like computer vision, computer games, robots, cars, medical systems, financial forecasting, information retrievals, medical diagnosis, natural language processing, and online advertising. Recent developments in technology and human requirements need more intelligent systems. Therefore, the importance of artificial intelligence is getting more important.

In our technologically based daily life, we generate and record data. This data continues to accumulate and grow, whether we like it or not. This is called 'Big Data' as a reference to its size. Although big data is constantly growing, it is not possible to make sense of this data as long as we do not implement the necessary methods. Many studies proposed by scientists to manipulate the data are being studied for their proper processing tools or methods.

In this chapter, we will investigate several machine learning techniques for wireless communications including deep learning (Brunato & Battiti 2005; Callado et al., 2009; Demestichas & Dimitrakopoulos 2006; Haykin 2005; Hussain et al., 2009; Pinto & Montez 2010; Wang 2007; Zhang et al., 2009; Zhang et al. 2013). Deep learning is a branch of artificial neural networks. Machine learning algorithms are divided into two parts as supervised learning and unsupervised learning according to their learning strategies. Linear regression, logistic regression, decision trees, k-nearest neighbors, Naïve Bayes, random forest and support vector machines algorithms are the examples for the supervised machine learning techniques. K-means clustering, hierarchical clustering, singular value decomposition, nonnegative matrix factorization and hidden markov models are some example algorithms for unsupervised learning. For most of the algorithms, before starting to execute learning algorithms, data preprocessing is taking an important place and has a big influence on the success of the accuracy of the method. The preprocessing part consists of feature extraction, data conversion, and feature selection.

Scientists and developers started to work on artificial neural networks many years ago. It is improved by analyzing the working principle of human brain. To make it clear we need to understand the terms about neural networks. Number of connections between neurons and speed of the data transmit are the reasons for why human brain was taken as inspiration source for learning. Biological neuron has dendrites, cell body and axon. Dendrites take signals, forward it to cell body to process it and send the processed data to axon to transmit it to other neurons. Similarly, artificial neurons get inputs with weight value that is generated particularly for each input. Inside the neuron, inputs and weights are multiplied and summed. The main point of the learning is about the determining and updating of the weights. Therefore, multilayer perceptrons and artificial neural networks with more layers created the deep learning concepts. Deep learning has the model that provides hierarchical learning. Each layer represents a different level of learning. If readers are interested in not only artificial neural networks and deep learning but also in other machine learning algorithms, they should know basic knowledge about calculus, linear algebra, probability theory, numerical computation, and optimization theory.

Rest of this chapter was organized as following. In section 2, some machine learning and big data processing methods, that used to solve necessary issue in wireless communication, have been categorized to reader for general overview. Some of methods that mentioned in section 2, has been explained briefly as subheadings in next chapter. Python programming language basics and setups were explained in last chapter to readers. After reading this chapter, the reader has knowledge about basic problems in wireless communication and the methods to solve these and how to use python programming language in basics.

MACHINE LEARNING AND BIG DATA PROCESSING APPROACHES IN WIRELESS COMMUNICATION

In wireless communication, there are lots of areas that use machine learning and big data processing approaches to handle such problems. Wireless communication has begun to enter our daily life since late of 20th century. But no significant increment of data traffic happened in wireless communication until past decade. Especially in the last decade, technology has become more effective than ever in areas such as education, business, and healthcare. Therefore, data and data traffic have expanded as never before. For instance, today's global IP data traffic per month is 150 peta byte while in 2016 that was 96 peta byte and experts estimate that this traffic will be able to reach 278 peta byte in 2021(Statistica, 2018). Another example IPv4 has not got enough address for machine that has connection to internet so this situation made that passing to IPv6 is necessary. Therefore, in the future it will be much more data and data traffic than now. So, the existing devices and methods are used to store data and communicate have begun to be inadequate. At this point, machine learning algorithms and big data approaches have begun to take over control. There are lots of subsections and algorithms in wireless communication that may be applied in terms of different purposes. Some of areas which use machine learning and big data approaches are discussed in this section. For instance, intrusion and anomaly detection, cognitive radio and modulation recognition approaches successfully has been used in wireless communication and signal processing based studies (Amos et al., 2013; Gohil et al., 2013; Hazar et. al., 2017; Rajasegarar et al., 2008; Tsagkari, 2008, Yu et al., 2008; Joel et. al., 2000; Roos et al., 2002; Tropp et al., 2006).

Wireless Sensor Networks (WSN) supply and evaluate communication dynamically to the several fields. WSN are being used in areas such as energy, distance of communication, and memory (Forster, 2007). When these conditions change, the structure of WSN also changes. To respond this in real-time, many approaches are used. In WSN there are lots of problems such as routing, clustering, localization, security, data integrity and aggregation (Alsheikh et al., 2014). Routing is a NP-hard type graph problem (Alsheikh et al., 2014). One of the reasons is the graph that is constituted by sensors, has dynamically changeable connections. Actually, this problem is similar to traveler salesman problem in some cases. To solve this problem many machine learning algorithms are used with some modifications. Firstly, regression algorithms like linear or polynomial and genetic algorithms are used for solution (Alsheikh et al., 2014; Forster, 2007; Pinto & Montez, 2010) (Pinto, Montez, Araújo, Vasques, & Portugal, 2014). Self-organizing map (SOM) is an unsupervised machine learning method based on artificial neural network. SOM is used many studies to assure maximum efficiency of the routing problem (Alsheikh et al., 2014). Another method is Q-learning, which is a reinforcement technique and with extra attributes, it becomes algorithmic pipelines named CLIQUE (Förster & Murphy, 2009), Q-MAP multicast in some research papers. Another problem is clustering to WSN. To achieve optimum communication between sensors is not easy. K-means, that simple clustering methods, principal component analysis (PCA), artificial neural networks are some of the methods used to cluster sensors (Jiang et al., 2017). Outlier detection problem is also one of the most researched topics in WSN. Researchers apply mostly different types of Support Vector Machine (SVM) methods to the networks of sensor (Alsheikh et al., 2014).

Cognitive Radio (CR) is a kind of intelligent radio device that can adapt itself dynamically to the environment. The mean of adaptation is that CR learns from the environment and then moves in the right way by programming itself. There are lots of issues that machine learning algorithms to be used, such as dynamic spectrum sharing or access, capacity maximization, re-configurability, generalized policy-making, and others (Clancy, Hecker, Stuntebeck, & O'Shea, 2007). But in general, there are two main categories such as decision making and classification. In decision making type of problems, CR has some knowledge from past state and has to make a decision about what it is going to do in the next step. To handle this situation, Markov states that based algorithms, reinforcement learning was applied (Clancy et al., 2007). In classification problems, almost every machine learning algorithms were used by the researcher for clustering, feature selection, classify to CR. K-means, mixture models for clustering, artificial neural networks, SVM, kernel based functions for classification or feature selection are some of methods (Wang et al., 2017) (Ding, Wu, Yao, Wang, & Chen, 2013) (Bkassiny, M., Li, Y. Jayaweera, 2013).

In a short term 5G to the internet and wireless communication will reach incredibly high level. There is a need of developing next generation wireless devices, policies, applications, methods with the 5G. In supervised learning methods such as regression model, SVM can be used to neighboring degree of sensors, linear or non-linear mapping of high dimensional network structure or classification. Bayesian based model such as mixture models, Hidden Markov Models (HMM) can be used to model large scale multi-input multi-output (MIMO) channel component. Principal and independent component analysis methods play significant role for anomaly, fault and intrusion detection of wireless or mesh network in 5G. Also some of reinforcement learning such as Q-learning, Markov decision process are solutions to make optimum decision in the 5G networks (Wang et al., 2017).

Nowadays, internet service providers (ISP) have to manage more instant data traffic. To solve this difficulty, ISP try to classify traffic between the networks. By classifying data traffic, not only many problems such as speed, security, intrusion detection, optimum pathway may solve more easily but also cyber-attacks can be detected more efficiently. Like other fields, machine learning techniques are the most promising solutions. Due to being core of internet traffic, IP is the main parameter of algorithms. For unsupervised methods, expectation maximization (EM), K-means algorithms show high accuracy to examples of successful network classification (Nguyen & Armitage, 2008). On the other hand, neural networks, genetic algorithms, linear discriminant analysis (LDA), Bayesian methods were used by being adapted situations. Wireless mesh networks (WMS) which may be considered as the next generation network, are the field for classification of traffic with machine learning methods. Decision trees such as C4.5, Naïve Bayes Classifier (Gu, Zhang, Xue, & Huang, 2011) and K-means are among the methods that are used for WMS (Nguyen & Armitage, 2008). Intelligent approaches in wireless networks block diagram can be seen from Figure 1.

Almost all machine learning algorithms were used in wireless communication area. But new and famous deep learning (DL) approach has not showed yet its full potential due to computational complexity. At most basic level, deep learning can be taught as very complex neural networks. This attribute enables high recognition accuracy. The size and amount of data created is too large compared to the past and DL has the ability to process these big data capacity. Today, DL is being used in many fields such as physical layer of wireless communication, channel decoding and detection (Wang et al., 2017).

In this part, we present machine learning algorithms in wireless network up to now. However, big data is a set of tremendous raw data that need to be constructing to a meaningful whole to be analyzed. Basically, it can be thought as combination of many different type normal dataset. Getting meaningful information, making predictions by analyzing with intelligent algorithms to content of big data, are some of new research area. Today, datasets are changing and expanding towards big data. Data volume that generated in the internet via individual information, distributed storage, social content, online business, cloud systems and others is growing extremely high speed day by day. For big data, lots of challenges exist wireless communication. A device connected to internet, produce different types of data such as

Figure 1. Intelligent approaches for wireless communication

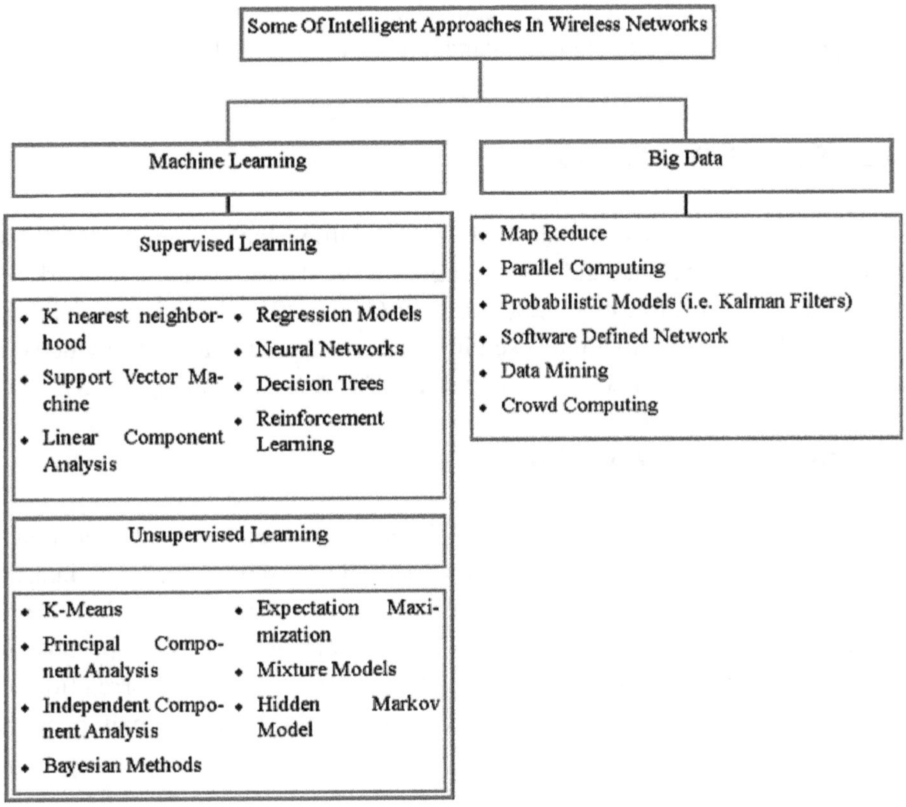

health, logistic, environmental and so many different types of devices produce all the types of these at the same time. Trying to process all of them is a huge challenge. Therefore, there are many big data analysis approaches such as Mapreduce, Bloom Filter, and Parallel Computing (Bi, Zhang, Ding, & Cui, 2015).

To manage big data in wireless communication optimally, some intelligent methods relating to machine learning are developed. These methods can be collected under different subtitles according to their purpose. Probabilistic models such as hidden Markov model, Kalman filters are mostly used to expose meaningful estimates from collected user data. Data mining that comes into daily life with big data may be most important concept. The aim of data mining is to discover embracing structures from unstructured data. It is very complex and difficult however during the recent years data mining is one of the most interested research area. This can be resembled to extreme dimension reduction. Besides, reinforcement algorithms, dimension reductions methods, distributed approaches, classification and clustering are also used by making them more compatible to big data processing (Bi et al., 2015).

An approach is that developing more effective and an intelligent data cache mechanism that is called data-aware cache management could reduce internet traffic. One of the best approaches is crowd computing (Bi et al., 2015). Internet users that are near could share their cached data to other users that need that data. Conventional routing and switching procedures are hardware based systems and they are insufficient for today's data traffic. Software defined networks (SDN) are a new approach that manages these processes with programmable units. SDN consist of three layers: control layer, data layer and management layer. Control and management layers are centered and routing devices has only data layer that is responsible only the routing. Programmable central unit is responsible to manage data traffic. This decoupling provides much more efficient, intelligent, flexible data traffic management. These are just the few of existing approaches that make big data management more effective in wireless communication.

MACHINE LEARNING METHODS

As we have discussed before in this chapter, there are two important points for wireless communication which are handling with big data and classification of the data to manage network correctly (Di & Joo, 2007; Kulkarni et al., 2011; Lee et al., 2007; Luo & Chang 2015; Mao 2009; Sandra 2013). In this section we want to explain popular methods used in wireless communication to reduce dimensions of a data set and classify the data in more detail (Alpaydin, 2014, Y. LeCun et al., 2015).

Principal Component Analysis (PCA)

PCA is one of the most popular machine learning methods that is used in a very large area in comparison to wireless communication. The main purpose of the method is to reduce the large datasets to small set. PCA is known with its process that provides reduction without losing information. It is called dimensionality reduction or data compression method. PCA is unsupervised method because it does not use the class label of data. PCA works based on finding the subsets have high correlations with principal component. Principal component is the rotation of the axes about data. The main principal component

Table 1. Example Dataset

	Data-1	Data-2	Data-3	Data-4	Data-5	Data-6
Feature 1	10	11	8	3	2	1
Feature 2	6	4	5	2.7	2.5	0.9

Figure 2. Data in coordinate system and average values of features (labeled as red)

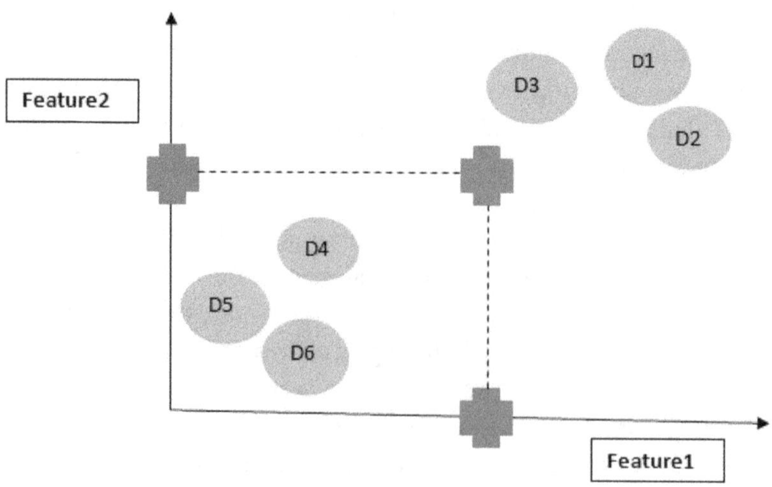

is the new axe that has the maximum variation of data. Other components can be listed by looking the variation amount. We have 3 steps to find principal components. First step is placing the data to coordinate system. We can explain the basic processes of PCA with an example data, which is shown in the Table 1 and Figure 2.

For the beginning, we have to put data to coordinate system. Example dataset in Table 1 have 2 features and 6 data. We can see the data in coordinate system in Figure 2.

After data is transferred in to coordinate system, we don't need original data in Table 1. Points labeled in red represent average values of Feature 1 and Feature 2. Using average values we calculate the center of the data that is labeled as blue in Figure 2. After the transformation, we shift the data to origin using center of the data. Shifting does not change the position of data relative to each other. Second step is the fitting a line to data using center as shown in Figure 3.

PCA finds the principal components by projecting the data on the line. It will measure the distances between data and line and try to minimize them. To be able to calculate the distance, we can use Pythagorean Theorem that is shown in Figure 4.

In the theorem shown above, we cannot change the distance a, because it's the distance between origin and data. Therefore, minimizing distance b or maximizing distance c can help us to find fitted line by rotating the red line. For each data in the system, using distances of data (c) sum of squared distances are calculated. PCA repeats the rotation procedure, until finding the largest sum of square distances between the projected points and the origin. The line that gives the largest distances is the first principal component and of course we can find other components after that.

If we will use PCA, we also need to know about eigenvector and eigenvalue. In PCA, first component is eigenvector. Moreover eigenvector is a direction and it has the biggest variance by comparing to other directions or vectors. Eigenvalue is a number that represents how much variance exists in the data or direction.

Figure 3. Line fitting to find principal components

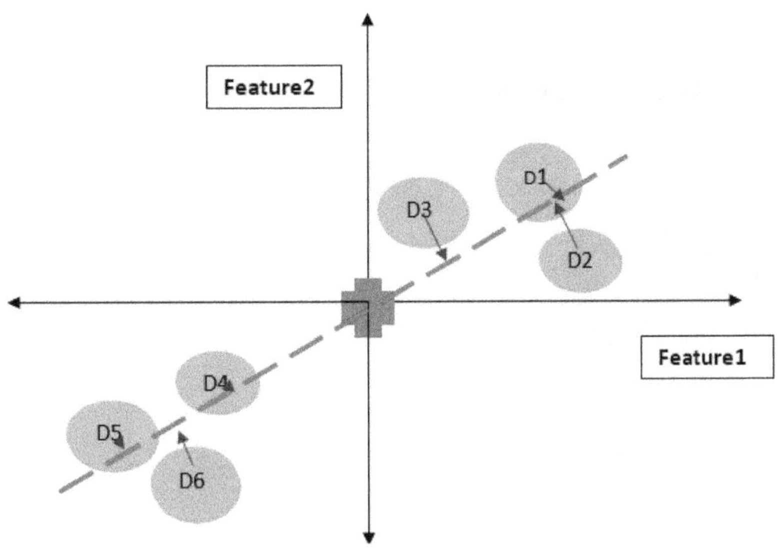

Figure 4. Pythagorean theorem used in principal component analysis

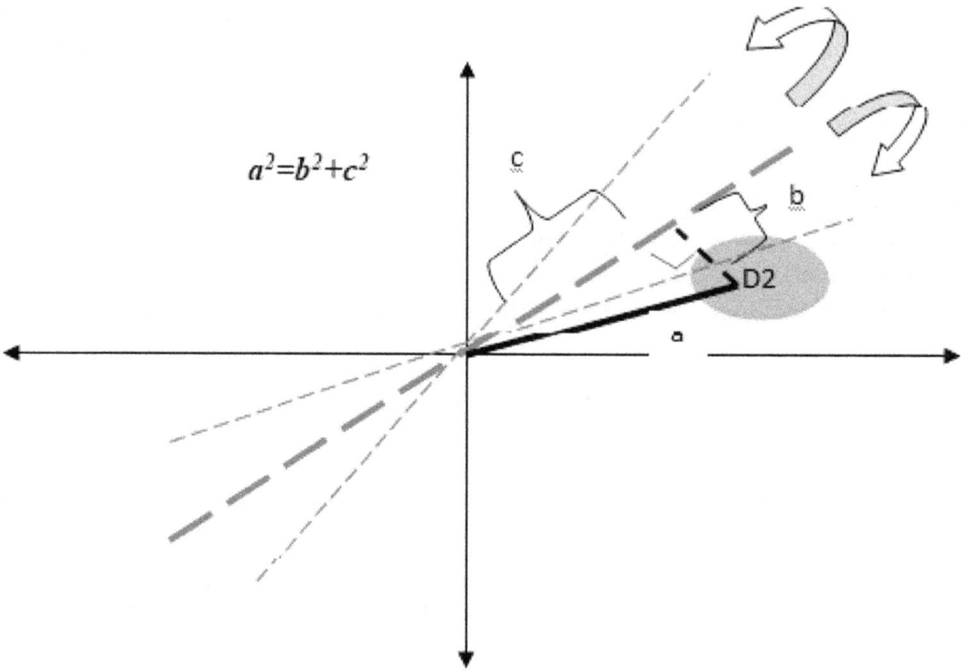

K-Nearest Neighbor (k-NN)

KNN is one of the popular and simple supervised classification methods. It is generally used for classification problems but there are studies about regression problems where KNN is used. First of all, we want to explain the algorithm basically. Assume that we have 2 classes in our dataset which are labeled as red and green and a data labeled as blue that is shown in Figure 5. We need to predict the class of the blue data.

Figure 5. Example data for KNN classification

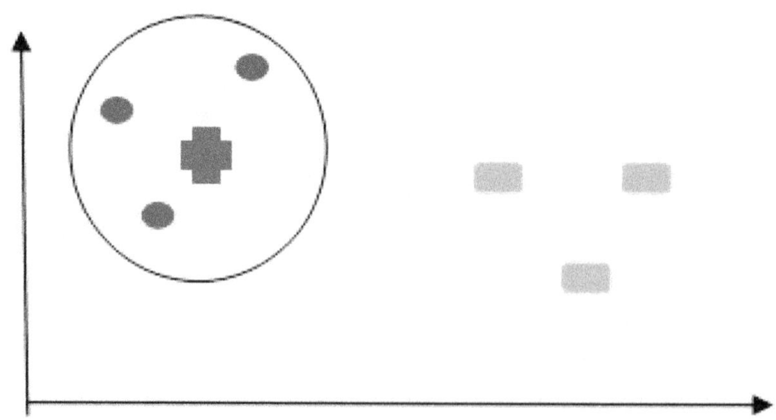

If we set k values as 3, KNN algorithm will calculate the distance between labeled data and find the closest 3 data. For the example above, closest 3 data belongs to red class, so blue labeled data's class can be set as red. However, unlike from the example above, the distribution of classified data is not always well separated according to the coordinate system. If the closest 3 data's classes don't belong to the same class, and then we have to choose the class that is mostly encountered. In general, KNN works better when the number of the input variables is not large. Because if it is high multi-dimensional input space, there can be similar data but placed far away from each other. On the other hand, if the value of K is increased, it is expected that error rate will increase.

Support Vector Machines (SVM)

Support vector machine method is one of the effective and basic ways for classification. It is known as a supervised learning method. For classification, it is enough to categorize classes into multiple part with borders. SVM method provides us to separate the classes with lines in an appropriate way. SVM defines how to draw the lines. As generally SVM is used for two-class classification problems.

First step is drawing lines that are closed to each class member and the lines must be parallel to each other which is shown in Figure 6. Second step is getting closer the lines and finding new joint border (w) line. Two lines and classes can be calculated as following formula 1.

$$\vec{w} * \vec{x} - b \geq 1, if y_i = 1 \tag{1}$$

$$\vec{w} * \vec{x} - b < -1, if i = -1$$

Vector w is the middle line between two closest class member and distance of w to origin can be calculated as $\frac{b}{w}$. Furthermore, distance between the lines can be found with the formula $\frac{2}{w}$.

Figure 6. Presentation of terms used in SVM

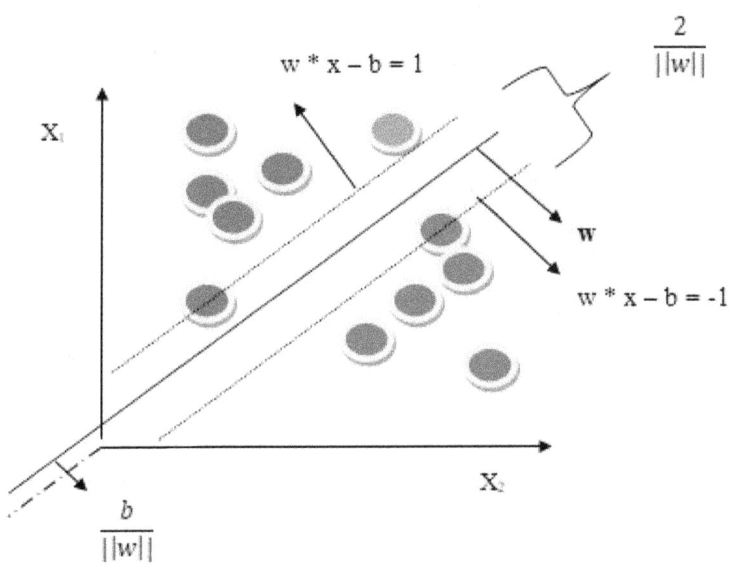

By taking the Figure 6 as an example, the data labeled with green will be classified according to equation 3.1. The data is located on the left side of the vector w, so 'y' which is stand for class label in the formula can be defined as 1. However, there may not be a line that can separate the classes in x-y plane. It can be a circular border line that is represented with $w = x^2 + y^2$ or other nonlinear formulas.

Artificial Neural Networks (ANN)

ANN is the most popular classification and prediction method in the literature. It is known with its successful prediction rate and speed. ANN is inspired from brain's working principle. In human brain, each neuron is working as separate processors that process in parallel. However, nowadays computers have only several processor cores. Processors in brain structure are slower than computer but the alternative of brain is high number of connection between neurons. Signals from other neurons are collected in the core of the neuron with axons. Collected signals are processed and transferred to other neurons with the connections. Using the brain structure, ANN method is improved as such a system with single command and multiple processors that each one has different parameters. ANN consists of inputs, weight of inputs, integration function, activation function and outputs. The structure of an artificial neuron can be seen from Figure 7.

Figure 7. Structure of an artificial neuron

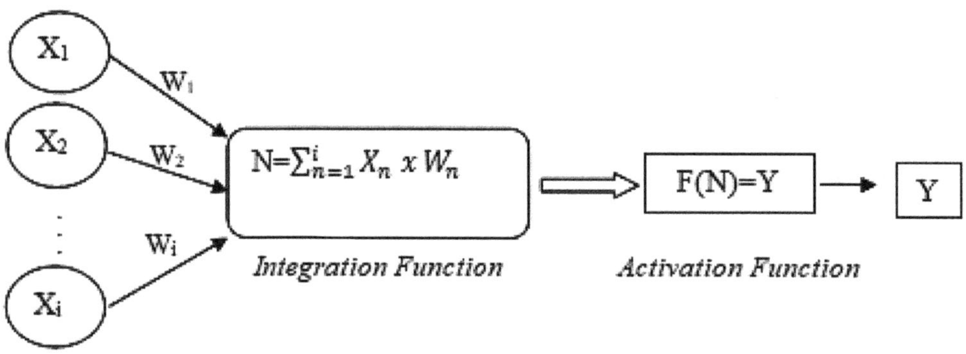

Figure 8. Structure of an artificial neural network

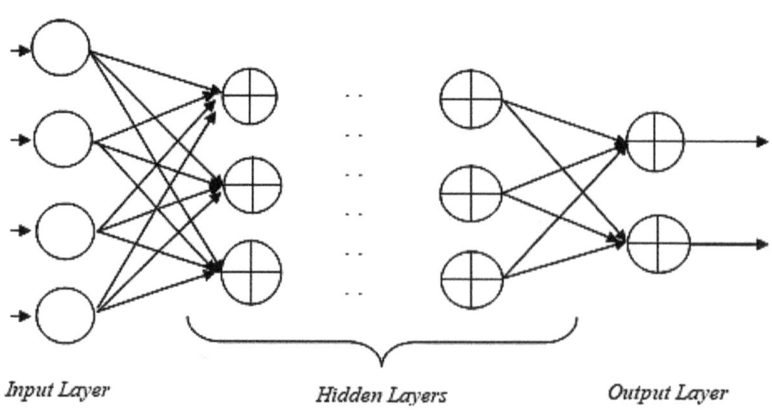

As we can see from the Figure 7, inputs (X_l) come from outside each one has own weight. The weights can be defined randomly at the beginning. After each output is calculated, by comparing to the expected output weights is updated until the error is very small. Inside the neuron, to calculate output Y, we have two functions namely integration and activation. Integration function is generally used as summation as shown in the Figure 7. Instead of it, multiplication, maximum, minimum or cumulative summation functions can be used. Output of integration function is the input for activation function which is generally used as sigmoid function and also linear function, step function or sinus functions. Basic properties of activation function are calculating the derivative easily and it is nonlinear.

A neural network has 3 different layers as shown in the Figure 8. Number of hidden layers can change up to different networks.

There is no standard to create an artificial neural network. An artificial neural network can be classified according to sequence of neurons, methods to calculate weights of inputs and learning speed. According to relations of neurons, we divide ANN into two parts as feed forward neural networks and feedback neural networks. Neurons in feed forward networks can take only previous neurons' output. On the other hand, an output of a layer can be input to next layers and also previous layers in feedback artificial neural networks. Feedback neural networks are also known as recurrent neural networks. Moreover, ANN can be classified into 3 parts according to learning type like supervised, unsupervised and reinforcement. In supervised ANN expected output and predicted output are used to calculate error and update the weight of the inputs. In contrast, unsupervised ANN does not use the expected output. It directly uses data similarity and makes clusters. Weights are updated to classify data according to similarity between them. Learning time is an important criterion for ANN. Static or dynamic learning has different effects on learning time. Static learning needs to train with some data to create network model. After the model and the suitable input weights are created, network can be used for classification. In dynamic learning, network training continues as long as network is used.

Moreover, in recent years convolutional neural networks (CNN) method is getting important. It has some additional layers different from traditional ANN that are convolutional layer, non-linearity layer, pooling layer and flattening layer. In general, CNN uses standard neural network for classification but to define some data and features other layers are used. Convolutional layer is used to perceive features of data. In the layer, to be able to find the features which are effective or not some filters are implemented. Non-linearity layer is generally used after convolutional layer to prevent ANN to be linear. Pooling layer is used to increase the number of calculations, parameters and the size of the filter. Pooling layer is not used very frequently because changing the convolutional layer is easier at the beginning. Flattening layer is the last layer before the simple ANN. Its purpose is to prepare the data as input.

Extreme Learning Machines (ELM)

ELM is known as a kind of ANN. It is also same with single hidden layer feed forward networks. The methods speed is very high and the results are successful. It is generally used for classification and regression problems. The method is using generally ANN structure, which has one or more hidden layers with tuned bias values. Its difference from the ANN is that in ELM the values are never updated again. The weights of the inputs are learned in a single step, there is no iterative calculation for parameters of the system. ELM is a method with high speed and generalization performance. The hidden layer is important for the method, because it is constructed only one time at the beginning with data projection methods.

Linear regression is also commonly used after random data projection step. One more advantage of ELM, in comparison to ANN, it does not need to process training data set to calculate random bias values.

Deep Learning

Deep learning is a kind of machine learning method that is generally used for such areas like image classification, video analysis, speech recognition and natural language processing (Y. LeCun et al., 2015). Deep learning uses many nonlinear sequential processing units for feature extraction and transformation. The method is generally used with graphical processing unit (GPU) because GPU provides powerful and parallel processing with high speed for big data. Before GPU, completing the complex processes with big data was not possible. Until 2015, a big amount of training data can be processed in short time with GPU support. Deep learning is used as deep neural network. Before GPU is used, ANN with CPU does not provide high performance for big data processes like natural language processing. Nowadays the number of neurons and layers can be easily increased to get better prediction rate with GPU support. Artificial intelligence is getting remarkable with this technological support.

As well known, deep learning architectures are convolution neural network, recurrent neural network, deep belief network (DBN) and Restricted Boltzmann machines (RBM). RBM is a fast learning algorithm that can learn the probability distribution over the input data set. It is a kind of Boltzman machines which includes 2 part graphs that have visible and hidden symmetrical connections. RBMs are mostly used for dimension reduction, classification, filtering, feature learning and modeling. On the other hand, DBN is also defined as stack of RBM. Each RBM layer is connected with both previous and next layers. In Figure 9, structure of DBM is given with its symmetric processing principle. A DBN consist of multiple RBM layer as symmetrically connected to each other. Last layer of RBM has classification or unsupervised clustering ability. DBN architectures are usually used for image recognition.

Figure 9. Structure of a Deep Belief Network (DBN) used in deep learning.

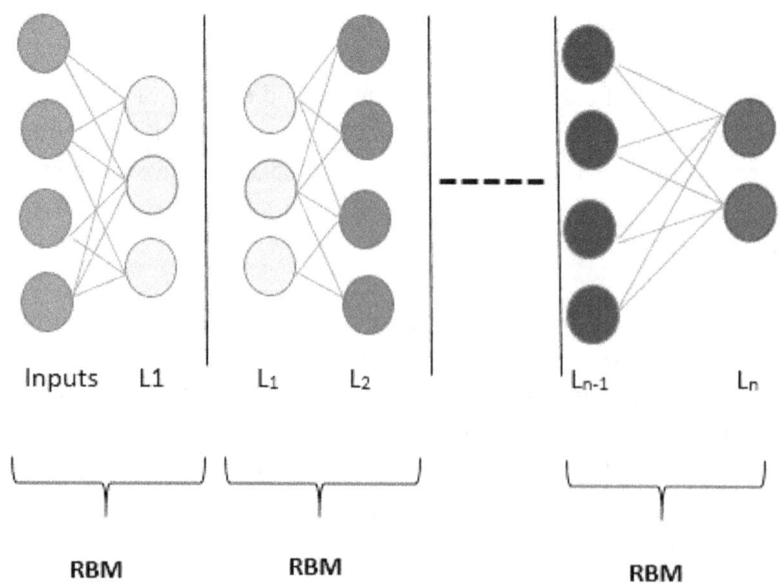

Figure 10. RBM structure with multiple hidden layers

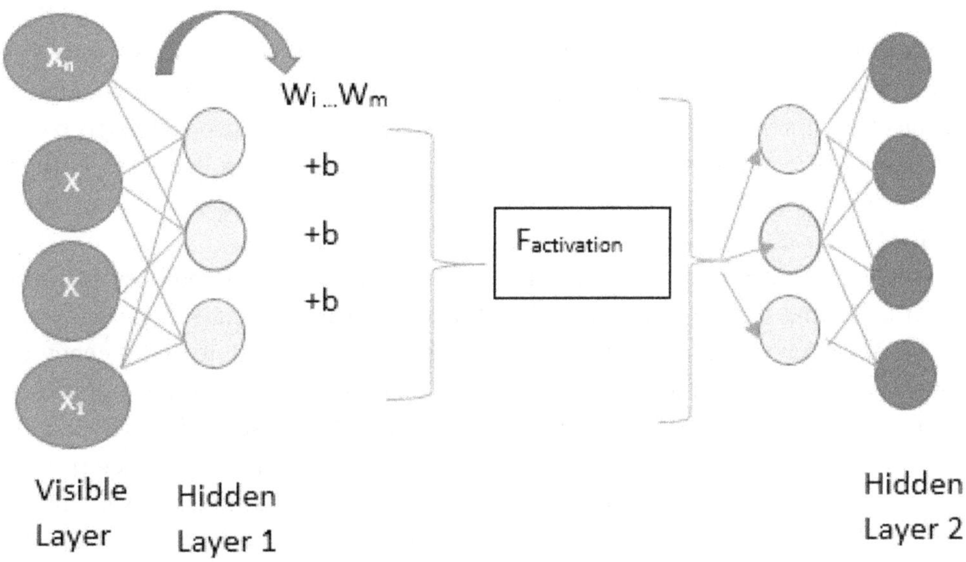

RBM which is used in DBN has two types of layer. First one is visible layer that includes inputs for hidden layers. In hidden layer, inputs from visible layer are processed with weights and other parameters. You can see the RBM structure with multiple hidden layers in Figure 10.

In RBM, similar to traditional ANN each layer's output Y is calculated with multiplying each neuron's weight with corresponding input then the value b is added as bias value. Before processing next layer, activation function produces the input by using previous output Y. As you can see, deep learning methods require powerful processors to handle with complex parallel processes and big data that are used in large networks with multiple layers and neurons. Now, we can use GPU in order to create more complex, fast and intelligent models.

INTRODUCTION TO PYTHON

The reason for the exponential increase in the development of computer science in recent years is that the algorithms, frameworks, software language have become open for community use. The subject we are dealing with is the sharing of researches, methods, software languages and frameworks with the community as open source. Machine learning and deep learning areas, which are the subdivisions of artificial intelligence, developed in this way and became the most popular research areas of our time. It is possible to do researches in these areas with many software languages (Matlab, Python, R, Java, C ++, Julia, Prolog, Lisp and others). But recently, there is a software language that stands out in machine learning and deep learning researches.

Why Do We Choose Python?

According to a research with many different criteria, Python is in top four of all software languages. When we look at machine learning and deep learning researches, it is the most widely used software language because its libraries, frameworks, algorithms open source, also language packages free to download. It is also recognized as one of the easiest languages to learn thanks to Python language's syntax, which is why Python is one of the most preferred software languages among researchers from different fields of science.

There are several libraries that we can use to study machine learning and deep learning in Python. Each of these libraries has their own advantages within different areas. But if we have to make a decision about which library we will use, the number of the contribution in Github and recency are important decision makers. From this point of view these are the examples of ML and DL libraries: Tensorflow, Scikit-learn, Theano, Caffe, Keras, Mxnet, PyTorch, CNTK. When researching with a complex data, it is crucial to visualize the data and instantly see some parts of the data, in order to determine the methods to be used in the research. Jupiter Notebook is one of the most popular frameworks in this field, it can be used with Python and allows you to see instantaneous output of the code you write and visualize your data.

Python Setup

As mentioned above, Python is a free and open source software language. Different versions can be easily downloaded and installed for each operating system from Python.org. You should download the correct python version that the libraries recommend you to use. If you are unsure, you can start by installing Python 3.6.x, the stable version of Python in almost every machine learning library. After downloading and setup Python, you can easily run python interpreter with writing *python* in your terminal. Python installation usually comes with a package manager named PIP. PIP finds the package that you want to use and does the installation for you with dependencies. Python packages and libraries can easily install with pip.

Every machine learning and deep learning libraries have detailed documentations, and numerous tutorials in their websites. Also, Python community is very helpful, you can ask your questions in different software knowledge markets and you can search for some python organizations that teach python for data science for latest libraries and trends.

FUTURE RESEARCH DIRECTIONS IN THE AREA / FUTURE TRENDS

In wireless communication area most of machine learning methods, especially deep learning approach, will be important. Moreover, several Internet of Things (IoT) based implementations and hardware solutions will be significant in this research topic. Our world will be more wireless in the future; therefore each new technical approach and standard will help for this purpose.

CONCLUSION

Machine learning techniques and after that, deep learning techniques are getting mostly used in recent years. When we look at the literature, we analyzed that the most popular machine learning techniques are PCA, k-NN, SVM, ANN, ELM and deep learning in wireless communication. The techniques will be more important in the near feature in communication, because data types and amount of the data are changing day by day. Saving, processing and converting data to knowledge will be remarkable for high quality communication and protecting the system. Recent developments about smart phones and other devices that use wireless communication like drones, wearable devices, machines with sensors needs reliable communication within Internet of Things (IoT) system. For this purpose, AI can provide secure systems and manage the data that is generated by the wireless system. In the near future, neural networks will be more popular for the IoT systems, because the systems are getting bigger with too many connected devices and it has dynamic structure which can be managed intelligently by the neural networks to be able to develop an intelligent system for communication, we gave information about python that is the most popular language for AI and other learning algorithms. Python is known with its speed and various affective packages in literature. We also recommend you to have knowledge about probability theory and mathematical background of the methods rather than using the ready functions to be able to create new unique techniques.

REFERENCES

Alpaydın, E. (2014). *Introduction to Machine Learning*. MIT Pres.

Alsheikh, M. A., Lin, S., Niyato, D., & Tan, H.-P. (2014). Machine Learning in Wireless Sensor Networks: Algorithms, Strategies, and Applications. *IEEE Communications Surveys and Tutorials, 16*(4), 1996–2018. doi:10.1109/COMST.2014.2320099

Amos, B., Turner, H., & White, J. (2013). Applying machine learning classifiers to dynamic android malware detection at scale. *2013 9th International Wireless Communications and Mobile Computing Conference, IWCMC 2013*, 1666–1671. 10.1109/IWCMC.2013.6583806

Bi, S., Zhang, R., Ding, Z., & Cui, S. (2015). Wireless communications in the era of big data. *IEEE Communications Magazine*, *53*(10), 190–199. doi:10.1109/MCOM.2015.7295483

Bkassiny, M., Li, Y., & Jayaweera, S. K. (2013). A survey on machine-learning techniques in cognitive radios. *IEEE Communications Surveys and Tutorials*, *15*(3), 1136–1159. doi:10.1109/SURV.2012.100412.00017

Brunato, M., & Battiti, R. (2005). Statistical learning theory for location fingerprinting in wireless LANs. *Computer Networks*, *47*(6), 825–845. doi:10.1016/j.comnet.2004.09.004

Callado, A., Kamienski, C., Szabo, G., Peter-Gero, B., Kelner, J., Fernandes, S., & Sadok, D. (2009). A survey on internet traffic identification. *IEEE Communications Surveys and Tutorials*, *11*(3), 1–16. doi:10.1109/SURV.2009.090304

Clancy, C., Hecker, J., Stuntebeck, E., & O'Shea, T. (2007). Cognitive radio networks. *IEEE Wireless Communications*, *14*(4), 47–52. doi:10.1109/MWC.2007.4300983

Demestichas, P., Dimitrakopoulos, G., Strassner, J., & Bourse, D. (2006). Introducing Reconfigurability and Cognitive Networks Concepts in the Wireless World. *IEEE Vehicular Technology Magazine*, *1*(June), 32–39. doi:10.1109/MVT.2006.283572

Di, M., & Joo, E. M. (2007). A Survey of Machine Learning in Wireless Sensor Networks. *Proceedings of the 6th International Conference on Information, Communications and Signal Processing (ICICS)*, 1–5. 10.1109/ICICS.2007.4449882

Ding, G., Wu, Q., Yao, Y., Wang, J., & Chen, Y. (2013). Kernel-Based Learning for Statistical Signal Processing in Cognitive Radio Networks. *IEEE Signal Processing Magazine*, *30*(4), 126–136. doi:10.1109/MSP.2013.2251071

Forster, A. (2007). Machine Learning Techniques Applied to Wireless Ad-Hoc Networks: Guide and Survey. *2007 3rd International Conference on Intelligent Sensors, Sensor Networks and Information*, 365–370. 10.1109/ISSNIP.2007.4496871

Förster, A., & Murphy, A. L. (2009). CLIQUE: Role-free clustering with Q-learning for wireless sensor networks. *Proceedings - International Conference on Distributed Computing Systems*, 441–449. 10.1109/ICDCS.2009.43

Gohil, A., Modi, H., & Patel, S. K. (2013). 5G technology of mobile communication: A survey. *2013 International Conference on Intelligent Systems and Signal Processing, ISSP 2013*, 288–292. 10.1109/ISSP.2013.6526920

Gu, C., Zhang, S., Xue, X., & Huang, H. (2011). Online Wireless Mesh Network Traffic Classification using. *Machine Learning*, *5*, 1524–1532.

Haykin, S. (2005). Cognitive radio: Brain-empowered wireless communications. *IEEE Journal on Selected Areas in Communications*, *23*(2), 201–220. doi:10.1109/JSAC.2004.839380

Hazar, M. A., Odabasioglu, N., Ensari, T., Kavurucu, Y., & Sayan, O. F. (2017). Performance analysis and improvement of machine learning algorithms for automatic modulation recognition over Rayleigh fading channels. *Neural Computing & Applications*, *29*(9), 351–360. doi:10.100700521-017-3040-6

Hussain, S., Schaffner, S., & Moseychuck, D. (2009). Applications of Wireless Sensor Networks and RFID in a Smart Home Environment. *2009 Seventh Annual Communication Networks and Services Research Conference*, 153–157. 10.1109/CNSR.2009.32

Jiang, C., Zhang, H., Ren, Y., Han, Z., Chen, K. C., & Hanzo, L. (2017). Machine Learning Paradigms for Next-Generation Wireless Networks. *IEEE Wireless Communications*, *24*(2), 98–105. doi:10.1109/MWC.2016.1500356WC

Kulkarni, R. V., Member, S., Förster, A., & Venayagamoorthy, G. K. (2011). Computational Intelligence in Wireless Sensor Networks : A Survey. *IEEE Communications Surveys and Tutorials*, *13*(1), 68–96. doi:10.1109/SURV.2011.040310.00002

LeCun, Y., Bengi, Y., & Hilton, G. (2015). Deep Learning. *Nature*, *521*(7553), 436–444. doi:10.1038/nature14539 PMID:26017442

Lee, M., Marconett, D., Ye, X., & Yoo, S. (2007). Cognitive network management with reinforcement learning for wireless mesh networks. *IP Operations and Management*, 168–179. doi:10.1007/978-3-540-75853-2_15

Luo, X., & Chang, X. (2015). A novel data fusion scheme using grey model and extreme learning machine in wireless sensor networks. *International Journal of Control, Automation, and Systems*, *13*(3), 539–546. doi:10.100712555-014-0309-8

Mao, G. (2009). *Localization Algorithms and Strategies for Wireless Sensor Networks*. Science And Technology; doi:10.4018/978-1-60566-396-8

Nguyen, T. T. T., & Armitage, G. (2008). A survey of techniques for internet traffic classification using machine learning. *IEEE Communications Surveys and Tutorials*, *10*(4), 56–76. doi:10.1109/SURV.2008.080406

Pinto, A. R., & Montez, C. (2010). Autonomic approaches for enhancing communication QoS in dense Wireless Sensor Networks with real time requirements. *Proceedings - International Test Conference*, 1–10. 10.1109/TEST.2010.5699288

Pinto, A. R., Montez, C., Araújo, G., Vasques, F., & Portugal, P. (2014). An approach to implement data fusion techniques in wireless sensor networks using genetic machine learning algorithms. *Information Fusion*, *15*(1), 90–101. doi:10.1016/j.inffus.2013.05.003

Rajasegarar, S., Leckie, C., & Palaniswami, M. (2008). Anomaly detection in wireless sensor networks. *IEEE Wireless Communications*, *15*(4), 34–40. doi:10.1109/MWC.2008.4599219

Roos, T., Myllymaki, P., Tirri, H., Misikangas, P., & Sievanen, J. (2002). A Probabilistic Approach to WLAN User Location Estimation. *International Journal of Wireless Information Networks*, *9*(3), 155–164. doi:10.1023/A:1016003126882

Sandra, P. R. (2013). A Survey on Recommendation Techniques in E-Commerce. *IEEE Communications Surveys and Tutorials*, *2*(12), 1586–1590.

Statistica. (2018). *Global IP Data Traffic from 2016 to 2021*. Retrieved from https://www.statista.com/statistics/499431/global-ip-data-traffic-forecast

Tropp, J. A., Gilbert, A. C., & Strauss, M. J. (2006). Algorithms for simultaneous sparse approximation. Part I: Greedy pursuit. *Signal Processing*, *86*(3), 572–588. doi:10.1016/j.sigpro.2005.05.030

Tsagkaris, K., Katidiotis, A., & Demestichas, P. (2008). Neural network-based learning schemes for cognitive radio systems. *Computer Communications*, *31*(14), 3394–3404. doi:10.1016/j.comcom.2008.05.040

Wang, T., Wen, C.-K., Wang, H., Gao, F., Jiang, T., & Jin, S. (2017, November). Deep learning for wireless physical layer: Opportunities and challenges. *China Communications*, *14*(11), 92–111. doi:10.1109/CC.2017.8233654

Wang, Y., Martonosi, M., & Peh, L. S. (2007). Predicting link quality using supervised learning in wireless sensor networks. *ACM SIGMOBILE Mobile Computing*, *11*(3), 71–83. doi:10.1145/1317425.1317434

Yu, Z., & Tsai, J. J. P. (2008). A Framework of Machine Learning Based Intrusion Detection for Wireless Sensor Networks. *2008 IEEE International Conference on Sensor Networks, Ubiquitous, and Trustworthy Computing (Sutc 2008)*, 272–279. 10.1109/SUTC.2008.39

Zhang, Y., Meratnia, N., & Havinga, P. (2009). Adaptive and Online One-Class Support Vector Machine-Based Outlier Detection Techniques for Wireless Sensor Networks. *2009 International Conference on Advanced Information Networking and Applications Workshops*, 990–995. 10.1109/WAINA.2009.200

Zhang, Y., Meratnia, N., & Havinga, P. J. M. (2013). Distributed online outlier detection in wireless sensor networks using ellipsoidal support vector machine. *Ad Hoc Networks*, *11*(3), 1062–1074. doi:10.1016/j.adhoc.2012.11.001

ADDITIONAL READING

Frey, B. J., Brendan, J. F., & Frey, B. J. (1998). *Graphical models for machine learning and digital communication*. MIT press.

Kibria, M. G., Nguyen, K., Villardi, G. P., Zhao, O., Ishizu, K., & Kojima, F. (2018). Big Data Analytics, Machine Learning and Artificial Intelligence in Next-Generation Wireless Networks. *IEEE Access: Practical Innovations, Open Solutions*, *6*, 32328–32338. doi:10.1109/ACCESS.2018.2837692

KEY TERMS AND DEFINITIONS

Artificial Intelligence: An attempt to recreate a living intellect, especially human intelligence, in a computer environment.

Artificial Neural Networks: A machine learning algorithm that is created by mimicking the information transmission and problem-solving mechanism in the human brain.

Deep Learning: The method for solving problems that have more probabilistic calculations based on artificial neural networks.

Machine Learning: The methods used to understand the patterns in the data and to obtain results from these patterns using various algorithms.

Open Source: The software that is shared freely with all users from source code to used methods.

Chapter 7
Machine Learning in Wireless Communication: A Survey

Neha Vaishnavi Sharma
Manipal University Jaipur, India

Narendra Singh Yadav
Manipal University Jaipur, India

ABSTRACT

As the circumstances are changing, mankind has turned out to be more inclined to snappy and speedier correspondence and access to information. The correspondence happens in numerous structures (e.g., presently, this correspondence is all the more a virtual substance than a physical one). So as to keep up fast correspondence, the coming age will depend on exceptionally tried and true, canny and self-learning/self-modifying correspondence organizers. In this context, this chapter reviews the most important machine learning techniques with the direct applicability in wireless ad-hoc systems. A guide of machine learning methods and their relevance is also provided. Different applications of ad-hoc wireless networks are discussed in terms of energy-aware communications, optimal node deployment and localization, resource allocation, and scheduling.

INTRODUCTION

As the circumstances are changing, mankind has turned out to be more inclined to snappy and speedier correspondence and access to information. The correspondence happens in numerous structures e.g. presently a days this correspondence is all the more a virtual substance than a physical one. So as to keep up fast correspondence the coming age will depend on exceptionally tried and true, canny and self-learning/self-modifying correspondence organize. In common situations applications are designed to work in enormous systems which are less inclined to failure. Addresses are allocated to hubs and hubs keep static directing tables with a specific end goal to have the capacity to discover the way to some

DOI: 10.4018/978-1-5225-7458-3.ch007

other hub and if some adjustment in topology happens, this data must be coursed through the whole system (Forster, 2007).

Alongside the quick advancement of versatile correspondence advances, countless quality remote administrations are required. As indicated by the report of Cisco VNI Global Mobile Data Traffic Forecast 2017, worldwide versatile information activity will increment about sevenfold in the vicinity of 2016 and 2021, and portable system association paces will increment triple by 2021. There is a major hole between the future necessities of remote administrations and current correspondences innovations, notwithstanding utilizing 4G/5G system. Step by step instructions to plan canny calculations/plans to make full utilization of the constrained remote assets is the subject of this extraordinary issue. As a vital teach, machine learning incorporates design acknowledgment and computational learning hypothesis in counterfeit consciousness and calculations to gain from the past and make expectations in confounded situations. It can be utilized to dissect the past/current radio circumstances and correspondence ideal models in remote interchanges, for example, range usage, channel limit, control level, reception apparatus designs, and heterogeneous connection properties, and help to create an ideal activity to enhance the Quality of Service (QoS).

MOTIVATION

Applications of ML in communications have a long history covering a wide range of applications. These comprise channel modeling and prediction, localization, equalization, decoding, quantization, compression, demodulation, modulation recognition, and spectrum sensing to name a few (O'Shea & Hoydis, 2017; Ibnkahla, 2000). Recently, some machine learning algorithms have been proposed for Wireless communications, intellectual radio systems, bioinspired systems, machine-to-machine interchanges, MIMO interface adaption, antenna selection, blockage control, etc.(Hinton, Osindero & Teh, 2006; Jiang, Zhang, Ren, Han, Chen, & Hanzo, 2017) . Machine learning has been a standout amongst the most dynamic research fields because of its incredible achievement in an extensive variety of spaces. Be that as it may, its effect on remote interchanges has so far been extremely constrained, despite the fact that the capability of machine learning in building best in class correspondence frameworks is expansive (Alsheikh, Lin, Niyato & Tan, 2014). The primary test is the means by which to figure the issues in correspondence frameworks as a legitimate machine learning model (Di & Joo, 2007).

AIM

The purpose of this chapter is to provide the students, researchers, network engineers and a wide range of readers with a guide to the fundamental concepts, machine learning approaches and algorithms required for a basic understanding and implementation to the Wireless communications. This chapter is both a survey for existing works and a guide for researchers willing to apply Machine learning to problems in ad hoc networking.

Section I is the introduction to the growth of Machine Learning with respect to Wireless communication. It also includes the aim and motivation of writing this chapter. Section II gives short descriptions of Computer Network, background or basics of wireless communication and its types. It also illustrates the relation between Communications and smart systems. Section III describes Machine learning concepts, its advantages and limitations. Section IV elaborates the distinctive properties of Machine learning while working with wireless communication network that must be considered in order to find the right Machine Learning algorithm which is further explained in section VI. It also presents different Machine Learning approaches applied to solve some network specific problems such as energy aware communication etc which are specified in section V.

WHAT IS A COMPUTER NETWORK?

A computer network is an interconnected collection of independent computers which aids communication in numerous ways. Apart from providing a good communication medium, sharing of available resources (programs and data on a computer are available to anyone on the network, regardless of the physical location of the computer and user), improved reliability of service (because of the presence of multiple computers), and cost-effectiveness (as small computers have a better price/performance ratio than larger ones) are some of the advantages of computer networking. In the early days, till even the early 1970s, computing and communication were considered to be two separate independent fields. The late 1970s and the early 1980s saw the merger of the fields of computing and communication. A modern-day computer network consists of two major components, namely, distributed applications and networking infrastructure. The distributed applications provide services to users/applications located on other computers. Common examples of distributed applications are Internet, electronic mail, air/train ticket reservation systems, and credit card authentication systems. The networking infrastructure provides support for data transfer among the interconnected computers where the applications reside. There are no restrictions on the physical media used for providing connectivity among the computers. Several physical media such as copper cable, optic fiber, and wireless radio waves are used in the present-day communication infrastructure.

Wireless Communication: A Background

Wireless Communication advancements have additionally encountered a broad improvement to fulfill the applications and administrations in the remote system. The blast of cutting edge remote applications, for example, assorted savvy terminal access, virtual reality, enlarged reality, and Internet of things, has impelled the advancement of remote correspondence into the fifth era to accomplish thousand crease limit, millisecond dormancy, and monstrous availability, along these lines influencing framework to outline a remarkably difficult errand. A few promising advances, for example, gigantic multi-input multi-yield (MIMO), millimeter wave (mm Wave), and ultra-densification organize (UDN) have been proposed to fulfill the previously mentioned requests.

Wireless communications has numerous structures like WSNs (Wireless Sensor Networks), WAN (Wireless Ad Hoc Networks), MANets (Mobile Ad Hoc Networks), VANets (Vehicular Ad Hoc systems) and nowadays quick creating AANETs (Aeronautical Ad Hoc Networks). Specially appointed remote systems are characterized as the class of remote systems that use multi-jump radio transferring and are fit for working without the help of any settled framework (thus they are likewise called foundation less systems). The nonappearance of any local organizer or base station makes the search an unpredictable one in contrast with cell systems.

Mobile Ad-Hoc Networks (MANETs)

Mobile ad-hoc networks (MANETs) are a subcategory of specially appointed systems; dynamic systems made out of remote empowered gadgets that frame a system among themselves without the requirement for any current foundation. MANETs are basically specially appointed systems where gadget versatility is reasonable. MANETs are regularly utilized as a part of situations where a particular undertaking should be expert or for asset sharing. MANETs are naturally conveyed in nature and are along these lines a perfect method for enabling different heterogeneous gadgets to pool assets. Because of their capacity to shape dynamic, impermanent systems without the requirement for coordinate client inclusion and start, MANETs offer incredible potential for future Internet frameworks as a method for sharing and broadening Internet get to.

Wireless Sensor Networks (WSN)

Wireless Sensor Networks (WSN) are a division for arrange that covers a few varieties in creations and sending of hubs. These systems are made out of little conveying hubs, which contain a detecting unit, remote correspondence module, processor, memory and a power supply, regularly a battery. The hubs that make these systems can gather scalars and they are likewise ready to speak with each other. The arrangement of hubs can be homogeneous or some of them may have exceptional attributes. Some WSN considers the utilization of a base station that has more computational power than different hubs. The base station has the duty to gather, process and store information sent by slave hubs.

Aeronautical Ad Hoc Networks (AANETs)

Aeronautical ad hoc networks (AANETs) can have a half breed topology of irregularly associated groups and work arrange. The half and half system topology intensified with profoundly unique nature of AANET prompts variable availability in the system. The availability in the system for guide aerial correspondence between flying machines is principally an element of speed of air vehicles, position of air vehicles, course of flight, scope of correspondence and clog.

The Vehicular Ad-Hoc Network

The Vehicular Ad-Hoc Network, or VANET, is an innovation that utilizations moves autos as hubs in a system to make a versatile system. VANET transforms each taking an interest auto into a remote switch or hub, permitting autos around 100 to 300 meters of each other to associate and, thus, make a system with a wide range. As autos drop out of the flag range and drop out of the system, different autos can participate, associating vehicles to each other with the goal that a portable Internet is made. It is assessed that the main frameworks that will coordinate this innovation are police and fire vehicles to speak with each other for wellbeing purposes.

Communication and Smart Systems

The wireless network administration nowadays is experiencing a noteworthy change. Individuals are moving from cell phones to shrewd frameworks broadly known as IoT (Internet of Things). Individuals need to remain associated with each gadget they claim and handle the things from anyplace. This expansion sought after of such gadgets has lead the whole correspondence convention to center around IoT gadgets and how this correspondence can be made more compelling continuously. This has prompted a sudden increment in arrange movement and will keep on rising by the day. Then, in tomorrow's wireless systems, huge volumes of information must be gathered, occasionally and progressively, over various monstrous detecting and wearable gadgets. Such monstrous short-bundle transmissions will prompt a generous activity over the remote uplink, which has customarily been significantly less congested than the downlink. The same remote system should likewise support cloud-based gaming, immersive virtual reality administrations, real time HD streaming, and traditional sight and sound administrations.

This change in perspective towards "smart" communication arranges that take advantage of the network data and organize information has in reality made current correspondence systems, and specifically portable systems to produce an immense measure of information at the system foundation level and at the client/client level. The information in the system contain an abundance of helpful data, for example, area data, and versatility and call designs. The vision of system administrators is to either empower new organizations through the provisioning of this information (or the data contained in it) to outside specialist co-ops and clients or to misuse the system information for in-house administrations, for example, arrange advancement and administration. So as to make the vision reality, there is a solid requirement for the improvement and usage of new machine learning strategies for huge information investigation in communication systems. The goal of these strategies is to remove valuable data from the system information while considering constrained communication assets, and after that to use this data for outer or in-house administrations.

MACHINE LEARNING

Machine Learning is an application of Artificial Intelligence that provides system the ability to automatically learn and improve from experiences without being explicitly programmed. The process of learning begins with observations or data, such as examples, direct experience or instructions. In order to look for patterns in data and make better decision in future the system needs to be trained with more and more examples. Machine learning methods are core part in many emerging applications of communication technology, e.g., smart cities (Tsai, Lai, Chiang & Yang, 2014) or internet of things (Werbos, 2011). Here, topics such as monitoring, fault prediction and scheduling are addressed with modern learning algorithms. The utilization of machine learning strategies in interchanges may give data about people that influence their security. In this manner, different protection saving ways to deal with information investigation have been as of late proposed e.g. (Agrawal & Srikant, 2000). Machine learning strategies are additionally broadly connected to handle security-related issues in communications, e.g., as a component of protection systems against spam assaults and infections (Guzella & Caminhas, 2009). The increasing convergence can be also observed in specific domains of communications such as image and video communication. While the immediate way to deal with planning pressure calculations utilizing auto encoders has given extremely restricted outcomes contrasted with the cutting edge, the utilization of machine learning as an improving segment for perspectives like video encoding, bit designation or different parts turned into a promising research direction (Zhang, Kwong, Wang, Yuan, Pan, Z & Xu, 2015). As most video signals are put away as compacted information, the theme of question acknowledgment and following in the packed space is additionally of high pertinence (Srinivasan, Lapuschkin,

Hellge, Müller & Samek, 2017). Video streaming is another application which profits by the utilization of learning algorithms. Regardless of the effective utilization of machine learning techniques in different correspondences applications, there are as yet numerous difficulties and inquiries that should be tended to. For example, the substantial size and high computational requests of present day machine learning calculations keep the vast scale utilization of these models in installed gadgets. Likewise 5G systems call for novel machine learning-based ways to deal with radio asset administration and system administration approaches that can adapt to vulnerabilities and deficient channel and system state data. Different issues concern unwavering quality, protection and security parts of machine learning models. The calculations and procedures utilized originate from numerous differing fields including insights, arithmetic, neuroscience, and software engineering.

The accompanying two traditional definitions catch the essence of machine learning:

1. The improvement of PC models for learning forms that give answers for the issue of information procurement and upgrade the execution of created frameworks.
2. The selection of computational techniques for enhancing machine execution by distinguishing and portraying textures and examples in preparing information (Langley, & Simon, 1995)
3. Applying these definitions to communication systems, we see that the guarantee of machine learning lies in exploiting authentic information to enhance the execution of sensor networks on given errands without the requirement for re-programming. All the more particularly, machine learning is important in systems administration applications for the accompanying principle reasons:
    a. Wireless networks usually monitor dynamic environments that change rapidly over time. For instance, a hub's area may change because of soil disintegration or ocean turbulence. It is desirable to create sensor networks that can adjust and work productively in such conditions.
    b. Wireless systems might be utilized for gathering new learning about inaccessible, unsafe areas (Paradis & Han, 2007) (e.g., well of lava ejection and waste water observing) in exploratory applications. Because of the unexpected conduct designs that may emerge in such situations, framework architects may create arrangements that at first may not work as expected. Framework designers would rather have robust machine learning algorithms that can align itself to recently obtained information.
    c. Wireless systems are normally conveyed in complicated conditions where specialists can't assemble precise numerical models to portray the framework conduct. In the meantime, a few errands in WSNs can be recommended utilizing basic scientific models yet may in any case require complex calculations to understand them e.g., the routing problems (Krishnamachari, Estrin & Wicker, 2002; Al-Karaki & Kamal, 2004). Under comparative conditions, machine learning gives low-many-sided quality appraisals to the framework display.
    d. Network creators regularly approach a lot of information yet might be not able concentrate critical Correlations in them. For instance, notwithstanding guaranteeing correspondence network and vitality supportability, the WSN application frequently accompanies least information scope prerequisites that must be satisfied by constrained sensor hardware resources. Machine learning techniques would then be able to be utilized to find critical relationships in the sensor information and propose improved sensor arrangement for maximum information scope.
    e. New uses and incorporations of WSNs, for example, in Cyber physical systems (CPS), machine-to-machine (M2M) interchanges, and Internet of things (IoT) technologies, have

Figure 1. Applications of machine learning in various areas of wireless communication

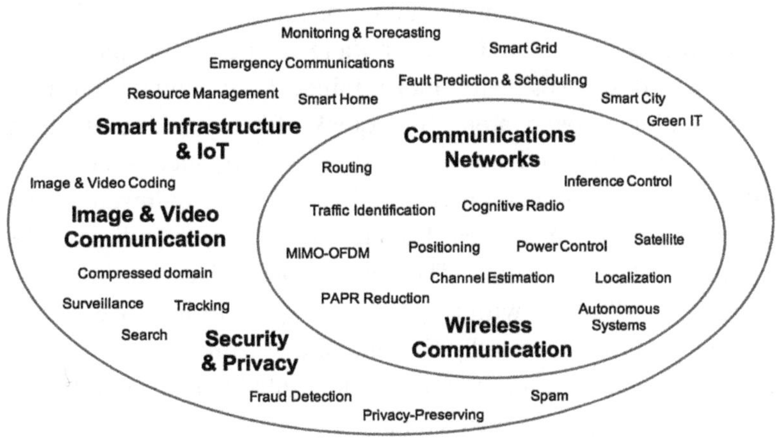

been presented with an inspiration of supporting more keen basic leadership and independent control. Here, machine learning is essential to separate the distinctive levels of reflections expected to play out the AI assignments with restricted human intervention (Bengio, 2009).

Limitations of Machine Learning

In any case, there are a couple of downsides and constraints that ought to be considered when utilizing machine learning procedures in wireless systems. Some of these are:

1. As a resource restricted structure, Wireless systems deplete a significant level of its vitality spending plan to foresee the exact theory and concentrate the agreement relationship among information tests. Along these lines, the planners ought to consider the exchange off between the calculation's computational prerequisites and the educated model's precision. In particular, the higher the required precision, the higher the computational prerequisites, and the higher vitality utilizations. Otherwise, the created frameworks may be utilized with incorporated and resource capable computational units to play out the learning task.
2. Generally, learning by cases requires a huge informational index of tests to accomplish the expected speculation capacities (i.e., genuinely little mistake limits), and the algorithm designer won't have the full control over the information plan process (Kulkarni, Lugosi & Venkatesh, 1998).

Machine Learning in Communications

Energy-Aware Communications

One of the real goals of numerous Wireless Networks research works is to enhance or optimize the execution of the whole system as far as vitality protection and system lifetime. The vast majority of the research exercises center on the outline of proficient directing convention at the system layer e.g. (Intanagonwiwat, Govindan & Estrin, 2000; Sohrabi, Gao, Ailawadhi & Pottie, 2000) choice of low-control modulation scheme at the physical layer e.g. (Chien, Elgorriaga, & McConaghy, 2001) or reception of energy sparing modular of activity at information interface layer e.g. (Sinha & Chandrakasan, 2001) to accomplish vitality mindfulness in WSNs. However, there is a detectable pattern that machine learning methods have been connected to investigate vitality mindful communication in WSNs. A significant number of the WSN applications depend intensely on quick, effective and solid information interchanges. Nonetheless, WSN communication protocols are characteristically inconsistent because of its inexactly coupled nature. In this way, communication conventions utilize circumstance mindful adjustment to recognize sound and vitality effective courses. The creators of present a directed learning approach for steering streamlining in Wireless Networks. The calculation utilizes machine learning procedures to enhance the circumstance mindfulness with a specific end goal to improve interchanges. Basically, it utilizes machine figuring out how to naturally find the relationships between info highlights (e.g., hub level and system level measurements, for example, cradle inhabitances, bundle length, hub remaining vitality) and yield (e.g., interface quality, ideal course). The machine learning procedure of is completed in four stages, to be specific

1. Feature choice and yield marking,
2. Sample gathering,
3. Offline preparing and
4. Online classification.

Feature choice and yield marking is the way toward picking the most suitable element vector that best speaks to the current issue. With regards to WSN, highlight vector could comprise of system parameters, for example, quick blurring, moderate blurring, movement design, flag quality and so forth. Yield marking is the way toward grouping yields utilizing area learning.

Sample gathering is the way toward gathering information for preparing reason. A backend server is likewise utilized as a part of the framework as an information gatherer and processor. This is known as brought together learning as far as engineering. Be that as it may, in some genuine applications, conveying a backend separate in WSNs isn't effective or even outlandish, therefore a decentralized/appropriated learning design is some of the time favored.

Offline preparing is the real learning process. The creators of (Wang, Martonosi & Peh, 2006) assesses two classifiers – choice tree (C4.5 algorithm) and rule-based learner. These learning calculations have a rich history; they may not be the best answer for WSNs as they adopt a brought together strategy and have no thought of asset requirements. Adapting overhead is just considered in the online characterization stage in light of the fact that disconnected preparing is finished by an asset rich backend server. In numerous WSN applications, adapting overhead is additionally present in the preparation stage, along these lines a more conveyed, proficient and versatile learning calculation limits the preparation overhead should be considered.

As a part of online preparation; the authors (Pandana & Liu, 2005) present a close ideal fortification learning system for vitality mindful sensor communications. The issue is detailed as normal throughput amplification per add up to expended vitality in sensor correspondences. A close ideal transmission technique was gotten in light of a fortification learning structure. This methodology picks the ideal balance level and transmission control while adjusts to the approaching activity rate, cradle and channel conditions. The motivation of utilizing the machine learning approach is that numerous arrangements of channel throughput augmentation require the state transition probability, which might be difficult to acquire. Machine learning, nonetheless, could get a close ideal transmission strategy in point-to-point situation, without inferring an intricate communication model.

Optimal Node Deployment and Localization

Sensor node deployment and localization are two interrelated issues in WSNs. With different node deployment strategies, the localization algorithms could also be totally different. For example, sensor nodes deployed manually may be localized using a walking GPS method. However, the GPS method would be too costly and time-consuming if sensor nodes are to be deployed randomly in large scale. In this case, parameters such as signal strengths, message delivering speeds, relative orientation can be used to estimate the sensor location using machine learning techniques (Di & Joo, 2007).

Deployment

The authors (Zhao & Liang, 2005) built up a fuzzy arrangement conspire for WSN observation application. The supposition for the greater part of the hub sending is that the environment is homogeneous. The impact of territory profile, (for example, obstacles, elevation, criticality of certain zone and so forth.) has been overlooked under such suspicion. By and by, these profiles might be essential in conveying a hub; for this situation, a more basic zone under reconnaissance may require more sensors. The arrangement issue given by (Zhao & Liang, 2005) has been planned as choosing an ideal node spread pattern to accomplish greatest data gain. To begin with, the region of intrigue is isolated into sub-zones, every one of which has its own particular territory profile. Second, in light of the territory profile, a fluffy rationale framework is utilized to register the quantity of hubs in each sub-region. Reproduction ponders demonstrate that the fluffy sending plan could accomplish a higher scope proportion and data pick up in the reconnaissance framework. The creator of (Shu & Liang, 2005) acquaints a fluffy enhancement calculation with proficiently modify sensor positions after an underlying arbitrary sending. The fundamental approach in (Shu & Liang, 2005) looks like that of (Zhao & Liang, 2005), both utilize fluffy rationale to get an ideal spread of hubs in the field to accomplish higher data pick up. Notwithstanding fluffy frameworks, technique in (Shu & Liang, 2005) likewise expect that hubs have certain versatility - one could change its position in view of some worldwide or nearby connection. A fluffy re-sending (or changing) conspire is done simply after the hubs have been arbitrarily sent in the field. The fluffy organization conspire presented in (Zhao & Liang, 2005) and (Shu & Liang, 2005) has an exceptionally constrained application, for example, field observation. The authors (Krause, Guestrin, Gupta & Kleinberg, 2006). present a general close ideal sensor arrangement display which boost data pick up while limits correspondence cost paying little mind to applications. This model isn't just relevant to reconnaissance WSNs yet additionally different WSNs applications also.

Localization

Location information is an important parameter in both networking and application domains of WSN. Accurate location estimation is a basic prerequisite for energy-aware routing and sensor event localization and reporting. There are generally two approaches to tackle the localization problem in WSN, namely hardware-based and probabilistic estimation based approaches. The authors (Dharne, Lee & Jayasuriya, 2006) propose using fuzzy logic to do away with strict probabilistic rules and set up heuristic fuzzy rules for localization in WSN. The algorithm uses a grid based approach in which a node's location is represented by its confidence level that it stands at a certain point in this grid.

The confidence level is computed based on the fuzzy logic system and the input variables of the fuzzy system are sensor measurements such as signal strength, time difference of arrival and etc. The authors (Tam, Cheng & Lui, 2006) adopted an evolutionary approach, namely a micro genetic algorithm and its extension to enhance the precision of existing localization methods. This is not a localization algorithm rather a post-optimizer for any other localization algorithms. It makes use of two genetic operators, namely the mutation and crossover operators, aiming at opportunistic decreases in the objective values obtained to mutate out current node location estimation, or crossover points for any pairs of existing chromosomes over successive generations for localization. Simulation results indicate that this post-optimizer improves the precision of various localization algorithms from 11.0% to 18.6% on average.

Resource Allocation and Task Scheduling

Resource Allocation and Task Scheduling are the real research challenges in the field of Wireless Communication as far as framework interaction is associated. Dissimilar to vitality mindful correspondence or ideal arrangement and confinement which streamlines a specific target capacity of a hub, enhancement issues planned under these two situations are from a more worldwide point of view, to be specific, how a gathering of sensor hubs could be overseen and booked to accomplish some framework goals, for example, a tradeoff between organize lifetime and data pick up.

The authors (van Norden, de Jong, Bolderheij & Rothkrantz, 2005) investigate three machine learning algorithms for undertaking planning for radar sensor systems and looks at the reenactment results. The algorithms utilized are fuzzy Lyapunov combination, hereditary algorithms and neural systems. The reenactment results demonstrate that GA outflanks the others. This examination was at first did to address the radar planning issue, be that as it may, the outcome could be instantly pertinent to Wireless Communication because of comparative framework engineering setting (Wireless Communication and radar sensor arrange). The authors (Foo, Lim, Zeng, Lam, Teo & Ng, 2000) and (Lim, Foo & Zeng, 2006) propose a versatile disseminated asset designation plot which determines moderately basic nearby activity to be performed by singular hubs in a Wireless Communication for administration of framework modes. Every hub adjusts its activity after some time in light of the status and input of its neighboring hubs. Despite the fact that the versatile activity is characterized locally, ideal worldwide conduct results from these nearby connections. The plan has been examined in two separate application situations, to be specific an acoustic Wireless Communication for field observation and camera arrange for movement checking. Reproduction results demonstrate that it gives a decent tradeoff between execution goals, for example, target following precision, scope zone, and system lifetime.

This area is predominantly to acquaint the perused with the calculations and systems that are perceived in this field. In addition, illustrations will be given to exhibit the way toward embracing machine learning in Wireless interchanges.

MACHINE LEARNING APPROACHES

Usually the wireless network programmers channelize machine learning as a collection of tools and algorithms that can be used to predict the up comings of the network whereas the experts in machine learning call it as a field covering vast themes and areas with pattern making. An understanding of such themes and patterns can be of great help for those studying machine learning. Application of machine learning to wireless communications can have tremendous benefits.

Existing Machine learning algorithms can be classified by their requirement in the context and most of them fall under the categories of Supervised, Unsupervised and reinforcement learning (Abu-Mostafa, Magdon-Ismail & Lin, 2012).

Supervised Learning

In Supervised Learning, a marked preparing set (i.e., predefined inputs and known yields) is utilized to fabricate the model. This model is utilized to speak to the learned relation between the input, output and system parameters. In this subsection, the major administered learning algorithms are examined with regards to Wireless Networks. In fact, supervised learning algorithms are widely used to comprehend a few difficulties in Wireless Networks, for example, security and Intrusion Detection on e.g. (Morelande, Moran & Brazil, 2008; Lu & Fu, 2009; Shareef, Zhu & Musavi, 2008) event detection and question preparing (e.g., Winter, Xu & Lee, 2005; Jayaraman, Zaslavsky & Delsing, 2010; Yu, Wang & Meng, 2005; Bahrepour, Meratnia, Poel, Taghikhaki & Havinga, 2010), media access control (e.g., Kim & Park, 2009; Shen & Wang, 2008; Kulkarni & Venayagamoorthy, 2009; Janakiram, Reddy & Kumar, 2006) security and intrusion detection and Qyality of Service (QoS), data integrity and fault detection (Branch, Giannella, Szymanski, Wolff & Kargupta, 2013; Kaplantzis, Shilton, Mani & Sekercioglu, 2007; Alsheikh, Lin, Niyato & Tan, 2014).

K-Closest Neighbor (k-NN)

This Supervised Learning algorithm orders a data sample (called a query point) in light of the marks (i.e., the output values) of the data samples. For instance, missing readings of a node can be anticipated utilizing the normal estimations of neighboring nodes inside particular measurement limits. There are a few capacities to decide the closest arrangement of nodes. A basic technique is to utilize the Euclidean distance between various nodes. K-closest neighbor does not require high computational power, as the capacity is figured in respect to nearby focuses (i.e., k-closest focuses, where k is a little positive whole number). This factor combined with the corresponded readings of neighboring hubs makes k-closest neighbor a reasonable appropriated learning algorithm for wireless networking. In the research by (Beyer, Goldstein, Ramakrishnan & Shaft, 1999) it has been demonstrated that the k-NN calculation may furnish off base outcomes while breaking down issues with high-dimensional spaces (more than 10-15 measurements) as the separation to various information tests ends up invariant (i.e., the separations to the closest and most distant neighbors are somewhat comparable). In wireless networking, the most critical use of the k-closest neighbor calculation is in the question processing subsystem (e.g. Jayaraman, Zaslavsky & Delsing, 2010; Yu, Wang & Meng, 2005).

Decision Tree (DT)

It is an arrangement technique for anticipating names of information by emphasizing the info information through a learning tree (Shareef, Zhu & Musavi, 2008). Amid this procedure, the element properties are contrasted relative with choice conditions to achieve a particular class. The writing is extremely

Figure 2. Example of node localization in wireless networks using supervised learning

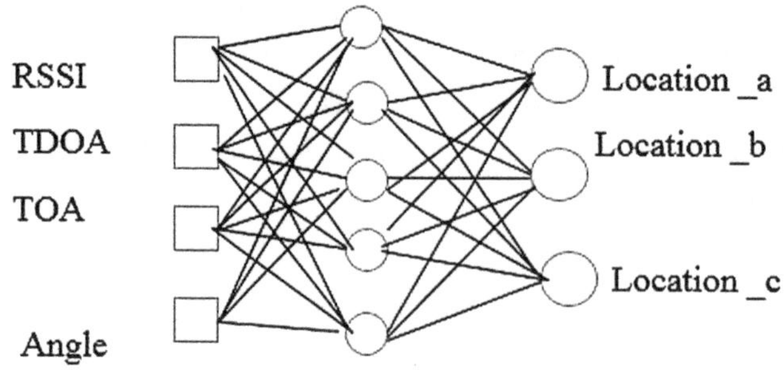

rich with arrangements that utilization DT calculation to determine diverse WSNs' plan challenges. For instance, DT gives a basic, yet effective strategy to distinguish connect unwavering quality in WSNs by recognizing a couple of basic highlights, for example, misfortune rate, debasement rate, mean time to disappointment (MTTF) and interim to reestablish (MTTR). Be that as it may, DT works just with directly divisible information and the way toward building ideal learning trees is NP-finished (winter, Xu & Lee, 2005).

Neural Systems (NNs)

This learning algorithm could be built by falling chains of choice units (e.g., perceptrons or radial based functions) used to perceive nonlinear and complex capacities. In Wireless Networks, utilizing neural systems in circulated behavior is as yet not all that inescapable because of the high computational necessities for taking in the system weights, and also the high administration overhead. Be that as it may, in unified arrangements, neural systems can take in numerous yields and choice limits at once (Lippmann, 1987) which makes them reasonable for comprehending a few system challenges utilizing a similar model.

We consider a sensor hub confinement issue (i.e., deciding hub's geological position) as an application case of neural system in WSNs. Hub confinement can be founded on engendering point and separation estimations of the got signals from grapple hubs. Such estimations may incorporate received signal strength indicator (RSSI), time of arrival (TOA), and time difference of arrival (TDOA) as showed in Figure 2. After directed preparing, neural system creates an expected hub area as vector-esteemed facilitates in 3D space. Related calculations to neural systems incorporate self-sorting out guide (or Kohonen's maps) and learning vector quantization (LVQ)

Support Vector Machines (SVMs)

It is a machine learning calculation that figures out how to arrange information focuses utilizing named preparing tests (Steinwart & Christmann, 2008). For instance, one approach for recognizing noxious conduct of a hub is by utilizing SVM to examine transient and spatial connections of information. To illustrate, given WSN's perceptions as focuses in the element space, SVM isolates the space into parts. These parts are isolated by as wide as conceivable edges (i.e., detachment holes), and new perusing will be characterized in view of which side of the holes they fall on. A SVM calculation, which incorporates advancing a quadratic capacity with straight limitations (that is, the issue of building an arrangement of

Figure 3. Example of simple clustering

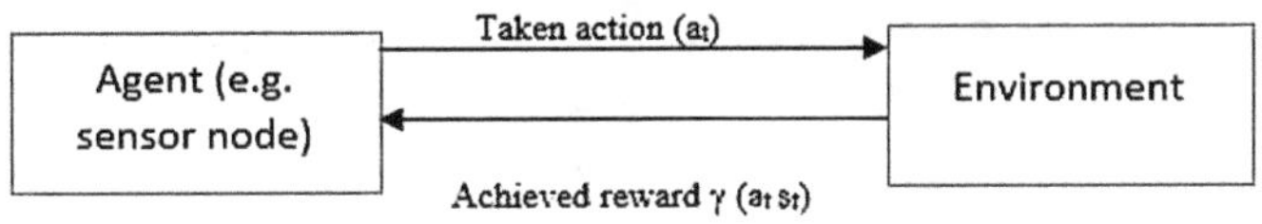

hyperplanes), gives an elective strategy to the multi-layer neural system with nonconvex and unconstrained improvement issue (Ayodele, 2010). Potential utilizations of SVM in WSNs are security.

Unsupervised Learning

Unsupervised students are not furnished with marks (i.e., there is no output vector). Essentially, the objective of an unsupervised learning calculation is to characterize the example set into various gatherings by researching the comparability between them. Not surprisingly, this subject of learning calculations is generally utilized as a part of hub grouping and information total issues. Without a doubt, this wide selection is because of information structures (i.e., no named information is accessible) and the coveted result in such issues.

K-Means Clustering

The K-means clustering: the research by (Kanungo, Mount, Netanyahu, Piatko, Silverman & Wu, 2002). is utilized to perceive information into various classes (known as bunches). This unsupervised learning calculation is broadly utilized as a part of sensor hub grouping issue because of its straight unpredictability and Fig. 3: Example of simple clustering explains how the data is clustered as per their classes and usage. The k-implies ventures to determine such hub bunching issue are:

1. Randomly pick k hubs to be the underlying centroids for various groups;
2. Label every hub with the nearest centroid utilizing a separation work;
3. Re figure the centroids utilizing the present hub participations and
4. Stop if the merging condition is legitimate (e.g., a predefined limit for the total of separations amongst hubs and their point of view centroids), generally backpedal to step 2.

Principal Component Analysis (PCA)

It is a multivariate strategy for information pressure and dimensionality lessening that expects to separate vital data from information and present it as an arrangement of new orthogonal factors called key segments (Jolliffe, 2011). As appeared in Fig. 4, the important parts are requested with the end goal that the principal segment compares to the most noteworthy change course of the information, et cetera for alternate segments. Thus, the slightest change parts can be disposed of as they contain the minimum data content. For instance, PCA lessens the measure of transmitted information among sensor hubs by finding a little arrangement of uncorrelated straight mixes of unique readings. Besides, the PCA technique disentangles the critical thinking by thinking about just couple of conditions in expansive variable issues

Figure 4. A simple 2D visualization pf principal component analysis algorithm

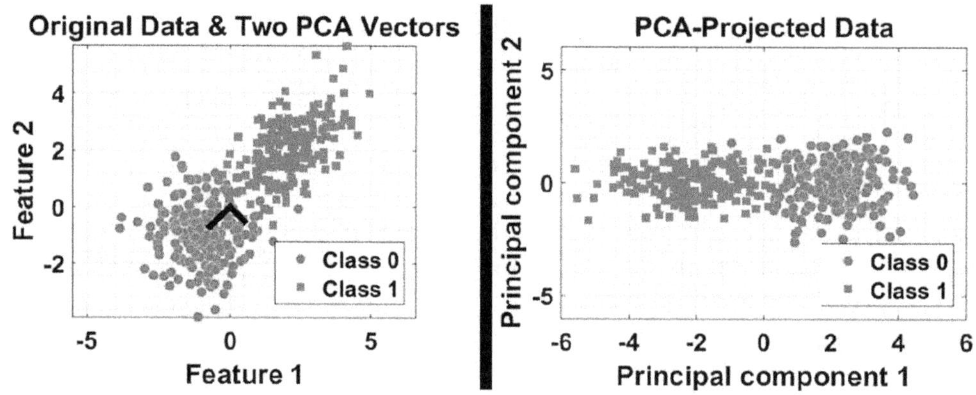

(i.e., tuning huge information into small information portrayal) (Feldman, Schmidt & Sohler, 2013). A careful exchange of the PCA hypothesis (e.g., the eigenvalue, eigenvector, and covariance lattice investigation) is given in (Jolliffe, 2011).

Reinforcement Learning

Reinforcement learning empowers a specialist (e.g., a sensor node) to learn by interfacing with its condition. The specialist will figure out how to take the best activities that maximize its long term compensates by utilizing its own involvement. The most understood Reinforcement learning procedure is Q-learning (Watkins & Dayan, 1992). As appeared in Fig. 5, a specialist frequently refreshes its accomplished prizes in light of the made a move at a given state. The future total reward (i.e., the Q-value) of playing out an activity at a given state a_t at a given state s_t is processed utilizing Eq. (1). $Q\ (s_{t+1};\ a_{t+1}) = Q\ (s_t;\ a_t) + \gamma(r\ (s_t;\ a_t) - Q\ (s_t;\ a_t))$ (1)where $r\ (s_t,\ a_t)$ signifies the quick reward of playing out an activity at a given state s_t, and γ is the learning rate that decides how quick learning happens (normally set to an incentive in the vicinity of 0 and 1). This algorithm can be effectively executed in a conveyed design like Wireless Networks, where every node tries to pick activities that are relied upon to augment its long haul rewards. Note that Q-learning has been broadly and effectively utilized as a part of Wireless Networks routing problems (e.g., Sun, Tatsumi & Zhao, 2002; Dong, Agrawal & Sivalingam, 2000; Forster & Murphy, 2007; Arroyo-Valles, Alaiz-Rodriguez, Guerrero-Curieses & Cid-Sueiro, 2007).

Figure 5. A visualization of Q-learning method

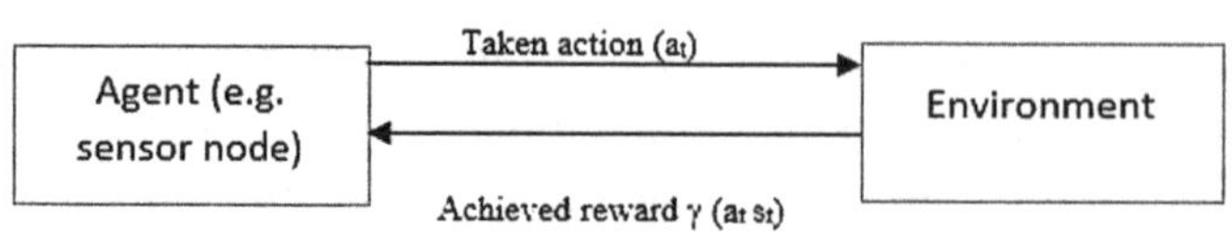

FINDING THE RIGHT ALGORITHM

Distinctive properties of Wireless ad hoc networks must be considered while choosing a decent ML approach.

- **Wireless Ad-Hoc Nature:** There is no settled foundation or a various leveled structure. The common remote medium sets extra limitations on the conceivable communication between the hubs and new issues emerge, like unbalanced connections.
- **Versatility and Topology Changes:** MANETs are normally exceptionally mobile situations, where new hubs always join the system, and existing ones either travel through the system or abandon it. Applications need to consider this and should have the capacity to adapt to the adjustable topology of the system. Then again, vitality isn't confined, since singular hubs are generally workstations or PDAs, which can be effortlessly recharged.
- **Energy Confinements:** On the other hand, WSNs are once in a while exceptionally portable, however exceedingly vitality limited. The fundamental situation incorporates a settled topology of sensor hubs, together with a set number of all the more capable base stations and no support or reviving is conceivable after organization. Accordingly, cost minimization and self-sufficient conduct are alluring.
- **Physical Appropriation:** Each hub in a MANET or a WSN is an independent computational unit, which speaks with its neighbors by means of messages. Information is likewise dispersed all through the hubs of the system and can be assembled on a focal station just with high correspondence costs. Thusly, calculations requiring worldwide data from the entire system turn out to be exceptionally costly. In this way, disreputability of the calculations is exceedingly suggested.

In dispersed ML algorithms, the arrangement is found in a shared way and every hub holds just its piece, specifically important to its activities (like the best next bounce in a routing problem).

FUTURE TREND OF MACHINE LEARNING IN WIRELESS COMMUNICATION

Machine learning based data handling in Wireless Networking is at its entering stage, when contrasted with conventional machine learning and Wireless Networks. As of now, examines predominantly center around applying machine learning methods to take care of a specific issue in Wireless Communication. Distinctive analysts will have diverse suppositions, application situations and inclinations in applying machine learning calculations. These distinctions speak to a noteworthy test in enabling scientists to expand upon each other's work with the goal that examination results will collect in the group. Subsequently, a typical design over the WSN machine learning group would be essential. It sets out two ways towards a brought together learning approach – dispersed learning in WSNs with a combination focus, where the attention is on how learning is affected when correspondence limitations constrain access to preparing information; and appropriated learning in WSN with in-arrange handling, where the emphasis is on how between sensor interchanges and neighborhood handling might be abused to empower correspondence proficient community learning.

SUMMARY

In this work, a review of the most imperative ML-based answers for the directing issues in Wireless ad-hoc systems (Wireless communication and Machine learning) is displayed. Furthermore, a guide on machine learning methods and their relevance to the depicted area is given. Future work will expand the rundown of displayed ML procedures and systems administration applications. Machine learning has been connected to steering issues in remote systems as right on time as 1994 (Boyan & Littman, 1994) and from that point forward discovers increasingly appropriateness. Extraordinary dispersed calculations have been created (Stone, 2000) or balanced (Stone & Veloso, 1999) to the wireless communication situation and an extensive range of conventional ML approaches has been connected to various issues, from support learning and swarm insight to neural systems. The nature of the works and their for the most part great outcomes contrasted with non-learning conventions plainly demonstrate the wide materialness of ML approaches and their points of interest in a versatile, temperamental condition. Also, a few ML systems have risen to be the most fitting for use in steering for remote specially appointed systems: support learning for vitality limited for the most part static Wireless Networks and swarm knowledge for profoundly portable, non-vitality confined MANETs. Both are completely appropriated, exceptionally tolerant to natural changes and simple to execute. Then again, just a couple of the conventions have achieved a more develop stage and have been tried in a genuine arrangement or have been widely assessed against different arrangements, both customary and ML-based. Likewise, now and again improper ML algorithms have been utilized, prompting higher correspondence expenses or lower quality outcomes. In rundown, future research in the territory needs to focus on encourage improvement and modification of the calculations to the specific situation and driving the answers for a develop convention state with true execution. Also, ML methodologies ought to be connected to a more extensive arrangement of issues and convention stack layers, with the objective of greater adaptability and freedom of the frameworks.

REFERENCES

Abu-Mostafa, Y. S., Magdon-Ismail, M., & Lin, H. T. (2012). Learning from data (Vol. 4). New York, NY: AMLBook.

Agrawal, R., & Srikant, R. (2000). Privacy-preserving data mining: Vol. 29. *No. 2*. ACM.

Al-Karaki, J. N., & Kamal, A. E. (2004). Routing techniques in wireless sensor networks: A survey. *IEEE Wireless Communications*, *11*(6), 6–28. doi:10.1109/MWC.2004.1368893

Alsheikh, M. A., Lin, S., Niyato, D., & Tan, H. P. (2014). Machine learning in wireless sensor networks: Algorithms, strategies, and applications. *IEEE Communications Surveys and Tutorials*, *16*(4), 1996–2018. doi:10.1109/COMST.2014.2320099

Arroyo-Valles, R., Alaiz-Rodriguez, R., Guerrero-Curieses, A., & Cid-Sueiro, J. (2007, December). Q-probabilistic routing in wireless sensor networks. In *Intelligent Sensors, Sensor Networks and Information, 2007. ISSNIP 2007. 3rd International Conference on* (pp. 1-6). IEEE. 10.1109/ISSNIP.2007.4496810

Ayodele, T. O. (2010). Types of machine learning algorithms. In *New advances in machine learning*. InTech.

Bahrepour, M., Meratnia, N., Poel, M., Taghikhaki, Z., & Havinga, P. J. (2010, November). Distributed event detection in wireless sensor networks for disaster management. In *Intelligent Networking and Collaborative Systems (INCOS), 2010 2nd International Conference on* (pp. 507-512). IEEE.

Bengio, Y. (2009). Learning deep architectures for AI. *Foundations and trends® in Machine Learning, 2*(1), 1-127.

Beyer, K., Goldstein, J., Ramakrishnan, R., & Shaft, U. (1999, January). When is "nearest neighbor" meaningful? In *International conference on database theory* (pp. 217-235). Springer.

Boyan, J. A., & Littman, M. L. (1994). Packet routing in dynamically changing networks: A reinforcement learning approach. In Advances in neural information processing systems (pp. 671-678). Academic Press.

Branch, J. W., Giannella, C., Szymanski, B., Wolff, R., & Kargupta, H. (2013). In-network outlier detection in wireless sensor networks. *Knowledge and Information Systems, 34*(1), 23–54. doi:10.100710115-011-0474-5

Chien, C., Elgorriaga, I., & McConaghy, C. (2001, August). Low-power direct-sequence spread-spectrum modem architecture for distributed wireless sensor networks. In *Proceedings of the 2001 international symposium on Low power electronics and design* (pp. 251-254). ACM. 10.1145/383082.383151

Dharne, A. G., Lee, J., & Jayasuriya, S. (2006, June). Using fuzzy logic for localization in mobile sensor networks: simulations and experiments. In *American Control Conference*, 2006 (pp. 6-pp). IEEE. 10.1109/ACC.2006.1656524

Di, M., & Joo, E. M. (2007, December). A survey of machine learning in wireless sensor netoworks from networking and application perspectives. In *Information, Communications & Signal Processing, 2007 6th International Conference on* (pp. 1-5). IEEE.

Dong, S., Agrawal, P., & Sivalingam, K. (2007, November). Reinforcement learning based geographic routing protocol for UWB wireless sensor network. In *Global Telecommunications Conference, 2007. GLOBECOM'07. IEEE* (pp. 652-656). IEEE. 10.1109/GLOCOM.2007.127

Feldman, D., Schmidt, M., & Sohler, C. (2013, January). Turning big data into tiny data: Constant-size coresets for k-means, pca and projective clustering. In *Proceedings of the twenty-fourth annual ACM-SIAM symposium on Discrete algorithms* (pp. 1434-1453). Society for Industrial and Applied Mathematics. 10.1137/1.9781611973105.103

Foo, M. C., Lim, H. B., Zeng, Y., Lam, V. T., Teo, R., & Ng, G. W. (2005, December). Impact of distributed resource allocation in sensor networks. In *Intelligent Sensors, Sensor Networks and Information Processing Conference, 2005. Proceedings of the 2005 International Conference on* (pp. 69-74). IEEE.

Forster, A. (2007, December). Machine learning techniques applied to wireless ad-hoc networks: Guide and survey. In *Intelligent Sensors, Sensor Networks and Information, 2007. ISSNIP 2007. 3rd International Conference on* (pp. 365-370). IEEE.

Forster, A., & Murphy, A. L. (2007, December). FROMS: Feedback routing for optimizing multiple sinks in WSN with reinforcement learning. In *Intelligent Sensors, Sensor Networks and Information, 2007. ISSNIP 2007. 3rd International Conference on* (pp. 371-376). IEEE.

Guzella, T. S., & Caminhas, W. M. (2009). A review of machine learning approaches to spam filtering. *Expert Systems with Applications*, *36*(7), 10206–10222. doi:10.1016/j.eswa.2009.02.037

Hinton, G. E., Osindero, S., & Teh, Y. W. (2006). A fast learning algorithm for deep belief nets. *Neural Computation*, *18*(7), 1527–1554. doi:10.1162/neco.2006.18.7.1527 PMID:16764513

Ibnkahla, M. (2000). Applications of neural networks to digital communications–a survey. *Signal Processing*, *80*(7), 1185–1215. doi:10.1016/S0165-1684(00)00030-X

Intanagonwiwat, C., Govindan, R., & Estrin, D. (2000, August). Directed diffusion: A scalable and robust communication paradigm for sensor networks. In *Proceedings of the 6th annual international conference on Mobile computing and networking* (pp. 56-67). ACM. 10.1145/345910.345920

Janakiram, D., Reddy, V. A., & Kumar, A. P. (2006, January). Outlier detection in wireless sensor networks using Bayesian belief networks. In *Communication System Software and Middleware, 2006. Comsware 2006. First International Conference on* (pp. 1-6). IEEE.

Jayaraman, P. P., Zaslavsky, A., & Delsing, J. (2010, June). Intelligent processing of k-nearest neighbors queries using mobile data collectors in a location aware 3D wireless sensor network. In *International Conference on Industrial, Engineering and Other Applications of Applied Intelligent Systems* (pp. 260-270). Springer. 10.1007/978-3-642-13033-5_27

Jiang, C., Zhang, H., Ren, Y., Han, Z., Chen, K. C., & Hanzo, L. (2017). Machine learning paradigms for next-generation wireless networks. *IEEE Wireless Communications*, *24*(2), 98–105. doi:10.1109/MWC.2016.1500356WC

Jolliffe, I. (2011). Principal component analysis. In *International encyclopedia of statistical science* (pp. 1094–1096). Berlin: Springer. doi:10.1007/978-3-642-04898-2_455

Kanungo, T., Mount, D. M., Netanyahu, N. S., Piatko, C. D., Silverman, R., & Wu, A. Y. (2002). An efficient k-means clustering algorithm: Analysis and implementation. *IEEE Transactions on Pattern Analysis and Machine Intelligence*, *24*(7), 881–892. doi:10.1109/TPAMI.2002.1017616

Kaplantzis, S., Shilton, A., Mani, N., & Sekercioglu, Y. A. (2007, December). Detecting selective forwarding attacks in wireless sensor networks using support vector machines. In *Intelligent Sensors, Sensor Networks and Information, 2007. ISSNIP 2007. 3rd International Conference on* (pp. 335-340). IEEE.

Kim, M. H., & Park, M. G. (2009). Bayesian statistical modeling of system energy saving effectiveness for MAC protocols of wireless sensor networks. In *Software Engineering, Artificial Intelligence, Networking and Parallel/Distributed Computing* (pp. 233–245). Berlin: Springer. doi:10.1007/978-3-642-01203-7_20

Krause, A., Guestrin, C., Gupta, A., & Kleinberg, J. (2006). *Near-optimal sensor placements*. Nashville, TN: IPSN. doi:10.1145/1127777.1127782

Krishnamachari, L., Estrin, D., & Wicker, S. (2002). The impact of data aggregation in wireless sensor networks. In *Distributed Computing Systems Workshops, 2002. Proceedings. 22nd International Conference on* (pp. 575-578). IEEE.

Kulkarni, R. V., & Venayagamoorthy, G. K. (2009, June). Neural network based secure media access control protocol for wireless sensor networks. In *Neural Networks, 2009. IJCNN 2009. International Joint Conference on* (pp. 1680-1687). IEEE.

Kulkarni, S. R., Lugosi, G., & Venkatesh, S. S. (1998). Learning pattern classification-a survey. *IEEE Transactions on Information Theory*, *44*(6), 2178–2206. doi:10.1109/18.720536

Langley, P., & Simon, H. A. (1995). Applications of machine learning and rule induction. *Communications of the ACM*, *38*(11), 54–64. doi:10.1145/219717.219768

Lim, H. B., Foo, M. C., & Zeng, Y. (2006, December). An adaptive distributed resource allocation scheme for sensor networks. In *International Conference on Mobile Ad-Hoc and Sensor Networks* (pp. 770-781). Springer. 10.1007/11943952_65

Lippmann, R. (1987). An introduction to computing with neural nets. *IEEE ASSP Magazine*, *4*(2), 4–22. doi:10.1109/MASSP.1987.1165576

Lu, C. H., & Fu, L. C. (2009). Robust location-aware activity recognition using wireless sensor network in an attentive home. *IEEE Transactions on Automation Science and Engineering*, *6*(4), 598–609. doi:10.1109/TASE.2009.2021981

Morelande, M. R., Moran, B., & Brazil, M. (2008, March). Bayesian node localisation in wireless sensor networks. In *Acoustics, Speech and Signal Processing, 2008. ICASSP 2008. IEEE International Conference on* (pp. 2545-2548). IEEE.

O'Shea, T., & Hoydis, J. (2017). An introduction to deep learning for the physical layer. *IEEE Transactions on Cognitive Communications and Networking*, *3*(4), 563–575. doi:10.1109/TCCN.2017.2758370

Pandana, C., & Liu, K. R. (2005). Near-optimal reinforcement learning framework for energy-aware sensor communications. *IEEE Journal on Selected Areas in Communications*, *23*(4), 788–797. doi:10.1109/JSAC.2005.843547

Paradis, L., & Han, Q. (2007). A survey of fault management in wireless sensor networks. *Journal of Network and Systems Management*, *15*(2), 171–190. doi:10.100710922-007-9062-0

Rastegardoost, N., & Etemadyrad, N. (2015). *Machine Learning in Wireless Sensor Networks*. Academic Press.

Shareef, A., Zhu, Y., & Musavi, M. (2008, February). Localization using neural networks in wireless sensor networks. In *Proceedings of the 1st international conference on MOBILe Wireless MiddleWARE, Operating Systems, and Applications* (p. 4). ICST (Institute for Computer Sciences, Social-Informatics and Telecommunications Engineering). 10.4108/ICST.MOBILWARE2008.2901

Shen, Y. J., & Wang, M. S. (2008). Broadcast scheduling in wireless sensor networks using fuzzy Hopfield neural network. *Expert Systems with Applications*, *34*(2), 900–907. doi:10.1016/j.eswa.2006.10.024

Shu, H., & Liang, Q. (2005, March). Fuzzy optimization for distributed sensor deployment. In *Wireless Communications and Networking Conference*, 2005 *IEEE* (Vol. 3, pp. 1903-1908). IEEE.

Sinha, A., & Chandrakasan, A. (2001). Dynamic power management in wireless sensor networks. *IEEE Design & Test of Computers, 18*(2), 62–74. doi:10.1109/54.914626

Sohrabi, K., Gao, J., Ailawadhi, V., & Pottie, G. J. (2000). Protocols for self-organization of a wireless sensor network. *IEEE Personal Communications, 7*(5), 16-27.

Srinivasan, V., Lapuschkin, S., Hellge, C., Müller, K. R., & Samek, W. (2017, March). Interpretable human action recognition in compressed domain. In *Acoustics, Speech and Signal Processing (ICASSP), 2017 IEEE International Conference on* (pp. 1692-1696). IEEE.

Steinwart, I., & Christmann, A. (2008). *Support vector machines*. Springer Science & Business Media.

Stone, P. (2000, June). *TPOT-RL applied to network routing*. ICML.

Stone, P., & Veloso, M. (1999, April). Team-partitioned, opaque-transition reinforcement learning. In *Proceedings of the third annual conference on Autonomous Agents* (pp. 206-212). ACM. 10.1145/301136.301195

Sun, R., Tatsumi, S., & Zhao, G. (2002, October). Q-map: A novel multicast routing method in wireless ad hoc networks with multiagent reinforcement learning. In *TENCON'02. Proceedings. 2002 IEEE Region 10 Conference on Computers, Communications, Control and Power Engineering* (Vol. 1, pp. 667-670). IEEE.

Tam, V., Cheng, K. Y., & Lui, K. S. (2006, November). A descend-based evolutionary approach to enhance position estimation in wireless sensor networks. In Null (pp. 568-574). IEEE.

Tsai, C. W., Lai, C. F., Chiang, M. C., & Yang, L. T. (2014). Data mining for Internet of Things: A survey. *IEEE Communications Surveys and Tutorials, 16*(1), 77–97. doi:10.1109/SURV.2013.103013.00206

van Norden, W., de Jong, J., Bolderheij, F., & Rothkrantz, L. (2005, July). Intelligent task scheduling in sensor networks. In *Information Fusion, 2005 8th International Conference on* (Vol. 2, pp. 8-pp). IEEE. 10.1109/ICIF.2005.1592013

Wang, Y., Martonosi, M., & Peh, L. S. (2006, May). A supervised learning approach for routing optimizations in wireless sensor networks. In *Proceedings of the 2nd international workshop on Multi-hop ad hoc networks: from theory to reality* (pp. 79-86). ACM. 10.1145/1132983.1132997

Watkins, C. J., & Dayan, P. (1992). Q-learning. *Machine Learning, 8*(3-4), 279–292. doi:10.1007/BF00992698

Werbos, P. J. (2011). Computational intelligence for the smart grid-history, challenges, and opportunities. *IEEE Computational Intelligence Magazine, 6*(3), 14–21. doi:10.1109/MCI.2011.941587

Winter, J., Xu, Y., & Lee, W. C. (2005, July). Energy efficient processing of k nearest neighbor queries in location-aware sensor networks. In *Mobile and Ubiquitous Systems: Networking and Services, 2005. MobiQuitous 2005. The Second Annual International Conference on* (pp. 281-292). IEEE.

Yu, L., Wang, N., & Meng, X. (2005, September). Real-time forest fire detection with wireless sensor networks. In *Wireless Communications, Networking and Mobile Computing, 2005. Proceedings. 2005 International Conference on* (Vol. 2, pp. 1214-1217). IEEE.

Zhang, Y., Kwong, S., Wang, X., Yuan, H., Pan, Z., & Xu, L. (2015). Machine learning-based coding unit depth decisions for flexible complexity allocation in high efficiency video coding. *IEEE Transactions on Image Processing*, *24*(7), 2225–2238. doi:10.1109/TIP.2015.2417498 PMID:25826804

Zhao, L., & Liang, Q. (2005, March). Fuzzy deployment for wireless sensor networks. In *Computational Intelligence for Homeland Security and Personal Safety, 2005. CIHSPS 2005. Proceedings of the 2005 IEEE International Conference on* (pp. 79-83). IEEE.

Section 2

Machine Learning and Artificial Intelligence Applications

Chapter 8
Guaranteeing User Rates With Reinforcement Learning in 5G Radio Access Networks

Ioan-Sorin Comşa
Brunel University London, UK

Sijing Zhang
University of Bedfordshire, UK

Mehmet Emin Aydin
University of West of England, UK

Pierre Kuonen
University of Applied Sciences of Western Switzerland, Switzerland

Ramona Trestian
https://orcid.org/0000-0003-3315-3081
Middlesex University London, UK

Gheorghiţă Ghinea
Brunel University London, UK

ABSTRACT

The user experience constitutes an important quality metric when delivering high-definition video services in wireless networks. Failing to provide these services within requested data rates, the user perceived quality is strongly degraded. On the radio interface, the packet scheduler is the key entity designed to satisfy the users' data rates requirements. In this chapter, a novel scheduler is proposed to guarantee the bit rate requirements for different types of services. However, the existing scheduling schemes satisfy the user rate requirements only at some extent because of their inflexibility to adapt for a variety of traffic and network conditions. In this sense, the authors propose an innovative framework able to select each time the most appropriate scheduling scheme. This framework makes use of reinforcement learning and neural network approximations to learn over time the scheduler type to be applied on each momentary state. The simulation results show the effectiveness of the proposed techniques for a variety of data rates' requirements and network conditions.

DOI: 10.4018/978-1-5225-7458-3.ch008

1. INTRODUCTION

The accelerated acquisition of powerful mobile devices is significantly contributing to the growing market of immersive multimedia applications. According to Cisco (2017), it is envisioned that by 2021 more than 80% of total mobile data will be represented by video traffic at different data rate requirements. In this context, the end-user Quality of Experience (QoE) will make the difference between network operators while providing these pretentious services (Trestian, Comsa, and Tuysuz, 2018). According to Ghinea, Timmerer, Lin, and Gulliver (2014), the concept of Multiple Sensorial Media (Mulsemedia) can enhance the user perceived QoE when experiencing poor video quality by incorporating additional senses such as: olfaction, wind, haptic, etc. However, the real factor that impacts the video quality degradation is denoted by the QoS provisioning schemes on the wireless interface that can differ from one operator to another. By providing higher video rates than the requested limit, the rate of packet drops is increased in order to keep the normal functionality of video decoders. On the other side, lower data rates for video services will increase the packet delays which have as a consequence a larger number of lost packets at the radio interface. Thus, guaranteeing certain data rates for video traffic is crucial in order to avoid the degradation of user perceived quality.

In 5th Generation (5G) of mobile communications standard, guaranteeing certain bit rate requirements is even stricter especially with the popularity of the new bandwidth hungry applications (i.e. high definition video, virtual reality traffic) (Elbamby, Perfecto, Bennis, and Doppler, 2018). This puts a significant pressure on Radio Resource Management (RRM) to provide these immersive services with very stringent QoS in multi-user scenarios (Li et al. 2017). Alongside of other RRM functions, the packet scheduler is in charge of allocating user data packets in frequency domain at each predefined Transmission Time Intervals (TTIs). According to Comşa (2014a), the scheduling process is conducted based on the scheduling rules aiming to maximize the satisfaction of particular QoS requirements. In literature, several scheduling rules are proposed to deal with the Guaranteed Bit Rate (GBR) objective. For example the scheduling rule proposed by Lundevall et al. (2004) is designed to work for WCDMA access networks and very low data rates of video services. Andrews, Qian, and Stolyar (2005) propose a scheduler for CDMA downlink networks in which a maximum number of 40 users are scheduled with the maximum rate of 160kbps. In the same type of access networks, the scheduler proposed by Kolding (2006) outperforms other GBR oriented schedulers for the considered networking scenarios. However, the proposed schedulers work appropriate only for particular scheduler states in terms of: access technologies, user rates, channel conditions, traffic load, etc. On one hand, these scheduling techniques must be upgraded for the novel access technologies imposed by 5G standard. On the other hand, the aim would be to use each of these scheduling rules on the best matching scheduler state in order to maximize over time the satisfaction of user rate requirements for various traffic types.

The proposed scheduling optimization problem must determine at each TTI the radio resources to be selected for all the active users as well as the most convenient scheduling rule to be followed in order to get the maximum possible GBR satisfaction outcome according to the scheduler momentary states. According to the selected scheduling rule, the resource allocation performs the frequency prioritization by favoring those users with poorer GBR satisfaction profile. If this frequency prioritization can be easily computed each TTI by simply calculating given scheduling metric for each user and radio resource, the scheduling rule selection must be a priori decided based on the momentary scheduler states. The idea is learn over time some preference values by applying random scheduling rules in many visits of scheduler states. But for the GBR-based maximization problem, this solution is unfeasible since the scheduler state space is continuous and multidimensional and hence, action-state values cannot be enumerated exhaustively. Thus, the only way is to approximate the best scheduling rule selection on each scheduler state. The aim is to learn these approximations in such a way that eventually, these solutions of the scheduling optimization problem would converge to optimal solutions. These approximations are represented by non-linear functions or neural networks whose weights are refined and adapted each TTI according to the selected scheduling rule. According to Van Hasselt (2011), the Reinforcement Learning (RL) principles and algorithms can be used to update these non-linear functions by reinforcing for each action function the corresponding time-based error.

This chapter proposes an innovative scheduler framework that aims to improve the user QoE of video services by maximizing the satisfaction of users' rates requirements under three different traffic type scenarios, such as: full buffer, Constant Bit Rate (CBR) and Variable Bit Rate (VBR). A new scheduling rule is proposed to maximize the satisfaction of GBR requirements. Then, the proposed scheduling framework makes use of RL algorithms to learn over time the policy of scheduling rules to be applied on each state. We study the impact of five actor-critic RL algorithms on the aforementioned traffic types with various channel and networking conditions. The scheduling process is conducted in Orthogonal Frequency Division Multiple Access (OFDMA) downlink systems. OFDMA is one of the access scheme candidates in 5G radio networks as proposed by FANTASTIG-5G (2016) European project.

The rest of this chapter is organized as follows: Section 2 provides a summary of related works; Section 3 presents the scheduling system model; Section 4 details the proposed RL framework with neural network approximations; Section 5 presents the simulation results for different traffic classes and network settings. Finally, Section 6 presents future trends and concludes this chapter.

2. RELATED WORK

The user QoE degradation of multimedia applications can be partially masked by introducing the mulsemedia services as stated by Yuan, Chen, Ghinea and Muntean (2014). According to Ghinea and Ademoye (2012), by adding olfaction sensorial components to the conventional video content, the user perceived experienced can be strongly improved if the perceived olfactory synchronization (Ademoye and Ghinea, 2009) is considered. Also, for some particular video content enhanced with olfaction sensory effects, the audio stream can be masked as suggested by Ademoye, Murray, and Ghinea (2016). An adaptation technique for the mulsemedia content based on the available bandwidth is proposed by Yuan, Ghinea and Muntean (2015). Comşa, Trestian and Ghinea (2018a) propose the concept of 360° mulsemedia in which, the QoE degradation of 360° video content due to fast head movements or poor network conditions can be masked if additional 360° sensory objects (i.e. wind, olfaction, heat) are perfectly mapped with the conventional content in both spatial and temporal domains. In order to satisfy the stringent QoS requirements of bandwidth-hungry 360° mulsemedia applications in multi-user scenario, a novel 5G scheduler framework is proposed by Comşa et al. (2018a) that makes use of the reinforcement learning technique.

The dynamic network conditions within the wireless environment play an important role in meeting the QoS requirements for a wide range of applications. In general, the location of the mobile users is important as the users located near the base station will experience better channel quality and can achieve higher data rates than those users located at the cell edge with poor channel quality (Trestian, Muntean, & Ormond, 2009; Trestian, Ormond, & Muntean, 2012; Vien, Akinbote, Nguyen, Trestian, & Gemikonakli, 2015). However, by enabling the available spectrum to the users with better channel conditions, other users are starved in receiving the requested data for longer time (Vien, Nguyen, Trestian, Shah, & Gemikonakli, 2016). Moreover, various adaptive multimedia mechanisms have been proposed in an attempt to cope with the QoS degradation and increase the QoE provisioning as presented by Zou, Trestian, and Muntean (2013, 2014, 2017).

To maximize the QoS satisfaction and implicitly the QoE provisioning of real time applications, Reinforcement Learning is used to learn over time the most convenient scheduling rule on each momentary state (Comşa, Aydin, Zhang, Kuonen, and Wagen, 2011; Comşa, Aydin, Zhang, Kuonen, and Wagen, 2012). The exploited scheduling rules are focused on satisfying various objectives, such as: user fairness satisfaction, GBR, delay and packet loss rates. The parameterization of Proportional Fair (PF) scheduling rule is used to attain dynamic trade-off between system throughput maximization and user fairness satisfaction according to the network conditions (Comşa, Zhang, Aydin, Kuonen, and Wagen, 2012; Comşa et al., 2014b, Comşa et al., 2014c). Neural networks are used to approximate the best parameterizations of Generalized PF (GPF) scheduling rule at each momentary state. Comşa et al. (2014b) use actor-critic RL algorithms to perform the simple parameterization of GPF rule. Various RL algorithms are used to update the weights of neural networks: some RL algorithms use fixed and discrete steps for GPF simple parameterization and update the neural networks for each step; another RL algorithm entitled continuous actor-critic RL algorithm makes use of only one neural network to output the continuous values that parameterize GPF rule at each TTI. When comparing with other state-of-the-art approaches, the learnt functions with these RL algorithms show gains higher than 10% when matching the user rates against the fairness satisfaction criterion. Moreover, Comşa et al. (2014c) use the same continuous actor-critic RL scheme for the double parameterization of the GPF scheduling rule by adapting the neural network that provides two continuous values at each TTI. The results indicate a gain larger than 5% when compared to actor-critic RL framework that uses the simple parameterization. Also, the double parameterization framework reveals a much higher capacity of recovering from unfeasible state space regions where the user fairness criterion is not satisfied.

Scheduling frameworks that use RL algorithms are also exploited to optimize the QoS objectives concomitantly. For instance, Comşa, De Domenico, and Ktenas (2017) proposes a 5G scheduler able to optimize the satisfaction of GBR, PLR and delay requirements at the same time for different types of traffic classes. The proposed actor-critic RL framework use a pool of three scheduling rule, each of them being oriented on one of the aforementioned particular objectives. When compared with simple scheduling rules, the proposed RL framework shows gains higher than 10% for different traffic types by monitoring the time when all users are satisfied from the viewpoint of all three objectives. In another work, Comşa, Trestian and Ghinea (2018b) optimize the satisfaction of both packet loss and delay objectives by using various RL algorithms. Different settings for the time windows are considered when computing the online packet loss rates for each user. Larger time windows may involve higher packet loss rates while shorter ones will decrease the probability of losing more data packets. Under various network conditions and time window settings, the learnt scheduling policies are able to outperform the standard scheduling rules when monitoring the number of TTIs with satisfied users from the viewpoint of packet loss and delay requirements.

This chapter considers the satisfaction of GBR requirements for different types of applications for downlink OFDMA scheduling. A variety of RL algorithms is considered to learn the best policy of scheduling rule for different settings of time windows that are used to compute the average user rates. Each RL consider a neural network for each scheduling rule from the pool. We aim to learn over time the best set of non-linear functions that can provide the selection of the best scheduling rule in each momentary state.

3. GBR BASED SCHEDULING MODEL

In this chapter we consider OFDMA downlink scheduling in which the available bandwidth is divided in equal Resource Blocks (RBs) at each TTI t. A RB is the smallest resource unit that can be allocated to one user based on the scheduling procedure. Then, we define by $\mathcal{B} = \{1, 2, ..., B\}$ the set of RBs corresponding to a given system bandwidth where B is the maximum number of RBs. Also, we define the User Equipment (UE) being defined by homogeneous traffic such as: full buffer, CBR and VBR. We denote by $\mathcal{I}_t = \{1, 2, ..., I_t\}$ the set of active users at TTI t where I_t is the maximum number of users. The set of active users $\mathcal{I}_t$ is time dependent since the total number of users I_t may change from one TTI to another.

The main role of the packet scheduler is to allocate each RB $j \in \mathcal{B}$ to user $i \in \mathcal{I}_t$ such that the satisfaction of GBR requirements is maximized. We define by $\underline{T_i}$ the GBR requirement for the average throughput of user $i \in \mathcal{I}_t$ characterized by a given traffic type. The GBR objective is satisfied for user $i \in \mathcal{I}_t$ at TTI t only if the average user throughput is greater than the GBR requirement $\underline{T_i}$. According to Comşa (2014a), the average user throughput can be calculated by using two types of filters, Exponential Moving Filter (EMF) and Median Moving Filter (MMF):

- EMF based average user throughput: we define by $\overline{T}_i[t]$ the average user throughput of user $i \in \mathcal{I}_t$ and calculated according to:

$$\overline{T}_i[t] = (1 - \beta) \cdot \overline{T}_i[t-1] + \beta \cdot T_i[t] \tag{1}$$

where β is the forgetting factor. When β takes low values, then the impact of new value throughput $T_i[t]$ is very low when computing its average value. On the other side, when β increases, then high oscillations are introduced when the scheduling performance is analyzed.

- MMF based average user throughput: the instantaneous user throughputs are stored for a given time window W. Then the computation of average user throughput $\overline{\overline{T}}_i[t]$ is determined each TTI according to the following formula:

$$\overline{\overline{T}}_i[t] = \frac{1}{W} \sum_{x=0}^{W} T_i[t-x] \tag{2}$$

When the time window W is high, the average user throughput is calculated based on the large number of instantaneous throughputs. In this case, the convergence of average user throughput $\overline{\overline{T}}_i[t]$ to its requirement $\underline{T_i}$ may take longer time to be achieved and the RRM outcome in terms of GBR satisfaction depends on very large number of observations. For very restrictive lengths of time windows, only some previous instantaneous throughputs are considered and then, the process can get significant oscillations and the RRM outcome in terms of GBR satisfaction is noisy. For this reasons, W must be carefully chosen in order to quantify best the scheduling rule applied on each state. In this sense, we propose to calculate the median moving window W as a function that depends on the number of active users I_t and the maximum number of schedulable users $I_{\max}$ such as:

$$W = \rho \cdot \left[I_t / I_{\max} \right] \tag{3}$$

where $I_{\max}$ represents the maximum number of users that can be scheduled at each TTI based on the signaling overhead constraints and system bandwidth and $\rho \in \mathbb{R}^+$ is the windowing factor.

According to some studies conducted by Comşa (2014a), the average user rates $\overline{T}_i\left[t\right]$ computed based on the exponential filter are used to compensate the channel fluctuations in OFDMA downlink scheduling, whereas the MMF based average user throughputs are used to measure the impact of each applied scheduling rule from the perspective of GBR objective. Thus, the setting of the windowing factor ρ plays a crucial role. When ρ is large, the probability of satisfying the GBR objective for each user increases since this performance is measured on the long term purpose by improving the system throughput and degrading the user fairness. But the long term GBR satisfaction does not guarantee the required amount of data in the short term downlink scheduling. Therefore, one purpose of this chapter is to find the optimal windowing factor which can permit to deliver the requested data rate on a short term purpose.

Let us define $\overline{\overline{T}}\left[t\right] = \left[\overline{\overline{T}}_1, \overline{\overline{T}}_2, ..., \overline{\overline{T}}_I\right]$ the momentary average throughput vector calculated based on MMF filter and $\underline{T} = \left[\underline{T}_1, \underline{T}_2, ..., \underline{T}_I\right]$ the corresponding vector of GBR requirements. It is said that, the GBR objective is satisfied if the instantaneous vector $\overline{\overline{T}}\left[t\right]$ satisfies the requirement vector for all active users. Furthermore, let us define $\mathcal{D} = \left\{1, 2, ..., D\right\}$ the set of scheduling rules oriented on GBR objective. At each TTI, each scheduling rule $d \in \mathcal{D}$ impacts differently in the GBR satisfaction of the average throughput vector $\overline{\overline{T}}\left[t\right]$. The aim is to apply at each momentary scheduler state the best scheduling rule $d^* \in \mathcal{D}$ in order to maximize the number of TTIs when the vector $\overline{\overline{T}}\left[t\right]$ satisfies the requirement vector $\underline{T}$. If the overall GBR satisfaction is not possible due to certain network conditions (i.e. poor channel conditions, high number of users), the aim would be to maximize the percentage of user throughputs from $\overline{\overline{T}}\left[t\right]$ that respects their GBR requirements from $\underline{T}$.

The scheduler should be aware about the benefit of allocating each RB $j \in \mathcal{B}$ to UE $i \in \mathcal{I}_t$ from the viewpoint of the GBR objective. This cost value is determined for each user by a concave and monotone utility function, defined as: $U_i : \mathbb{R} \rightarrow \mathbb{R}$, $U_i\left(\overline{T}_i\right) = F_i\left(\overline{T}_i\right) \cdot G\left(\overline{\overline{T}}_i\right)$, where F_i is a function that depends on the average user rate calculated with exponential moving filter in order to compensate the channel oscillations and $G\left(\overline{\overline{T}}_i\right)$ is the utility weight function designed to take as input the average user rate $\overline{\overline{T}}_i$

and output the priority of user $i \in \mathcal{I}_t$ to be scheduled on a given RB at TTI t. In fact, a scheduling rule defines the type of utility function with the condition that the same utility is assigned for all active users within one TTI. Then, the utility function of user $i \in \mathcal{I}_t$ can be defined as: $U_{d,i}\left(\overline{T}_i\right) = F_i\left(\overline{T}_i\right) \cdot G_{d,i}\left(\overline{\overline{T}}_i\right)$ in which we keep the same function F_i for all scheduling rules while the weight function $G_{d,i}$ depends on the selected scheduling rule $d \in \mathcal{D}$ at each TTI.

3.1 Optimization Problem

By using the Taylor's expansion (Comşa, 2014a), the proposed short-term optimization problem aims to allocate the radio resources as well as to decide the scheduling rule to be applied at each TTI, such as:

$$\max_{b,c} \sum_{d=1}^{D} \sum_{i=1}^{I_t} \sum_{j=1}^{B} b_{d,i}\left[t\right] c_{i,j}\left[t\right] \cdot F_i^{'}\left(\overline{T}_i\right) \cdot G_{d,i}\left(\overline{\overline{T}}_i\right) \cdot r_{i,j}\left[t\right]$$
$$s.t. \begin{cases} \sum_d b_{d,i}\left[t\right] = 1, \; i = 1,\ldots,I_t, & (a) \\ \sum_i b_{d^*,i}\left[t\right] = I_t, \; d^* \in \mathcal{D}, & (b) \\ \sum_i b_{d^{\otimes},i}\left[t\right] = 0, \; d^{\otimes} \in \mathcal{D}, \forall d^{\otimes} \neq d^*, & (c) \\ \sum_i c_{i,j}\left[t\right] \leq 1, \; j = 1,\ldots,B, & (d) \\ b_{d,i}\left[t\right] \in \left\{0,1\right\}, \; \forall d \in \mathcal{D}, \forall i \in \mathcal{I}_t, & (e) \\ c_{i,j}\left[t\right] \in \left\{0,1\right\}, \; \forall i \in \mathcal{I}_t, \forall j \in \mathcal{B}, & (f) \end{cases} \tag{4}$$

where, the optimization problem uses the first derivative of utility function $U_{d,i}^{'}\left(\overline{T}_i\right) = F_i^{'}\left(\overline{T}_i\right) \cdot G_{d,i}\left(\overline{\overline{T}}_i\right)$. The function F_i must be chosen to attenuate the channel variations such as: $F_i\left(\overline{T}_i\right) = \log\left(\overline{T}_i\right)$, and then, $F_i^{'}\left(\overline{T}_i\right) = 1/\overline{T}_i$. In (4), $r_{i,j}\left[t\right]$ is the achievable rate of user $i \in \mathcal{I}_t$ for RB $j \in \mathcal{B}$ at TTI t, being calculated as $r_{i,j}\left[t\right] = N_{i,j}^{bits}\left[t\right]/0.001$ (Comsa, 2014a), where $N_{i,j}^{bits}\left[t\right]$ is the maximum number of bits that can be transmitted if RB $j \in \mathcal{B}$ would be allocated to user $i \in \mathcal{I}_t$. Basically, $N_{i,j}^{bits}\left[t\right]$ is determined based on the Channel Quality Indicator (CQI) which is transmitted by each UE to the base station at certain number of TTIs. However, if $r_{i,j}\left[t\right] \cdot F_i^{'}\left(x_i\right) \approx 1$, then the optimization problem is focused more on the GBR objective since those users with higher values of $G_{d,i}\left(\overline{\overline{T}}_i\right)$ are preferred to be scheduled. Also, if $G_{d,i}\left(\overline{\overline{T}}_i\right) = 1$, then the optimization problem is focused more on scheduling those users with higher $r_{i,j}\left[t\right] \cdot F_i^{'}\left(\overline{T}_i\right)$ values, and thus, the fairness objective is addressed.

The optimization problem from (4) is non-linear programming model and the set of constraints is convex. In the optimization problem, $b_{d,i}\left[t\right] \in \left\{0,1\right\}$ is the scheduling rule assignation variable: $b_{d,i}\left[t\right] = 1$ when scheduling rule $d \in \mathcal{D}$ is assigned to user $i \in \mathcal{I}_t$ and $b_{d,i}\left[t\right] = 0$, otherwise. The time-based RB allocation variable is $c_{i,j}\left[t\right] = 1$ when the RB $j \in \mathcal{B}$ is allocated to user $i \in \mathcal{I}_t$, and $c_{i,j}\left[t\right] = 0$, otherwise.

Constraints (a) denote that the only one scheduling rule or utility function can be associated for each user $i \in \mathcal{I}_t$. The selected scheduling rule $d^* \in \mathcal{D}$ must be same for each user $i \in \mathcal{I}_t$, fact which is denoted by constraints (b) and (c). Constraints (d), allocates at most one user (if the corresponding data queues are not empty) per each RB $j \in \mathcal{B}$. Moreover, constraints (e) and (f) make the problem combinatorial. If the type of utility function is a priori decided, then the allocation metric becomes:

$$m_j[t] = \arg\max_{i \in \mathcal{I}_t} \left\{ G_{d,i}\left(\overline{\overline{T}}_i\right) \cdot r_{i,j}[t] / \overline{T}_i \right\} \tag{5}$$

where $m_j[t]$ indicates that RB $j \in \mathcal{B}$ is allocated to user $m \in \mathcal{I}_t$, $\forall m \neq i$ at TTI *t*. In this case, $c_{m,j}[t] = 1$ and the rest of variables are $c_{i,j}[t] = 0$, $\forall \{m, i\} \in \mathcal{I}_t$, $\forall i \neq m$. The same computation (5) is repeated for each RB until the entire spectrum is allocated. Once the resource allocation is performed, then the Modulation and Coding Scheme (MCS) is determined for each allocated user according to their CQIs. Also, the Transport Block (TB) is calculated in order to define the number of bits to be sent to each allocated user.

3.2 Conventional Utility Functions

According to (5), the performance of the scheduling process depends on the exploited rule $d \in \mathcal{D}$ that assigns the unique weight function $G_{d,i}$. In literature, three scheduling rules are proposed to optimize the GBR objective, such as:

- **Barier Function (BF)** ($d = 1$)**:** proposed by Lundevall et al. (2004) with the utility weight defined as follows:

$$G_{1,i}\left(\overline{\overline{T}}_i\right) = 1 + \omega_1 \cdot \exp\left[-\omega_2 \cdot \left(\overline{\overline{T}}_i - \underline{T}_i\right)\right] \tag{6}$$

where ω_1 and ω_2 are BF parameters. Basically, if user $i \in \mathcal{I}_t$ has its average throughput $\overline{\overline{T}}_i$ much below of its GBR requirement, then the allocation probability on RB $j \in \mathcal{B}$ increases exponentially as defined by (6).

- **Minimum/Maximum (mM)** ($d = 2$)**:** proposed by Andrews et al. (2005) stores an additional token counter for each user and the weight function is defined as follows:

$$G_{2,i}\left(\overline{\overline{T}}_i\right) = \exp\left(\omega_3 \cdot TC_i[t]\right), TC_i[t] = \max\left\{0, TC_i[t-1] + \underline{T}_i - T_i\right\} \tag{7}$$

where ω_3 is the corresponding parameter and $T_i = \overline{\overline{T}}_i(W = 1, x = 0)$ is the instantaneous user throughput. Actually, mM rule introduces a timer when defining the weight function in the sense that, if the user instantaneous throughput is higher than its requirement, then that user can be still scheduled for certain period until the cooldown period starts and $TC_i[t] < 0$.

- **Required Activity Detection (RAD) ($d = 3$):** Kolding (2006) proposes a simple weight function defined as:

$$G_{3,i}\left(\overline{\overline{T}}_i\right) = \underline{T_i} / \overline{\overline{T}}_i \tag{8}$$

where, the user with the lowest average rate compared to its requirement has the greatest chance to be scheduled on that resource block.

Studies conducted by Comsa (2014) reveal different behavior of these scheduling rules when the network conditions are variable in terms of: downlink channel type, number of active users, arrival bit rates and GBR requirements.

3.3 Proposed Utility Function

The scheduling rule proposed in this section is entitled Lagrange Multiplier (LM) since the proposed weight function take into account the following multiplier $\lambda_i[t]$ which is determined according to the following equation:

$$\lambda_i[t] = \lambda_i[t-1] + \omega_4 \cdot \left(\underline{T_i}[t] - T_i[t]\right) \tag{9}$$

where ω_4 is the forgetting factor which corresponds to the Lagrange multiplier computation for the GBR objective and $T_i = \overline{\overline{T}}_i\left(W = 1, x = 0\right)$ is the instantaneous user throughput at TTI *t*. Then, the proposed weight function is:

$$G_{4,i}\left(\overline{\overline{T}}_i\right) = \log\left(\omega_5 + \lambda_i\right) \tag{10}$$

By setting proper values for ω_5, the user with the lowest multiplier has much less chances to be scheduled for the upcoming scheduling instants, whereas users with higher multipliers are much likely to be served by using the logarithmic scale. Even when users are experiencing very high fading oscillations $\left(r_{i,j}[t] \cdot F_i'\left(\overline{T}_i\right) >> 1\right)$, the optimization problem can be still focused on the GBR requirement satisfaction due to the logarithmic form of the weight function.

4. REINFORCEMENT LEARNING FRAMEWORK

The solution to the optimization problem exposed in (4) aims to find at each TTI *t*, the best rule assignation and resource allocation variables $\left\{b_{d,i}[t], c_{i,j}[t]\right\} \in \left\{0,1\right\}$ such that the GBR satisfaction condition $\overline{\overline{T}}_i \geq \underline{T_i}$ is met for each active user $i \in \mathcal{I}_t$. To decide jointly both decision variables would be practically impossible since the overall problem is combinatorial and each solution must be provided within

Figure 1. Proposed system model

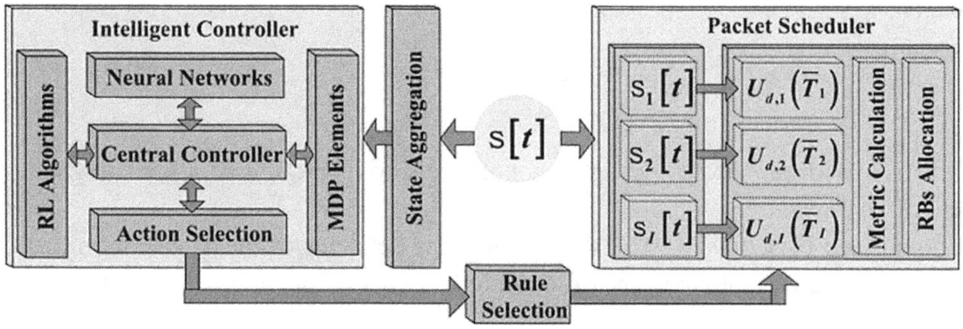

one TTI. In order to implement the proposed optimization framework in real-time schedulers, we propose a sub-optimal solution that can be defined as follows: in the first stage, the scheduling rule assignation variable must be decided whereas, in the second stage, the resource allocation can be performed according to the metrics' calculation provided in (5). In this sense, we propose a reinforcement learning framework able to learn over time the most appropriate scheduling rule to be applied on each scheduler state.

Figure 1 presents the proposed RL framework, where an intelligent controller learns to take proper scheduling decisions based on the momentary scheduler states. By $\mathcal{S}$ we define the measurable, continuous and multi-dimensional scheduler state space, and $s\left[t\right] \in \mathcal{S}$ is a vector that defines the momentary scheduler state at TTI t. The OFDMA packet scheduler (Fig. 1, right side) takes as input at each TTI two types of information: *a)* the momentary state $s \in \mathcal{S}$ and *b)* the utility function decided by the controller. Then, the utility weights are determined based on one of the equations (6)-(10) followed by

the metrics calculations and resource allocations that are performed according to (5). In contrast to the packet scheduler, the controller behaves as a *black box* that needs to decide the rules to be followed by the entire scheduling procedure based only on the momentary scheduler states.

At TTI t, the scheduler controller observes the momentary state $s\left[t\right] \in \mathcal{S}$ and takes a decision regarding the type of scheduling rule that must be used by the scheduler. At TTI t+1, a new state $s\left[t+1\right] = s' \in \mathcal{S}$ is observed and a reward value is obtained from the RRM entity in order to evaluate the performance of the applied scheduling rule in the previous state $s\left[t\right] \in \mathcal{S}$. Here, the reward value gives a general measure for the GBR satisfaction including all active users. The controller needs to experience high number of states' transitions in order to learn over time which is the best scheduling rule to be applied on each momentary state. This is basically the way how the RL principle works (Sutton and Barto, 2017): for many visits of the state-action pairs, the reward values are stored and discounted; then, the controller just follows the action that has the highest accumulated reward on that state. Due to the actual circumstances of the scheduler state space (continuous and multi-dimensional), the state-action values cannot be enumerated exhaustively. In this sense, we propose to learn only the approximation of the best scheduling rule to be applied on each momentary scheduler state.

According to Figure 1, the dimension of scheduler state space has to be reduced to a fixed dimension in order to enhance the learning performance and reduce the framework complexity. Then, we propose the use of neural networks to approximate the scheduling rule decision for each momentary state. A given neural network takes as input the momentary scheduler state and outputs the preference value of selecting its corresponding rule on that state. The previous state, action, reward and current state are stored in the Markov Decision Process (MDP) at every iteration. For each GBR oriented scheduling rule, we use one neural network for its approximation. According to the selected rule in the previous state, the corresponding neural network is updated in the next state once the reward value is given. The neural network weights corresponding to the selected scheduling rule are updated according to the information provided by the MDP entity and the type of RL algorithm which is used. This stage when the neural networks are updated is entitled *learning*. During the learning stage, the action selection block may choose to improve or to exploit what the neural networks have learnt so far. If the exploitation step is chosen, then the rule with the highest neural network output is selected to perform the scheduling procedure. If the improvement step is decided, then a different scheduling rule may be selected according to some probabilities. The action selection has the role of enhancing the learning outcome by enlarging the state space exploration. The learning stage continues until some convergence criteria are met. When the exploitation stage is performed, only the scheduling rule with the highest neural network output is selected on each momentary state.

4.1 State Space

The scheduler state space is divided in two disjoint sub-spaces: *a)* controllable sub-space $\mathcal{S}_C$ that evolves based on the selected scheduling rules and *b)* uncontrollable sub-space $\mathcal{S}_U$ which is rather stochastic and it cannot be predicted in general. Then, the obtained scheduler state space is $\mathcal{S} = \mathcal{S}_U \cup \mathcal{S}_C$. We define by $c\left[t\right] \in \mathcal{S}_C$ the momentary controllable state and by $z\left[t\right] \in \mathcal{S}_U$ the momentary uncontrollable state at TTI *t*. Consequently, the momentary scheduler state is defined as: $s = \left[c\left[t\right], z\left[t\right]\right] \in \mathcal{S}$. By momentary uncontrollable state we mean the CQI reports from all users and the arrival rates in data queues. To improve the learning performance, the vector of GBR requirements $\underline{T} = \left[\underline{T}_1\left[t\right], \underline{T}_2\left[t\right], \ldots, \underline{T}_I\left[t\right]\right]$ is randomly changed at different time periods. The number of active user is dynamically changed in the learning stage in order to increase the generality of the neural network approximations. Then, these parameters are also included in the uncontrollable scheduler state. The controllable state comprises the following elements: $c = \left[\overline{T}, \overline{\overline{T}}, \phi, q\right]$, where $\overline{T} = \left[\overline{T}_1\left[t\right], \overline{T}_2\left[t\right], \ldots, \overline{T}_I\left[t\right]\right]$ is the instantaneous vector with average user rates calculated based on the EMF filter, $\overline{\overline{T}} = \left[\overline{\overline{T}}_1\left[t\right], \overline{\overline{T}}_2\left[t\right], \ldots, \overline{\overline{T}}_I\left[t\right]\right]$ is the vector of average user rates determined based on the MMF filter, $\underline{T} = \left[\underline{T}_1\left[t\right], \underline{T}_2\left[t\right], \ldots, \underline{T}_I\left[t\right]\right]$, and $\phi = \left[\phi_1\left[t\right], \phi_2\left[t\right], \ldots, \phi_I\left[t\right]\right]$ is the vector of differences between user rates and GBR requirements calculated for each use with the following formula: $\phi_i = \max\left\{0, \underline{T}_i\left[t\right] - \overline{\overline{T}}_i\left[t\right]\right\}$. Finally, $q = \left[q_1\left[t\right], q_2\left[t\right], \ldots, q_I\left[t\right]\right]$ is the vector with queue lengths for each active user. We consider for this study, that each active user has one active data queue.

4.2 Action Space

It is defined by $\mathcal{A}=\{a_1, a_2, ..., a_D\}$ the finite set of controller actions. If the selected action at TTI t is $a[t]=d$, then the scheduling rule $d \in \mathcal{D}$ is selected to conduct the scheduling procedure on that TTI.

4.3 Reward Function

In the reward computation, the average user throughput with MMF filter is considered only for each active user. From the user perspective, the reward is calculated as a normalized value of the difference between the average user throughput with MMF and its GBR requirement as expressed by the following equation:

$$r_{1,i}\left(\overline{\overline{T}}_i[t]\right)=\left(\overline{\overline{T}}_i[t]-\underline{T}_i[t]\right)/\underline{T}_i[t],\ \forall i \in \mathcal{I}_t \tag{11}$$

If the reward is $r_{1,i}\left(\overline{\overline{T}}_i[t]\right)>0$, then user $i \in \mathcal{I}_t$ is considered to be satisfied from the viewpoint of the GBR constraint. For reasons related more to the convergence of neural networks, the reward function for each active user is modeled by using the following equation:

$$r_{2,i}\left(\overline{\overline{T}}_i[t]\right)=\begin{cases} r_{1,i}\left(\overline{\overline{T}}_i[t]\right), & if\ r_{1,i}\left(\overline{\overline{T}}_i[t]\right)<0 \\ 1, & if\ r_{1,i}\left(\overline{\overline{T}}_i[t]\right)\geq 0 \end{cases} \tag{12}$$

The intrinsic scheduler reward value is obtained by summing the rewards from (12) at each TTI such that:

$$r_{3,i}\left(\overline{\overline{T}}[t]\right)=\sum_{i=1}^{I_t} r_{2,i}\left(\overline{\overline{T}}_i[t]\right) \tag{13}$$

The proposed reward function evaluates the performance of applying action $a[t]\in\mathcal{A}$ in state $s[t]\in\mathcal{S}$. The reward value for pair (s,a) is received in state $s'[t+1]\in\mathcal{S}$. It would be interesting to measure if there is any improvement of the intrinsic reward value $r_{3,i}\left(\overline{\overline{T}}[t+1]\right)$ received at TTI t+1 compared to the previous intrinsic reward value $r_{3,i}\left(\overline{\overline{T}}[t]\right)$ at TTI t. If we further consider now that $s'[t+1]\in\mathcal{S}$ is the current state and $s[t]\in\mathcal{S}$ is the previous one, then the global reward function is calculated as follows:

$$r\left(\overline{\overline{T}}[t+1],\overline{\overline{T}}[t]\right)=\begin{cases} 1, & if\ r_{3,i}\left(\overline{\overline{T}}[t+1]\right)=1 \\ r_{3,i}\left(\overline{\overline{T}}[t+1]\right)-r_{3,i}\left(\overline{\overline{T}}[t]\right), & otherwise \end{cases} \tag{14}$$

When the reward is $r = 1$, then the reward is maximized and all active users satisfy the GBR requirements. On the other side, when $r \geq 0$, then the reward is moderate and the scheduling rule applied in the previous state was a good action. However, when $r < 0$, then the controller gets a punishment and the applied action may not be a good option. For instance, in some networking conditions (when the traffic load is high and stringent GBR requirements) whatever the applied rule is, the reward is still negative. This makes part of noisy data which cannot be avoided when running real time RL frameworks. However, the scheduling rule that provides the minimum punishment value should be selected. The commitment of the proposed solution is to maximize the number of TTIs when the reward is maximized and minimize as much as possible the amount of punishment rewards.

4.4 Value Functions

The action selection block from Figure 1 implements the scheduling policy that maps momentary states over the action space. In learning stage, the scheduling policy is determined according to some probability functions in order to enhance the exploration capability of RL framework. In the exploitation stage, the output values provided by each neural network are used as distributions to rank the best scheduling rule to be applied every state. However, the scheduling policy is defined as a function $\pi : \mathcal{S} \times \mathcal{A} \rightarrow \left[0,1\right]$ and determined based on the following formula (Sutton and Barto, 2017):

$$\pi\left(d \mid s\right) = \mathbb{P}\left(a\left[t\right] = d \mid s\left[t\right] = s\right) \tag{15}$$

where $\mathbb{P}\left(\cdot\right)$ is the probability of selecting action $a\left[t\right] = d$ by π when the actual state is $s\left[t\right] = s$.

The interest would be to measure the value of the initial state from where the whole learning stage starts. This means that we need a sum of discounted rewards from state to state starting from, let us say, $s\left[0\right] = s$ when given the policy π. By using notations from machine learning domain, this measurement is obtained by the value function defined as $V^{\pi} : \mathcal{S} \rightarrow \mathbb{R}$ and calculated as expressed below (Sutton and Barto, 2017):

$$V^{\pi}\left(s\right) = \mathbb{E}_{\pi}\left[\sum\nolimits_{t=0}^{\infty} \gamma^{t} \cdot R_{t+1} \mid s\left[0\right] = s\right] \tag{16}$$

where $\left(\gamma^{t} \cdot R_{t+1}; t \geq 0\right)$ is the accumulated reward value from state to state being discounted by the $\gamma \in \left[0,1\right]$ and $s\left[0\right]$ is considered a random state such that $\mathbb{P}\left(s\left[0\right] = s\right) > 0$. If we also assume that the first action of the learning process is randomly chosen and $\mathbb{P}\left(a\left[0\right] = d\right) > 0$ keeps valid for all scheduling rules $d \in \mathcal{D}$, then we get the action-value function $Q^{\pi} : \mathcal{S} \times \mathcal{A} \rightarrow \mathbb{R}$ determined as follows (Sutton and Barto, 2017):

$$Q^{\pi}\left(s,d\right) = \mathbb{E}_{\pi}\left[\sum\nolimits_{t=0}^{\infty} \gamma^{t} \cdot R_{t+1} \mid s\left[0\right] = s, a\left[0\right] = d\right] \tag{17}$$

and the rest of scheduling rules that follows state $s[0] = s$ are chosen according to policy π.

The purpose is to be able to access the value of these functions in between consecutive states in order to update them based on the received rewards. According to Comşa et al. (2018b), the following recursions are obtained for equations (16) and (17), respectively:

$$V^{\pi}(s) = r\left(\overline{\overline{T}}[t+1], \overline{\overline{T}}[t]\right) + \gamma \cdot V^{\pi}(s') \tag{18.a}$$

$$Q^{\pi}(s,d) = r\left(\overline{\overline{T}}[t+1], \overline{\overline{T}}[t]\right) + \gamma \cdot V^{\pi}(s') \tag{18.b}$$

The target of the proposed RL framework is to train the neural networks in order to obtain near-optimal value and action-value functions. However, the value function $V^{*}(s)$ is optimal when the values provided in each state have the highest expectable return such as $V^{*}(s) = \max_{\pi} V^{\pi}(s)$. Similarly, the optimal action-value function $Q^{*}(s,d)$ is the highest expected return when the scheduling process starts from $s \in \mathcal{S}$ and the applied scheduling rule is $d \in \mathcal{D}$. When both functions are optimal, then the policy itself is optimal and the scheduling rule to be applied in momentary state $s \in \mathcal{S}$ can be extracted by using the following formula:

$$d^{*} = \arg\max_{d' \in \mathcal{D}} \pi(d' \mid s) \tag{19}$$

According to the optimality conditions, the equations expressed in (18.a) and (18.b) respectively, become:

$$V^{*}(s) = r\left(\overline{\overline{T}}[t+1], \overline{\overline{T}}[t]\right) + \gamma \cdot V^{*}(s') \tag{20.a}$$

$$Q^{*}(s,d) = r\left(\overline{\overline{T}}[t+1], \overline{\overline{T}}[t]\right) + \gamma \cdot V^{*}(s') \tag{20.b}$$

If we also consider that the optimal value function in state $s' \in \mathcal{S}$ is $V^{*}(s') = \max_{d' \in \mathcal{D}} Q^{*}(s', d')$ (Comşa et al., 2018b), then the above equations can be rewritten as follows:

$$V^{*}(s) = r\left(\overline{\overline{T}}[t+1], \overline{\overline{T}}[t]\right) + \gamma \cdot \max_{d' \in \mathcal{D}} Q^{*}(s', d') \tag{21}$$

For the purpose of the scheduling process oriented on GBR satisfaction, the optimality of these functions is not guaranteed since the state-action pairs are not exhaustively stored and enumerated. The aim is to find the best approximations of these functions by implementing the neural networks. During the learning stage, the value and action-value non-linear functions must be updated in each state according to the received rewards. The way of how thee weights of these functions are updated gives in fact the type of RL algorithms which is used. Actually, based on equations (20.a), (20.b) and (21), each RL algorithm has its own strategy to perform the learning stage.

4.5 State Space Aggregation

Under its original form, the momentary scheduler state is variable due to the dependence of controllable parameters on the number of active users. Also, the CQI state depends as well on the number of RBs and implicitly on the system bandwidth. For these reasons, the scheduler state space needs to be aggregated in order to reduce the state dimension to fixed value regardless the number of active users and system bandwidth. The aggregation scheme aims to reduce the complexity of neural networks and speed-up the learning process. Comşa (2014) proposes an aggregation technique of CQI state space under three layers: *1)* the pre-processing layer is responsible of eliminating the system bandwidth dependence; *2)* classification layer performs the Radial Basis Function (RBF) classification for each user CQI to a given cluster of preprocessed CQI and *3)* regression layer applies statistical models to extract the most relevant features. Therefore, the CQI state space dimension is reduced from $I_t \times B$ to 4. Also, we use the proposal given by Comşa et al. (2018b) to reduce the dependency on I_t of the controllable elements $c\left[t\right] \in \mathcal{S}_C$ and user arrival rates. We use the mean and standard deviation function based on likelihood estimators for each of these vectors that are depending on the number of users. This way, we obtain a reduction of this space dimension from $I_t \times 5$ to 2×5. The vector of GBR requirements does not need to be processed, since the same requirement is set for all active users. Thus, by encompassing the compression techniques expressed above, the obtained state is $v\left[t\right] \in \overline{\mathcal{S}}$ that is an aggregate version of $s\left[t\right] \in \mathcal{S}$ with a much lower dimension.

4.6 Function Approximations

The proposed RL framework makes use of $D+1$ neural networks: the neural network used to approximate the optimal value function is used as a critic regarding the selected actions on each states whereas other D neural networks are used to approximate the optimal action value functions for each considered scheduling rule. Then, let $\overline{V^*} : \overline{\mathcal{S}} \rightarrow \mathbb{R}$ be the value neural network and $\overline{Q^*} : \overline{\mathcal{S}} \times \mathcal{A} \rightarrow \mathbb{R}$ action-value neural networks represented based on the following equations:

$$\overline{V^*}\left(v\right) = N\left[\theta_t, \psi\left(v\right)\right] \quad (22.a)$$

$$\overline{Q^*}\left(v, d\right) = N_d\left[\theta_t^d, \psi\left(v\right)\right] \quad (22.b)$$

where $\left\{N, N_1, \ldots, N_D\right\}$ are the neural networks that model the non-linear functions, $\left\{\theta_t, \theta_t^1, \ldots, \theta_t^D\right\}$ are the weights at TTI t for each neural network, and $\psi\left(v\right)$ is the feature vector with non-linear transformations. Figure 2 illustrates the insights of the proposed RL framework when the learning stage is conducted

Figure 2. Proposed RL framework

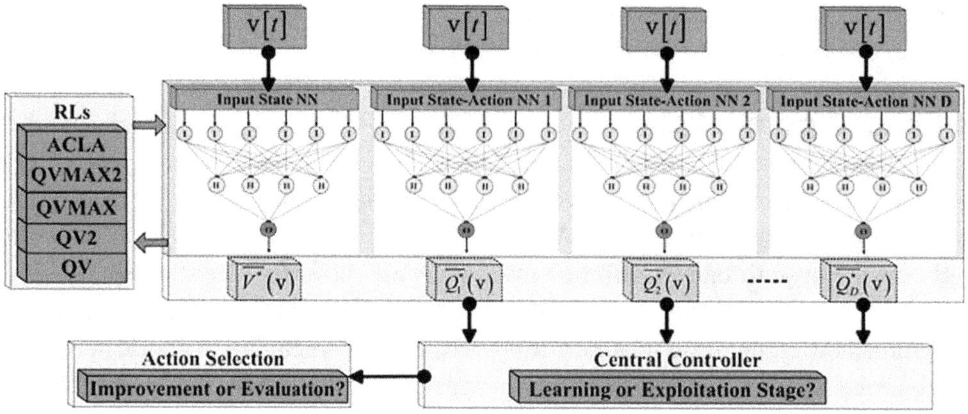

based on five types of RL algorithms. At each TTI *t*, the momentary state $v[t] \in \overline{\mathcal{S}}$ is propagated through all neural networks. If the action selection block decides to exploit the non-linear functions, then the scheduling rule corresponding to the neural network with the highest output is selected. Otherwise, a random action is chosen to perform the scheduling task. This state and action are saved in the MDP entity. At TTI *t*+1 when the reward is received the neural networks are updated according to the performed RL algorithm that uses the temporal difference values given by one of the equations (20.a), (20.b), and (21). After the learning stage is finished, the exploitation stage uses only the action value neural networks that rank the best scheduling rule to be applied in each compressed scheduler state.

Each neural network needs to define its structure before the learning stage is performed. A neural network is defined by L number of layers and M_l number of hidden nodes for each layer $l \in \{1,2,...,L\}$. The number of hidden nodes for input layer M_1 is the dimension of the compressed state $v[t] \in \overline{\mathcal{S}}$ and $M_L = 1$ since a single output value is provided by each neural network. A given vector of weights consists a number of $L-1$ matrices of weights to be updated such as: $\theta_t = [w_{1,2}, w_{2,3}, ..., w_{L-1,L}]$, where the matrix between layers $l-1$ and l is: $w_{l-1,l} = \{w_{j,k}; j = 1,...,M_{l-1}, k = 1,...,M_l\}$. So, when running the learning stage, these weights for each neural network are updated according to the data provided by the MDP entity in three main stages: *a)* forward the momentary states $\{v[t], v'[t+1]\} \in \overline{\mathcal{S}}$ through the neural networks; *b)* determine the error to be reinforced in each neural network according to the type of RL algorithm; *c)* back-propagate the errors to update the weights for each involved neural network.

4.6.1 Forward Propagation

We denote by $v_+^{(l-1)}$ the compressed scheduler state being propagated through $l-1$ layers including the biased points and by $v^{(l)}$ the scheduler state at the output of layer l. Then, the non-linear transformation between layers $l-1$ and l becomes (Comşa et al., 2018b):

$$v^{(l)} = \psi_l\left(w_{l-1,l}^T \times v_+^{(l-1)}\right) \tag{23}$$

where $v_+^{(l-1)}$ denotes the input values of $v^{(l-1)}$ plus the bias point. The result of propagating the input state $v[t] \in \overline{\mathcal{S}}$ through the entire structure looks like (Comşa et al., 2018b):

$$v^{(L)} = \overline{V^*}(v) = \psi_L\left(w_{L-1,L}^T \times \ldots \times \psi_l\left(w_{l-1,l}^T \times \ldots \times \psi_1\left(v^{(1)}\right)\right)\right) \tag{24}$$

Actually in the learning stage, the momentary state is propagated through a number of $D+1$ neural networks and the obtained values are: $\left\{v^{(L)}, v_1^{(L)}, \ldots, v_D^{(L)}\right\}$.

4.6.2 Error Calculation Based on RL Algorithm

Over the learning stage, we aim to learn a set of $D+1$ neural networks. But at each TTI, only the value neural network and only one action value are updated at each TTI, where the updated action-value neural network corresponds to the selected rule in the previous state. Let us consider that the framework is in the current state $v' \in \overline{\mathcal{S}}$ at TTI t+1 and the selected action in the previous state $v \in \overline{\mathcal{S}}$ is $a[t] = d$. Then, at TTI t+1 only the weights $\left\{\theta_t, \theta_t^d\right\}$ are updated. This is done by reinforcing the corresponding errors that are able to evaluate the quality of scheduling decisions at each TTI. We define by $e_{t+1}\left(\theta_t, v', v\right)$ the value error calculated at TTI t+1 by using the value neural network with weights θ_t. Similar, $e_{t+1}^d\left(\theta_t^d, v', v\right)$ is the action value error corresponding to the neural network with weights θ_t^d. These errors measure the difference between the target values obtained by using equations (20.a), (20.b) and (21) reported to propagated values $\left\{v^{(L)}, v_1^{(L)}, \ldots, v_D^{(L)}\right\}$ in state $v \in \overline{\mathcal{S}}$. In this sense, different RL algorithms are used to learn the best non-linear functions able to take proper scheduling decisions for different traffic types and network settings:

1. 1. QV (Wiering and van Hasselt, 2009): the target functions for both value and action-value functions are determined based on equations (20.a) and (20.b), and the errors are calculated as follows:

$$e_t\left(\theta_t, v', v\right) = \left[r\left(\overline{\overline{T}}[t+1], \overline{\overline{T}}[t]\right) + \gamma \cdot \overline{V^*}(v')\right] - \overline{V^*}(v) \tag{25.a}$$

$$e_{t+1}^d\left(\theta_t^d, v', v\right) = \left[r\left(\overline{\overline{T}}[t+1], \overline{\overline{T}}[t]\right) + \gamma \cdot \overline{V^*}(v')\right] - \overline{Q^*}(v, d) \tag{25.b}$$

2. 2. QV2 (Wiering and van Hasselt, 2009): the action value error is similar to (25.b), the target value functions is determined based on (20.a), and the value error is calculated as follows:

$$e_t\left(\theta_t, v', v\right) = \left[r\left(\overline{\overline{T}}\left[t+1\right], \overline{\overline{T}}\left[t\right]\right) + \gamma \cdot \overline{V^*}\left(v'\right)\right] - \overline{Q^*}\left(v, d\right) \tag{26}$$

3. 3. QVMAX (Wiering and van Hasselt, 2009): the action value error is similar to (25.b), the target for the value function is determined based on (21) and the corresponding error is determined as:

$$e_t\left(\theta_t, v', v\right) = \left[r\left(\overline{\overline{T}}\left[t+1\right], \overline{\overline{T}}\left[t\right]\right) + \gamma \cdot \max_{d' \in \mathcal{D}} \overline{Q^*}\left(v', d'\right)\right] - \overline{V^*}\left(v\right) \tag{27}$$

4. 4. QVMAX2 (Wiering and van Hasselt, 2009): the action value error is similar to (25.b), the target value function is determined based on (21) and the corresponding error is calculated as follows:

$$e_t\left(\theta_t, v', v\right) = \left[r\left(\overline{\overline{T}}\left[t+1\right], \overline{\overline{T}}\left[t\right]\right) + \gamma \cdot \max_{d' \in \mathcal{D}} \overline{Q^*}\left(v', d'\right)\right] - \overline{Q^*}\left(v, d\right) \tag{28}$$

5. 5. ACLA (Van Hasselt and Wiering, 2009): the value error is determined similar to (25.a); if this error is $e_t\left(\theta_t, v', v\right) \geq 0$, then the action $a\left[t\right] = d$ applied in state $v \in \overline{\mathcal{S}}$ is a good option. Otherwise, the probability of selecting this action must be decreased. The action-value error is determined as follows:

$$e_{t+1}^d\left(\theta_t^d, v', v\right) = \begin{cases} 1 - \overline{Q^*}\left(v, d\right), & \textit{if } e_{t+1}\left(\theta_t, v', v\right) \geq 0 \\ -1 - \overline{Q^*}\left(v, d\right), & \textit{otherwise} \end{cases} \tag{29}$$

Regardless the chosen RL algorithm to conduct the learning stage, both value and action-value errors must decrease over time. At each TTI, the instantaneous errors must be reinforced in the corresponding neural networks in order to refine the weights $\left\{\theta_t, \theta_t^d\right\}$, process entitled back propagation.

4.6.3 Back Propagation

Both errors $\left\{e_{t+1}, e_{t+1}^d\right\}$ are back-propagated layer by layer at TTI t+1. For simplicity, we explain the back-propagation concept for the value error while the action-value error follows the same computations. Let us define the vector of errors $e_{t+1}^{(l-1)} = \left[e_1^{(l-1)}, e_2^{(l-1)}, \ldots, e_{M_{l-1}}^{(l-1)}\right]$ that is propagated to output of layer $l-1$. These errors are back-propagated from layer l by using the following formula (Van Hasselt, 2011):

$$e_{t+1}^{(l-1)} = w_l^T \times \Delta^T\left(\Psi_l', e_{t+1}^{(l)}\right) \tag{30}$$

where $\Psi_l^{'} = \left[\psi_{l,1}^{'}, \psi_{l,2}^{'}, \ldots, \psi_{l,N_l}^{'}\right]$ and $\Delta\left(\Psi_l^{'}, e_{t+1}^{(l)}\right) = \left[\psi_{l,1}^{'} \cdot e_1^{(l)}, \psi_{l,2}^{'} \cdot e_2^{(l)}, \ldots, \psi_{l,M_l}^{'} \cdot e_{1,M_l}^{(l)}\right]$. By following equation (30), the errors are back-propagated to the input layer. We proceed in this way to update each particular matrix of weights between all layers by using the gradient descent principle. We denote by $w_{l-1,l}^{t} = \left\{w_{l-1,l}^{t}; j = 1, \ldots, M_{l-1}, k = 1, \ldots, M_l\right\}$ the matrix of weights between layers l and $l-1$ that needs to be updated. For these computations, we need the propagated vector $v^{(l-1)}$ to the output of layer $l-1$ and the error at the input of layer l $\Delta\left(\Psi_l^{'}, e_{t+1}^{(l)}\right)$ (if we browse the neural network from input to output layer). Then, the updating process of weight $w_{l-1,l}^{t}$ takes the following form (Van Hasselt, 2011):

$$w_{j,k}^{t+1} = w_{j,k}^{t} + \eta_t \cdot v_j^{(l)} \cdot \psi_{l,k}^{'} \cdot e_k^{(l)} \tag{31}$$

where $\eta_t \in \left[0, 1\right]$ is the learning rate and $v_j^{(l)}$ is the output value of node j of layer l.

4.7 Action Selection in Learning Stage

In the learning stage, after updating the neural weights, an action has to be decided for state $v'\left[t+1\right] \in \overline{\mathcal{S}}$. The action selection block from Figure 2 aims to enhance the exploration of the scheduler state space by selecting the improvements or exploitation steps based on some probabilities. Two types of probability distributions are studied: ε- greedy and Boltzmann. The former one aims to select the action $a\left[t+1\right] \in \mathcal{A}$ based on the following policy function (Van Hasselt, 2011):

$$\pi\left(d' \mid v'\right) = \begin{cases} \varepsilon_{t+1}\left(d'\right), & if \quad \varepsilon \le \varepsilon_{t+1} \\ v_d^{'(L)}, & otherwise \end{cases} \tag{32}$$

where $\varepsilon_{t+1}\left(d'\right) \in \left[0, 1\right]$ is the random variable for action $a\left[t+1\right] = d'$, ε_{t+1} is the probability that decides between improvement and exploitation steps, and $v_d^{'(L)}$ is the output value of action-value neural network in state $v'\left[t+1\right] \in \overline{\mathcal{S}}$. Parameter ε can be fixed or can evolve according to some functions. Lower ε values imply more improvements steps since random actions are likely to be selected. On the other hand, higher ε values imply more exploitation steps. The Boltzmann distribution considers only the output of the action-value neural networks. The potentially good actions for momentary state $v' \in \overline{\mathcal{S}}$ are determined by using the following policy function (Van Hasselt, 2011):

$$\pi\left(d' \mid v'\right) = \frac{\exp\left(v_{d'}^{'(L)} / \tau\right)}{\sum_d \exp\left(v_d^{'(L)} / \tau\right)} \tag{33}$$

where $\tau \in \left[0, \infty\right]$ is the temperature factor that sets how greedy the action should be. Lower τ values imply a more greedy scheduling policy that aim to select the scheduling rule corresponding to the neural network with the highest output. When τ is very large, the selection is more random.

5. SIMULATION RESULTS

The purpose of the obtained simulation results can be divided into two categories: (a) to study the performance of QV, QV2, QVMAX, QVMAX2 and ACLA RL algorithms for different traffic types and network settings and b) to evaluate the performance of learnt scheduling policies compared to state-of-the art schedulers. Three types of traffic are considered: infinite buffer, CBR and VBR. For each traffic type, the learnt scheduling policies are studied in terms of different settings of the windowing factor ρ in order to find the optimal filter length when computing the average user rates that maximizes the percentages of TTIs when all active users are 100% satisfied from the viewpoint of the GBR objective. The pool of scheduling rules consist the following techniques: BF, mM, RAD and LM as expressed by equations (6)-(10). For each configuration (traffic type and windowing factor), the learning stage is launched only once with the same time period. Ten simulations with different random user positions are launched for the exploitation stage. The results are averaged and the standard deviations are presented for each RL algorithm.

5.1 Parameter Settings

The considered system bandwidth is 100 MHz with the maximum number of $B = 100$ RBs. The retransmission scheme is an ARQ algorithm with maximum 5 retransmissions. Data packets failing to be retransmitted within this limit are declared lost. During learning and exploitation stages, the number of active users is randomly changed at each 1000 TTIs in the domain of $I_t \in [15;120]$. All users are moving in the macro-cell scenarios with 120kmph by using the random direction mobility model for the learning and exploitation stages in order to experience as many CQI observations as possible. For the interference model, we consider a cluster with 7 cells, and the simulation model runs only on the central cell, with others being used to provide the interference levels. The training stage runs for 500s by using the same user-network-application conditions for all five RL algorithms. The exploitation stages are launched in 10 different simulations of 95s each, and the results are averaged

The full buffer traffic model will consider full data queues at any time. However, those data packets which are waiting in the queue for more 300ms are dropped and implicitly are declared lost. The rates of CBR traffic that are also changed in the same interval of 1000 TTIs with the same values for all active users belonging to the following set $\{32;64;128;256;512;1024\}$. The VBR traffic generates the packet sizes and arrival rates according to Pareto and Geometric distributions (Comsa 2014a), respectively. The GBR requirements for each traffic type are switched for each active user randomly in the same interval of 1000 TTIs from the set of rate requirements $\underline{T}_i[t] \in \{32;64;128;256;512;1024\}$ kbps. In this way, the environment is very dynamic and the proposed RL framework is able to experience a variety of network conditions.

Based on some extensive simulation results, the optimum number of maximum schedulable uses at each TTI is set to $I_{\max} = 10$. It is also observed that for very low windowing factors, the GBR requirements cannot be satisfied by any involved scheduling rule. Very high windowing factors will increase the time when the GBR requirements are satisfied by the proposed scheduling policies and they are not able to react and take appropriate decisions for such large horizons of user throughputs. However, we would like to find the most suitable value for the windowing factor for each traffic class such that the GBR satisfaction is maximized. The rest of scheduler parameters are provided in Table 1.

Table 1. OFDMA RL framework parameters

Parameters Name	Description/Values
System Bandwidth/Cell Radius	20 MHz/1000m
User Speed/Mobility Model	120kmph/Random Direction
Channel Model	Jakes Model
Path Loss / Penetration Loss	Macro Cell Model / 10 dB
Interfered Cells/Shadowing	0/8dB
Carrier Frequency/DL Power	2GHz/43dBm
Frame Structure	FDD
CQI Reporting Mode	Full-band, periodic at each TTI
PUCCH Model	Errorless
Scheduler Type	BF, mM, RAD, LM
$\{\omega_1;\omega_2;\omega_3;\omega_4\}$	$\{1.25;13.1\cdot10^{-5};10.1;2\}$
Traffic Type	Homogeneous: Infinite Buffer, CBR, VBR
Max. Number of schedulable users	10
RLC ARQ	Acknowledged Mode (Max. 5 retransmissions)
AMC Levels	QPSK, 16-QAM, 64-QAM
Target BLER	10%
Number of Users	Variable: 15-120
RL Algorithms	QV, QV2, QVMAX, QVMAX2, ACLA
Controller Timescale	1 TTI
Number of layers for neural networks	3
Number of Hidden Nodes	100
Exploration/Exploitation Periods	500s/95s
Windowing Factor	$\{2.5;4.0;5.5\}$
Dynamic GBR Constraints	$\underline{T_i}=\{32;64;128;256;512;1024\}\,kbps$
Maximum HoL Delay	300ms
CBR Traffic Type	Data Rates based on GBR Constraints $\lambda=\{32;64;128;256;512;1024\}\,kbps$
VBR Traffic Type	Packet size: Pareto Distrib. $(x=35.5;\alpha=1.1)$ Arv. Rate: Geometric Distrib. $(\mu=1.5;\sigma=1.93)$

Table 2. RL algorithms' parameters

RL Algorithm	Learning Rates for Action Value Functions	Learning Rates for Value Functions	Discount Factor (γ)	Exploration Type (ε,τ)
QV	0.001	0.00001	0.99	Boltzmann $(\tau = 10)$
QV2	0.001	0.00001	0.95	Boltzmann $(\tau = 10)$
QVMAX	0.001	0.00001	0.99	Boltzmann $(\tau = 10)$
QVMAX2	0.001	0.00001	0.95	Boltzmann $(\tau = 10)$
ACLA	0.0001	0.0001	0.99	Greedy $(\varepsilon = 5 \cdot 10^{-4})$

The controller parameters are represented by learning rates, discount factors, exploration parameters (ε,τ) and the configuration parameters of neural networks. For the given time period of learning, these parameters need to be properly set in order to minimize the set of errors $\{e, e^1, e^2, ..., e^d\}$. Table 2 illustrates the most suitable parameters for each RL algorithm. Also, the neural networks need to be parameterized before launching the learning stage in terms of $\{L, M_1, ..., M_L\}$. When the neural network is too flexible (involving large number of layers and hidden nodes), the RL framework complexity is higher, the learning is slower and the neural networks may overfit the input data. On the other side, reducing the number of hidden layers and nodes will underfit the input data in the sense that some parts from the scheduler state space will not be very well represented by the neural network functions. In both cases, the value error starts to increase starting at certain points in the learning stage. Different configurations of neural networks are tested by using extensive simulation results and considering the under-fitting, over-fitting and system complexity trade-off, the optimal setting for each neural network when addressing the GBR objective is: $L = 3$ and $M_2 = 100$. The same configuration is used for value and action-value neural networks. Also, the activation function for input and output layers $\{\Psi_1, \Psi_3\}$ is linear and for the hidden layer Ψ_2 is tangent hyperbolic.

The learning performance is studied for each RL algorithm which is used to update the neural network weights by measuring the percentage of TTIs when the reward is punishment, moderate and maximized, respectively. Let $p(-1 < r < 0)$ be the mean percentage of TTIs when the reward is punishment, $p(0 \le r < 1)$ is the mean percentage of TTIs when the reward is moderate and finally, we denote by $p(r = 1)$ the mean percentage of TTIs when the reward is maximized and consequently, the GBR requirements are satisfied by all active users. The aim is to find those scheduling policies that are able to provide the highest amount of $p(r = 1)$ and the lowest percentage of $p(-1 < r < 0)$. In the exploitation stage we also measure the mean percentage of TTIs when certain percent of users satisfies the GBR requirements. Let $q\%$ be the percentage of users that satisfy the GBR requirements and $p\left(q\%\right)$ the mean percentage of TTIs when the GBR objective is met for $q\%$ of active users. The exploitation stage is running in the same conditions as the learning stage and the simulation conditions are similar to all involved scheduling and RL algorithms.

5.2 Full Buffer Traffic

For the full buffer traffic model, the impact of the learnt policies is studied only in terms of dynamic GBR requirements and number of users since the data queues are always full. When $\rho = 2.5$, QV, QV-MAX and ACLA show the best performance from the viewpoint of $p\left(r = 1\right)$ maximization (as seen in Fig. 3.a). From the viewpoint of $p\left(-1 < r < 0\right)$, ACLA and QVMAX are the best options. In Figure 3.b, the performance of classical scheduling rules, ACLA and QVMAX policies is presented when measuring the mean percentage of TTIs when users are satisfied in the range percentage of $q \in \left[0,100\right]\%$. The proposed LM scheduling rule outperforms other rules for the entire domain of q. Also, the scheduling policies learnt with ACLA and QVMAX algorithms provide similar results when compared to LM scheduling rule. However, a small performance degradation can be observed for ACLA when the range between $p\left(40\%\right)$ and $p\left(80\%\right)$ is considered. Figure 4 presents the scheduling performance when $\rho = 4$. ACLA and QVMAX assures the best percentages of maximized rewards as presented in Figure 4.a. When comparing with other schedulers, ACLA and QVMAX policies together with the LM scheduling rule perform the best. When the windowing factor increases to $\rho = 5.5$, only ACLA algorithm is able to assure the best percentage of maximized rewards (Fig. 5.a). From the perspective of the mean percentage of TTIs with satisfied users, LM scheduling rule and ACLA policy are the best options. From Figures (3.a), (4.a) and (5.a), it can be observed that the mean $p\left(100\%\right)$ gradually increases according to the windowing factor. We conclude that $\rho = 5.5$ is the optimum windowing factor for the full buffer traffic type and the best choice is ACLA policy by providing a gain higher than $p\left(100\%\right) > 40\%$.

5.3 CBR Traffic

The GBR constraints are set based on the arrival rates that are switched at each 1000 TTIs for all active users. In Figure 6.a, ACLA and QVMAX obtain the best amount of maximized rewards $p\left(r = 1\right)$. QVMAX provides a slightly lower amount of punishments but a higher percentage of TTIs for $p\left(0 \le r < 1\right)$. These differences are quantified in Figure 6.b where it can be observed a slight degradation of provided services when exploiting the neural network trained by QXMAX. However, these policies are able to outperform the conventional scheduling rules when measuring the mean percentage of TTIs when users are satisfied in the interval of $\left[p\left(60\%\right); p\left(100\%\right)\right]$. If the windowing factor increases to $\rho = 4$, the same policies are able to provide the best performance when $p\left(r = 1\right)$ is measured (Fig. 7.a), while QV gets the lowest amount of punishments but at the price of much higher percentage of TTIs with moderate rewards. Figure 7.b show the advantage of using ACLA policy that is able to follow the distribution of percentages with satisfied users as follows: between $p\left(20\%\right)$ and $p\left(50\%\right)$, ACLA follows the LM rule that is the best choice for the considered interval; for $\left[p\left(50\%\right);p\left(100\%\right)\right]$, the policy provided by ACLA is able to outperform other candidates. As seen in Fig. 7.a, QVMAX obtains higher percentage of TTIs with punishment and moderate rewards when compared to ACLA policy. This difference is quantified in Fig. 7.b where ACLA policy outperforms QVMAX. When $\rho = 5.5$, ACLA and QV policies obtain the highest amount of maximized rewards and the lowest percentage for the punishments rewards (Fig. 8.a). From Fig. 8.b. it can be observed that ACLA performs the best when measuring the percentages of TTIs when more than $q > 60\%$ of users satisfy the dynamic GBR requirements. When comparing to Fig. 7.b, ACLA policy with $\rho = 5.5$ is able to obtain higher percentage of TTIs when the considered

Figure 3a. Mean percentages of TTIs for punishment, moderate and maximum rewards when the windowing factor is $\rho = 2.5$

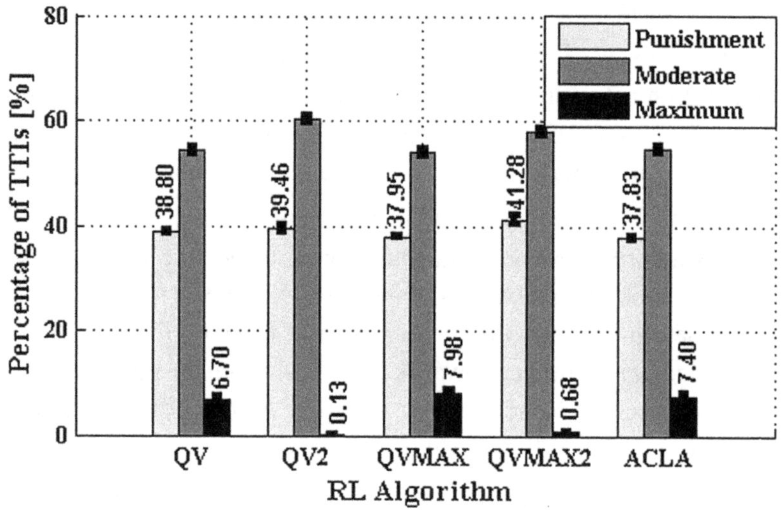

Figure 3b. Mean percentages of TTIs for percentages of satisfied users when the windowing factor is $\rho = 2.5$

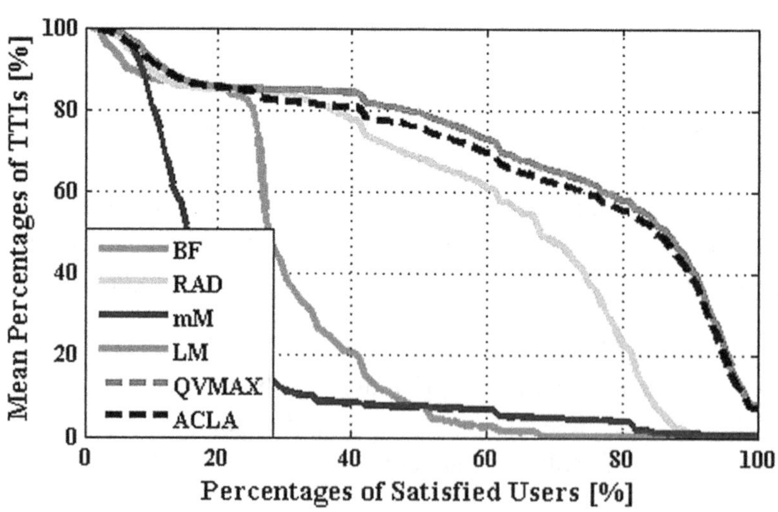

interval is $q \in \left[70;90\right]\%$, but with lower amount of TTIs when 100% of active users are satisfied from the viewpoint of the GBR objective. Then, for CBR traffic type, the optimum value for the windowing factor is $\rho = 4$ and ACLA algorithm learns the best.

5.4 VBR Traffic

The VBR traffic is generated by considering variable arrival rates for each user at each TTI. Basically, the variability of networking conditions is higher than in other scenarios considered for full buffer and

Figure 4a. Mean percentages of TTIs for punishment, moderate and maximum rewards when the windowing factor is $\rho = 4$

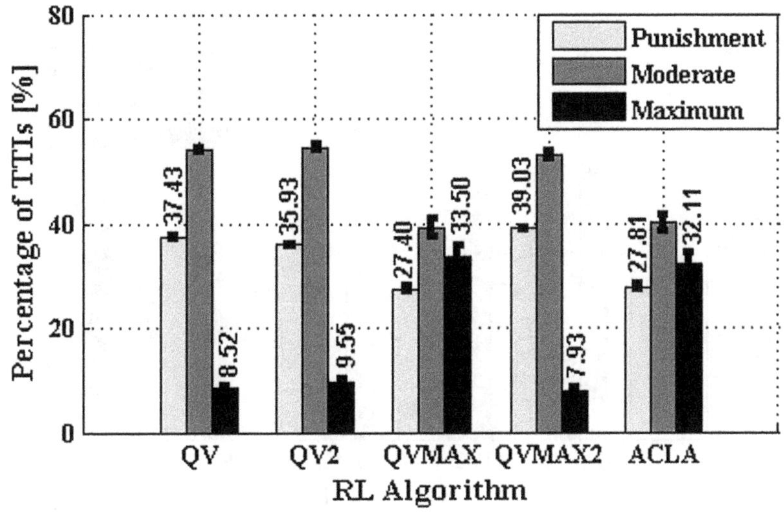

Figure 4b. Mean percentages of TTIs for percentages of satisfied users when the windowing factor is $\rho = 4$

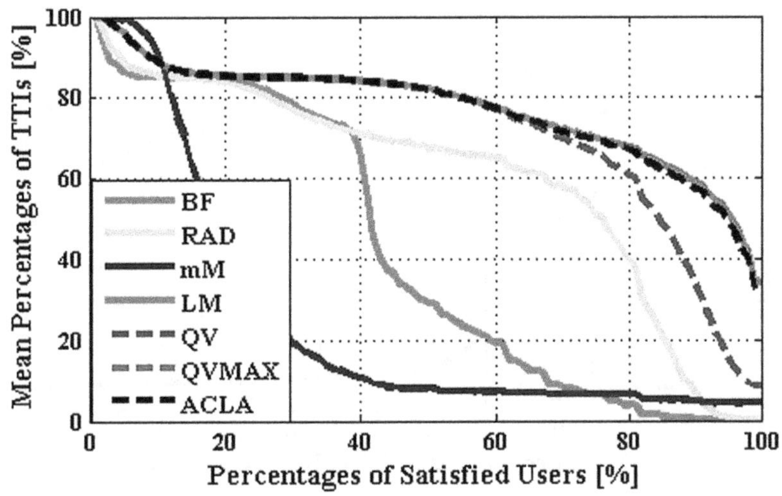

CBR traffic types. For lower windowing factors like $\rho = 2.5$, QV2 policy provides the highest percentage of TTIs when the reward is maximized while ACLA is getting the lowest percentage of punishments (as seen in Fig. 9.a). When comparing to other scheduling rules, ACLA achieves a gain greater than 10% when considering the range of $\left[p(90\%); p(100\%)\right]$. When increasing the windowing factor to $\rho = 4.0$ (Fig. 10.a), the QV policy is the best option from both viewpoints of $p(r = 1)$ and $p(-1 < r < 0)$, respectively. ACLA is the second best policy since QVMAX has the highest percentage of $p(0 \leq r < 1)$ which means that, this policy is not able to provide satisfactory results for larger percentages' range of user satisfaction. Figure 10.b shows the performance evaluation of QV and other conventional rules. QV

Figure 5a. Mean percentages of TTIs for punishment, moderate and maximum rewards when the windowing factor is $\rho = 5.5$

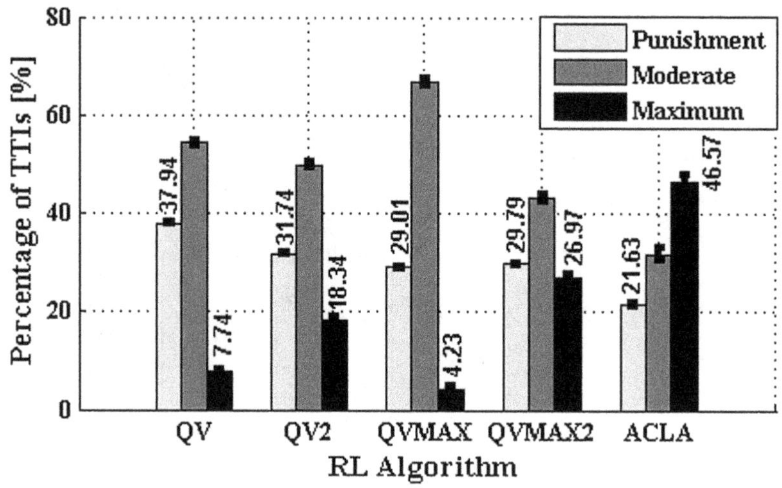

Figure 5b. Mean percentages of TTIs for percentages of satisfied users when the windowing factor is $\rho = 5.5$

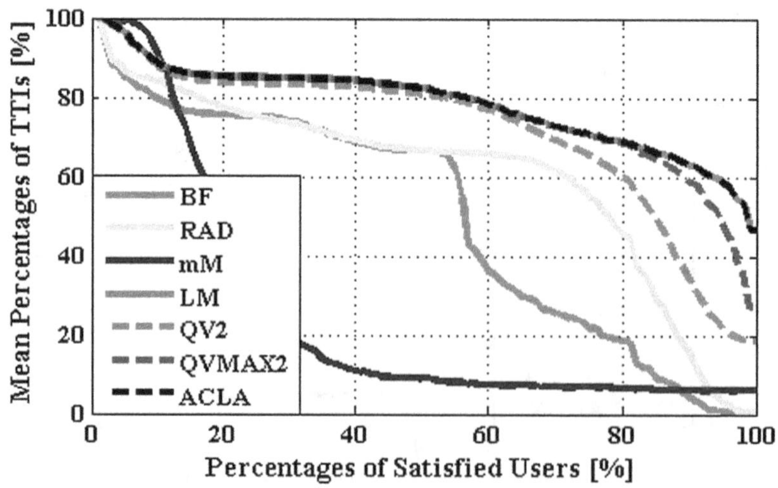

policy is able to outperform the proposed LM scheduling rule when $p(100\%)$ is measured, but this policy provides a slight degradation when certain satisfaction intervals are considered. When measuring the performance of the obtained scheduling policies (Fig. 11.a), only ACLA and QV achieve satisfactory results when compared to other policies. However, as seen from Figure 11.b, ACLA is able to outperform the LM rule from the viewpoint of $p(100\%)$, but at the price of a degradation for the percentages of TTIs when the considered range of satisfied users is $q \in [70;90]\%$. This is in contradiction with the CBR traffic type, where ACLA is able to provide much better results. So, for windowing factors

Figure 6a. Mean percentages of TTIs for punishment, moderate and maximum rewards when the windowing factor is $\rho = 2.5$

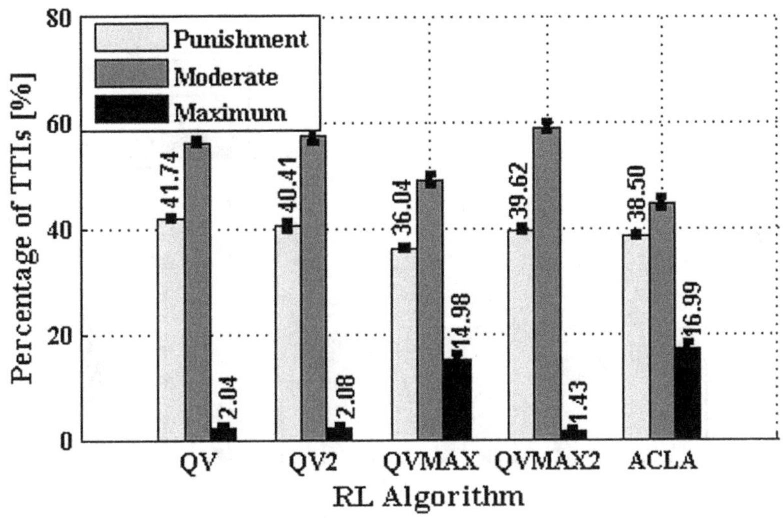

Figure 6b. Mean percentages of TTIs for percentages of satisfied users when the windowing factor is $\rho = 2.5$

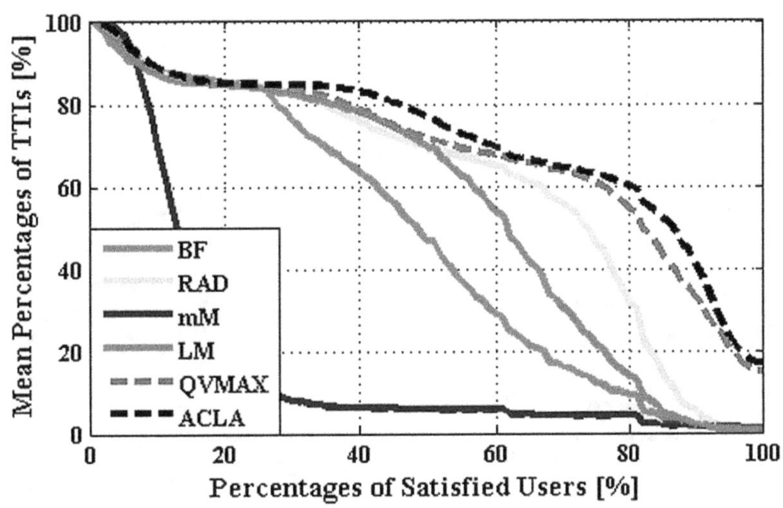

larger than $\rho = 4$, the scheduling policies are not able to follow the trajectory imposed by other scheduling rules. This is explainable since for larger horizons of time when computing the average rates, the scheduling policies are not able to predict the best scheduling rule decisions when the variable arrival rates are considered. Then, the best performance for the GBR satisfaction with VBR traffic is obtained when the windowing factor is $\rho \leq 4$ and the best RL algorithms must be chosen between QV and ACLA.

Figure 7a. Mean percentages of TTIs for punishment, moderate and maximum rewards when the windowing factor is $\rho = 4$

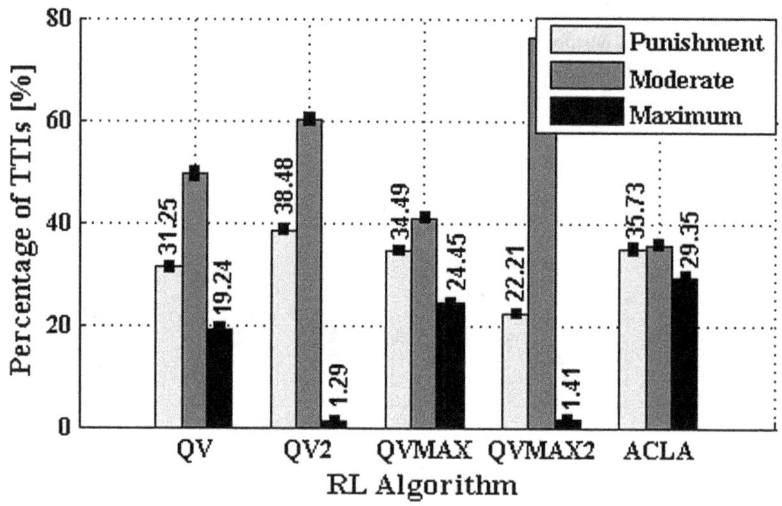

Figure 7b. Mean percentages of TTIs for percentages of satisfied users when the windowing factor is $\rho = 4$

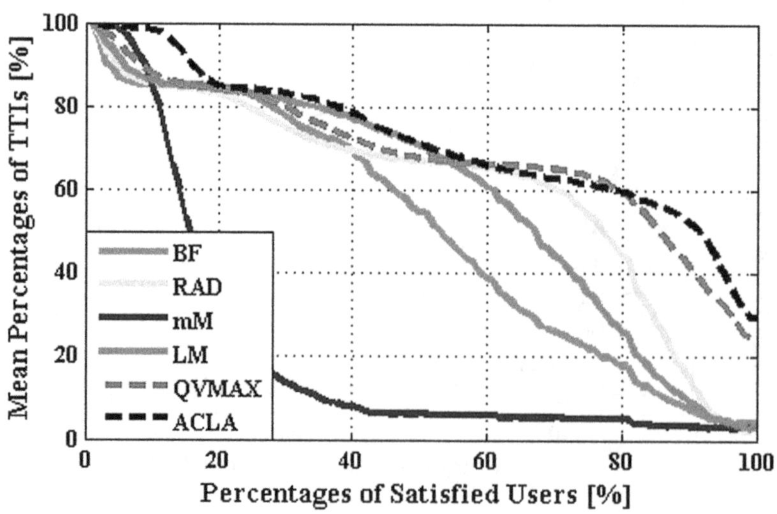

5.5 General Remarks

Alongside the windowing factor parameterization, the type of simulated traffic has a great influence on the performance of the exploited policies. In this sense, the proposed LM scheduling rule outperforms other candidate schemes from the viewpoint of the percentages of TTIs when the active users are satisfied in proportion of 100% for the full buffer traffic model. Only the policy trained by ACLA algorithm can follow the trajectory imposed by LM by indicating a gain of 5% to 40% when measuring $p\left(100\%\right)$ performance metric. For this traffic type, the optimum windowing factor is $\rho = 5.5$. With the CBR

Figure 8a. Mean percentages of TTIs for punishment, moderate and maximum rewards when the windowing factor is $\rho = 5.5$

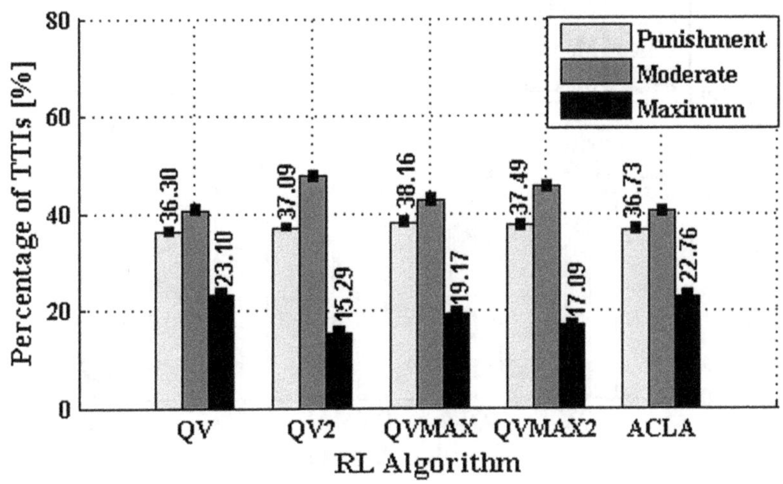

Figure 8b. Mean percentages of TTIs for percentages of satisfied users when the windowing factor is $\rho = 5.5$

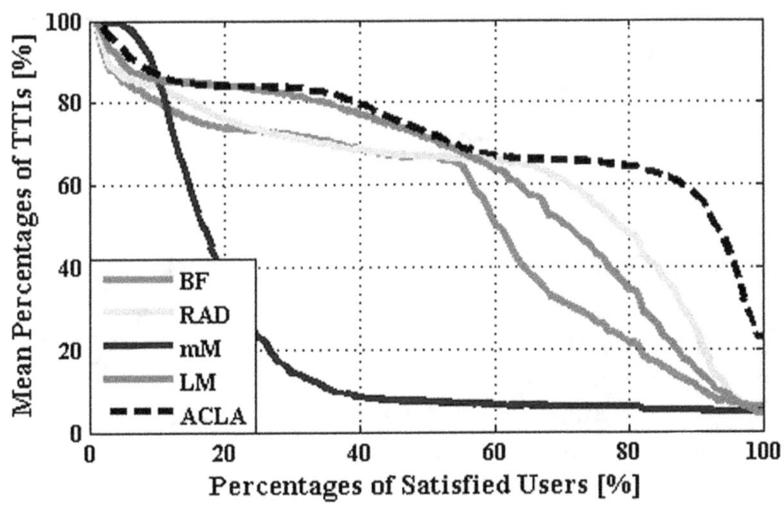

traffic type, the combination of different scheduling rules improves the GBR satisfaction of all active users by about 15% to 20% when ACLA policy is exploited. But the optimum windowing factor is reduced to $\rho = 4$. When the VBR traffic type is scheduled, the proposed scheduling policies are able to provide the best results only if the windowing factor is $\rho \leq 4$. For larger windowing factor values, the exploited scheduling policies are able to provide the best percent of TTIs when all active users are satisfied but at the price of degrading the performance when the interval is $\left[p(65\%);p(95\%)\right]$.

Figure 9a. Mean percentages of TTIs for punishment, moderate and maximum rewards when the windowing factor is $\rho = 2.5$

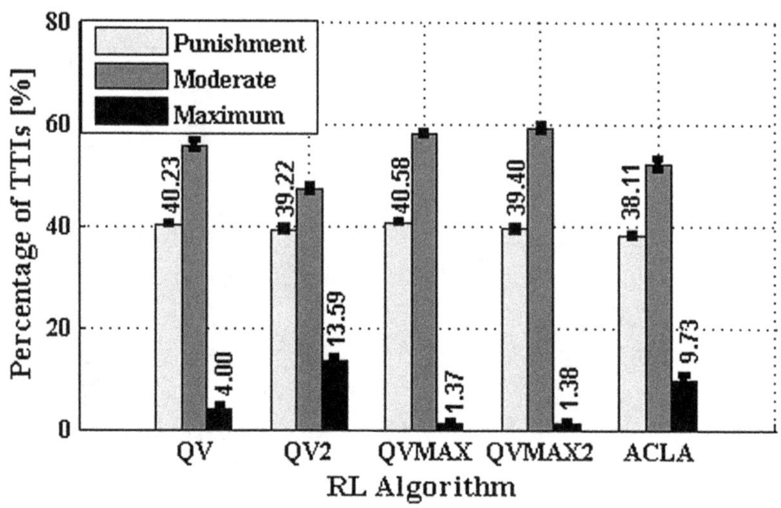

Figure 9b. Mean percentages of TTIs for percentages of satisfied users when the windowing factor is $\rho = 2.5$

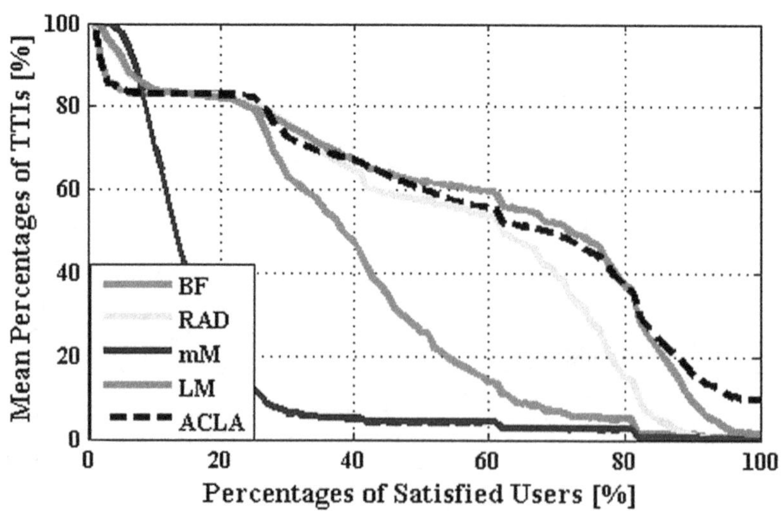

6. CONCLUSION AND FUTURE TRENDS

This chapter proposes an innovative RL framework able to improve the GBR satisfaction for different types of traffic such as: full buffer, CBR and VBR. These traffic characteristics were chosen in order to cover a wide range of applications with various arrival rates and GBR requirements (e.g. video, VoIP, FTP, web browsing). By taking into account the dynamic channel conditions and traffic loads within the learning stages, the stability of the learnt policies are evaluated in terms of various RL algorithms. Each RL algorithm aims to select in each momentary scheduler state the most suitable scheduling rule that can get the highest GBR satisfaction outcome. In order to reduce the complexity of the proposed

Figure 10a. Mean percentages of TTIs for punishment, moderate and maximum rewards when the windowing factor is $\rho = 4$

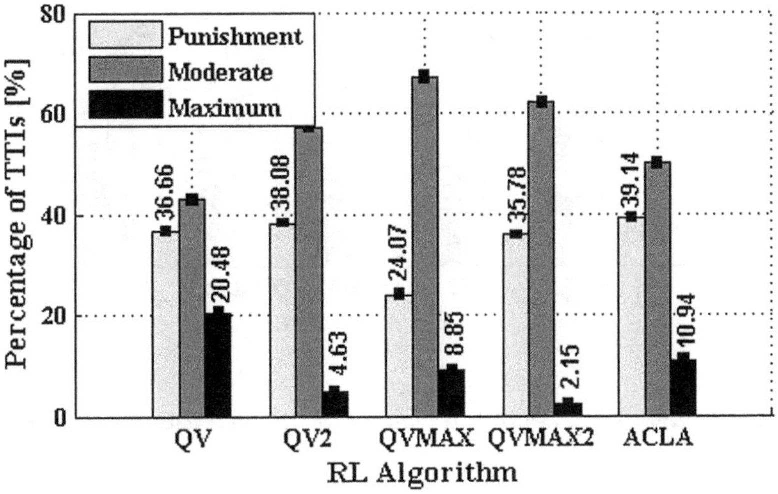

Figure 10b. Mean percentages of TTIs for percentages of satisfied users when the windowing factor is $\rho = 4$

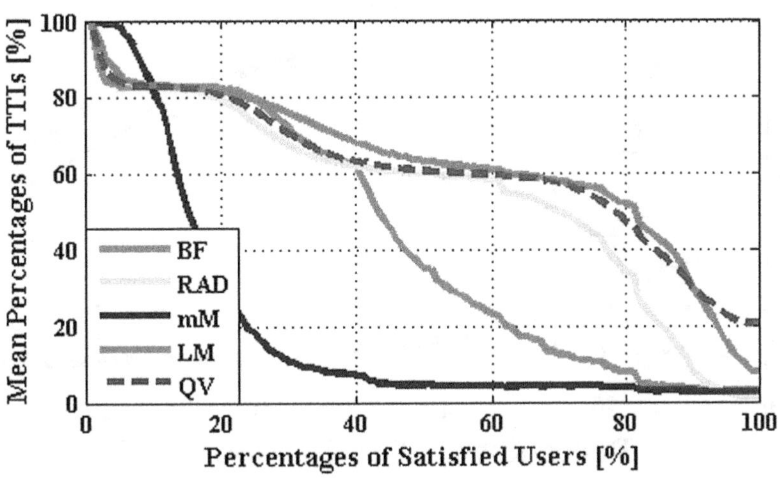

RL framework due to the very high dimension of the scheduler state space, the scheduling decisions are approximated by using non-linear neural networks. The simulation results indicate that the proposed RL framework is able to outperform other conventional scheduling rules when setting optimum time windows of average user rates for each category of involved traffic.

Figure 11a. Mean percentages of TTIs for punishment, moderate and maximum rewards when the windowing factor is $\rho = 5.5$

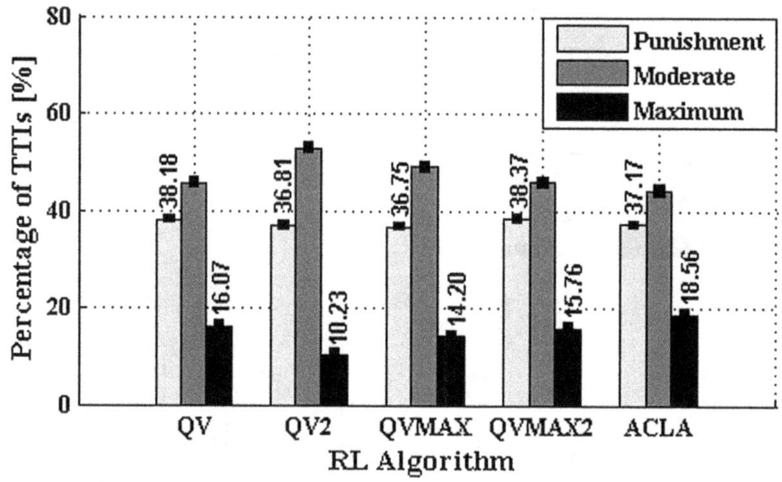

Figure 11b. Mean percentages of TTIs for percentages of satisfied users when the windowing factor is $\rho = 5.5$

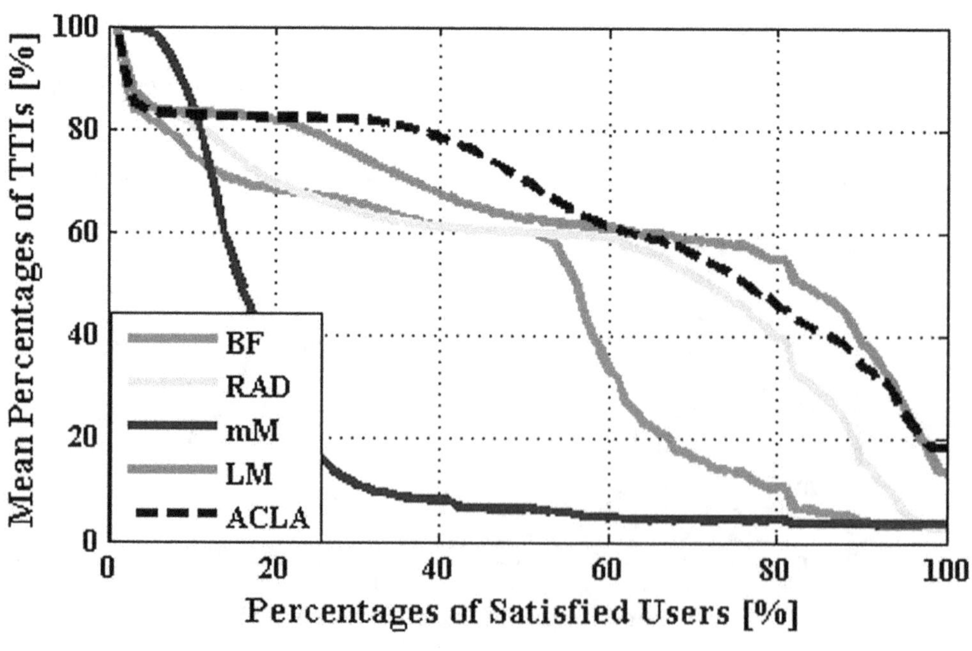

Possible future research direction would be to consider a heterogeneous dynamic environment which integrates various traffic types. In this scenario, the idea would be to train different scheduling policies based on different traffic types by using slave controllers whereas the scheduling decision on the traffic classes is performed by a master controller. Under a more realistic radio access network scheduling scenarios, the traffic heterogeneity implies different QoS requirements. Considering that the reward functions for different objectives represent the sum of normalized sub-rewards from different traffic classes, the traffic heterogeneity will require different reward functions for different traffic priority classes. Consequently, in this context, the controller state space dimension will become much larger than the classical state space with homogeneous traffic.

REFERENCES

Ademoye, O. A., & Ghinea, G. (2009). Synchronization of olfaction-enhanced multimedia. *IEEE Transactions on Multimedia, 11*(3), 561–565. doi:10.1109/TMM.2009.2012927

Ademoye, O. A., Murray, N., Muntean, G. M., & Ghinea, G. (2016). Audio masking effect on inter-component skews in olfaction-enhanced multimedia presentations. *ACM Transactions on Multimedia Computing Communications and Applications, 12*(4), 51. doi:10.1145/2957753

Andrews, M., Qian, L., & Stolyar, A. (2005). Optimal Utility based Multi-User Throughput Allocation Subject to Throughput Constraints. In *Proceedings IEEE 24th Annual Joint Conference of the IEEE Computer and Communications Societies*, (pp. 2415- 2424), Miami, FL: IEEE. 10.1109/INFCOM.2005.1498527

Cisco. (2017). *Cisco Visual Networking Index: Global Mobile Data Traffic Fore-cast Update, 2016-2021 White Paper* Retrieved from http://www.cisco.com/c/en/us/solutions/collateral/service-provider/visual-networking-index-vni/mobile-white-paper-c11-520862.html

Comşa, I.-S. (2014a). *Sustainable Scheduling Policies for Radio Access Networks Based on LTE Technology*. Bedfordshire, UK: University of Bedfordshire.

Comşa, I. S., Aydin, M., Zhang, S., Kuonen, P., & Wagen, J.-F. (2011). Reinforcement learning based radio resource scheduling in LTE-advanced. In *Proceedings of 17th International Conference on Automation and Computing* (pp. 219 - 224). Huddersfield, UK: Academic Press.

Comşa, I. S., Aydin, M., Zhang, S., Kuonen, P., & Wagen, J.-F. (2012). Multi Objective Resource Scheduling in LTE Networks Using Reinforcement Learning. *International Journal of Distributed Systems and Technologies, 3*(2), 39–57. doi:10.4018/jdst.2012040103

Comşa, I.-S., Aydin, M., Zhang, S., Kuonen, P., Wagen, J.-F., & Lu, Y. (2014b). Scheduling Policies Based on Dynamic Throughput and Fairness Tradeoff Control in LTE-A Networks. In *Proceedings of IEEE Conference on Local Computer Networks (LCN)* (pp. 418-421). Edmonton, AB, Canada: IEEE. 10.1109/LCN.2014.6925806

Comşa, I.-S., De Domenico, A., & Ktenas, D. (2017). QoS-Driven Scheduling in 5G Radio Access Networks - A Reinforcement Learning Approach. In *Proceedings of IEEE Global Communications Conference (GLOBECOM)* (pp. 1-7). Singapore: IEEE. 10.1109/GLOCOM.2017.8254926

Comşa, I.-S., Trestian, R., & Ghinea, G. (2018a). 360° Mulsemedia Experience over Next Generation Wireless Networks - A Reinforcement Learning Approach. In *Proceedings of Tenth International Conference on Quality of Multimedia Experience (QoMEX)* (pp. 1-6). Cagliari, Italy: Academic Press. 10.1109/QoMEX.2018.8463409

Comşa, I.-S., Zhang, S., Aydin, M., Chen, J., Kuonen, P., & Wagen, J.-F. (2014c). Adaptive Proportional Fair Parameterization Based LTE Scheduling Using Continuous Actor-Critic Reinforcement Learning. In *Proceedings of IEEE Global Communication Conference (GLOBECOM)* (pp. 4387 - 4393). Austin, TX: IEEE. 10.1109/GLOCOM.2014.7037498

Comşa, I.-S., Zhang, S., Aydin, M., Kuonen, P., Lu, Y., Trestian, R., & Ghinea, G. (2018b). Towards 5G: A Reinforcement Learning-based Scheduling Solution for Data Traffic Management. *IEEE eTransactions on Network and Service Management*, 1–15.

Comşa, I.-S., Zhang, S., Aydin, M., Kuonen, P., & Wagen, J. (2012). A Novel Dynamic Q-Learning-Based Scheduler Technique for LTE-advanced Technologies Using Neural Networks. In *Proceedings of IEEE Conference on Local Computer Networks (LCN)* (pp. 332 – 335). Clearwater, FL: IEEE.

Elbamby, M. S., Perfecto, C., Bennis, M., & Doppler, K. (2018). Toward Low-Latency and Ultra-Reliable Virtual Reality. *IEEE Network*, *32*(2), 78–84. doi:10.1109/MNET.2018.1700268

FANTASTIC-5G. (2016). *D4.1: Technical Results for Service Specific Multi-Node/MultiAntenna Solutions.* Retrieved from http://fantastic5g.com/wp-content/uploads/2016/06/FANTASTIC-5G_D4.1_ Final.pdf

Ghinea, G., & Ademoye, O. (2012). The sweet smell of success: Enhancing multimedia applications with olfaction. *ACM Transactions on Multimedia Computing Communications and Applications*, *8*(1), 2. doi:10.1145/2071396.2071398

Ghinea, G., Timmerer, C., Lin, W., & Gulliver, S. R. (2014). Mulsemedia: State of the Art Perspectives and Challenges. *ACM Transactions on Multimedia Computing Communications and Applications*, *11*(17), 1–23.

Hasselt, H. P. (2011). *Insights in Reinforcement Learning Formal Analysis and Empirical Evaluation of Temporal-Difference Learning Algorithms.* University of Utrecht.

Kolding, T. E. (2006). QoS-Aware Proportional Fair Packet Scheduling with Required Activity Detection. In *Proceedings of IEEE Vehicular Technology Conference*, (pp. 1-5). Montreal, Quebec, Canada: IEEE. 10.1109/VTCF.2006.277

Li, Y., Pateromichelakis, E., Vucic, N., Luo, J., Xu, W., & Caire, G. (2017). Radio Resource Management Considerations for 5G Millimeter Wave Backhaul and Access Networks. *IEEE Communications Magazine*, *55*(6), 86–92. doi:10.1109/MCOM.2017.1601118

Lundevall, M., Olin, B., Olsson, J., Wiberg, N., Wanstedt, S., Eriksson, J., & Eng, F. (2004). Streaming applications over HSDPA in Mixed Service Scenarios. In *Proceedings of IEEE 60th Vehicular Technology Conference VTC2004-Fall*, (pp. 841-845). Los Angeles, CA: IEEE. 10.1109/VETECF.2004.1400139

Sutton, R. S., & Barto, A. G. (2017). *Reinforcement Learning: An Introduction.* London: IMT Press. doi:10.1007/978-1-4899-7687-1_720

Trestian, R., Comşa, I.-S., & Tuysuz, M. F. (2018). Seamless Multimedia Delivery within a Heterogeneous Wireless Networks Environment: Are We There Yet? *IEEE Communications Surveys and Tutorials*, *20*(2), 945–977. doi:10.1109/COMST.2018.2789722

Trestian, R., Muntean, G., & Ormond, O. (2009). Signal Strength-based Adaptive Multimedia Delivery Mechanism. In *IEEE 34th Conference on Local Computer Networks, 2009. LCN 2009* (pp. 297–300). IEEE. 10.1109/LCN.2009.5355092

Trestian, R., Ormond, O., & Muntean, G.-M. (2012). On the Impact of Wireless Network Traffic Location and Access Technology on Mobile Device Energy Consumption. In *Proceedings of the 2012 IEEE 37th Conference on Local Computer Networks (LCN 2012)* (pp. 200–203). Washington, DC: IEEE. 10.1109/LCN.2012.6423608

Van Hasselt, H., & Wiering, M. A. (2009). Using Continuous Action Spaces to Solve Discrete Problems. In *Proceedings of International Joint Conference on Neural Networks* (pp. 1149–1156). Atlanta, GA: Academic Press. 10.1109/IJCNN.2009.5178745

Vien, Q. T., Akinbote, T., Nguyen, H. X., Trestian, R., & Gemikonakli, O. (2015). On the coverage and power allocation for downlink in heterogeneous wireless cellular networks. In *2015 IEEE International Conference on Communications (ICC)* (pp. 4641–4646). IEEE. 10.1109/ICC.2015.7249055

Vien, Q. T., Nguyen, H. X., Trestian, R., Shah, P., & Gemikonakli, O. (2016). A Hybrid Double-Threshold Based Cooperative Spectrum Sensing over Fading Channels. *IEEE Transactions on Wireless Communications*, *15*(3), 1821–1834. doi:10.1109/TWC.2015.2496598

Wiering, M. A., & van Hasselt, H. (2009). The QV Family Compared to Other Reinforcement Learning Algorithms. In *Proceedings of IEEE International Symposium on Adaptive Dynamic Programming and Reinforcement Learning* (pp. 101–108). Nashville, TN: IEEE. 10.1109/ADPRL.2009.4927532

Yuan, Z., Chen, S., Ghinea, G., & Muntean, G. M. (2014). User quality of experience of mulsemedia applications. *ACM Transactions on Multimedia Computing Communications and Applications*, *11*(1s), 15. doi:10.1145/2661329

Yuan, Z., Ghinea, G., & Muntean, G. M. (2015). Beyond multimedia adaptation: Quality of experience-aware multi-sensorial media delivery. *IEEE Transactions on Multimedia*, *17*(1), 104–117. doi:10.1109/TMM.2014.2371240

Zou, L., Trestian, R., & Muntean, G.-M. (2013). DOAS: Device-Oriented Adaptive Multimedia Scheme for 3GPP LTE systems. In *2013 IEEE 24th International Symposium on Personal Indoor and Mobile Radio Communications (PIMRC)* (pp. 2180–2184). IEEE.

Zou, L., Trestian, R., & Muntean, G.-M. (2014). eDOAS: Energy-aware device-oriented adaptive multimedia scheme for Wi-Fi offload. In *2014 IEEE Wireless Communications and Networking Conference (WCNC)* (pp. 2916–2921). IEEE.

Zou, L., Trestian, R., & Muntean, G. M. (2017). E3-DOAS: Balancing QoE and Energy-Saving for Multi-Device Adaptation in Future Mobile Wireless Video Delivery. *IEEE Transactions on Broadcasting*, 1–15.

ADDITIONAL READING

Moysen, J., & Giupponi, L. (2018). From 4G to 5G: Self-organized network management meets machine learning. *Computer Communications, 129*, 248–268. doi:10.1016/j.comcom.2018.07.015

Polese, M., Jana, R., Kounev, V., Zhang, K., Deb, S., & Zorzi, M. (2018). Machine Learning at the Edge: A Data-Driven Architecture with Applications to 5G Cellular Networks. Retrieved from https://arxiv.org/abs/1808.07647

Vien, Q., Ogbonna, N., Nguyen, H. X., Trestian, R., & Shah, P. (2015). Non-Orthogonal Multiple Access for Wireless Downlink in Cloud Radio Access Networks. In Proceedings of European Wireless 2015; 21th European Wireless Conference (pp. 1–6).

Zou, L., Trestian, R., & Muntean, G.-M. (2013). A utility-based priority scheduling scheme for multimedia delivery over LTE networks. *In 2013 IEEE International Symposium on Broadband Multimedia Systems and Broadcasting (BMSB) (pp.* 1–7*).* 10.1109/BMSB.2013.6621745

KEY TERMS AND DEFINITIONS

Machine Learning: Category of algorithm that uses statistical techniques to analyze data and make predictions.

Multi-Objective Optimization: Mathematical optimization problems involving more than one objective function to be optimized simultaneously.

Radio Resource Management: Function of a mobile communication systems used for establishing, maintaining, and releasing of radio resources.

Reinforcement Learning: Training/learning method aiming to automatically determine the ideal behavior within a specific context based on rewarding desired behaviors and/or punishing undesired one.

Unsupervised Learning: Type of machine learning algorithm that is capable of learning a function that best represents a model from datasets consisting of unlabeled input data.

Chapter 9
Online Learning and Heuristic Algorithms for 5G Cloud-RAN Load Balance

Melody Moh
https://orcid.org/0000-0002-8313-6645
San Jose State University, USA

ABSTRACT

The rapidly evolving 5G cellular system adapts cloud computing technology in its radio access networks (RAN), namely Cloud RAN or CRAN. CRAN enables better scalability, flexibility, and performance that allows 5G to provide connectivity for the vast volume of IoT devices. This chapter presents two major research results addressing the load balance (LB) problem in CRAN. First, the authors propose a generic online learning (GOL) system; GOL integrates reinforcement learning (RL) with deep learning method for an environment not fully visible, changing over time, while receiving feedbacks of fluctuating delays. Simulation results show that GOL successfully achieves the LB objectives of reducing both cache-misses and communication load. Next, they study eight practical LB based on real cellular network traffic characteristics provided by Nokia Research. Experiment results of these algorithms on queue-length analysis show that the simple, light-weight queue-based LB is almost as effectively as the much more complex waiting-time-based LB.

INTRODUCTION

Advances in cellular network technology continue to develop to address increasing demands from the growing number of Internet of Things (IoT) devices. A research done by the United Nation's panel on global sustainability stated that, by 2050, 70% of the world population will live in urban areas, which only cover 2% of the entire Earth surface, yet are responsible for 75% of the greenhouse gas emissions (United Nations, 2012). Based on this understanding, the concept of Smart Communities therefore needs the solutions and practices to advance the development and to allow the sustainability of urban environments. In particular, the use of Information and Communication Technologies (ICT) will provide the

DOI: 10.4018/978-1-5225-7458-3.ch009

necessary backbone, not only for maintaining existing services but for enabling new ones; the Internet of Things (IoT) is among the most useful, hopeful ICT technologies for this purpose (Casana & Redondi, 2017). Many expect that the growth of IoT will bring a better quality of life and enable numerous new opportunities.

The rapid growth of IoT also raises new challenges in resource constrained wireless networks, which are the fundamental backbone for IoT services. The emerging 5G communication systems aim to address this challenge by providing ubiquitous, scalable connectivity for vast, exponential growing IoT devices as well as existing cellular applications (Palattela, 2016; Saxena, Roy, Sahu, & Kim, 2017). 5G cellular systems uses emerging technologies such as Millimeter Wave, massive MIMO (Multiple-Input Multiple-Output), and CRAN (Cloud Radio Access networks). These would allow massive connectivity, resource pooling, and energy efficiency needed to support rapid growth of IoT needed for smart urban areas. Among them, the CRAN model is characterized by centralized computation and resource pooling; it therefore promises great scalability and flexibility to support the connectivity of a large volume of devices (Saxena et al., 2017; Checko et al., 2015; Karneyenka, Mohta, & Moh, 2017). More background in CRAN will be described in the background section.

Intelligent systems (IS) have the capability to observe their surrounding environments through their sensory input channels, and interact with the environments via their output channels, through which their actions directly or indirectly affect the environments. Also, a realistic environment is dynamic: with time it changes itself, as well as the elements it depends on.

It is often built to achieve a set of goals. IS in each time-frame requires to make a decision based on the observed data of its surrounding environment. For problems in domains such as robotics, drones, and network controls where the decision (action) is made in real-time, the system is required to adapt in real-time based on the feedback it received. Feedback can be interpreted as some changes in observed data from an environment based on actions of the system. The feedback of a particular action, however, may reach to the system with some delay, or the system may have no knowledge of mapping between observed feedback and its previous actions.

Reinforcement learning (RL) is one of the best learning methods for real-time decision-making (Sutton & Barto, 1998). RL learns from interaction with the environment via recognitions and actions to achieve a goal. On each interaction step, the agent (system), based on the state of the environment chooses an action that alters the state of the environment, and a reward or punishment is then provided to the agent as the desirability of the chosen action. RL algorithms are very useful for solving a wide variety of problems especially when the model is not known in advance; they have been used to solve various problems related to mobile networking and messaging (Muñoz, Barco, Ruiz-Avilés, de la Bandera, & Aguilar, 2013; Shahriari & Moh, 2016; Wu, Lu, Huang, & Zheng 2015).

Deep learning (DL), inspired by the mechanism of human vision, recently attracted more and more attention due to its good performance in many fields such as speech recognition, computer vision, and natural language processing (Hinton & Salakhutdinov, 2006; Krizhevsky, Sutskever, & Sutskever, 2012; Farabet, Couprie, Najman, & LeCun, 2013). As DL has a hierarchical structure, thus it can be applied to any classical learning algorithm.

W proposed GOL (Generic Online Learning), an IS that is able to adapt to dynamic environments without any prior knowledge of the environment or its action policy. GOL forms its abstract definition of state-action space with three hierarchical, deep, layered network consisting of entities. We claim that three layers are sufficient for online-learning. The self-organizing hierarchical state-action space system interacts with a dynamic, partially visible environment based on the input and output data of system. Then the system receives a set of variably delayed feedbacks from the environments to adjust its internal structure.

GOL has been applied for real-time control in 5G networks (Song, Yoo, & Song, 2016; Karlsson et al., 2016). More specifically, it is used for load balancing (LB) of processing user equipment (UE) events in a CRAN (Cloud – Radio Access Networks) BBU (Baseband Units) pool, which consists of several VM (Virtual Machines) with each VM having a cache of limited size. GOL will observe arriving user types and event types, and place each arriving UE event to an appropriate VM, with the goal to balance the load so as to minimize both cache-miss and communication load.

The second part of the paper investigates eight LB algorithms for CRAN based on the real-network traffic and system characteristics provided by Nokia researchers (A. Rottinghaus, D. Baretto, R. Van Egeren, & R. Edward, 2017). It follows the same high-level architecture used in studying cache management for CRAN (Chin & Moh, 2017a, 2018). This is a continuation of our on-going work on online learning for smart cities and 5G (Shahriari & Moh 2016, Shahriari, Moh, & Moh, 2017), CRAN (Karneyenka, Mohta, & Moh, 2017; Shahriari, Moh, & Moh, 2017; Tsai & Moh, 2017a, 2017b, 2018), cloud and fog computing (Huang, Wu, & Moh, 2014; Reguri, Kogatam, & Moh, 2016; Kaur & Moh, 2018; Choudhari & Moh, 2018; Gao & Moh, 2018), and mobile networks and SDN (Sathyanarayana & Moh, 2016; Su & Moh, 2018).

This chapter is an extension of two conference papers (Shahriari, Moh, & Moh, 2017; Tsai & Moh, 2017b). The chapter is organized as follows. Sections 2 and 3 describe the background and related studies of reinforcement and deep learnings and of CRAN and LB, respectively. They are followed by architecture, learning algorithm, and performance evaluation of GOL, and of eight LB algorithms. The chapter ends with remarks on conclusion and future work.

BACKGROUND AND RELATED WORKS ON REINFORCEMENT LEARNING AND DEEP LEARNING

All expert systems use a variation of learning methods depending on the problem they are solving. The common notion in all learning methods is to find a set of patterns that describes the problem space. For instance, in classifying images, the learning model learns the feature-patterns in input images for each category, or even for object recognition the system extracts and learns a set of feature-patterns for each trained object. Here we focus mainly on learning methods that adapt or modify themselves based on training data or some feedback, such as classification and RL.

Reinforcement Learning

Standard RL theories are based on the concept of Markov decision process (MDP). A MDP is denoted as a tuple S, A, R, P , where *S* is the state space, *A* is the action space, *R* is the reward function (feedback), and *P* is the state transition probability, as shown in Figure 1.

The goal of RL is to learn the optimal policy π^* , so that the expected sum of discounted reward of each state will be maximized

Figure 1. Interaction of IS (agent) with environment with RL

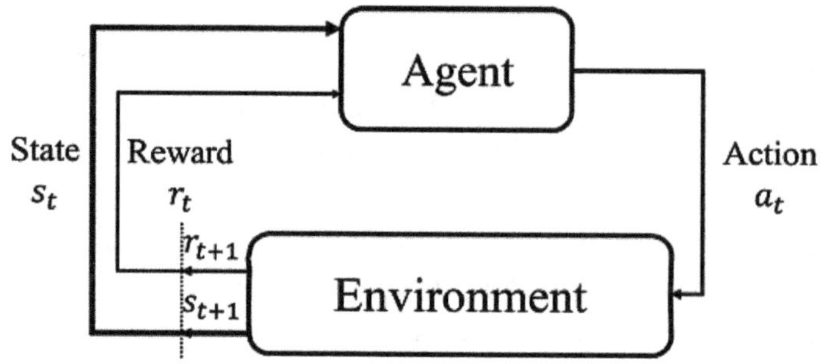

$$J_{\pi^*} = \max_{\pi} J_{\pi} = \max_{\pi} E_{\pi}\left[\sum_{t=0}^{\infty} \gamma^t r_t\right] \tag{1}$$

where $\gamma \epsilon \left[0,1\right)$ is the discount factor which, if set to zero, makes the agent "opportunistic" about current reward while agent with γ equal to 1 strives for a long-term high reward. r_t is the reward at time-step t. $E_{\pi}\left[\cdot\right]$ stands for the expectation with respect to the policy π and the state transition probabilities, and J_{π} is the expected total reward. A value function $Q\left(s,a\right)$ represents the estimate of expected return attainable from executing action a in state s. Its computation repeatedly sweeps through the state-action space of MDP. The value function of each state-action pair is updated according to:

$$Q\left(s,a\right) \leftarrow \sum_{s'} p\left(s' \middle| s,a\right)\left[r\left(s,a,s'\right) + \gamma \max_{a'} Q\left(s',a'\right)\right] \tag{2}$$

until the largest change Δ in the value of any state-action pair is smaller than a preset constant threshold, where $p\left(s' \middle| s,a\right)$ is the probability of state transition from s to s' after executing action a, and $r\left(s' \middle| s,a\right)$ is the corresponding reward. After the algorithm converges, the optimal policy is followed by simply taking the greedy action in each state s as

$$a^* = \arg\max_{a} Q^*\left(s,a\right), \left(\forall s \in S\right) \tag{3}$$

As for model-free cases where the agent has no prior knowledge of the environment, *Q*-learning (an RL algorithm) can achieve optimal policies from delayed rewards. At a certain time step t, the agent observes the state s_t, and then chooses an action a_t. After executing action a_t, the agent receives a reward r_{t+1} and gets into the next state s_{t+1}. Then the agent will choose the next action a_{t+1} according to the best-known knowledge and learned policy.

Let α_t be the learning rate where α_t equal to zero makes the agent incapable of learning anything while α_t equal to one makes it consider only the most recent information. The one-step updating rule of *Q*-learning can be described below:

$$Q(s_t, a_t) = (1 - \alpha_t) Q(s_t, a_t) + \alpha_t \left(r_{t+1} + \gamma \max_{a'} Q(s_{t+1}, a') \right) \tag{4}$$

While the *Q*-learning algorithm chooses the best action based on the state-action pair with highest *Q* value, the learning algorithm used in this paper, SARSA (State-Action Reward State-Action) (Sutton & Barta, 1998), chooses actions by ε *-greedy* policy, and the updating algorithm is described as follows,

(5)where is the *Q* value selected with -greedy policy at for . In -greedy policy the action is chosen based on highest *Q* value with probability of 1- otherwise it is chosen randomly, where . The -greedy policy allows the agent to explore the state-action space more and avoids fast convergence to local optimal solution.

To deal with the changing environment, an agent has to learn non-stationary knowledge incrementally and adapt to the new environment gradually. In (Wang, Chen, Li, Dong, & Tarn, 2016), they have proposed an incremental learning method for RL in a dynamic environment, where RL agent detects change (drift) in the environment and adapts based on the drifts with dynamic programming.

For realistic agents, such as robots, drones, and network controls that deal with continuous state-action space, the original *Q*-learning or other RL methods do not have a good performance. Fuzzy RL has been applied to various problems with continuous state-action space (Jouffe, 1998; Err & Deng 2006), as fuzzy inference system has proven to be extremely effective on problems for domains with continuous space such as control systems.

Deep-Learning Classification and RL

Deep learning (DL) has been inspired by the mechanism of human vision, and has recently gained increasing attention due to its great performance in many fields (Hinton et al., 2006; Krizhevsky et al., 2012; Farabet et al., 2013). The intention of DL is to discover more abstract representations in higher levels (Bengio & Delalleau, 2011). It involves a class of models to hierarchically learn high-level features of input data with a deep hierarchical architecture. DL has the general formulation as follow,

$$f(\boldsymbol{x}) \approx g_1\left(g_2\left(\ldots\left(g_n(\boldsymbol{x})\right)\ldots\right)\right) \tag{6}$$

where **x** is the input, $g_i\,(i = 1, \ldots, n)$ is the operation on the ith layer, and $f(\boldsymbol{x})$ is the new representation of $\boldsymbol{x}$. The input of a higher layer is the output of its previous layer in the DL models. In this way, it can progressively lead to more abstract and complex features at deeper layers. More abstract features are generally invariant to most local changes of the input. Commonly used DL models include deep belief networks (Hinton, Osindero, & The, 2006), deep Boltzmann machines (Salakhutdinov & Hinton, 2009), stacked auto-encoders (SAE) (Bengio, Lamblin, & Popovici, 2007), and convolutional neural networks.

As DL has a hierarchical structure, thus it can be applied to any classical learning algorithm such as classification (Zhou & Wei, 2016). Also, most importantly, recently DL has been applied to RL (Zhao & Du, 2016; Mnih et al., 2016) on the transformation of state-action space as well as utilization of deep network to find the best action policy in a very dynamic environment, such as games (Mnih et al., 2013) where the agent played a game to master it after abundant iterations.

BACKGROUND AND RELATED WORK ON 5G, CRAN, AND LOAD BALANCING

In this section, we first describe background information including the basic 5G/LTE-A architecture and CRAN model. This is followed by related works in LB.

5G, CRAN, and UEC

5G Architecture

The basic architecture of 5G/LTE-A (Long Term Equivalence – Advance) consists of two parts: The Radio Access Network (RAN) shown on the left side of Figure 2 (Alcatel, 2009) and the Evolved Packet Core (EPC) on the right side.

CRAN

A RAN consists of eNodeB and User Equipment (UE). Traditional eNodeB consists of remote radio head (RRH) and baseband unit (BBU) with RRH on every eNodeB, as shown on the left side of Figure 3 (Checko, 2015). While RRH is mainly to transmit and to receive wireless signals as well as to amplify signals for transmission, BBU is responsible for transforming IP packets into digital baseband signal, processing baseband signal from RRH, plus management of user mobility and Quality of Services (QoS)

Figure 2. Basic 5G/LTE-A architecture

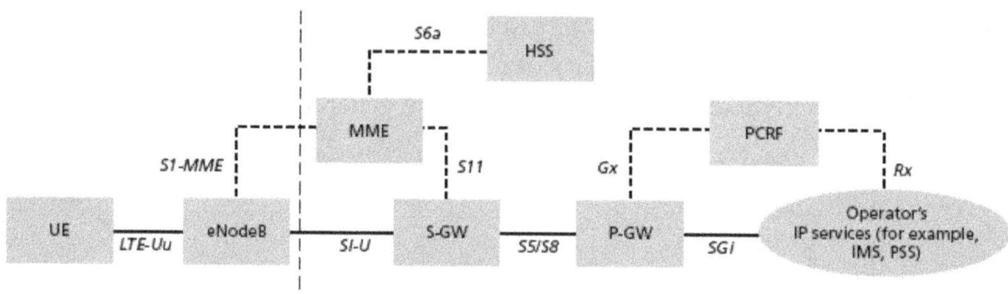

Figure 3. RAN (left) and CRAN (right)

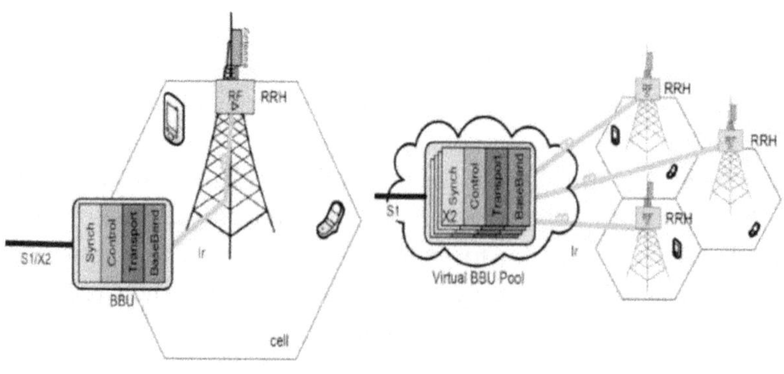

(Watanabe * Machida, 2012). In CRAN, the BBUs of a group of eNodeB's are pooled together, as shown on the right side of Figure 3. The CRAN model therefore reduces power consumption, increases scalability, and lowers delay. It also offers other advantages such as high resource utilization, and better mobility and radio interference management (Checko, 2015). These are crucial for 5G supporting IoT services in smart communities. More background information on 5G and CRAN is described in Section 2.

User Equipment Context (UEC)

Cellular networks use *User Equipment Context (UEC)* to represent the control information of a user's call/session, including the user's ID, subscription including QoS information, and state data of the current session or event. UEC records therefore are stored in the BBU, and need to be quickly accessed, read, and written.

An example of the *benefits of CRAN* would be handover (HO) support. HO occurs in cellular networks when a user moves from one eNodeB to another. Its purpose is to transfer the UEC record to the current eNodeB the user is located. In the traditional LTE network, UECs are stored in a local BBU of a eNodeB; thus, it takes time to transfer UECs between the two eNodeB. In CRAN, because BBUs are pooled, many HO steps are now internal processes of the BBU pool(Liu et al., 2012); this significantly reduces HO latency (Karneyenka et al., 2016). Furthermore, having an active user's UEC in a BBU cache is very important to maintain acceptable user experience levels (Tsai & Moh, 2017b, 2018). Efficient LB for UEC records to be stored and processed in one of the VMs within the BBU pool would also reduce latency and improve user experience (Shahriari et al., 2017).

Load Balance

There have been related studies on LB for emerging 5G and for SDN systems. Elshaer, Boccardi, Dohler, and Irmer (2015) proposed LB algorithm for micro and small cells; the algorithm increased throughput for UE and reduced up-link signal to noise ratio. Earlier, Muñoz, Barco, Ruiz-Avilés, Bandera and Aguilar (2013) used fuzzy logic controller combined with fuzzy rule-based reinforced learning System to solve congestion problem for enterprise LTE femtocells. There are a few related studies on LB for SDN (Sathyanarayana et al., 2016; Koushika & Selvi, 2014; Zhang & Guo, 2014). Our work focused on the CRAN component of 5G networks, we investigate LB algorithms, identified those that are effective to reduce queuing and service times, thus also decrease congestion and increase system throughput.

Most recently, LB has been studied for coded storage systems in cloud computing architecture. Hu, Wang, Liu, Niu, and Huang (2017) proposed proactive, intelligent LB mechanisms. They made optimal decisions either per request or statistically across requests. Experiments showed that the proposed schemes significantly reduced both median and tail latencies. Through queue-length analysis, our work has evaluated several basic, practical LB algorithms, and demonstrated that some of them are able to reduce the median latency (in terms of both waiting time and overall service time) within a CRAN system.

REINFORCEMENT LEARNING ARCHITECT WITH SELF-ORGANIZED STATE-ACTION SPACE

This section presents the GOL architecture, including the environment, the input and output channels, the GOL entities, the GOL hierarchical structure, the generic entity network, and the learning algorithms, as presented in the following subsections.

Environment

As the environment is dynamic, the GOL (agent) perceives it via a set of sensory input channels. Also, the perception of the agent from the environment can be partial, some of the input data can be missing or corrupted. The input data enters the system as a stream in real-time.

Channels

Channels are the critical part of GOL structure, as it forms its perception of its surrounding environment based on the input data it receives, and interacts with the environment via output channels.

Input Channel (IC), is the source of input data stream to GOL, which it perceives its environment through ICs. As an example, if the agent is a robot then visual and auditory data can be defined as ICs.

Output Channel (OC), is the mediator between agent and environment for interaction. For the robot example above, its arms, fingers, and legs can be defined as OCs.

Both the IC's input data and the OC's output data format is a stream of numerical data. The OCs' output will be further translated into an actual action based on the type of channel and the agent we are building.

GOL Entities

There are two types of entity: Numerical Entity (EN) and Generic Entity (GN), including Medium-Entity (EM) and High-Entity (EH). Recall that the structure of GOL has three layers; the ENs form the first layer, and GN forms the other two. As entities in the second and third layers are similar in nature, we distinguish them into two groups for clarity. We define GN of the second layer as *Medium-Entity* (EM) and for the third layer as *High-Entity* (EH). Note that each entity at each layer is the abstract definition of its lower layer entities.

Numerical Entity (EN)

It is the representation of numerical value with some loss in the granularity of value. Note that EN can be a fuzzy number (Sussner, Esmi, & Barros, 2016). In this paper, for simplicity instead of using fuzzy numbers to reduce the number of ENs, we do the conversion with an acceptable loss in granularity of number up to two decimal digits. For example, numbers such as 102, 110, and 100 are all mapped to the same EN.

Medium Entity (EM)

It represents a pattern of observed ENs for each channel over time. As in each layer, we store a limited number of entities, then each EM is formed from the history of ENs. Thus, Each EN has one-to-many connections to several EMs as those EM are built based on ENs in the first layer.

High Entity (EH)

It represents a *high-level abstract pattern* of observed EMs for a set of ICs and OCs over time. Each EH is formed from an EMs' history of some chosen channels. EH is the high-level description of observed input and output data over time. As granularity of EH is lower than EMs, then we expect to have less number of EHs compare to Ems. Thus, each EM has one-to-many connections to several EHs as those EH are built based on EMs in the second layer.

GOL Structure

GOL has a hierarchical structure with three layers made of two different types of entities.

First Layer

The recent input data and previous executed output data are entering the system as a stream. As the data are numerical data, the system converts each input data to an EN. Figure 4 depicts the first layer, where each node represents an EN. It has limited memory for the data stream of OC and IC in the format of ENs. New data stream enters the system from right side, so the old data are pushed to left side.

Figure 4. First layer of GOL where each node represents an EN

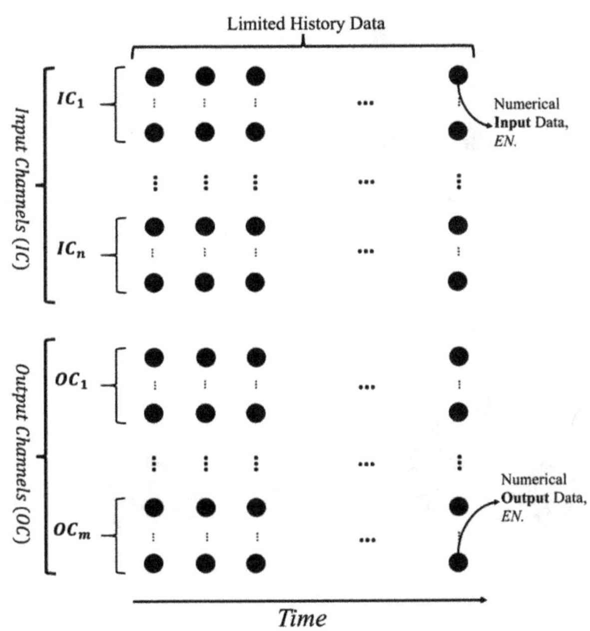

GOL includes its previous executed actions (output data) because it finds the EH of the observed environment and its interaction with the environment to adapt to the best policy in time. Thus, EH describes the environment and GOL's actions in time for an adaptation of GOL.

First layer stores EN of IC and OC data stream for the past m_1 observation, where we consider $10 \leq m_1 \leq 50$ and in this paper, we set m_1 to 50. So, when there is m_1 ENs in memory of each channel in the first-layer and new data is observed then GOL forget the oldest data.

Second Layer

It consists of EMs of each channel as a stream. As later we will describe the GOL learning algorithm, GOL finds the best four EMs set (4EM) matching the current data of each channel in the first layer, in the current time frame. As shown in Figure 5 and Figure 6, there are at most four EMs per each channel in the first layer, note that they are input data to the second layer, as shown in Figure 7.

In Figure 7, for the second layer and like the first layer, it stores the last m_2 observed data stream, and we set m_2 to 20. However, as it is evident from Figure 7, the granularity of time for second layer is lower than first layer, as each 4EM set describes the pattern of data in first layer for a specific time-frame, then history of these 4EMs that shapes the second layer is representative in a different time frame, or

Figure 5. EMs either generated from or have high similarity with ENs in the first layer

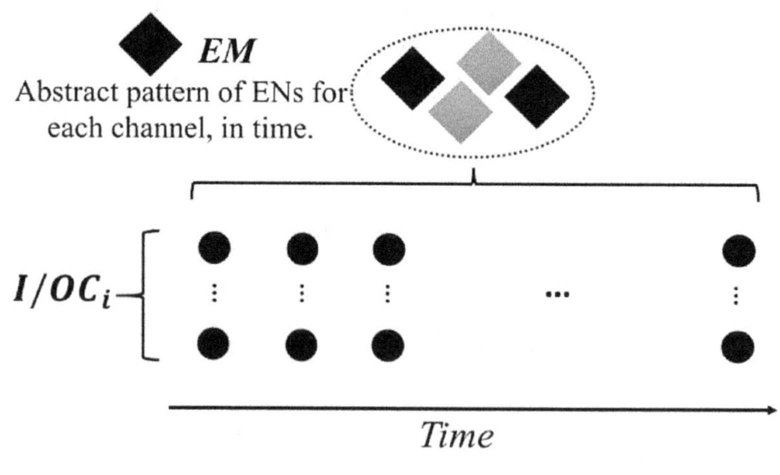

Figure 6. A group of selected EMs based on ENs in the first layer as input data for the second layer

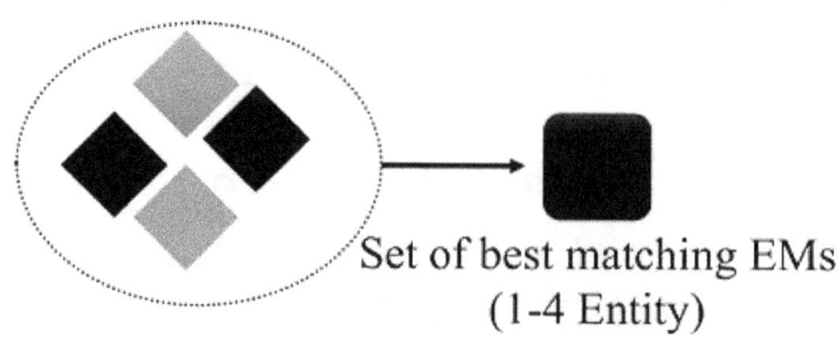

Figure 7. Second layer of GOL

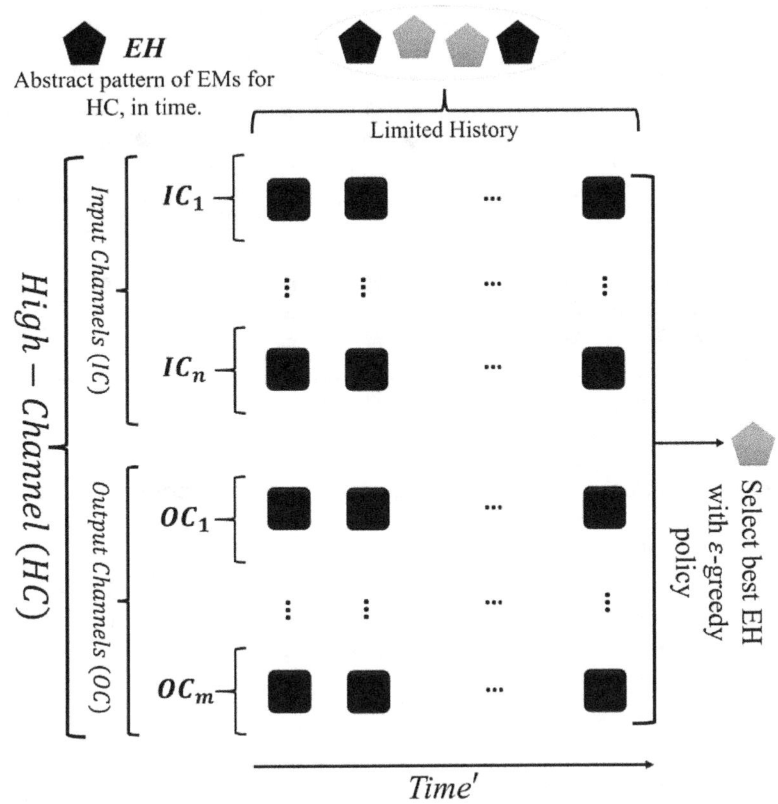

more precisely a longer time-frame. Also, according to Fig. 5, GOL perceives both sets of ICs and OCs as a high-level channel (HC) in the second layer. Note that in this layer both input or output data have lost their meaning, they are both considered as input data.

Third Layer

It is the highest layer in GOL where EHs are stored; this layer has no other functionality.

GN (Generic Entity) Network

As it is described, EHs and EMs are formed in the first and the second layer based on their lower layer entities. Both EH and EM have the same general structure; however, we represent them separately for clarity.

EM Network

It represents a pattern of ENs over time for an IC or OC. EM network is structured in three sections: (1) Past, (2) Present, and (3) Future. Figure 8 depicts the EM network structure in detail. The purpose of each node in EM network is illustrated in Figure 9. Each main node in EM network represents an entity

Figure 8. EM network structure and its sections for an IC or OC

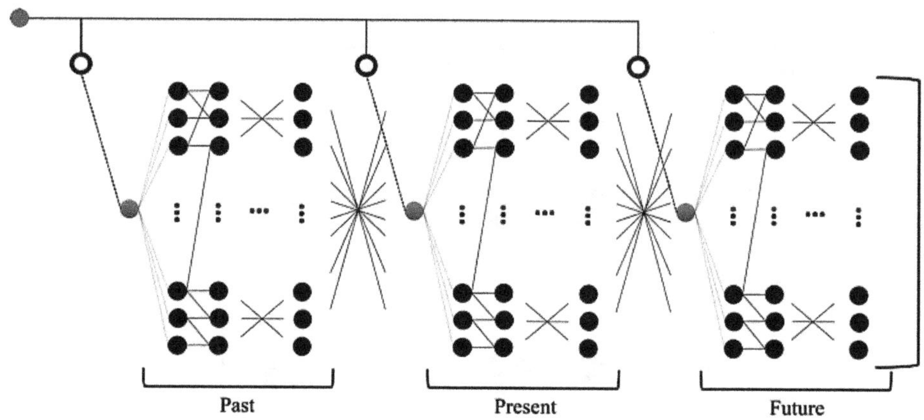

Figure 9. EM network nodes description, including eNs and section nodes

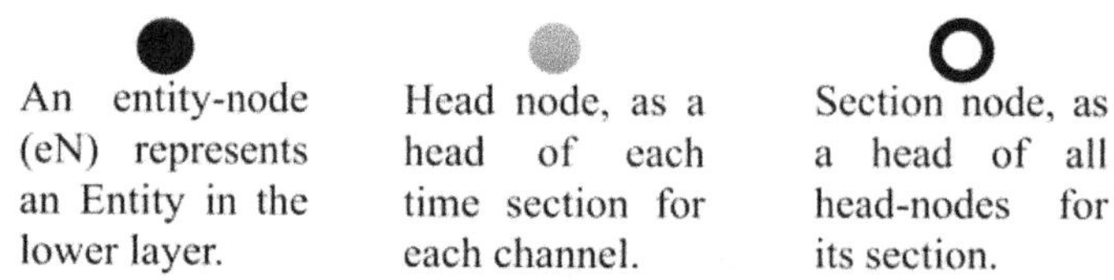

in the lower layer. EM network is separated into three sections: The present section is used to match each EM with the current data in a channel of the first layer, the past section is used to validate the best matching EMs based on their present network entities, and the future section is used to predict the coming pattern or select the best possible action.

According to Figure 9, each entity-node (eN) in entity's network that represents an entity in a lower layer has an immediate connection to the eN in front of it in the next time step. Also, each eN has random connections with its neighbor nodes in the next time step in the network. For simplicity in this paper, the connections between nodes have no extra data such as weight. Also, the structure of network will remain constant during learning. The eNs in the different sections are connected to each other. Each section does not detach eN's connections in network structure, the main purpose of head-nodes and section-nodes, as shown in Figure 9, are to organize the network, make evaluation and prediction computationally more convenient.

EH Network

It represents HC (High Channel). Thus it includes all EMs of selected ICs and OCs. Illustrated in Figure 10, the EH network structure is similar to EM's network, except the fact that each section has two head-nodes. One head-node for ICs' entities and another one for OC's entities. Note that the connections between eNs of each part (IC and OC) exist as the EH network describes the pattern for all input data and output data in a time-frame.

Figure 10. EH network structure and its sections for selected ICs and OCs

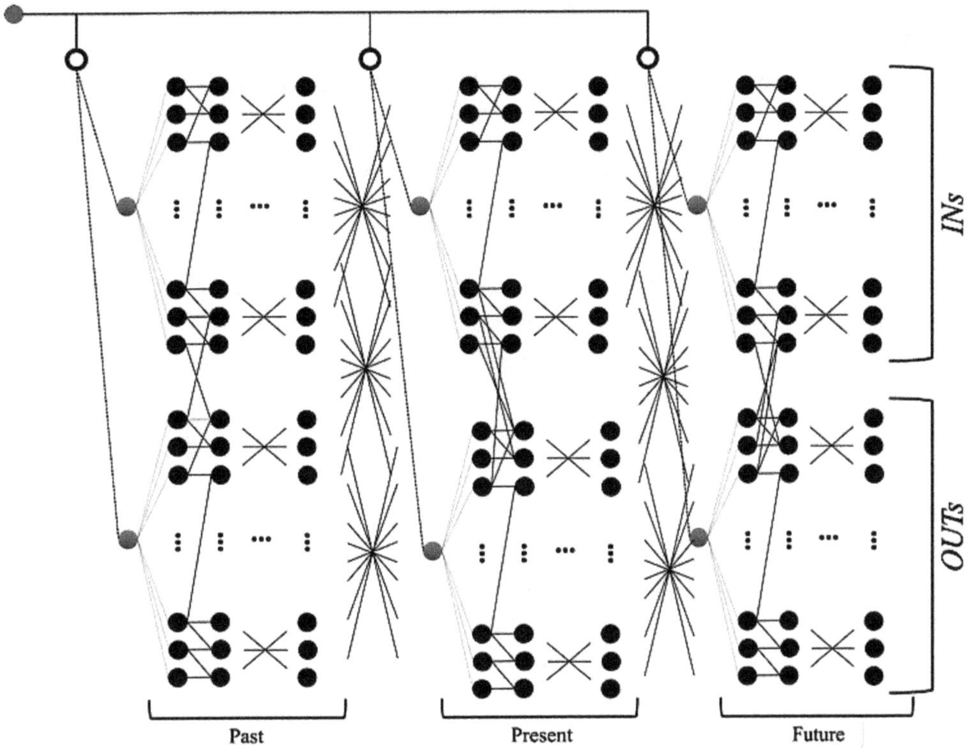

In both the EM and EH network structures, the number of layers in the present section is higher than that of the other two sections. In this paper, the ratio is as follows, 3:6:1 (past: present: future), with the majority given to the present section.

GOL Learning Algorithm

As an RL system, GOL adapts based on the feedbacks it received in response to the actions it has executed. So, GOL needs to first find the best action based on the received data; then it adapts based on the observed feedbacks.

Finding the Best EH

To find the best action, GOL must first find the best EH that describes the recently observed data and self-taken actions. Note that the system needs to have at least one EH to start making decisions. A system with no EH is called a *crude* system.

When GOL is crude, it means the first layer or/and second layer do not have enough data to form any EMs and consequently an EH. During crude state, the system makes random actions based on the nature of its OC. When the first EH is formed, then it is *refined* and it can make decisions and adapt. In a period of crudeness, the first layer forms new EMs based on ENs if no matching exists for the observed ENs. Note however that even during a refined state and even with the existence of an EM with a good match of ENs in the first layer, the system forms a new EM with a small probability of 0.005. Note also that the system behaves similarly in forming EH in the second layer, with the probability of forming a new EH is 0.0005.

Algorithm 1 illustrates the scheme for finding the best action based on the current ENs in the first layer.

Algorithm 1. Finding the best action based on ENs in first layer

```
Input: ENs of the first layer.
1. A set of sensory input data from each channel arrives; however, some sen-
   sory data may be missing or corrupted.
2. GOL finds ENs of each input data and output data, of its previous actions,
   for each channel.
3. GOL adds the current ENs to the memory of first entities layer of each
   channel, first-layer (FL).
4. Get the list of all EMs (cEM) that current ENs defines them.
5. For every EM in cEM, match their network with entities in FL for their re-
   spective channel. Referring to Algorithm 2.
6. Pick the best 4 possible matches based on ε-greedy policy for each channel
   (4EM) in cEM after the comparison.
7. Add 4EM of each channel to the second layer (SL).
8. Find all the best matching EHs (cEH) based on the entities in the second
   layer. Referring to Algorithm 2.
9. Pick the best EH (bEH) in cEH with ε-greedy policy as the best solutions or
   the observed patterns.
10. Make an action based on bEH.
   a. Pick all last EMs that are OUT EMs in bEH.
   b. Now for each out-channel pick the best action based on ε-greedy policy.
Output: The best action for each OC.

* Value of ε is set to 0.15 for all the algorithms in this paper.
```

Algorithm 2, the algorithm to find the best EH based on the entities in the lower layer, is illustrated below. GOL stores the last 10 EHs and their ENs that resulted in action from algorithm 1. Afterward, it uses them in the adaptation step, illustrated in Algorithm 2.

Algorithm 2. Finding the best matching higher entities based on the lower layers

```
Input: Entities of first or second layer.

1. IF system is crude THEN build an entity. Even IF system is refined THEN
   with small probability make a new entity.
```

```
2. Start from PRESENT section of the entity's network and level by level match
   the nodes' entities with entities in FL/SL. The number of matches is the
   number of activations of the PRESENT section.
3. Validate the entity based on its PAST section, then calculate the final ac-
   tivation value of each entity:
   a =PRESENT activation
   a'=PAST activation
   IF a' < a THEN sign = -1
   ELSE sign = 1
   a_e, entity activation, is calculated as:
   a_e = a × 0.7 + a'× sign × 0.3
4. Select the 4 of the best matching entities based on ε-greedy policy.

Output: A list of matching entities, EMs or EHs, depending on the given layer.
```

Adaptation

After finding the best action for each OC, the system interacts with the environment. Simultaneously, the system is receiving feedbacks from the environment; however, the system does not know which executed action should receive a particular feedback as feedbacks are arriving with various delays.

Below we have the adaptation algorithm based on the feedbacks and SARSA adaptation algorithm. After applying the current feedback for adaptation of GOL structure, the system keeps the current feedbacks with a probability of 0.2 to use them in the next adaptation iteration.

GOL Adaptation Algorithm Based on Feedbacks

Input: Stream of feedbacks.

GOL collects reinforcement feedbacks it receives from environment between arrivals of new data from channels.

GOL store EHs from *Algorithm 1* in a limited-sized list, LEH.

1. For each feedback (fb) in the list of collected feedback.
   a. fb affects all entities in LEH (with size of $L = 10$) with effect-ratio (er), er_k, where k is the index of entity in LEH. k is the ranked position of action-entity (EH); $k = 0$ represents the most recent action taken.
   b. er_k may follow one of the following equations,
      i. Linear ($er_k = 1 - k / L$)
      ii. Exponential ($er_k = e^{-k^2}$) where $er_k \leq 1$ for all $k \geq 0$, linear equation is used for simplicity. Each out-channel and each portion of that channel can have a distinguish adaptation policy.
   c. Adaptation follows SARSA algorithm, which at the time of adaptation system finds the best action based on current data and uses Eq. (5) with $\alpha = 0.15$ and $\gamma = 0.85$.

Note that feedback can be another form of input data, but here we consider them as a reinforcement data – not an input data which may include some feedbacks not given to them system directly and are observable through input channels; they are indirect feedbacks that we ignore them in this research.

SIMULATION OF THE 5G NETWORK LOAD-BALANCER WITH AN ENVIRONMENT OF DYNAMIC PARAMETERS

In the following, we briefly describe the problem of LB in a 5G network. This is followed by description of simulation setting (user and event types), LB architecture (LB, VM, and S or Cloud Storage), and GOL variables (IC, OC, punishment, and rewards) described in the subsequent subsections.

Provisioning of a high data rate wireless connectivity in rural and remote areas is one of the goals for perspective 5G wireless networks. Unlike the previous generation, 5G network is service-driven rather than technology-driven (Song et al., 2016). The concept of the "broadband access everywhere" is a pillar of 5G service requirements, and the target performance of 50 Mbps everywhere is considered regarding the experienced user throughput (Karlsson et al., 2016).

We apply GOL to event load balancer (LB) for a 5G mobile network; note that GOL, in general, may be applied to any cloud load balancer. We consider the load balancing issue in BBU pools in a C-RAN: As UE events arriving to the LB, GOL is applied to place each UE event to one of the VMs so as to balance the computing load of the VMs.

More specifically, each VM is associated with a limited-sized cache for storing the incoming UE event data. Any cache miss will cause the user profile data to be downloaded to the cache from the (secondary) cloud storage supporting the BBU pool, and any cache overflow will also cause the oldest user event data to be evicted from the cache and uploaded back to the cloud storage. Also, from time to time there is a need of synchronization between VM cache and cloud storage (upload to cloud storage). Thus, the goals of GOL are to place UE events to the VMs so as to minimize total cache-misses as well as the communication load between VM cache and the cloud storage.

User Types

In our simulation, each user has a unique ID, called userID. Also, we consider four types of users, ranked in descending order of priority levels: premium-mobile, premium-stationary, basic-mobile, and basic-stationary. Premium users are assumed to be far less in number compared with basic users.

Event Types

Any mobile user (i.e., UE) constantly generates new events, including the following event types:

1. **Attach Request:** An event to register user equipment (UE) with a tower.
2. **Tracking Update:** An event to notify the tower and cell about geo-location movement and movement between cells that cover different frequencies.
3. **Service Request:** An event from UE to the tower that contains a request message for some data.
4. **Paging Request:** An event from tower to the Internet to fetch the requested data. (In this paper we do not consider this event as it is not generated by mobile user.)

5. **Handover Request:** An event that handover a UE to another tower.
6. **Detach Request:** An event for termination of the connection with a tower.

Load Balancer (LB)

LB is responsible for assigning each arrival UE event to a virtual machine (VM) to store the event data. The size of event data is 2Kbytes. LB keeps track of assigned VM per user based on user-ID, frequency of each user-type, frequency of each event-type, and frequency for each pair of (user-type, event-type). Figure 11 depicts the LB and VMs.

Virtual Machine (VM)

There are R VMs (R = 3) in the simulation. Each VM has an individual cache, that (Cache-Size) can store 700 UE event data, as shown in Fig. 12. Within a cache, in our design each user type has its separate cache-queue; the size of each separate queue can change in time as new users are arriving. Also, when there are 700 events stored in a VM cache and a new one arrives then the VM removes the oldest one from one of the queues randomly. As new UE event arrives, the old ones are pushed down the queue.

Cloud Storage (CS)

To maintain data persistency and fault tolerance, a cloud storage, referring to Fig. 11, is used. Each VM synchronizes its event data with CS based on the synchronization-time interval (ST) of the event data, so that the CS would have its updated copy. Note that a small ST would provide more rapid synchronization. GOL algorithm would decide the ST for each user event.

Figure 11. The structure of LB and R number of VMs with CS for persistency of data

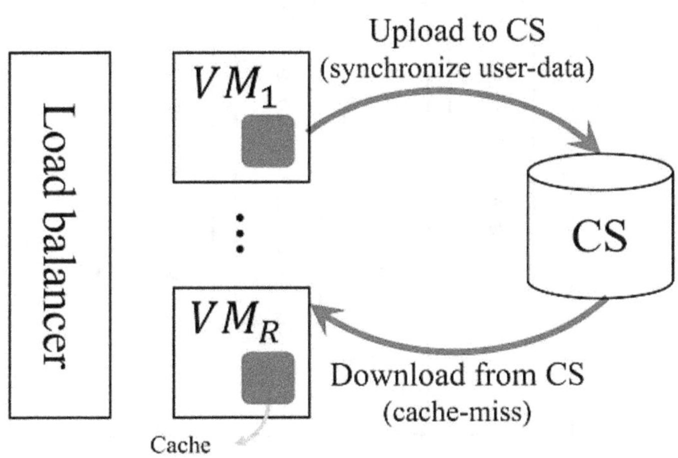

ICs and OCs for LB

LB uses GOL to find the best possible VM for each event. Thus, first we need to define the ICs and OCs for the LB problem, as follows:

$$IC_0 = \begin{cases} frequency\,of\,observing\,this\,user_{type} \\ frequency\,of\,observing\,this\,event_{type} \\ frequency\,of\,observing\left(user_{type}, event_{type}\right) \\ user_{type} \\ event_{type} \end{cases} \tag{7}$$

$$IC_i = \begin{cases} queue\,size\,for\,each\,user_{type}\;in\,VM_i \\ average\,of\,ST\;for\,each\,queue\,in\,VM_i \end{cases} 1 \le i \le R \tag{8}$$

$$OC_j = \begin{cases} ST\;for\,each\,queue\,in\,VM_j \\ value\,of\,VM_j\left(Q\,value\right) \end{cases} 1 \le j \le R \tag{9}$$

To simulate a realistic environment, in the simulation we randomly set values of some input data in ICs (except IC_0) to -1 to make the environment partially invisible to the LB. $\left|\mu_{\Delta Q}\right| < 2 \times \sigma_{\Delta Q}$

Figure 12. The internal structure of a VM and user-queues

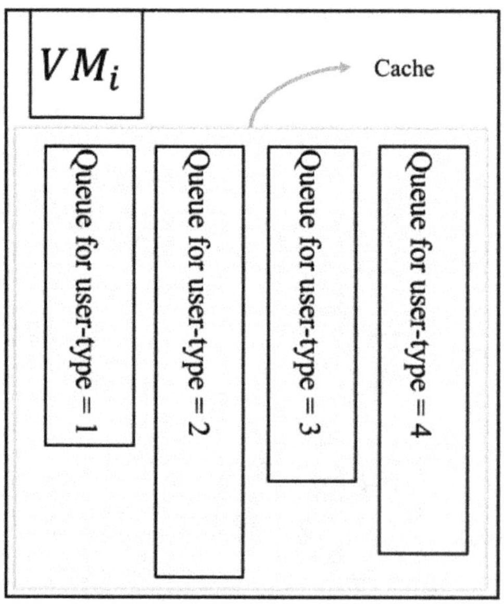

System Objectives, Punishment, and Reward

System Objectives

The LB has two objectives:

1. Minimize the total cache-misses for all VMs as LB tries to assign each event data to a right VM.
2. Minimize the communication load between CS and VMs. When a cache-miss happens, the assigned VM download from the CS the stored (existing) user profile. Also, a VM synchronizes (upload data to CS) according to ST.

For GOL to minimize cache-misses and communication load between VMs and CS, we need to define punishment and reward for each case.

Punishment

Cache-Miss Punishment

It affects the value of selected VM, Eq. (9). For each action entity (EM) we calculate the cache-miss punishment as follow:

$$p_{cm}^{i} = -er_{k} \times \left(100 - 10 \times u_{t}\right) \tag{10}$$

where p_{cm}^{i} is the cache-miss punishment of the action taken for ith VM in (8) (jth VM in (9)), u_t is the user-type, for $u_t = 1$ refers to the top priority user (premium-mobile) receiving the maximum punishment for a cache-miss, and er_k is as described in the *Adaptation Algorithm*.

Cloud-Load Punishment

- **Upload:** When the type of cloud-load is *upload* to CS, it influences the ST value of user-data with the following punishment:

$$p_{cu}^{i} = -er_{k} \times \left(5 \times ds\right) \tag{11}$$

where p_{cu}^{i} is the upload punishment due to synchronization of UE event with the CS, and ds is the upload data size, equals to 2KB.

- **Download:** When the type of cloud-load is *download* from CS, then we have a cache-miss and punishment is defined as:

$$p_{cd}^{i} = -er_{k} \times ds \tag{12}$$

where p_{cd}^{i} is the download punishment and ds is as defined above, or 0 if the user profile does not exist in CS (i.e., it is a new user).

VM Miss-Match Punishment

It happens when LB assigns a new incoming event to a VM different than an existing event of the same user. Thus, the system has an inner VM load. The punishment for inner-load is defined as follow,

$$p_{lv}^{i} = -er_{k} \times \left(0.1ds\right) \tag{13}$$

where p_{lv}^{i} is inner-VMs communication load due to miss-match.

Reward

The system receives rewards for each event that it processes and they are as follows:

VM Value Reward

It is defined as follow:

$$r = 200 - 3 \times \left(n_{fb} - n_{ST}\right) \tag{14}$$

$$r_{v}^{i} = er_{i} \times \left(r_{rnd2} + \frac{r}{2}\right) \tag{15}$$

where r_{v}^{i} is the reward for VM value of the executed action, n_{fb} is number of feedbacks that GOL has received between processing two events, as feedbacks are delayed the system can receive more than one feedback between arrival of two user-events, n_{ST} is the number of upload punishments that effected ST of user-data, and r_{rnd2} is a random value between 0 and $r / 2$.

VM ST Reward

For each user-type is defined as follow:

$$r' = cm - 10 \times n_{ST} \tag{16}$$

$$r_{ST}^{i} = er_{i} \times \left(r_{rnd2}^{'} + \frac{r'}{2}\right) \tag{17}$$

where r_{ST}^{i} is the reward of synchronization time of users in all queues, note that a longer synchronization time would reduce uploads to CS; $r_{rnd2}^{'}$ is similar to r_{rnd2} in (15). In (16), cm is equal to 1,000; the intuition for selection of its value is that we estimated that n_{ST} could have a maximum value of 100, as during our experiment we observe that value of n_{ST} fluctuates between 3 to 5.

Table 1. Environment simulation parameters

Parameter	Description
N_{users}	Number of mobile user in environment is set to 3,000.
$User_{type}$	Number of user-types is set, with 1 as premium-mobile user.
$Event_{type}$	Number of event-types is set to 5.
R	Number of VMs is set to 3.
Cache-Size	Maximum number of user events that each VM can cache the event data is set to 700.
N_{events}	Number of events is set to 50,000 for the entire simulation.
Epoch	Simulation time frame; the time duration of 20 events.

As the system receives feedbacks between processing two consecutive events, it stores them. After processing all the feedbacks, with a probability of 0.2 it keeps all currently observed feedbacks for processing in the iteration, since, due to the nature of feedbacks that are delayed feedbacks.

PERFORMANCE EVALUATION OF REINFORCEMENT LEARNING

Simulation Parameters

The parameters for the environment are summarized in Table 1.

Load Balancing Methods

We evaluate and compare the performance of GOL with two LB methods.

Memorized-BestVM (MB)

For each new user event arriving to the system, the LB selects the best VMs based on the cache-queue size depending on the user-type, and select its ST randomly from range of [300-15,000] milliseconds (ms) and memorized where the event data of this user is stored. Afterward, for all the events of the user, it sends them to the same VM but each time chooses their ST randomly. Thus, in this case, we never have VM miss-matches, but we still have cache-misses.

Memorized-BestVM With Fixed ST (MB-F)

Is similar to MB except with a fixed ST set to 1,000 ms (1 second).

Performance in Cache-Misses

Figures 13-16 depict the simulation results for all three LB methods; they are further summarized in Table 2.

Fig. 13 compares the three LB methods for average number of cache-misses per VM per epoch (20 events), during the entire simulation. It is clear that GOL has achieved significantly lower cache misses than the other two LB methods. Per event, the cache misses are 11.15%, 11.0%, and 4.2%. In another word, comparing with the other two methods, GOL has reduced the total number of cache-misses to 37.67% when compared with the MB method (Table 2).

To see more clearly the cache-miss performance of the three methods, Figure 14 and 15 depict the growth of total cache-misses for MB and GOL, respectively, for all three VMs. It is clear that while the total number of cache-misses grows linearly with simulation time in epochs for MB algorithm (Fig. 14), the growth in GOL is far slower (Fig. 15).

Figure 13. Comparison for average number of cache-misses per VM per epoch (1 epoch = 20 events)

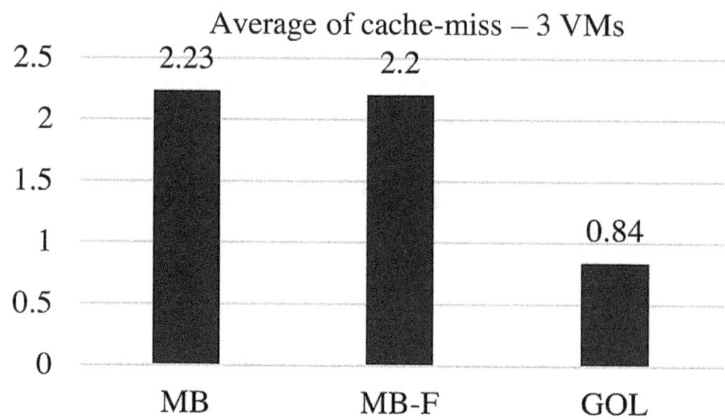

Figure 14. Total cache-miss vs. time (epochs) for MB

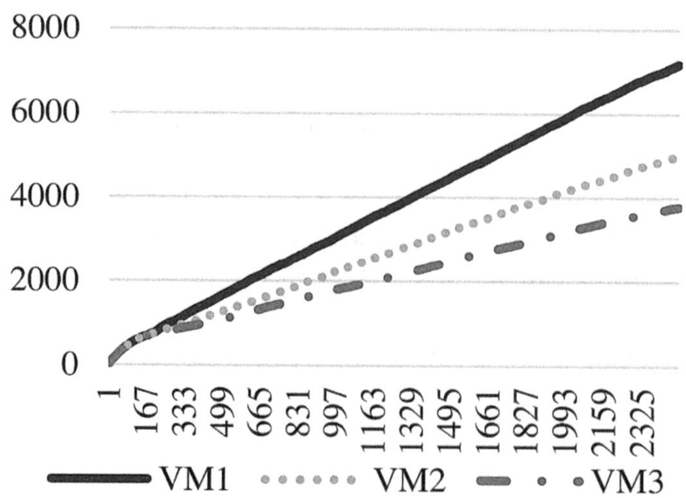

Figure 15. Total cache-miss vs. time (epochs) for GOL

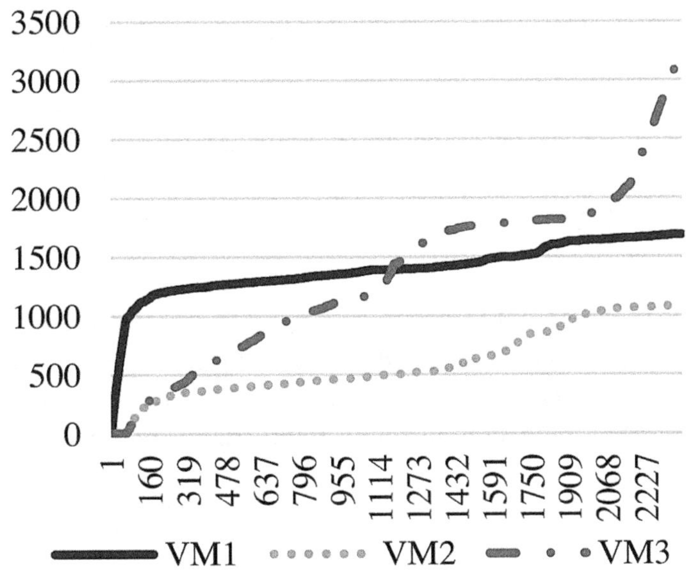

Performance in Communication Load

Next, we look at the communication load between VMs and CS, as illustrated in Fig. 16. It is clear that GOL has significantly reduced both upload and download data, to 40.47% and 75.77%, respectively, when compared with the MB algorithm (referring to Table 2).

Figure 16. Total upload and download data (megabytes) for three algorithms

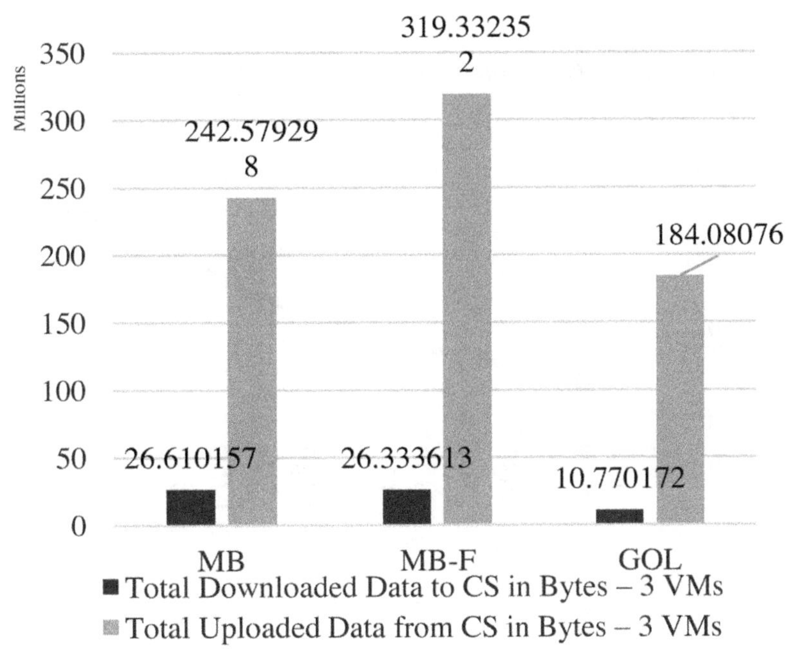

Table 2. Results of Simulation of All Three Load Balancing Methods

	Average of Cache-Miss per VM per Epoch	Total Download Data to CS in Bytes	Total Upload Data From CS in Bytes	Average of ST (ms)
MB	2.23	26,610,157	242,579,298	5,244
MB-F	2.2	26,333,613	319,332,352	976.1
GOL	**0.84**	**10,770,172**	**184,080,760**	**18,502.6**
GOL/MB	**37.67%**	**40.47%**	**75.88%**	**3.52**

It is clear from Table 2 that, by using the proposed punishment and reward scheme, GOL has kept a longer ST, and thereby successfully achieved the system objectives: far less cache misses and far less upload and download data between VM and CS. It also confirms that the static structure of GOL with three layers has the capability to adapt to dynamic environments without having prior knowledge or pre-defined state-action space.

Computation Overhead

While it is clear that the proposed GOL architecture has effectively achieved the two system objectives (minimizing cache-misses and communication load), it is important to understand the computation overhead of the GOL method. As EH is an abstract description of the observed input and output data over time, the number of EH represents the total number of patterns to describe the entire environment, and the system's interaction with that environment.

Figure 17 illustrates the growth of the number of EHs over simulation time. Observe that the total number of EHs grows to 51 during the entire simulation. This small number of EHs proves that the proposed system has a good performance on finding the limited number of high-level patterns that are sufficient to describe a dynamic and chaotic environment, with feedback policies expressed by (15) and (17).

Figure 17. Growth for number of EHs vs. time (epochs)

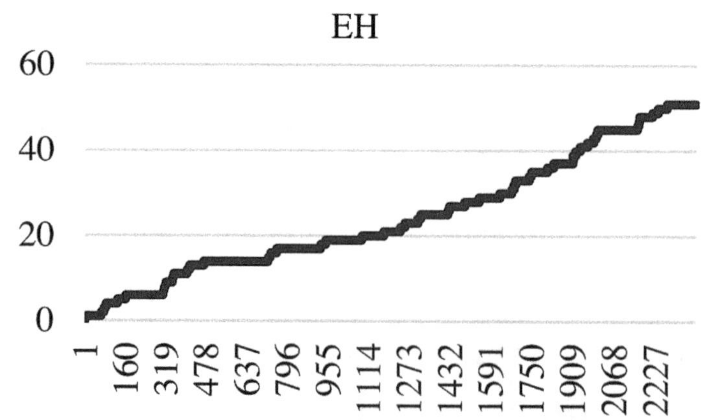

HEURISTIC ALGORITHMS OF PRACTICAL CRAN LOAD BALANCING

To allow for carrier-grade real-time services while supporting rapidly growing IoT communications, we consider LB in 5G CRAN, as depicted in Figure 18. On the left is the control core of eNodeB, which hosts layer-2 (L2) control applications, which in tradition maintains stateful information of all the callers. There are multiple L2 control applications running on the same host, each receiving user requests for a cell and/or a frequency band. User requests are represented by UEC. To leverage cloud technology, coming to the BBU pool of CRAN (shown on the right of Figure 18), these requests, instead of processed by a single machine for each cell, would be pooled together and processed by several Virtual Machines (VM) hosted by a server. Note also that each VM in the host is assumed to have its own local cache. The local cache associated with each VM stores copies of UEC for fast, latency-sensitive access (Tsai et al., 2018).

Observed from Figure 18 that, as each UEC record arrives at the control core of a 5G eNodeB (the left box), it is important that the UEC be assigned to an appropriate VM, to achieve a balanced load among the VMs, so that the time needed for accessing and processing the UEC is minimized. Note also that cache management is another important task for the CRAN model depicted in Figure 18, which has been addressed in an earlier work (Tsai et al., 2018).

Figure 18. High level diagram of LB in CRAN

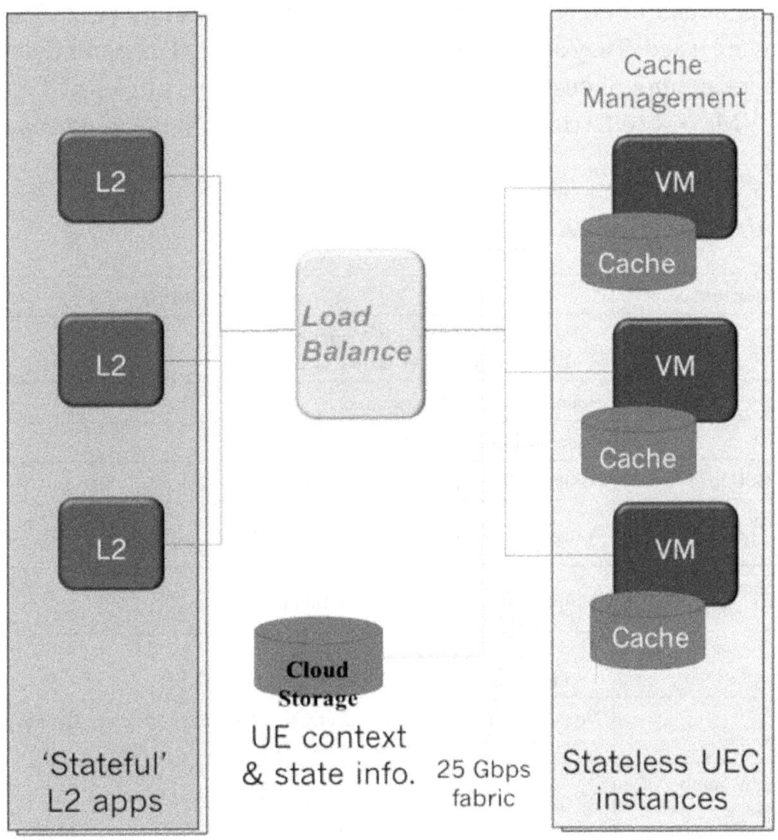

Table 3. 5G UE events break down

UE Events	Event Arrival Rate Events/eNB/second	Relative CPU Load
UE state transitions	750	1.00
Handovers	100	0.82
Tracking area updates (TAU)	30	1.24
Paging	500	0.26
Attach/detach	25	2.31
Actual Event Total	1400	

While there have been many complex, intelligent LB algorithms proposed for various distributed computing and networked systems (Muñoz et al., 2013; Koushika et al., 2014; Zhang et al; 2014; Elshaer et al., 2015; Sathyanarayana et al., 2016; Hu et al., 2017), this work focuses on evaluating several basic, practical LB algorithms that may be readily deployed in actual 5G networks. To achieve this goal, it is important to understand the characteristics of real network traffic load, in terms of UEC, to be processed for CRAN in 5G networks. Table 3 summarizes the major UE event types need to be processed by CRAN, and their associated traffic load and CPU load, provided by Nokia Research (A. Rottinghaus, D. Baretto, R. Van Egeren, & R. Edward, 2017).

Based on Table 3, a total of eight LB algorithms are studied, as summarized in Table 4. First, five basic LB algorithms are listed. They are followed by three additional algorithms that were based on some ad-hoc design with an attempt of customizing to the traffic characteristics listed on Table 3; assuming that the number of VMs = 4 (referring to Table 4, again agreed with the real network setting provided

Table 4. Eight LB Algorithms for CRAN

Acronym	Algorithm	Description
Five Basic LB Algorithms (Listed in Ascender Order of Complexity)		
RR	Round-robin	Round-robin of VM assignment.
RND	Random	Random VM assignment.
CPU	Based on CPU load	Assign to VM with min. cumulative CPU load.
S_{queue}	Based on shortest queue size	Assign to VM with shortest queue length.
S_{wait}	Based on shortest waiting time	Assign to VM with shortest waiting time.
Three Additional LB Algorithms (Listed in Ascender Order of Complexity)		
Ded	Dedicated	3 types of events with high occurrence (UE State transition, Handovers, Paging) each has own dedicated VM; all others share one VM.*
Access	Based on min. number of UEC accesses	Assign to VM with min. number of UEC accesses.
Q_{CPU}	Based on queued time at CPU	Split all event types into 4* categories according their relative CPU load. The highest CPU load category will be assigned to VM with shortest waiting time and vice versa.

Note: *assuming no. of VM = 4

Table 5. Experiment Parameter Values

Parameter	Value
No. of VM	4
VM cache size	2,500MB
UEC record size	200 KB
UEC arrival process	Poisson
λ, mean UEC arrival rate	1,400 UEC/second
Total no. of distinct users	25,000
Total simulation time	5 minutes

by Nokia Research Research (A. Rottinghaus, D. Baretto, R. Van Egeren, & R. Edward, 2017). Note that the algorithms are listed in the ascending order of algorithmic complexity.

PERFORMANCE EVALUATION OF PRACTICAL CRAN LB HEURISTIC ALGORITHMS

Experiment Setup

CloudSim (Calheiros, Ranjan, Beloglazov, Rose, & Buyya 2011)is used for the experiment. One host and 4 VMs are set up to represent the BBC pool for one 5G eNodeB. Other simulation parameters are summarized in Table 5, which is based on the real cellular network setting provided by Nokia Research (A. Rottinghaus, D. Baretto, R. Van Egeren, & R. Edward, 2017).

Since the main goal of the LB algorithms is to reduce the latency of IoT communications and to improve user experience, the following two performance criteria are considered:

- **Waiting Time:** The time a UEC record spent in a VM queue while waiting to be processed by the VM.
- **Total Service Time:** Total time a UEC record spent in a VM, including waiting time plus the time the VM processes the UEC.

Performance Results

We evaluate the eight LB algorithms by queue-length analysis. In particular, we measure the waiting time and service time experienced by each UEC record, and present the results in different formats, as shown in the following subsections.

Eight LB Algorithms: Probability Density Function (PDF) and Average Values

The PDF of waiting time and total service time experienced by UEC records under the 8 different algorithms are shown in Figures 19 and 20, respectively. Their average values and standard deviation are also summarized in Tables 6 and 7, respectively.

Figure 19. PDF of queue size (no. of UEC records)

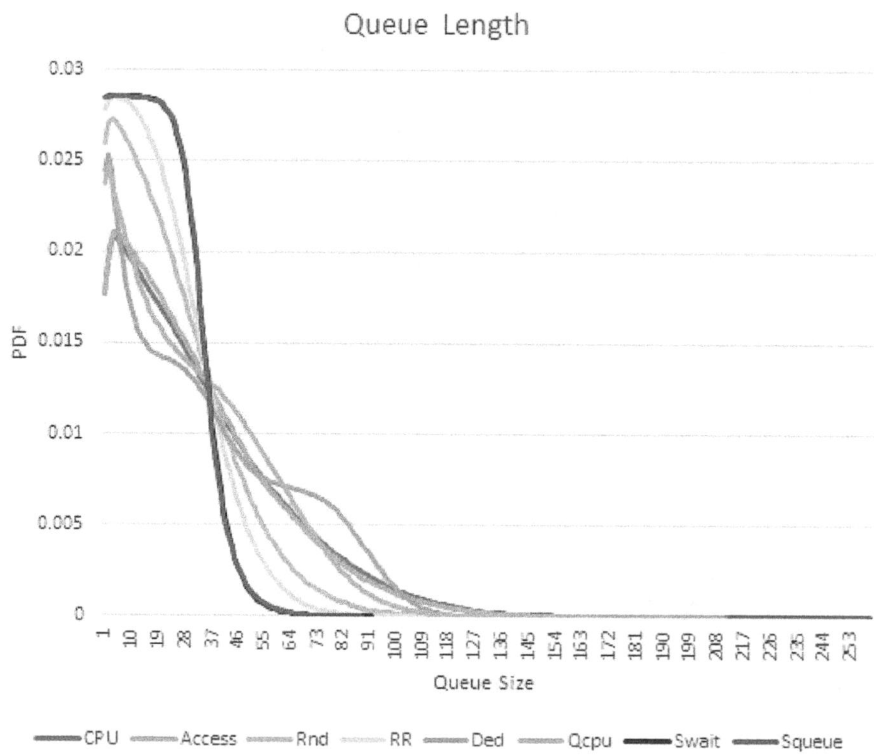

Figure 20. PDF of total service time (normalized with CPU load to process UE state transition)

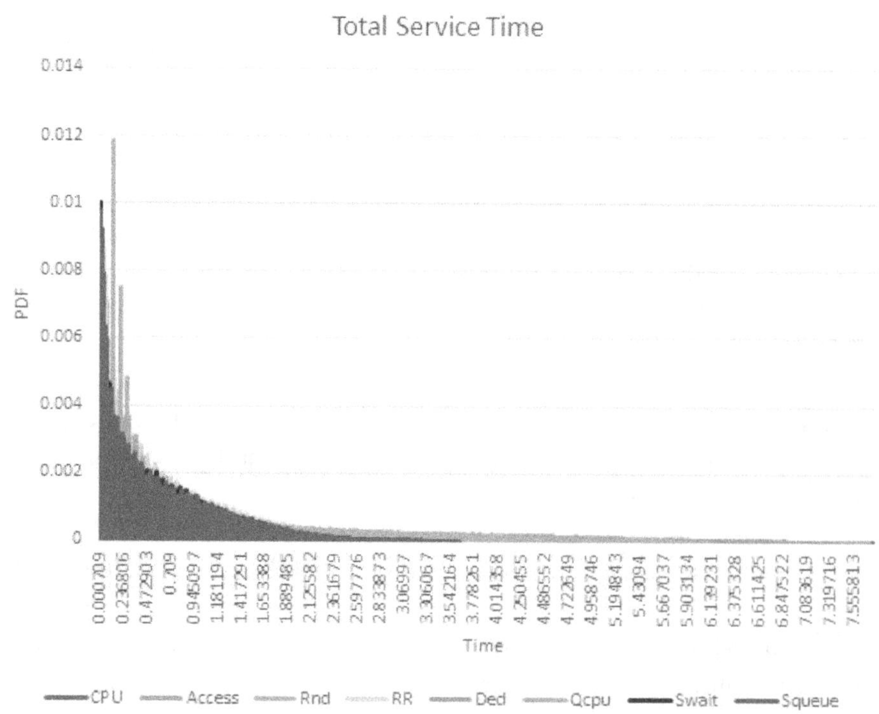

Table 6. Average and standard deviation of queue size (in terms of no. of UEC records)

	RR	Rnd	CPU	S_{queue}	S_{wait}	Ded	Access	Q_{cpu}
Avg	21.192	24.387	34.970	18.995	18.990	36.245	33.495	31.801
StdDev	1.007	1.119	3.022	0.5321	0.5545	0.888	2.122	1.9065

Table 7. Average and standard deviation of total service time (msec)

	RR	Rnd	CPU	S_{queue}	S_{wait}	Ded	Access	Q_{cpu}
Avg	1.163	1.366	1.913	0.920	0.894	2.103	1.862	1.759
StdDev	0.221	0.309	0.482	0.153	0.149	0.420	0.480	0.316

Observed that the LB algorithms that are based on queue size or waiting time (such as S_{queue} and S_{wait}) achieve the best results – with very short queue size and short total service time, while LB algorithms that are based on CPU loads or access counts (such as Access, CPU, Qcpu) suffer from long queue size and long total service time. The results of Round-robin and Random algorithms are somewhere in between.

The relative result between queue-length/waiting time-based algorithms and CPU-load-based algorithms is understandable: since the main goal for these LB algorithms is to improve both IoT communications and user experiences by reducing latency through balancing the VM loads, LB algorithms that are based on queue size or waiting time can better achieve balance in queue size or waiting time, thereby reducing the waiting time and total service time.

Five Basic LB Algorithms: Cumulative Density Function (CDF) and Box Plots

In order to demonstrate the relative merits of the LB algorithms, this section focuses on the five basic algorithms (refer to Table 4):

- **CPU:** Based on CPU load
- **RND:** Random
- **RR:** Round-robin
- **S_{queue}:** Based on shortest queue size
- **S_{wait}:** Based on shortest wait time

The CDF results are shown in Figures 21 and 22, respectively for queue length and waiting time. It is clear that S_{queue} and S_{wait} both achieve the best results, followed by RR, RND, and lastly CPU. Comparing S_{queue} with S_{wait}, S_{queue} performs seems to perform as good as S_{wait} (see also Tables 6 and 7). This is very encouraging since it is much simpler to implement S_{queue} than to implement S_{wait}.

Finally, we show the box plots of the queue length and total service time of these five basic LB algorithms, in Figures 23 and 24, respectively. It is very clear that that S_{queue} and S_{wait} both achieve very strong results, followed by RR, RND, and lastly by CPU. The box plots illustrate not only the median value, but also the maximum, minimum, first and third quartiles. These plots allow a clear view of the distribution of queue size and total service time. They not only display that S_{queue} and S_{wait} both attain very

Figure 21. CDF of queue size (in terms of no. of UEC records)

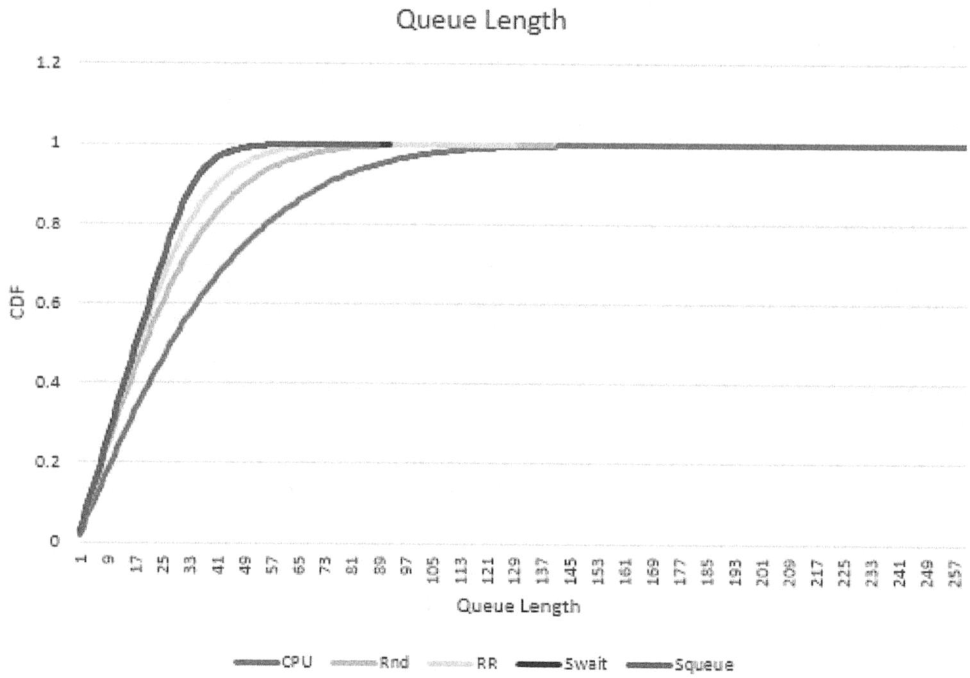

Figure 22. CDF of total service time (msec)

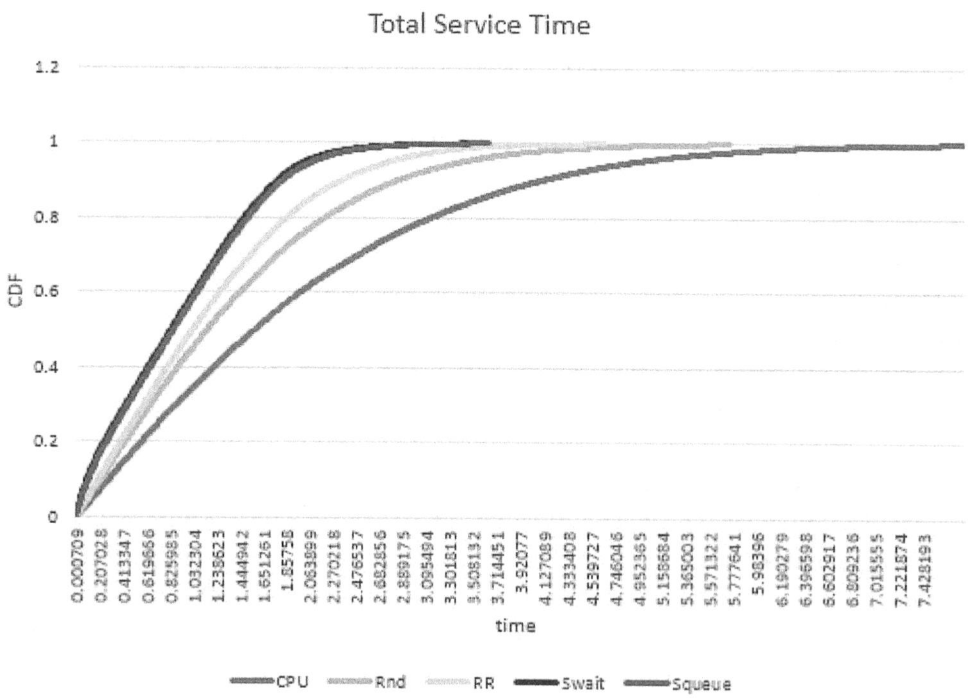

Figure 23. Box plot of queue size (in terms of no. of UEC records)

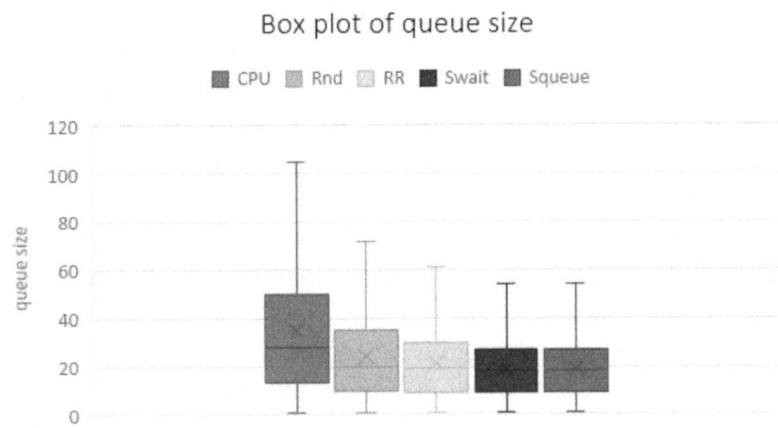

Figure 24. Box plot of total service time (msec)

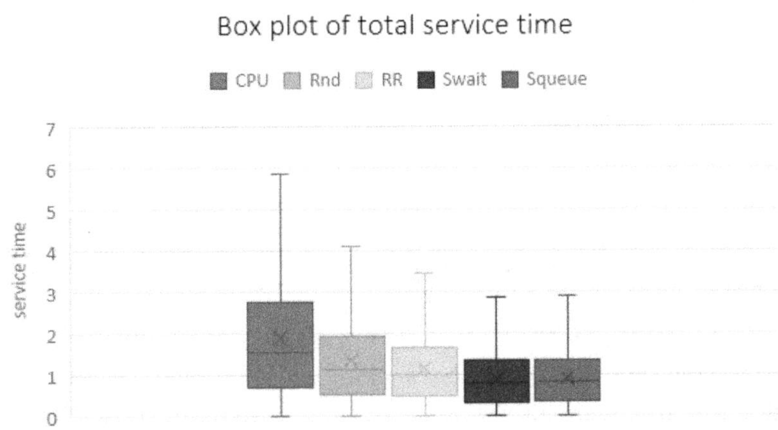

short queue size and total service time, but their values are with smaller variance. Thus, they achieve excellent load balance among VMs, which allows each UEC record to experience short delay, and with short variance among the delay values.

It is very interesting to compare the two best-performed algorithms: S_{queue} and S_{wait}. Note that S_{queue} is based on queue size while S_{wait} is based on waiting time. Clearly S_{queue} is simpler to implement than S_{wait}. Yet, comparing with queue size (referring to Table 4), the average value achived by S_{queue} is only .026% longer than that by S_{wait}. This is a very encouraging, showing there if often a trade-off between complexity and performance, yet a much simpler implementation (S_{queue}) can perform almost as good as the more complicated ones (S_{wait}).

SUMMARY

The emerging 5G cellular technologies aim to provide scalable, reliable communication for rapidly growing of IoT devices. This chapter has studied the effectiveness of LB algorithms for CRAN that is a critical component of 5G system. Two main research results are included. First, we proposed GOL, an intelligent system based on RL algorithms with a simple three deep-layered structure, for online learning of a dynamic and partially visible environment with delayed feedbacks. The GOL architecture has been illustrated in detail, and its learning algorithms presented. The proposed method has been applied for load balancing in 5G CRAN networks; the problem domain has been mapped to the GOL architecture with a proper design of punishments and rewards for its learning algorithm. Simulation results have demonstrated that the proposed method has the capability to adapt to some optimal solutions over time by forming high-level and abstract features from the observed data. Comparing with two other methods, GOL is able to achieve significantly better performance in terms of reducing both cache-misses and communication load.

Next, eight LB algorithms have been carefully developed and evaluated, based on real cellular network characteristics provided by Nokia Research. The results have demonstrated how well these eight algorithms are able to reduce latencies experienced by IoT communications in 5G system. Among these LB algorithms, S_{queue} (based on queue length) and S_{wait} (based on waiting time) have achieved the shortest waiting time and shortest total service time. The performance of S_{queue} is only 0.026% worse than S_{wait}, yet it is much simpler to develop.

FUTURE RESEARCH DIRECTIONS

Future research may gear towards several directions: First, improvement of GOL architecture. This would include the following: 1) using fuzzy numbers and fuzzy RL, 2) the connections between nodes in entities network can adapt in time or add some weights to form a better structure for finding better actions, and 3) extend the GOL architecture to other application domains such as control of robots, drones and other autonomous systems, and resource management in 5G C-RAN (Karneyenka et al., 2017).

The next future research direction would be working on improving CRAN. This would consist of integrating LB with cache management in CRAN (Tsai et al., 2018); from there developing a multi-constrained optimization problem for CRAN in terms of latency, network bandwidth, CPU processing, and service qualities; applying predictive traffic modeling; and combining with other aspects of CRAN resource management (Karneyenka et al., 2017; Su et al., 2018).

The third research direction would be developing fog and edge computing algorithms (Choudhari et al., 2018; Gao et al., 2018) that are the future of cloud computing as supported by 5G technologies, based on machine learning methods. These methods would include both supervised and unsupervised ones, such as those discussed for secured IoT applications (Raju et al., 2018). Along this direction, one would also work on how cloud, fog, and edge computing architectures would support distributed machine learning methods.

ACKNOWLEDGMENT

This work is supported in part by SJSU RSCA and by Nokia Foundation. The author appreciates former students Mr. Behrooz Shahriari and Mr. Chin Tsai for their master research work, and appreciates many helpful discussions with Mr. Alan Rottinghaus, David Baietto, Rick Van Egeren and Rob Edward of the Department of Radio Access Networks, Nokia Research.

REFERENCES

Alcatel, L. (2009). *The LTE network Architecture—A comprehensive tutorial." Strategic Whitepaper*. Lucent Alcatel.

Behrooz & Moh. (2016). Intelligent mobile messaging for urban networks: Adaptive intelligent mobile messaging based on reinforcement learning. *Proceedings of 12th IEEE International Conference on Wireless and Mobile Computing, Networking and Communications (WiMob)*, 1–8.

Bengio, Y., & Delalleau, O. (2011). On the expressive power of deep architectures. In *Algorithmic Learning Theory* (pp. 18–36). Berlin, Germany: Springer. doi:10.1007/978-3-642-24412-4_3

Bengio, Y., Lamblin, P., Popovici, D., & Larochelle, H. (2007). Greedy layerwise training of deep networks. Proc. Adv. Neural Information Processing System, 153–160.

Calheiros, R. N., Ranjan, R., Beloglazov, A., De Rose, C. A. F., & Buyya, R. (2011). CloudSim: A toolkit for modeling and simulation of cloud computing environments and evaluation of resource provisioning algorithms. *Software, Practice & Experience*, *41*(1), 23–50. doi:10.1002pe.995

Casana, M., & Redondi, A. E. C. (2017). IoT communicatin technologies for smart city. In V. Angelakis & ... (Eds.), *Designing, Developing, and Facilitating Smart Cities*. Springer Publishing. doi:10.1007/978-3-319-44924-1_8

Checko, A., Christiansen, H. L., Yan, Y., Scolari, L., Kandaras, G., Berger, M., & Dittmann, L. (2015). Cloud RAN for mobile networks – A technology overview. *IEEE Communications Surveys and Tutorials*, *17*(1), 405–426. doi:10.1109/COMST.2014.2355255

Choudhari, T., & Moh, M. (2018). Prioritized Task Scheduling in Fog Computing. In *Proc. of 2018 Annual ACM Southeast Conference*. Richmond, KY. ACM Press. 10.1145/3190645.3190699

Elshaer, H., Boccardi, F., Dohler, M., & Irmer, R. (2015). Load & Backhaul Aware Decoupled Downlink/Uplink Access in 5G Systems. In *Proceedings of IEEE International Conference on Communications (ICC)*, pp. 5380–5385. 10.1109/ICC.2015.7249179

Err, M.-J., & Deng, C. (2004). *Online tuning of fuzzy interference systems using dynamic fuzzy Q-learning. IEEE Trans. on Systems, Man, and Cybernetics, Part B*, 34, 1478–1489.

Farabet, C., Couprie, C., Najman, L., & LeCun, Y. (2013). Learning hierarchical features for scene labeling. *IEEE Transactions on Pattern Analysis and Machine Intelligence*, *35*(8), 1915–1929. doi:10.1109/TPAMI.2012.231 PMID:23787344

Gao, L., & Moh, M. (2018). Joint Computation Offloading and Prioritized Scheduling in Mobile Edge Computing. In *Proceedings of IEEE International Conference on High Performance Computing and Simulation*. Orleans, France: IEEE.

Hinton, G., & Salakhutdinov, R. (2006). Reducing the dimensionality of data with neural networks. *Science*, *313*(5786), 504–507. doi:10.1126cience.1127647 PMID:16873662

Hinton, G. E., Osindero, S., & Teh, Y. (2006). A fast learning algorithm for deep belief nets. *Neural Computation*, *18*(7), 1527–1554. doi:10.1162/neco.2006.18.7.1527 PMID:16764513

Hu, Y., Wang, Y., & Liu, B. (2017). Latency reduction and load balancing in coded storage systems. In *Proceedings of ACM Symposium on Cloud Computing*. Santa Clara, CA. ACM Press. 10.1145/3127479.3131623

Huang, J., Wu, K., & Moh, M. (2014). Dynamic Virtual Machine Migration Algorithms Using Enhanced Energy Consumption Model for Green Cloud Data Centers. In *Proceedings of the International Conference on High Performance Computing and Simulation (HPCS)*. Bologna, Italy. IEEE Press. 10.1109/HPCSim.2014.6903785

Hwang, K. (2017). *Cloud Computing for Machine Learning and Cognitive Applications*. The MIT Press.

Jouffe, L. (1998). *Fuzzy Inference System Learning by Reinforcement Methods. IEEE Trans. on Systems, Man, and Cybernetics, Part C* , 28, 338–355.

Karlsson, A., Al-Saadeh, O., Gusarov, A., Renuka, V. R. C., Tombaz, S., & Sung, K. W. (2016). Energy-efficient 5G deployment in rural areas. *Proceedings of IEEE 12th International Conference on Wireless and Mobile Computing, Networking and Communications (WiMob)*, 1–7.

Karneyenka, V., Mohta, K., & Moh, M. (2017). Location and Mobility Aware Resource Management for 5G Cloud Radio Access Networks. In *Proc. of the Int. Conf. on High Performance Computing and Simulation (HPCS)*. Genoa, Italy: IEEE Press. 10.1109/HPCS.2017.35

Kaur, G., & Moh, M. (2018). Cloud Computing Meets 5G Networks: Efficient Cache Management in Cloud Radio Access Networks. In *Proc. of 2018 Annual ACM Southeast Conference*. Richmond, KY: ACM Press. 10.1145/3190645.3190674

Koushika, A. M., & Thamarai Selvi, S. (2014). Load Balancing Using Software Defined Networking in cloud environment. In *Proceedings of 2014 International Conference on Recent Trends in Information Technology (ICRTIT)*. IEEE Press. 10.1109/ICRTIT.2014.6996164

Krizhevsky, A., Sutskever, I., & Hinton, G. E. (2012). Imagenet classification with deep convolutional neural networks. Proc. Advance Neural Information Processing System, 25, 1097–1105.

Liu, L., Yang, F., Wang, R., Shi, Z., Stidwell, A., & Gu, D. (2012). Analysis of handover performance improvement in cloud-RAN architecture. In *Proc. 7th International ICST Conference on Communications and Networking in China*. IEEE Press.

Mnih, V., Badia, A. P., Mirza, M., Graves, A., Harley, T., Silver, D., & Kavukcuoglu, K. (2016). Asynchronous Methods for Deep Reinforcement Learning. In *Proceedings of the 33rd International Conference on Machine Learning*. New York, NY: W&CP.

Mnih, V., Kavukcuoglu, K., Silver, D., Graves, A., Antonoglou, I., Wierstra, D., & Riedmiller, M. (2013). Playing Atari with Deep Reinforcement Learning. *Proceedings of NIPS Deep Learning Workshop*. arXiv:1312.5602

Muñoz, P., Barco, R., Ruiz-Avilés, J. M., de la Bandera, I., & Aguilar, A. (2013). *Fuzzy Rule-Based Reinforcement Learning for Load Balancing Techniques in Enterprise LTE Femtocells* (Vol. 62). IEEE Transactions on Vehicular Technology.

Palattella, M., Dohler, M., Grieco, A., Rizzo, G., Torsner, J., Engel, T., & Ladid, L. (2017). Internet of Things in the 5G era: Enables, architecture, and business model. *IEEE Journal of Selected Advanced Topics in Communications.*

Raju, R., & Moh, M. (2018). Machine learning techniques for security of Internet of Things (IoT) and fog computing systems. In *Proceedings of IEEE International Conference on High Performance Computing and Simulation.* Orleans, France: IEEE.

Reguri, V. R., Kogatam, S., & Moh, M. (2016). Energy Efficient Traffic-Aware Virtual Machine Migration in Green Cloud Data Centers. In *Proc. the 2nd IEEE Int. Conf. on High Performance and Smart Computing (HPSC).* IEEE Press. 10.1109/BigDataSecurity-HPSC-IDS.2016.55

Salakhutdinov, R., & Hinton, G. E. (2009). Deep Boltzmann machines. *Proc. Int. Conf. AI Statististics.*, 448–455.

Sathyanarayana, S., & Moh, M. (2016). Joint Route-Server Load Balancing in Software Defined Networks using Ant Colony Optimization. In *Proceedings of the International Conference on High Performance Computing and Simulation (HPCS).* Innsbruck, Austria. IEEE Press. 10.1109/HPCSim.2016.7568330

Saxena, N., Roy, A., Sahu, B., & Kim, H. (2017, February). Efficient IoT gateway over 5G wireless: New design with prototype and implementation results. *IEEE Communications Magazine.*

Shahriari, B., & Moh, M. (2017). Generic Online Learning for Partial Visible Dynamic Environment with Delayed Feedback - Online Learning for 5G C-RAN Load-Balancer. In *Proc. of the International Conference on High Performance Computing and Simulation (HPCS).* Genoa, Italy. IEEE Press. 10.1109/HPCS.2017.36

Song, J., Yoo, T., & Song, P. J. (2016). Mobility level management for 5G network. *Proceedings of International Conference on Information and Communication Technology Convergence (ICTC)*, 940–943.

Su, G., & Moh, M. (2018). Improving Energy Efficiency and Scalability for IoT Communications in 5G Networks. In *Proc. of 12th ACM Int. Conf. on Ubiquitous Information Management and Communication (IMCOM).* Langkawi, Malaysia: ACM. 10.1145/3164541.3164547

Sussner, P., Esmi, E., & Barros, L. C. (2016). Controlling the width of the sum of interactive fuzzy numbers with applications to fuzzy initial value problems. *Proceedings of IEEE International Conference on Fuzzy Systems*, 1453–1460.

Sutton, R. S., & Barto, A. G. (1998). *Reinforcement Learning: An Introduction.* MIT Press.

Tsai, C. & Moh, M. (2018). Cache Management for 5G Cloud Radio Access Networks. In *Proc. of 12th ACM Int. Conf. on Ubiquitous Information Management and Communication (IMCOM).* Langkawi, Malaysia: ACM Press. 10.1145/3164541.3164559

Tsai, C., & Moh, M. (2017a). Abstract: Cache Management and Load Balancing for 5G Cloud Radio Access Networks. In *Proc. ACM Symposium on Cloud Computing.* Santa Clara, CA: ACM Press. 10.1145/3127479.3132690

Tsai, C., & Moh, M. (2017b). Load Balancing in 5G Cloud Radio Access Networks Supporting IoT Communications for Smart Communities. In *Proc. of 2017 IEEE International Symposium on Signal Processing and Information Technology (ISSPIT)*. Bilbao, Spain: IEEE. 10.1109/ISSPIT.2017.8388652

United Nations. (2012). United Nations Secretary-General's high-level panel on global sustainability. In *Resilient People, Resilient Planet: A future worth choosing*. Author.

Wang, Z., Chen, C., & Li, H.-X. (2016). A novel incremental learning scheme for reinforcement learning in dynamic environments. *Proceedings of 12th World Congress on Intelligent Control and Automation (WCICA)*, 2426–2431. 10.1109/WCICA.2016.7578530

Watanabe, K., & Machida, M. (2012). Outdoor LTE Infrastructure Equipment (eNodeB). *Fujitsu Scientific and Technical Journal*, *48*(1), 27–32.

Wu, J. W., Liu, J., Huang, Z., & Zheng, S. (2015). Dynamic Fuzzy *Q*-Learning for Handover Parameters Optimization in 5G multi-tier networks. *Proceedings of International Conference on Wireless Communications & Signal Processing (WCSP)*, 1–5. 10.1109/WCSP.2015.7341220

Zhang, H., & Guo, X. (2014). SDN-based load balancing strategy for server cluster. In *2014 3rd Int. Conference on Cloud Computing and Intelligence Systems (CCIS)*. IEEE Press.

Zhao, W., & Du, S. (2016). *Spectral–Spatial Feature Extraction for Hyperspectral Image Classification: A Dimension Reduction and Deep Learning Approach* (Vol. 54). IEEE Transactions on Geoscience and Remote Sensing.

Zhou, Y., & Wei, Y. (2016). *Learning Hierarchical Spectral–Spatial Features for Hyperspectral Image Classification* (Vol. 46). IEEE Transactions on Cybernetics.

Chapter 10
Machine Learning-Based Subjective Quality Estimation for Video Streaming Over Wireless Networks

Monalisa Ghosh
Indian Institute of Technology Kharagpur, India

Chetna Singhal
Indian Institute of Technology Kharagpur, India

ABSTRACT

Video streaming services top the internet traffic surging forward a competitive environment to impart best quality of experience (QoE) to the users. The standard codecs utilized in video transmission systems eliminate the spatiotemporal redundancies in order to decrease the bandwidth requirement. This may adversely affect the perceptual quality of videos. To rate a video quality both subjective and objective parameters can be used. So, it is essential to construct frameworks which will measure integrity of video just like humans. This chapter focuses on application of machine learning to evaluate the QoE without requiring human efforts with higher accuracy of 86% and 91% employing the linear and support vector regression respectively. Machine learning model is developed to forecast the subjective quality of H.264 videos obtained after streaming through wireless networks from the subjective scores.

INTRODUCTION

This chapter provides insight into the emerging applications of machine learning for catering the needs of the next generation wireless networks. It is expected that the next generation wireless networks must reinforce ultra-reliable, low-latency communication in addition to bestowing high video quality to users playing real time streaming or ongoing download applications. Several deployment of LTE femto cells, WiFi hotspots not only makes accessing faster, but also increases the number of users enormously. In next generation wireless networks, substantial volumes of data has to be gathered periodically and

DOI: 10.4018/978-1-5225-7458-3.ch010

in real time across several sensing devices. The wireless uplink will confront growth of traffic due to large number of short-packet transmissions, opposed to general view of uplink being less congested compared to downlink. Tomorrow's wireless network must support cloud-based gaming, real-time HD streaming and several multimedia services. Infact, this will lead to varying networking environment whose applications and their heterogeneous quality-of-service (QoS) along with need for efficient network resource utilization, while maintaining reasonable cost endorse a fundamental change in the way in which next generation wireless networks are modeled, designed and optimized. Upcoming trends also aim at solutions that optimize the Quality of Experience (QoE) of the end users. This necessity for meeting the QoE can be realized by implementing the fundamental notions of machine learning across network infrastructure and the end-user equipments. This chapter presents basic architecture, training procedure, as well as the associated challenges and opportunities for communication problems that can be addressed using machine learning.

Recent studies are coming up that address the current state-of the-art applications of machine learning for QoE evaluation. The work by Grazia et al., (2018) devised a machine learning approach to carry out QoE-deployed video admission control and resource management algorithms. They developed a multi-stage learning system that combines the unsupervised learning of video features from the size of H.264-encoded video frames with a supervised classifier trained to automatically extract the quality-rate characteristics of unknown video sequences. They have developed a multi-stage machine learning system that blends the unsupervised learning of features obtained from video from the size of H.264-encoded video frames with a supervised classifier trained to spontaneously withdraw the quality-rate characteristics of unknown video sequences. This characterization of QoE is then used to govern parallel video transmissions via a shared channel in order to assure a minimum quality level conveyed to the destined users. The discussions by Casa & Wassermann (2017) presented a new proposal to multi-dimensional Quality of Experience (QoE) prediction incase of mobile video by use of machine learning models. A high dimensional input space is used by them inorder to model influence of buffering and beginning delay on QoE. The proposed models are trained and tested on publicly available mobile video dataset. The work by Bampis, Li, Katsavounidis, & Bovik (2018), deploy diverse recurrent dynamic neural networks that carry out continuous time subjective QoE prediction. They trained a variety of recurrent neural networks and non-linear autoregressive models to estimate the QoE using different available subjective QoE database. Another work by Gao, Kwong, & Jia (2017) devise a joint framework consisting of machine learning and game theory (MLGT) for inter frame coding tree unit (CTU) level bit allocation and rate control (RC) optimization in high efficiency video coding (HEVC). At first, the authors have devised a support vector machine-based multi classification scheme inorder to improve the prediction accuracy of CTU-level rate-distortion (R-D) model. Second, they have proposed a mixed R-D model-based cooperative bargaining game theory for bit allocation optimization. The minimum utility is adjusted by the reference coding distortion and frame-level quantization parameter (QP) change.

With extensive utilization of internet in today's world, it can be foreseen that several multimedia applications like video teleconferencing, video on demand, video streaming over the wireless, online streaming sites like YouTube, Hotstar etc. will occupy a huge proportion in the next generation wireless networks. Videos undergo ample distortions during any of these processes i.e. processing, compression or transmission which results in degradation of visual quality before reaching the destined users. Fluctuating bandwidth is a distinctive feature of the wireless channels that corrupts the transmitting bitstream. The recent advances in video coding and compression techniques make it feasible to deliver high quality video services to the end users. Majority of the video coding and compression standards deploy block-based coding schemes and motion compensation techniques. Due to application of these techniques, the decoded videos are subjected to one or multiple compression artifacts i.e. blurriness, false edges, color bleeding, jagged motion, chrominance mismatch, ringing and flickering. Study of video quality metrics provides insight into the video distortions. Hence, proper error correction and concealment techniques can be applied to ameliorate the quality of the video.

MEETING THE QOE AND QOS PARAMETERS

The issue of video quality evaluation is essential for video service providers from the perspective of better service delivery and maintaining acceptable performance levels across common network impairments. Human beings visual perception of the video gives assessment of video quality. Since network-related video impairments negatively influence quality of experience (QoE), so its evaluation is utmost important. Various video quality evaluation methods help in quality monitoring; thus maintaining quality of service (QoS) requirements and improving overall system performance.

NEED FOR REDUCING HUMAN EFFORTS IN SUBJECTIVE VIDEO QUALITY ASSESSMENT

As per the Recommendation of ITU-R BT.500-11 (2000), obtaining subjective scores from several groups of non-expert assessors for each and every video is a tedious and expensive process. Assessors need to be made familiar with assessment methods, type of impairment, grading scale etc. Discussions by Seshadrinathan, Soundararajan, Bovik, & Cormack (2010) show the presence of correlation between several objective quality metrics and human mean opinion scores. The dependence of several objective metrics on subjective scores can be made use for applying machine learning algorithms for forecasting the subjective scores. So, in this chapter a machine learning model is proposed which predicts the subjective scores

Many robust error detection mechanisms for H.264 and H.263++ encoding schemes are discussed by Farrugia & Debono (2008, 2009). Their methodology employs check sum of transport protocols and subjective scores for dividing the corrupted and uncorrupted Macro blocks (MBs) in a frame. Numerous error concealment techniques are used to mask the corrupted MBs. Only few corrupted MBs result in quality loss of the video. Some corrupted MBs incorporate extremely little distortions; thus removing them will adversely affect the video quality. The process involves classification of MBs in the scale of 1 to 5, followed by highly corrupted MBs detection. Subsequently, the content in these extremely corrupted MBs is used for training the Support Vector Machine (SVM).

The rest of the chapter is organized as follows. The next section discusses the significance of video quality and different assessment methods associated with it. These include subjective quality as well as objective quality methodologies. A brief description of the objective metrics i.e. PSNR, SSIM, MS-SSIM, SVD and MOVIE is also presented. Another section includes discussion on main motive behind proposing the machine learning model with a view to forecast the subjective scores of videos obtained through wireless transmission from the objective metrics. A short description of different wireless transmission methods used to obtain the distorted video sets is provided. The complete process involved in DMOS scores calculation is outlined. The further section provide details of the process involved in implementing the machine learning model. The process starts with correlation computation between the objective and subjective scores, followed by predictive model training and performance analysis of different regression models. A discussion on some of the open challenges is provided. The final section presents scope for future works and conclusion.

IMPORTANCE OF VIDEO QUALITY AND ITS ASSESSMENT METHODS

Quality of a video is assessed after passing the video through video transmission system. Several assessment techniques give an estimate of perceived degraded video compared to reference video. This methodology can be subjective as well as objective. Subjective quality assessment methods gather opinion scores of the quality from the users. On the contrary, Objective quality assessment methods make use of spatial as well as temporal information of the video.

Subjective Quality Methodology

Subjective video quality methods involve conducting the subjective tests, directly asking the human subjects for their point of view regarding the perceived quality. Choice of subjects should be such that they are not skilled in this field. An atmosphere that is isolated from the common environment is provided for viewing the videos aiming for better results. Subjective quality assessment is carried out as per Recommendation of ITU-R BT.500-11 (2000). The assessor has to give an unbiased score of his choice to video sequences. The options range from excellent to bad. Different scores and their corresponding opinions is given in Table 1. Approximately, 20-40 subjects are required to give their opinion scores. This methodology is much time and resource consuming.

Objective Quality Methodology

1. 1. **Peak Signal to Noise Ratio (PSNR):** PSNR is most easily interpreted through Mean Squared Error (MSE) between the original video and distorted videos. Generally, PSNR is a basic standard for objective video quality assessment metric. The MSE is given as

$$MSE = \frac{\sum_{l=1}^{L}\sum_{m=1}^{M}\left[S\left(l,m\right)-s\left(l,m\right)\right]^{2}}{L\times M}$$

Table 1. Quality and corresponding opinion scores of a video

Video Quality	Opinion Score
Excellent	5
Good	4
Fair	3
Poor	2
Bad	1

$L \times M$ is the pixel resolution of the video. $S(l,m)$ is the original image. $s(l,m)$ is the distorted image. The expression for PSNR is given as,

$$PSNR = 20\log_{10}\frac{255}{MSE} \tag{1}$$

2. **Structural Similarity Index (SSIM):** SSIM is one of the approaches for quality evaluation of still images. Structural information comprises of those attributes in an image which depicts the construction of objects in the scene, independent of mean luminance and contrast (Wang, Bovik, Sheikh, & Simoncelli, 2004). A good perception of image distortion is given by the computation of structural information change, considering that human visual system is adjusted to obtain structural information from viewing field. The SSIM is calculated on luminance component of the video frame-by-frame basis. Then, the mean of all the frame level gives the overall SSIM index for complete video sequence. SSIM index is expressed as

$$SSIM = \frac{\left(2\mu_y\mu_z + d_1\right)\left(2\sigma_{yz} + d_2\right)}{\left(\mu_y^2 + \mu_z^2 + d_1\right)\left(\sigma_y^2 + \sigma_z^2 + d_2\right)} \tag{2}$$

where Y, Z are reference and distorted frames. d_1, d_2 assume some constant values. σ_y, σ_z are the standard deviations of luminance intensities for reference and test videos.

3. **Multi-Scale Similarity Index (MS-SSIM):** The details of the image at various resolutions can be incorporated conveniently using the multi-scale method. Prolongation of single scale approach in SSIM leads to MS-SSIM (Wang, Lu, & Bovik, 2004). In MS-SSIM, original and test image are taken as inputs. Low pass filtering is performed by the system. The filtered image is then down sampled by a factor of two. MS-SSIM is represented as,

$$MS - SSIM\left(y,z\right) = \left[l_m\left(y,z\right)\right]^{\alpha}\prod_{m=1}^{M} c_m\left(y,z\right)^{\beta}\left[S_m\left(y,z\right)^{\gamma}\right] \tag{3}$$

where

$$l_m\left(y,z\right) = \frac{2\mu_y\mu_z + d_1}{\left(\mu_y{}^2 + \mu_z{}^2 + d_1\right)},\ c_m\left(y,z\right) = \frac{2\sigma_y\sigma_z}{\sigma_y{}^2 + \sigma_z{}^2 + d_2},\ S_m\left(y,z\right) = \frac{\sigma_{yz} + d_3}{\sigma_y\sigma_z + d_3}$$

4. **M-Singular Value Decomposition Metric (SVD):** M-SVD is one of the parameters used for image quality assessment that is established on SVD concept (Shnayderman, Gusev, & Eskicioglu, 2006). M-SVD is expressed as a function of distance between singular values of the reference and distorted image block. SVD is expressed as,

$$SVD = \sqrt{\sum_{j=1}^{n}\left(F_j - F_j^{'}\right)^2} \tag{4}$$

where $F_j, F_j^{'}$ are the singular values of original and distorted blocks, n is the number of blocks.

5. **Motion-Tuned Video Integrity Evaluation (MOVIE):** MOVIE records perceptually significant distortions all along the motion trajectories; thus combining motion information into the video quality assessment process. In this way, MOVIE enhances spatial artifacts measurement in videos (Seshadrinathan, & Bovik, 2010). MOVIE uses Gabor filter family to decompose the original and distorted videos into the spatio-temporal bandpass channels. MOVIE model is distinct in the sense that it utilizes neuroscience based models as to how brain involves in visual information processing. In a video, spatial and temporal distortions are captured using the "Spatial MOVIE Index" and "Temporal MOVIE Index" model respectively. Spatial MOVIE and Temporal MOVIE indices are integrated to give overall MOVIE index.
6. **Spatial Motion-Tuned Video Integrity Evaluation Index:** Different spatial distortions include blurring, false contouring, ringing, mosaic patterns, noise and so on. Apart from primarily processing the spatial distortions, Spatial MOVIE index captures temporal distortions to a very little extent.
7. **Temporal Motion-Tuned Video Integrity Evaluation Index:** Various distortions like mosquito noise, ghosting, smearing, motion compensation mismatch etc. constitute the temporal distortions. For temporal video quality assessment, motion statistics is calculated from original video in the pattern of optical flow fields. The optical flow fields obtained is then combined with the spatio-temporal Gabor decomposition of original and test videos to evaluate temporal quality. Temporal MOVIE index primarily captures temporal distortions.
8. **Spatial Perceptual Information (SI):** Spatial perceptual information determines the quantity of spatial characteristics of a picture. The scenes with are more spatially complex usually contain higher spatial information. SI does not convey entropy measurement (Recommendation ITU-T P.910, 2008). SI calculation depends on Sobel filter. At given time n, every video frame $\left(Fr_n\right)$ is initially filtered using Sobel filter $\left[Sobel\left(Fr_n\right)\right]$. In every Sobel filtered frame, the computation of standard deviation is performed over the pixels $\left(Std_{space}\right)$. This process is continued for every frame contained in the video sequence. This in turn leads to creation of a time series of SI of the scene. Out of this time series, the maximum value is selected to depict the SI content of the scene $\left(\max m_{time}\right)$. This complete procedure can be represented in equation as

$$SI = \max m_{time}\left\{Std_{space}\left[Sobel\left(Fr_n\right)\right]\right\} \tag{5}$$

9. **Temporal Perceptual Information (TI):** Temporal perceptual information gives a measure of temporal changes taking place in a video sequence. Sequences with high motion usually contain higher temporal information. Like SI, TI also does not convey entropy measurement (Recommendation ITU-T P.910, 2008). TI computation depends on motion difference feature $\left[Motion_n\left(j,k\right)\right]$. This motion difference evaluates the difference between pixel values (of luminance plane) that are at

identical locations in space, but at consecutive times or frames. $Motion_n(j,k)$ can be expressed as function of time n as

$$Motion_n(j,k) = Fr_n(j,k) - Fr_{n-1}(j,k) \tag{6}$$

where $Fr_n(j,k)$ denotes pixel location at j^{th} row and k^{th} column of nth time frame. $Fr_{n-1}(j,k)$ denotes pixel location at j^{th} row and k^{th} column of $(n-1)^{th}$ time frame. TI is expressed as-

$$TI = \max m_{time}\left\{Std_{space}\left[Motion_n(j,k)\right]\right\} \tag{7}$$

where $\max m_{time}$ gives maximum value of time, Std_{space} gives standard deviation over space. The SI and TI values for the sample videos are given in Table 2.

OBJECTIVE OF PROPOSING THE MACHINE LEARNING MODEL

This chapter introduces a machine learning model that is developed to forecast the subjective scores of videos obtained through wireless transmission from the objective metrics. This particular model reduces tremendous human efforts needed in accessing the video. Obtaining a prediction of the subjective scores provides an estimate of QoE. Flow diagram depicting the model is given in Figure 1. The process of training this model involves providing objective and subjective scores that can predict subjective scores of new test sequence.

Figure 1. Flow diagram of proposed machine learning model

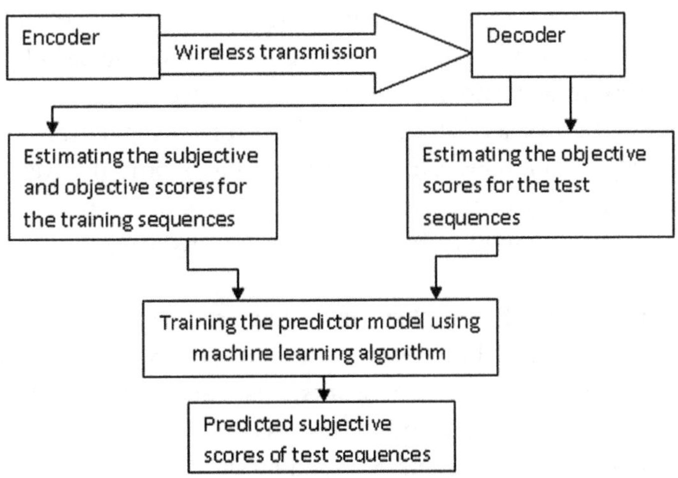

Ten different sets of video sequences consisting of natural sceneries, pedestrian area etc. have been used in this study. These video sequences are obtained after transmission through wireless networks collected from LIVE video quality database (2003). Each set consists of one reference video sequence and eleven distorted video sequences. The distortions are obtained in the following ways-

1. **H.264 Compression:** There are four test videos per reference which are H.264 compressed. The compression rate ranges from 200 kbps to 5 Mbps. The H.264 compressed videos were created using the JM reference software (Version 12.3) which was made accessible by Joint Video Team (JVT) (available online- H.264/AVC Software Coordination) (2007).
2. **Transmission Over IP Networks:** Different applications like video conferencing, Video on Demand require transmission of videos over IP networks. The database consists of three "IP" test videos per reference which were obtained by simulating IP losses on H.264 compressed videos. The JM reference software from H.264/AVC Software Coordination [available online] was used to create H.264 compressed videos with compression range varying between 0.5 and 7 Mbps. S. Wenger in "H.264/AVC over IP" gives a detailed study of transmitting H.264 videos over IP networks. Generally, the packet losses occur in IP network because of buffer overflow at intervening nodes in congested network. The test videos obtained after transmission of compressed H.264 videos over error prone IP networks contain between 1 and 4 slices per frame and 1 slice per packet. Use of one slice per frame leads overhead reduction caused by IP headers with compromise in terms of robustness. Use of 4 slices per frame enhances robustness to error, but in turn decreases the compression efficiency.
   The IP error patterns with four different loss rate of 3%, 5%, 10% and 20% is used which is provided by VCEG (Video Coding Experts Group). After obtaining the H.264 compressed packetized video; the test videos were created by dropping the packets defined in error pattern. Two packets were not dropped inorder to allow decoding- (i) the first packet carrying Instantaneous Data Refresh (IDR) and (ii) the last packet (as decoder fails to detect this last packet). The H.264 bit stream obtained was further decoded employing H.264 Software coordination. Further, losses were suppressed deploying built-in error concealment mechanism (mode 2-motion copy) given in H.264/MPEG-4 AVC Reference software manual (2007).
   Several observed artifacts in transmission over IP networks rely on different factors such as (i) whether the Intracoded frame (I frame) or Predicted frame (P frame) is lost (ii) the ordering of Flexible Macroblock (iii) whether each frame transmission takes place in 1 packet or 4 packets. Loss of I frame results in more distortions of the video sequence compared to P frame. Transmission of complete frame as one slice leads to remarkable distortions than transmission using four slices. Regular as well as dispersed modes of Flexible Macroblock ordering was used in the simulations. More distortions was observed in regular mode. In dispersed mode, the frames were subsampled by 2 along both rows as well as columns. Four packet groups were composed using these subsampled frames.
3. **Transmission Over Wireless Networks:** The following assumptions are made while simulating transmission of H.264 compressed videos over wireless networks. While simulating transmission of H.264 compressed videos over wireless networks; it is assumed that packet loss occurs although it contains single bit error. As a consequence of this assumption, probability of loss of longer packet is more compared to shorter sized packets. So, sending shorter sized packets is advantageous. Video stream is encoded employing multiple slices for each frame. Each packet comprises roughly 200

Table 2. Description of the video sets

Video	Description	Camera Movement	Video Duration	Frame Rate	SI	TI
Blue sky (bs)	Blue sky and some trees	Circular camera motion	8.68	25	125.6266	26.0787
Mobile and Calendar (mc)	Toy train moving horizontally and a calendar moving vertically in the background	Recorded by camera pan	10	50	109.2197	0.0063
Pedestrian area (pa)	Few people walking about in a street intersection	Recorded by still camera	10	25	51.5969	7.4527
Park run (pr)	A person running across a park and comes to stand still.	Recorded by camera pan	10	50	146.5951	0
River bed (rb)	A river bed containing some pebbles and water	Recorded by still camera	10	25	104.4118	17.2233
Rush hour (rh)	Rush hour traffic on a street	Recorded by still camera	10	25	85.7463	7.9567
Sunflower (sf)	A bee moving over a sunflower in close-up	Recorded by still camera	10	25	49.1202	4.4439
Shields (sh)	A person walking across a display pointing at shields	Camera pans, then becomes still and zooms in	10	50	111.1766	0.0063
Station (st)	A railway track, a train and some people walking across the track	Recorded by still camera	10	25	107.9593	0.4670
Tractor (tr)	A tractor moving across some fields	Recorded by camera pan	10	25	53.6444	0.8260

bytes per packet providing equal liability to bit errors. The errors in wireless networks are produced using bit error patterns and software made available by VCEG. The packet error rate varies between 0.5-10%. Decoding as well as error concealment mechanism is similar to IP simulations. Observed artifacts in wireless transmission also rely on whether it results in loss of I or P frames. Also Flexible Macroblock Ordering contributes to observed artifacts.

Figure 2a. Video samples: bs

Figure 2b. Video samples: mc

Figure 2c. Video samples: pa

Figure 2d. Video samples: pr

Figure 2e. Video samples: rb

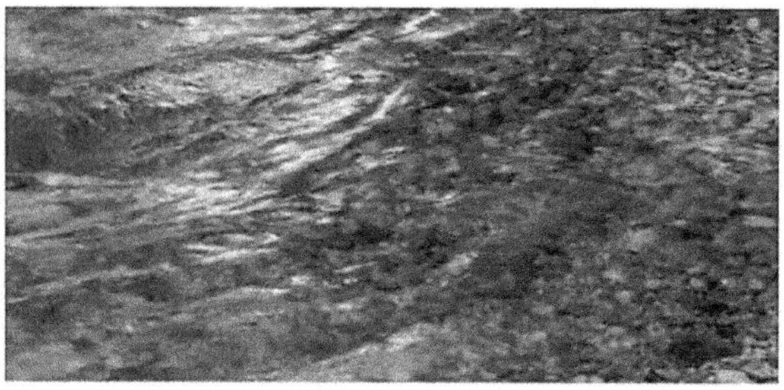

Figure 2f. Video samples: rh

Figure 2g. Video samples: sf

Figure 2h. Video samples: sh

Figure 2i. Video samples: st

Figure 2j. Video samples: tr

In total, hundred and ten distorted video sequences (11 distorted sequences for each video sample) are H.264 compressed, produced by simulating losses carried by H.264 compressed video stream in a wireless environment. Spatial resolution of each video is 768×432. Detailed description of the video sets is provided in Table 2 and snapshot of video samples is shown in Figure 2.

Subjective Quality Metric

Subjective quality tests were carried out with the help of 38 human subjects. The opinion scores were computed according to ITU- R Absolute category rating (ACR) scale. There were different sessions when the tests were conducted. Over all sessions, the mean opinion score was calculated for a video. These opinion scores obtained from assessors were transformed to difference opinion scores with a view to remove assessor preferences for any particular reference videos. The difference opinion score is computed as

$$D_{pqs} = r_{pq_{ref}s} - r_{pqs} \tag{8}$$

where r_{pqs} represent the rate (score) allocated by assessor p to video q in session $s = \{1,2\}$.

This is termed as hidden reference removal. The difference scores obtained per session was subsequently converted to Z-scores.

$$z_{pqs} = \frac{D_{pqs} - \mu_{ps}}{\sigma_{ps}} \tag{9}$$

where $\mu_{ps} = \frac{1}{T_{ps}} \sum_{q=1}^{T_{ps}} D_{pqs}$ and $\sigma_{ps} = \left(\frac{1}{T_{ps} - 1} \sum_{q=1}^{T_{ps}} \left(D_{pqs} - \mu_{ps} \right)^2 \right)^{1/2}$. T_{ps} denotes the number of test videos watched by assessor p in session s.

In the process of test conduction, the assessor sees each video contained in database only once- either in first or second session. The matrix $\{z_{pq}\}$. results after combining the Z-scores of both the sessions. $\{z_{pq}\}$ corresponds to Z-scores given by assessor p to video q where q varies from 1 to T_{ps}. $T_{ps} = 110$ distorted videos collected from LIVE Video Quality Database (LIVE VQD) (2003).

A procedure is laid down in ITU-R BT 500.11 to reject scores from unreliable assessors. The recommendation of ITU-R BT 500.11 at first governs if the scores given by an assessor are normally distributed by evaluating kurtosis of the scores. When the kurtosis lies between the values of 2 and 4, the scores are deemed to be normally distributed. Incase of normally distributed scores, an assessor gets rejected when greater than 5% of scores allocated by him falls beyond the range of two standard deviations from mean scores. Generally, those subjects that are constantly pessimistic or optimistic in judging the quality are retained.

Linear rescaling was done to make the Z-scores remain in the range of [0,100]. Under the assumption that Z-scores given by an assessor follow standard Gaussian distribution, 99% of the scores fell in the range of [-3,3]. The authors in ("Study of subjective and objective quality assessment of video") performed rescaling following linear mapping from [-3,3] to [0,100] using

$$z'_{pq} = \frac{100\left(z_{pq} + 3\right)}{6} \tag{10}$$

For each video, the DMOS was calculated as mean of rescaled Z-scores from rest subjects left after rejection. Only 29 subjects were left after subject rejection $\left(R = 29\right)$.

$$DMOS_q = \frac{1}{R}\sum_{p=1}^{R} z'_{pq} \tag{11}$$

In In this study, the DMOS scores were collected from LIVE VQD. LIVE VQD aimed to sample the extent of visual quality in almost uniform manner. The histogram of DMOS scores acquired from LIVE VQD is shown in Table-3. Table-3 shows out of total videos, the number of videos with good and poor quality are almost equal. Table-3 depicts that the DMOS scores obtained from LIVE VQD lies in the range of [30,90]. This range seems apparently small to subjects dealing with evaluating subjective scores employing Double Stimulus Continuous Quality Scale (DSCQS) method. The authors- Seshadrinathan, Soundararajan, Bovik, & Cormack (2010) applied the DSCQS method in VQEG phase 1 study, where a rating in the scale of [0,100] was given by the subjects to original and test videos. The difference between the rating of original and test videos was calculated as DMOS. On the contrary, LIVE VQD applies a sole stimulus paradigm with hidden reference removal and DMOS is calculated taking the average of Z-scores over all sessions; not as difference between ratings of original and test videos. Transforming difference scores to Z-scores is essential to justify for differences in applying ratings by the subjects. Under the assumption that Z-scores given by an assessor follow standard Gaussian distribution, 99% of the scores fell in the range of [-3,3] which is analogous to DMOS score in range of [0,100]. DMOS range of [30,90] in LIVE VQD is equivalent to range of [-1.2,1.92] on average Z-scores, which is roughly 86% of area of standard normal distribution.

Table 3. Distribution of DMOS scores over the video sequences

DMOS Scores	Number of Videos
30-40	10
40-50	27
50-60	35
60-70	25
70-80	10
80-90	3

Objective Quality Metric

PSNR, SSIM, MS-SSIM, SVD, Spatial MOVIE, Temporal MOVIE and MOVIE index constitute the objective quality metrics. Out of them, PSNR, SSIM, MS-SSIM and SVD are computed on luminance component at each frame with respect to reference video in similar set. Concatenations of frames result in formation of sequence files. Calculating the mean of corresponding metric across all frames gives the metrics of that video sequence. Spatial MOVIE, Temporal MOVIE and MOVIE index were estimated for all test (distorted) video sequence with reference to original video sequence. For implementing the machine learning model, dataset was composed of all the hundred and ten distorted video sequences. The subjective and objective scores of distorted videos were incorporated in the data set. The objective quality metrics was computed for distorted videos of blue sky, park run, shields, tractor and so on. Distribution of PSNR for complete dataset of videos is shown in Figure 3.

Figure 3a. PSNR distribution for complete dataset of videos: PSNR boxplot

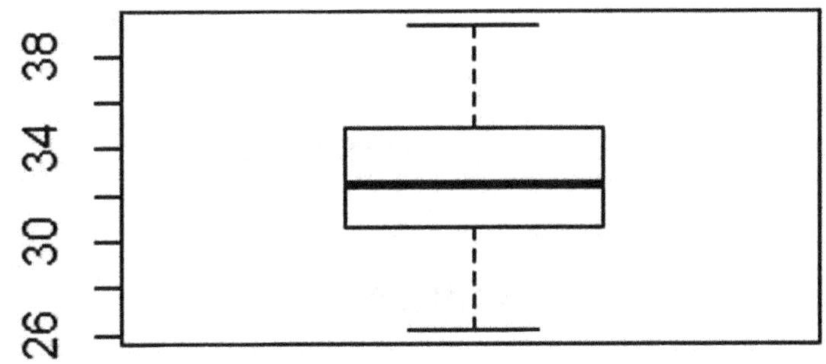

Figure 3b. PSNR distribution for complete dataset of videos: Scatter plot of PSNR~DMOS

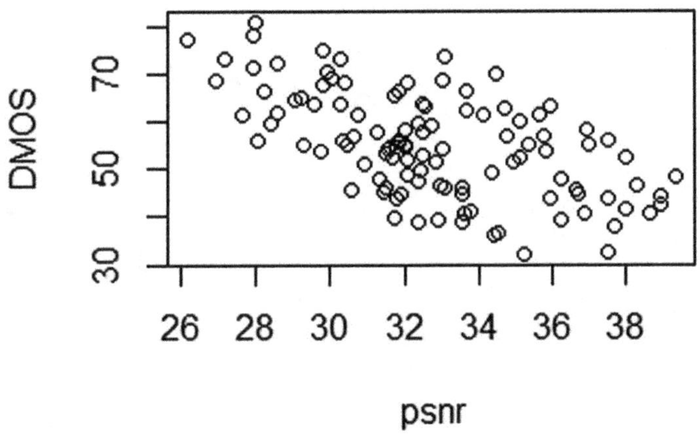

IMPLEMENTING THE MACHINE LEARNING MODEL

The dataset for study of proposed machine learning model contains subjective and objective scores of 110 test videos. Keeping in mind the fact that the model intends to predict the subjective score i.e. DMOS, it is at first necessary to make out the dependence of each and every objective score on the subjective score. Distribution of objective metrics incase of river bed is shown in Figure 4.

Figure 4a. Objective metrics for test videos in river bed (rb): SSIM and MS-SSIM

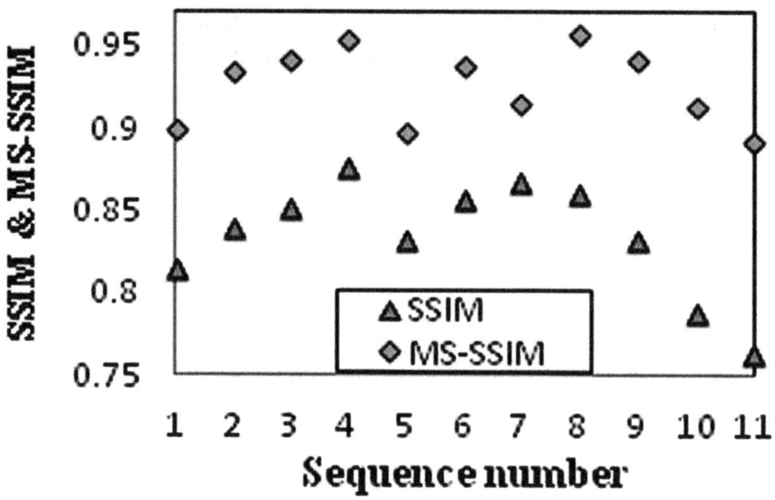

Figure 4b. Objective metrics for test videos in river bed (rb): PSNR and SVD

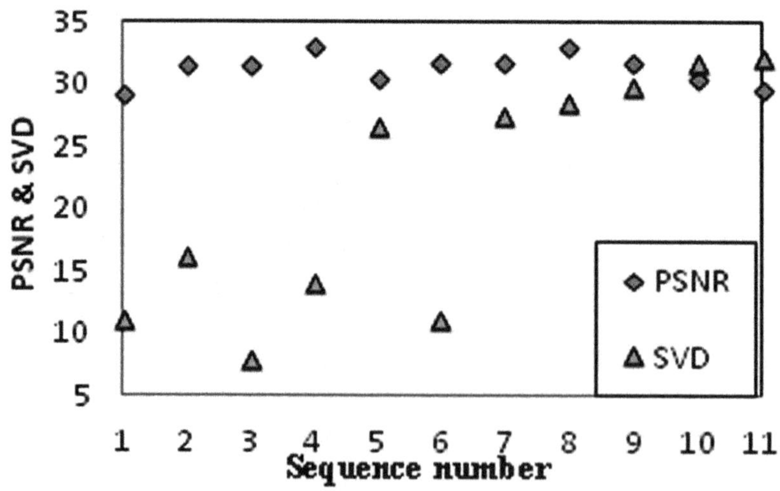

Figure 4c. Objective metrics for test videos in river bed (rb): Spatial, Temporal and MOVIE index

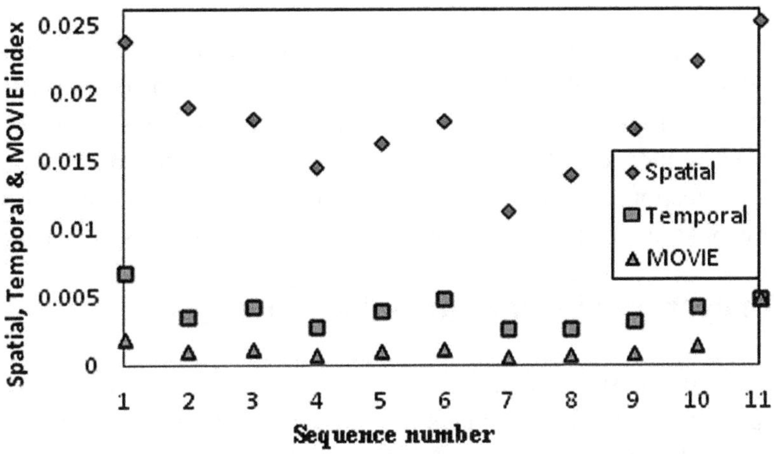

So, the following steps need to be followed to execute the model-

1. Finding out the association (correlation) between every objective score and subjective score.
2. Training the predictive model.
3. Applying the Linear regression model to predict the response variable i.e. DMOS.
4. Applying the Support Vector regression model to predict the response variable i.e. DMOS.
5. Performance comparison of Linear and Support Vector Regression models.

Computing Correlation Between Objective Scores and Subjective Score

Correlation test is carried out to assess the association between objective scores and the subjective score. The correlation coefficient shows the dependency of DMOS on a specific objective score. Correlation coefficient gives the statistical relationship between PSNR and DMOS, SSIM and DMOS, MS-SSIM and DMOS, Spatial MOVIE and DMOS etc. The computed correlation coefficient values are depicted in Table 4.

Table 4. Correlation between Objective scores and DMOS

Objective Score~DMOS	Correlation Coefficient
PSNR	-0.5847
SSIM	-0.4583
MS-SSIM	-0.6674
SVD	-0.3906
Spatial MOVIE	0.7607
Temporal MOVIE	0.5660
MOVIE index	0.3355

Training the Predictive Model

The complete dataset is separated into training and test datasets. The predictive model is trained using 90% of the dataset, while rest 10% of the dataset is used for evaluating the performance of trained model.

This predictive model comprises of two variables- predictor and response variable. The predictor variables include the objective metrics- PSNR, SSIM, MS-SSIM, SVD, Spatial MOVIE, Temporal MOVIE and MOVIE index. The response variable contains the subjective score- DMOS. Linear regression and support vector regressions are used for predicting the response variable considering its quantitative nature.

Performance Analysis of Linear Regression Model

Linear regression is one of the simplest approach for predicting the quantitative response. A linear regression model helps in determining the values of the coefficients used in the representation with the aid of available data (Tibshirani, James, Witten, & Hastie, 2013). Linear regression considers that there exists a linear relationship between the response variable ' y ' and the predictor variables ' x_1 ' and ' x_2 ' of the sort $y = K + Lx_1 + Mx_2$, where K, L and M are the coefficients of the given equation. After the training set is fed, linear regression builds a linear equation for prediction on the test set.

The response variable–DMOS is predicted from objective metrics using a simple linear regression model constructed in R studio. The contents in training dataset are used to train the model and its performance on the test dataset is obtained as shown in Table 5. Mean absolute error and variance of linear regression model is 5.6904 and 15.6630 respectively. The adjusted R –squared value obtained for this model in R-studio is 0.5716.

Performance Analysis of Support Vector Regression (SVR) Model

The support vector machine (SVM) is an augmentation of the support vector classifier which is the outcome of expanding the feature space in a distinct way by applying kernels. Support vector machine (SVM) provides an approach for classification whereas the support vector regression finds application in regression problem (Tibshirani et al., 2013). SVR provides a quantitative approach than a qualitative approach. In case of non-linear SVR, the kernel functions modify the data into higher dimensional feature space for attaining the linear separation.

Table 5. Performance comparison of Linear Regression and Support Vector Regression

Parameters	Linear Regression	Linear Regression With Bagging	Support Vector Regression (SVR)	SVR With Bagging
Mean absolute Error	5.6904	5.8544	5.0842	6.2757
Standard deviation of error	3.9576	3.8252	3.0530	3.7334
Root mean square error	6.9314	6.9933	5.9304	7.3022
Variance of error	15.6630	14.6325	9.3213	13.9388
Mean absolute percentage error(%)	10.9747	11.3196	10.0237	11.7442

The response variable is predicted using a support vector regression model constructed in R studio. SVR is trained using training instances and tested on the test dataset. In this study, a polynomial kernel is used to construct the SVR model because of its better performance compared to others. Mean absolute error obtained here is 5.0842 and variance of 9.3213.

Performance Comparison of Linear Regression and SVR

In every simulation, the mean absolute error value varies due to the random sampling of the train and test sequences. With a view to analyse the actual performance of these models, Bagging (Bootstrap Aggregation) is performed on these models. The bootstrap is used to quantify the doubt related to a particular estimator. Bootstrap estimates the actual performance of Linear regression and SVR. Linear regression and SVR performance comparison and their corresponding bagged models is depicted in Table 5. It can be noted that SVR has a better performance compared to linear regression.

K-Fold Cross Validation

Sometimes it leads to over fitting while trying to fit the model with training instances. This degrades the performance of the model while testing. This problem can be overcome using the K-fold Cross validation, a resampling technique that helps to avoid over fitting (Rodriguez, Perez, & Lozano, 2010). In this process, observation set is randomly divided into k folds of almost equal size. The starting fold is the validation set and method is fit on the rest $k-1$ folds.

In this study, 10 fold cross validation is applied. Figure 5 shows the corresponding fold number and mean absolute error value of linear regression (LR) and support vector regression (SVR). Mean absolute error of 7.414 and 6.275 is obtained using LR and SVR respectively on all folds.

Performance of SVR on an Unseen Sequence

In order to test the unbiased performance of SVR on any particular video sequence, some tests were carried out. SVR was trained on starting 9 tractor (tr) videos and its performance on 10th tr video was evaluated. This gives mean absolute error of 5.6388 and variance of 9.905. It is seen that the performance is almost similar to SVR in section 4.5. Thus, unbiased performance of SVR is proved.

Figure 5. Variation of error in LR and SVR in 10-fold cross validation

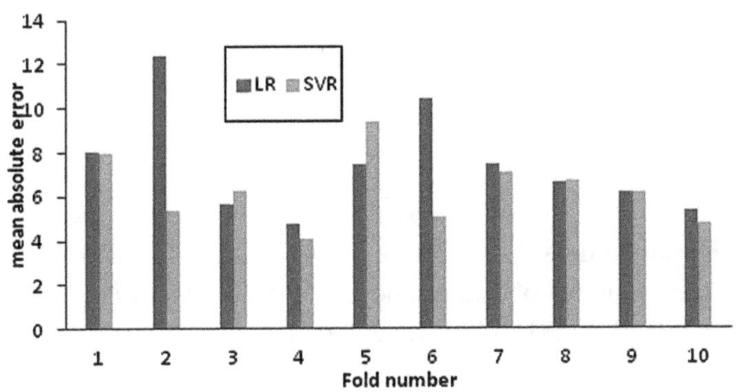

OPEN CHALLENGES

The study presented in this chapter was intended to reduce the efforts involved in video quality evaluation by the use of machine learning model. Although few objective metrics have been considered in this chapter, still other major QoE metrics are not taken into account that could have affected sharp quality changes. The identification of a function that quantifies the quality degradation due to change in SSIM, PSNR, SVD etc can be a research challenge. Any small change in any of the objective metrics will affect the QoE immensely. The amount of trade off that can be obtained by reducing the SSIM characteristics, thus providing space for more numbers of end users to be served can be looked into. The process of allowing new users to be served at the cost of minimal degradation of the objective metrics is still an open area of challenge. A model can be devised that must be optimized in terms of maximum users served at the cost of minimum quality degradation in the overall system. Steps can be taken forth inorder to further improve the accuracy of the machine learning models. Feature set can be extended to include the initial delay, jitter and rebuffering events to predict the QoE. Brief analysis of complexity involved in the end user side for implementing the machine learning model can be considered.

CONCLUSION AND FUTURE DIRECTIONS

This chapter has presented a novel machine learning model that is applied to evaluate the Quality of Experience of a video sequence from its objective quality metrics. Infact, this provides an insight of how machine learning finds applications in next generation wireless networks. Averaging of quality assessment techniques at frame level for a video gives the objective score for that video. Certain observations were deduced from this study. Spatial MOVIE was found to have highest correlation with subjective scores. Training objective scores with machine learning algorithms provides better evaluation of subjective quality. To check the accurate performance of Linear and Support vector regression model, bootstrap aggregation was performed. The performance of support vector regression was better compared to linear regression. Further, it was noticed that this model operates well on unseen sequences which tells that the model is unbiased by a particular set of sequences.

In future, several objective features that have good correlation with subjective scores can be simulated to increase the predictive capacity of our model. Exploiting the spatial and temporal information of video sequences and deploying them for enhancing video quality can be carried out. Many no- reference quality assessments metric like video intrinsic integrity and distortion evaluation oracle (VIIDEO) can be included as features. Apart from this, numerous full reference quality assessment metrics can also be incorporated. Using the principal component analysis to remove those features (from multiple available features) that may adversely affect the response variable.

REFERENCES

Bampis, C. G., Li, Z., Katsavounidis, I., & Bovik, A. C. (2018). Recurrent and Dynamic Models for Predicting Streaming Video Quality of Experience. *IEEE Transactions on Image Processing*, *27*(7), 3316–3331. doi:10.1109/TIP.2018.2815842 PMID:29641409

Casa, P., & Wassermann, S. (2017). Improving QoE prediction in mobile video through machine learning. *8th International Conference on the Network of the Future (NOF)*, 1-7. 10.1109/NOF.2017.8251212

Farrugia, R. A., & Debono, C. J. (2008). A Robust Error Detection Mechanism for H.264/AVC Coded Video Sequences Based on Support Vector Machines. *IEEE Transactions on Circuits and Systems for Video Technology*, *18*(12), 1766–1770. doi:10.1109/TCSVT.2008.2004919

Farrugia, R. A., & Debono, C. J. (2009). A Support Vector Machine Approach for Detection and Localization of Transmission Errors Within Standard H.263++ Decoders. *IEEE Transactions on Multimedia*, *11*(7), 1323–1330. doi:10.1109/TMM.2009.2030651

Gao, W., Kwong, S., & Jia, Y. (2017). Joint Machine Learning and Game Theory for Rate Control in High Efficiency Video Coding. *IEEE Transactions on Image Processing*, *26*(12), 6074–6089. doi:10.1109/TIP.2017.2745099 PMID:28858792

Grazia, M. D. F. D., Zucchetto, D., Testolin, A., Zanella, A., Zorzi, M., & Zorzi, M. (2018). QoE Multi-Stage Machine Learning for Dynamic Video Streaming. *IEEE Transactions on Cognitive Communications and Networking*, *4*(1), 146–161. doi:10.1109/TCCN.2017.2784449

H.264/AVC Software Coordination. (2007). Retrieved from http://iphome.hhi.de/suehring/tml/

H.264/MPEG-4AVC Reference Software Manual. (2007). Retrieved from http://iphome.hhi.de/suehring/tml/JM(JVT-X072).pdf

Int. Telecommun. Union. (2000). *Methodology for the Subjective Assessment of the Quality of Television Pictures*. ITU-R Recommendation BT.500-11, Tech. Rep.

LIVE -Labaratory for Image and Video Quality Engineering, an Image Quality Assessment Database. (2003). Retrieved from http://live.ece.utexas.edu/research/quality/subjective.htm

Rodriguez, J. D., Perez, A., & Lozano, J. A. (2010). Sensitivity Analysis of k-Fold Cross Validation in Prediction Error Estimation. *IEEE Transactions on Pattern Analysis and Machine Intelligence*, *32*(3), 569–575. doi:10.1109/TPAMI.2009.187 PMID:20075479

Seshadrinathan, K., & Bovik, A. C. (2010). Motion tuned spatio-temporal quality assessment of natural videos. *IEEE Transactions on Image Processing*, *19*(2), 335–350. doi:10.1109/TIP.2009.2034992 PMID:19846374

Seshadrinathan, K., Soundararajan, R., Bovik, A. C., & Cormack, L. K. (2010). Study of Subjective and Objective Quality Assessment of Video. *IEEE Transactions on Image Processing*, *19*(6), 1427–1441. doi:10.1109/TIP.2010.2042111 PMID:20129861

Shnayderman, A., Gusev, A., & Eskicioglu, A. M. (2006). An SVD-based grayscale image quality measure for local and global assessment. *IEEE Transactions on Image Processing*, *15*(2), 422–429. doi:10.1109/TIP.2005.860605 PMID:16479812

Telecommunication Standardization Sector of, ITU. (2008). Subjective video quality assessment methods for multimedia applications. [Tech. Rep.]. *ITU-T P*, *910*, 4–5.

Tibshirani, R., James, G., Witten, D., & Hastie, T. (2013). Linear Regression, Support Vector Machines, Tree-Based Methods. In An Introduction to Statistical Learning with Applications in R (pp. 59-102, 344-354, 316-321). New York, NY: Springer.

Wang, Z., Bovik, A. C., Sheikh, H. R., & Simoncelli, E. P. (2004). Image quality assessment: From error visibility to structural similarity. *IEEE Transactions on Image Processing*, *13*(4), 600–612. doi:10.1109/TIP.2003.819861 PMID:15376593

Wang, Z., Lu, L., & Bovik, A. C. (2004). Video quality assessment using structural distortion measurement. *Signal Processing Image Communication*, *19*(2), 121–132. doi:10.1016/S0923-5965(03)00076-6

Chapter 11
Adaptive Principal Component Analysis-Based Outliers Detection Through Neighborhood Voting in Wireless Sensor Networks

Ekaterina Aleksandrova
University of Glasgow, UK

Christos Anagnostopoulos
University of Glasgow, UK

ABSTRACT

This chapter introduces statistical learning methods and findings of a group decision-making algorithm in internet of things (IoT) and edge computing environments. The discussed methodology locally detects outliers in an on-line and adaptive mode. It is driven by three perspectives—opinion, confidence, and independence—and exploits the incremental principal component analysis using the power method for eigenvector and eigenvalue estimation and Knuth and Welford's online algorithms for variance estimation. The methodology is implemented and evaluated over real contextual data in a wireless network of environmental sensors from where appropriate conclusions are drawn about the capabilities and limitations of the proposed solution in IoT environments.

INTRODUCTION

In the growing world of the 'Internet of Things', more and more computing and sensing devices are joining the wireless sensor networks (WSNs) at the network edge allowing on-line contextual data processing and localized knowledge exchange. However, keeping a WSN stabilised only a part of it is having a fully-usable environment. In order to make the exchanged contextual data logically consistent to other computing and sensing devices, contextual knowledge has to be locally processed and only suf-

DOI: 10.4018/978-1-5225-7458-3.ch011

ficient statistics are sent to the relevant destination, e.g., sink nodes, micro-data centres, edge gateways, and IoT applications. This would require devices to have the computing capabilities and resources for making on-line calculations and adaptive transformations of the gathered contextual data, presenting a high energy cost for the whole WSN. Such concerns in the past have led to the idea of a centralised computing power in a dematerialised system with enormous capacities and simple architectures, also named as the 'Cloud' (Lopez, 2015).

Cloud computing has made the current state of IoT environments a possibility by completely transforming the data cycle but at certain costs. Having all contextual data transferred to a centralised data centre involves a significant communication overhead. Furthermore, the required computational power for processing the data varies from device to device, as well as the storage capacity for preserving the results (Sharma & Wang, 2017), nevertheless, that increases the need for more powerful data centre cores. Nowadays, such problems have brought the highlight back to the 'edge' of the network given that modern devices have gained considerable computational abilities. The new concept allows devices with smaller capabilities than data centres but powerful enough to process any incoming contextual information from any closely located other devices/sensors and if needed pass the data to the core of the system.

Maintaining consistency in the exchanged data requires not only lossless contextual data transfer but also identifying and correcting corrupted and/or intentionally altered pieces of data and/or dimensions. Such pieces of data deviate from the rest and are known in the literature as "outliers" (Deng, 2016). The existence of outliers in distributed environments and WSNs demands for the development of a powerful, local and adaptive machine learning technique for outlier detection within neighbours/cliques of a WSN, which does not introduce higher network bandwidth or energy consumption (Deng,2016) (Anagnostopoulos & Hadjiefthymiades, 2014) (Harth & Anagnostopoulos,2017).

Aim

The aim of this chapter is to propose and analyse an on-line adaptive statistical learning algorithm for outlier detection in wireless sensor networks based on the Principle Component Analysis technique via different voting policies among sensors belonging in the same WSN neighbourhood/clique. The objective is to incrementally and locally learn the minimum sufficient statistics over each sensing and computing device in a WSN such that, based on the opinion of the neighbouring nodes via voting, the outlier detection process is revised to further improve the true positive rate.

Problem

In a summarised form outliers (also called anomalies) are defined as "statistical deviations from the majority of existing data" (Deng, 2016). And when there is such a tremendous amount of contextual and time-series data collected in a present-day IoT environment, there is also a very high probability that a part of that data is corrupted by faulty devices, or it reflects an extreme change in the scanned environment or is alternated with malicious intentions. Having such anomalies would result in unexpected future decisions based on the processed data.

The existence of outliers in a distributed system calls for finding a powerful technique for outlier detection. But in order to be most beneficial from the added extra computations, the algorithm has to be efficient, without a substantial increase in the network bandwidth or power consumption (Deng, 2016), (Anagnostopoulos, Hadjiefthymiades, & Georgas, 2012) (Anagnostopoulos & Hadjiefthymiades, 2014),

(Harth, Anagnostopoulos, & Pezaros, 2018), (Harth et al., 2017) but with the preserved ability to correctly identify anomalies in real-time.

Proposed Solution and Hypothesis

Hypothesis

The intended outcome is to reject or fail to reject the hypothesis whether a local opinion on an outlier decision gathered by more than one sensing device in a given WSN neighbourhood could positively affect the local decisions on anomaly/novelty and outlier identification.

The proposed methodology in this chapter is based on the established 'voting with a majority rule' in order to resolve a common problem of classifying a piece of data as an outlier or not within a specified WSN neighbourhood of sensing devices based on their local minimum sufficient statistics of the current data-subspaces' eigenvectors. The voting system over local pieces of knowledge (by adaptively maintaining the current eigenbasis) would allow a close-ranged network of wireless devices to form a local binary decision on anomalies in each sensor's data pool. Such a technique is expected to create a light local data processing at the 'edge' level of the network before information is sent for further processing to the Cloud.

The current methodology explores a local contextual decision-making algorithm driven by three different perspectives: confidence-based decision making, independent-based decision making, and opinion/social-based decision making for outlier detection. The contextual information processing algorithm analyses and incrementally obtains the first eigenvector of the contextual data stream, which is further used for driving the outlier decision making process by sharing pieces of knowledge with the neighbouring WSN nodes.

Context Outline

The paper is organised as follows: Section II reports on related work and previous research on the topic. Section III provides a context behind the data used for the experiments, together with the performed multivariate contextual analysis. Section IV explains the methodologies and techniques used to build the proposed solution for local outlier detection in Section V. Section VI presents a description on each voting policy applied to the algorithm and the analysis of their performance. Section VII demonstrates the evaluation techniques and results for each policy given a dynamically estimated threshold. Sections VIII and IX discuss future research plans based on the conclusions drawn from the results of the conducted research and performed experiments.

BACKGROUND

Cloud computing has made the current state of Internet of Things environments a possibility by completely transforming the data cycle but at certain costs. Having all data transferred to a centralised data centre involves a significant communication overhead. Furthermore, the required computational power for processing the data greatly varies from device to device, and so does the storage capacity for preserving the results (Sharma & Wang, 2017). Such issues create a reminder of the computational abilities already available in the modern personal devices (Lopez, et al., 2015). And since switching back to completely decentralised systems based on past experience is not an option, the recently found solution is to shift completely from the "core", right to the "edge" of the network in a new concept accordingly named edge-computing (Lopez, et al., 2015). This allows devices with smaller capabilities than data centres but powerful enough to process parts of or the whole incoming data from its surrounding devices/sensors and if needed pass it to the core system.

Wireless Sensor Networks are a sub-problem of the global IoT environments where solutions are aiming for appropriate data reductions in order to prolong the lifetime of the systems (Anastasi, Conti, Di Francesco, & Passarella, 2009) (Intanagonwiwat, Estrin, Govindan, & Heidemann, 2002) (Yao & Gehrke, 2002) (Wei, He, He, & Liu, 2017) (Cheng, Zhihuang, Xiong, & Yang, 2016) (Cheng, Feng, Shi, & Chen, 2018). Such networks consist of "a sensor node (sources), a processing node (relays) and a sink node (consumers)" (Anagnostopoulos & Hadjiefthymiades, 2014). The environment is obseved through the source and then sent to the next node where the data is processed. The processing occurs in the relay nodes which stand as a neighbourhood central node where all data from nearby nodes are gathered, processed, compressed and then transferred either to the next relay node or to the sink node. (Anagnostopoulos & Hadjiefthymiades, 2014). In this context the relays are devices with higher computational but lower storage capabilities, which supports the idea behind the architecture of Edge Computing systems.

MULTVARIATE CONTEXTUAL DATA

Wireless Sensor Network

In order to experiment and improve the anomaly detection algorithm, there was a need for a suitable multidimensional raw dataset. In 2014 the Berkeley Intel Research Lab collected sensor data from 54 environment readers (Bodik, et al., 2004), which were timestamped and made publicly available in their raw form, making them perfect for outlier analysis. In Figure 1 is a map with the location of each sensor on the floorplan of the Berkeley Lab.

The sensors were equipped with 4 different weather boards and the results are these real-valued multivariate vectors in the form of $x = [x_1, x_2, x_3, x_4]$, where x_1 is temperature, x_2 is humidity, x_3 is light and x_4 is voltage. The values were recorded in the corresponding units: x_1 in degrees Celsius, x_2 in percentages reflecting temperature corrected relative humidity, x_3 in Lux units and x_4 expressed in volts. Each vector $\boldsymbol{x}$ is locally sensed by each sensing device at a discrete time instance $t = 1, 2, 3, \ldots$ where each instance is with duration of 31 seconds (Bodik, et al., 2004). In order to simplify the sorting of the data, the provided columns for date and time were combined in a timestamp, forming a 5th dimension of the data.

Multivariate Contextual Data Analysis

Performing an analysis on the dataset was faced by a difficulty regarding missing data due to connectivity issues or data truncation. Examining the lost data showed that 4 of all sensors $[5, 15, 17, 28]$ have missing datapoints in their first 1000 rows which are intended as a training dataset in the algorithm.

Figure 1. Wireless Sensor Network, sensor arrangement Berkeley Intel Research Lab
Source: Bodik, et al., 2004

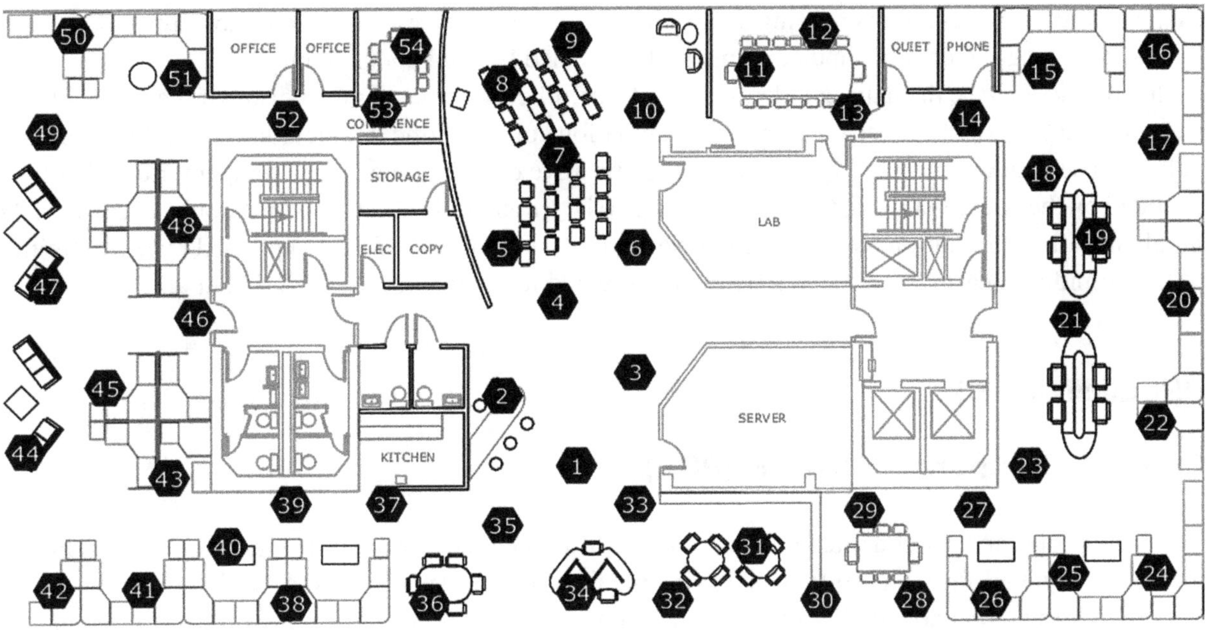

Figure 2. Scatterplots for all 5 dimensions (time, temperature, humidity, light, voltage) in 24h

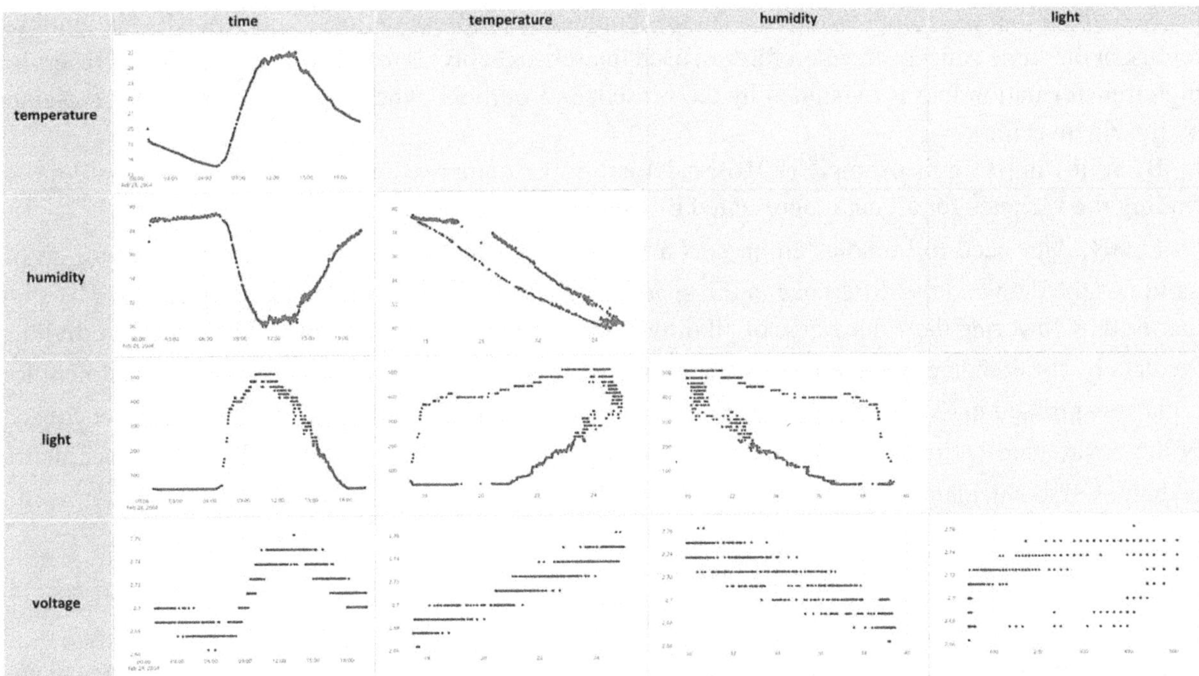

Further analysis showed that only sensors 5 and 28 have more than 20% missing data, making the other two sensors usable for the purpose of training the model. This was considered in a later stage of the chapter, where any datapoints missing a value in one or more dimensions were removed from the training dataset using the data manipulation methods provided by the Pandas library.

In order to analyse the data and identify their form, all dimensions for the period of the first 24 hours were plotted in 2-by-2 scatterplots for the combinations of date-time, temperature, humidity, light and voltage. The resulted plots in Figure 2 show high positive correlation between temperature and voltage as stated by the source of the data (Bodik, et al., 2004). Another observation which could be made is the highly negative correlation between the temperature and humidity, which is explained by the definition of the humidity measurement as "temperature corrected relative humidity" (Bodik, et al., 2004).

PRELIMINARIES

Principle Component Analysis (PCA)

The basis of outlier detection is identifying extreme differences occurring in an environment. A technique introduced by Karl Pearson (1901) is the Principle Component Analysis (PCA) which gives the ability of compressing data knowing the variance and reconstructing the data given the produced eigen-space with small reconstruction error (Deng, 2016).

A dataset is transformed using PCA such that it is represented by a "reduced number of 'effective' features" (Haykin, 1999), preserving the value of the content and allowing multivariate data to be compressed given that the compression loss (in this chapter, "compression loss" and "transformation loss" represent the same concept and therefore are used interchangeably) is tolerable (Smith, 2002). Observing high transformation loss is explained by the presence of outliers, whose variance cannot be explained by the eigenvector.

By following Deng's example (2016), calculating the compression rate for all sensors is done by finding the variance for all data represented by the covariance matrix Σ in Table 1.

Firstly, data need to be adjusted based on the mean vector $\bar{x} = [\bar{x}_1, \bar{x}_2, \bar{x}_3, \bar{x}_4]$. Depending on the nature of the data and the difference in the sparsity between the dimensions, the data can also be standardised by lowering the value range of all dimensions to a common one. This is calculated by dividing the data by the standard deviation vector $\sigma = [\sigma_1, \sigma_2, \sigma_3, \sigma_4]$ in (2) Table 1. The overall adjustment formula is expressed in (3) in Table 1. The covariance matrix representation of the whole dataset consists of the covariance $\text{cov}(x_i, x_j)$ between each dimension forming a $\Sigma_{4\times4}$ matrix. The covariance matrix is then used to calculate the transformation matrix by solving the equality of:

$$\Sigma \cdot [q_1, q_2 ... q_p] = [\lambda_1, \lambda_2 ... \lambda_p] \cdot [q_1, q_2 ... q_p], \text{ where } p \leq d$$

and q_i is the corresponding transformation matrix consisting of the sorted by significance eigenvectors and λ_i is the corresponding eigenvalues vector. The transformation matrix can then be applied to the adjusted data, which will produce a compressed matrix with a compression degree depending on the specified number of dimensions p to be taken from the transformation matrix:

$$x_{compressed} = [q_1, q_2 \dots q_i]^T \cdot x'$$

The decompression is performed in a similar way by applying the same transformation matrix on the compressed data.

$$x_{decompressed} = x_{compressed} \cdot [q_1, q_2 \dots q_i]$$

The transformation cost is then computed by finding the difference between the original data and the decompressed matrix.

$$\left\| x' - x_{decompressed} \right\|$$

Given that the authors chose to have the first $N = 1000$ datapoints as training data, PCA was performed for each sensor using the first N datapoints. Following the methodical steps of finding the eigenvector for each sensor's first N points, the authors derive the feature vector resulting in the eigenvectors sorted by significance (principle component with corresponding highest to corresponding lowest eigenvalue).

Table 1. Variance and covariance formulas

Variable	Explanation	Formula
d	Number of dimensions	$\bar{x} = \frac{\sum_{1}^{n} x_i}{n}$ (1)
n	Number of data points	
x	A single multivariate data point i	$\sigma = \sqrt{\frac{\sum_{1}^{n} (x_i - \bar{x})^2}{n-1}}$ (2)
$\bar{x}$	Mean vector	
σ	Standard deviation vector	$x' = \frac{x - \bar{x}}{\sigma}$ (3)
x'	Adjusted and standardized vector	
$\text{cov}(x_i, x_j)$	Covariance between dimensions i and j	$\Sigma_{i,j} = \text{cov}(x_i, x_j) = \frac{\sum_{1}^{n} x'_i x'^T_j}{n-1}$ (4)
Σ	Covariance matrix for all data points	$\Sigma = \begin{pmatrix} \text{cov}(x_1, x_1) & \text{cov}(x_1, x_2) & \cdots & \text{cov}(x_1, x_d) \\ \text{cov}(x_2, x_1) & \text{cov}(x_2, x_2) & \cdots & \text{cov}(x_2, x_d) \\ \vdots & \vdots & \vdots & \vdots \\ \text{cov}(x_d, x_1) & \text{cov}(x_d, x_2) & \cdots & \text{cov}(x_d, x_d) \end{pmatrix}$ (5)

Figure 3. Compression error rate for all sensor data using 1, 2, 3 and 4 principal components

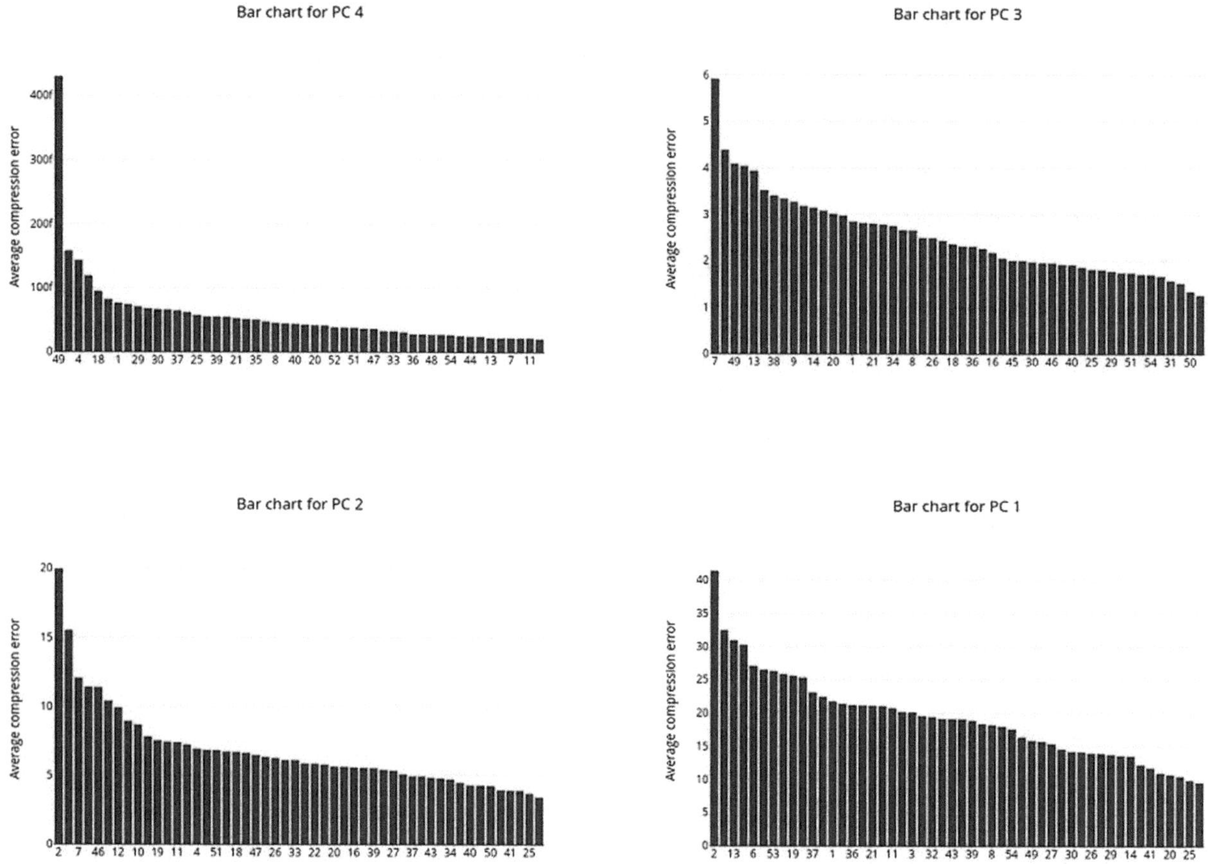

Figure 4. Percentage of variance explanation using 1, 2 and 3 principal components

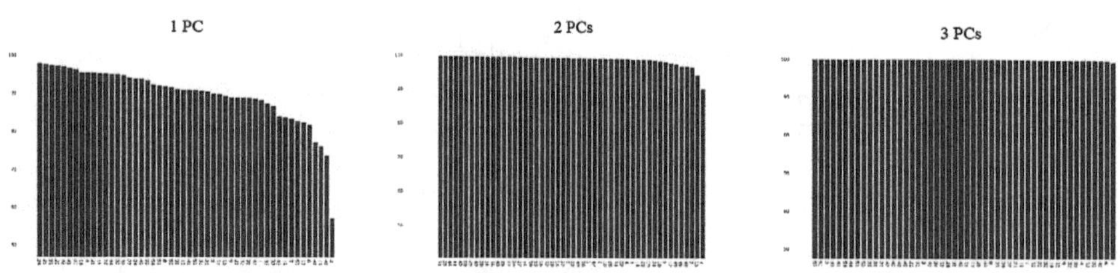

Given the K-dimensional feature vector, where K is also the number of dimensions of the initial data, the authors take the relative four principle components and compress the initial data, then decompress and find the reconstruction cost. This is also known as "compression error rate" which is presented in Figure 3. Analysing the histogram returns that the average compression error using 1 principle component is 11.3%, in other words 1st PC explains on average 89.7% of the data in Figure 4.

The Power Method

Following the analysis of the compression error rate, it was concluded that the data could be explained just by using the first eigenvector (first principle component). Due to the dependency between the eigenvector and the eigenvalues the authors conducted further analysis on the eigenvalues and drew conclusions on the eigenvectors. As seen in Figure 5 the difference between the 1st and 2nd eigenvalue is significant for all (excluding sensors 5,15,17,28 due to missing values) sensors indicating that the first eigenvector for all sensors is strongly dominant.

Taking the significant difference between the first and second eigenvalue into account, it is appropriate to consider a more powerful and yet less computationally expensive technique such as the power method (Edelman, 2018). The algorithm, as seen in the pseudo-code snippet following Holmes's iterative power method (2002), starts with choosing an initial random vector q_0 or an already known approximation to the dominant eigenvector with shape $(n \times 1)$ where n is the number of dimensions of the covariance matrix. For this case the initial vector is set to all ones. The next step is to calculate a new estimation of the eigenvector z_i by computing the dot product on the already known covariance matrix of the data Σ and the previously estimated eigenvector q_{i-1}. The following step is computing the Euclidean norm of the newly calculated vector $\|z_i\|$, which is the square root of the inner product of the eigenvector with its transpose.

This is an estimation of the eigenvalue λ_i corresponding to that eigenvector z_i, which is then used to normalise the estimated eigenvector by dividing the vector by the Euclidean norm. The estimated

Figure 5. Eigenvalue difference between 1st and 2nd eigenvalue

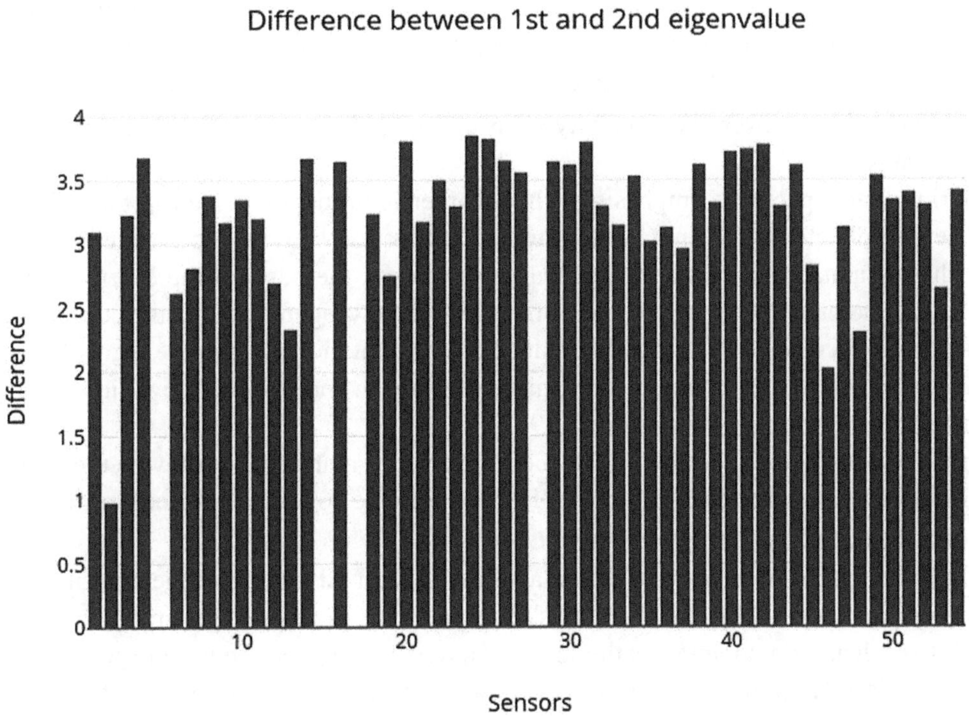

vector is used as the initial eigenvector on the next iteration of the algorithm. All steps are repeated until the eigenvalue converges to an initially set difference ϵ between the eigenvalues in two consecutive iterations. The smaller the difference the more accurate the estimations are but at the cost of more performed calculations. In this case, the ϵ is set to 0.001, which allows the eigenvalue to converge in less than 5 iterations.

Algorithm 1. The Power method

Set an initial all-ones vector $q_0 = (d \times 1)$
Set $i = 1$
While $|\lambda_i - \lambda_{i-1}| > \epsilon$ do
$z_i = \Sigma \cdot q_{i-1}$
$\lambda_i = \|z_i\|$
$q_i = \frac{z_i}{\lambda_i}$
$i = i + 1$
End While

Online Variance Estimation

As seen by the structure and the used variables, the computation of the mean and standard deviation vectors requires going through each datapoint in the whole set of length n, which creates a time complexity for both algorithms of $O(d \times n)$. Similarly, the algorithm for computing the covariance value between two dimensions depends on their corresponding mean and standard deviation, given that they are already computed, the result is a complexity of $O(d \times n)$. Performed for each pair of dimensions in the covariance matrix, the overall complexity sums up to $O(d^2 n)$ where d is the number of dimensions for the given dataset.

Considering that aspect, the emphasis of this chapter is on data received in the form of a stream, where the newly collected data is not kept in storage but instead sent for processing instantaneously.

Due to the computational cost for one covariance matrix, the basis of the eigenvector calculation, adding a new datapoint to the initial data and computing the new eigenvector would cost the same amount of immense power. In order to considerably reduce the computation cost for the algorithm, Welford has developed a very good online approach of estimating the mean and standard deviation given the current estimations and the new datapoint (Welford, 1962).

Similarly, the algorithm for computing the new covariance matrix relies only on the previous covariance matrix and the newly received datapoint. This reduces the computational cost of the algorithm and the time complexity changes from $O(d^2 n)$ to $O(1)$.

Being independent from the size of the data allows a great advantage to the algorithm for its usage in devices with smaller computational abilities. This eases the process of generating a new eigenvector with regards to each new datapoint from the stream. Given the last generated eigenvector representing the data without the last datapoint, compared to the new eigenvector considering the new datapoint shows

Table 2. Complexities

$\bar{x} \Rightarrow O(d \times n)$	$\sigma \Rightarrow O(d \times n)$	$\mathrm{cov}(X_i, X_j) \Rightarrow O(d \times n)$	$\Sigma \Rightarrow O(d^2 n)$

Table 3. Online Variance and Covariance formulas

Variable	Explanation	Formula
n	Number of data points processed until now	$\bar{x}' = \bar{x} + \frac{1}{n+1}(x^* - \bar{x})$ (1)
d	Number of dimensions	
q	Eigenvector	$\sigma' = \sigma + \frac{1}{n+1}(x^* - \sigma)$ (2)
λ	Eigenvalue	$x^{*\prime} = \frac{x^* - \bar{x}'}{\sigma'}$ (3)
x^*	New multivariate data point	
$\bar{x}'$	Estimated mean vector	$\Sigma_{i,j}{}' = \mathrm{cov}'(x_i, x_j) = \frac{\sum_1^n (x_i^{*\prime} x_j^{*\prime T})}{n-1}$ (4)
σ'	Estimated standard deviation vector	$\Sigma' = \Sigma + \frac{1}{n+1}(x^{*\prime} \cdot x^{*\prime T} - \Sigma)$ (5)
$x^{*\prime}$	Adjusted and standard deviation vector	
$cov'(x_i, x_j)$	Estimated new covariance between dimensions i and j	
Σ'	Estimated new covariance matrix given the new data point	

to what extent the new data deviate from the already known data. The difference between the two vectors shows whether the datapoint could be considered an outlier or not. This raises the question of how much divergence is considered normal and where the reason to doubt the nature of the datapoint starts.

SOLUTION FUNDAMENTALS

Online Outlier Detection Algorithm

The basis of the outlier algorithm considered in this paper lays on the identification of a group of sensors in a similar environment. For the Berkeley Lab dataset, the most intuitive approach for grouping sensors is by location, assigning each sensor a group of specific number of nearest "living" neighbourhood sensors, a group of odd or even number of sensors, which share the same environment and communicate directly.

The number of sensors is highly dependent on the voting rules (policies) of the "neighbourhood". If the algorithm is configured to group an odd number of sensors then for the sensor being processed in the group, there is an even number of "neighbours" voting on the nature of the data. This is possible to result in a tie. One possible resolution is applying a policy which establishes equal weights to all votes, no matter if the vote is from the source of the data or not. If the number of sensors in a neighbourhood is configured to be an even number, then there is one sensor as a source of the data and an odd number of voting neighbours. This is possible to result in a tie given that all votes have equal weights. However, if the source and the neighbourhood's votes have an unequal significance, that would allow exploring different rules for the voting system and finding the most optimal policy for a particular case.

All votes may also have different weights due to the nature of the outlier detection algorithm in each device. Each sensor is allowed to rely on different techniques for its core algorithm, which would create diversity depending on whether it is a model-based, as performed in this chapter, or a model-free approach (Dayan & Berridge, 2014).

After identifying a group in a common environment, the rest of the proposed outlier detection Algorithm 2 consists of two more core components - training a model given an already collected dataset and labelling unknown incoming data from a stream. Creating a model for a single sensor in a "neighbourhood" begins by collecting an initial set of N datapoints forming a sample of valid training data, required to have 0 dimensions with missing or out of the domain values. A valid dataset is standardised and adjusted according to the mean vector and the standard deviation vector following the PCA algorithm in section Principle Component Analysis. The next step is computing a covariance matrix based on the given sample and then producing an eigenvector explaining the data in the eigenspace.

In order to identify a datapoint as an outlier, the degree of its divergence from the whole data needs to be above a certain limit. This is represented by the difference between the trained model eigenvector and the new eigenvector calculated with regards to the new datapoint. Such difference is addressed as a threshold model estimation and is further explained in section Eigenvector Angle Threshold Estimation.

The second phase of the algorithm is applying the created model on unknown data received in the form of a stream. Each time a multivariate datapoint is sensed by the source sensor, it is sent in a transformation pipeline. In order to inspect the effect on the variance on the whole dataset with respect to the new datapoint, the pipe follows the PCA/power method algorithm. First, the datapoint is standardised and adjusted according to the online variance and covariance techniques in section Online Variance Estimation and then a vector is produced representing the new data in the eigenspace.

The final step is deciding on the nature of the datapoint (outlier/normal) following an initially specified policy for the "neighbourhood". In order to compare and exploit the most valuable voting policy, the source sensor has the ability to label its data as an outlier or not based on its own decision algorithm

Algorithm 2. Local Outlier Detection Algorithm

Set *defineNeighbourhood*() ▷ phase 0: forming a sensor neighbourhood
For *sensor* in *neighbourhood* do ▷ phase 1: training a model
$data \leftarrow gatherNValidDatapoints()$
$meanVector \leftarrow calculateMeanVector(data)$
$standardDeviationVector \leftarrow calculateStandardDeviationVector(data)$
$adjustedMatrix \leftarrow (data - meanVector) \div standardDeviationVector$

```
covarianceMatrix ← calculateCovarianceMatrix()
eigenVector ← powerMethod(covarianceMatrix, ϵ)
normalize(eigenVector)
trainThresholdModel()
▷ Section Eigenvector Angle Threshold Estimation
End For
While streaming do   ▷ phase 2: processing unknown data
datapoint ← fromStream
newMeanVector ← calculateOnlineMeanVector(datapoint)
newStandardDeviationVector ← calculateOnlineStandardDeviationVector(datapoint)
newAdjustedMatrix ← (datapoint − newMeanVector) ÷ newStandardDeviationVector
newCovarianceMatrix ← calculateOnlineCovarianceMatrix()
newEigenVector ← powerMethod(newCovarianceMatrix, ϵ)
normalize(newEigenVector)
angle ← getAngleBetween(oldEigenVector, newEigenVector)
getLocalVote(angle)
For i = 0 to neighbours do
sensors[i] ← getVote(datapoint)
End For
decision ← applyPolicy(localVote, groupVote)
If decision == NONOUTLIER then
updateModel()   ▷ Section Online Model Update
Else
SIGNAL OUTLIER
End If
End While
```

and also consults with the rest of the sensors in the "neighbourhood". Each "neighbour" must provide its vote, whether the datapoint should be labelled as an outlier or not.

Eigenvector Angle Threshold Estimation

One approach of measuring the extent of divergence between two vectors is computing the Euclidean distance. The Euclidean distance represents "the geometric distance in the multidimensional space" (Hill & Lewicki, 2006). However, in different cases the results vary based on the type of the given data. Euclidean distance could be used in fields where the value domain of each variable is known in advance, such as image processing (Wang, Zhang, & Feng, 2005). However, it is desired that this outlier algorithm is applicable to any kind of data, therefore, such measurement is unsuitable for this case.

Another technique for estimating the difference between two eigenvectors is to compute the angle between them. That is done by performing the dot product on the old and new eigenvector. This results in the cosine of the angle in-between, which using inverse cosine returns the measurement of the angle. In Deng's paper on outlier detection (2016) this technique is mentioned as a "correlation score of the eigenvector" with regards to the inspected datapoint. This is another point of view on the eigenvectors, which does not address the difference of the data as deviation but as a measurement of correlation of the new datapoint to the initial data.

Knowing a technique to measure the difference between the new and the known data is only the first step to a good outlier detection algorithm. The next step is creating a good heuristic for an appropriate threshold. The initial experiments were conducted using a constant value of 45°. This showed poor results due to the data having less significant deviation than expected, however, such results contributed to reducing the computational costs of the algorithm.

Another technique experimented with was using the first N datapoints as training data for the eigenvector model and the same quantity of next datapoints, creating N eigenvectors. Calculating the difference between each new eigenvector and the model resulted in a significant amount of information on the variance of the angles. Given that information the next step was to use the mean of the newly computed correlation scores and set that as an outlier threshold. This showed an improvement of the detected anomalies but was not sufficient.

In order to have satisfactory results of the experiments, it was decided to use an already established statistical heuristic of finding outliers using the statistical boxplot chart (NIST/SEMATECH). The two formulas presented a technique for setting a threshold for "mild" and "extreme" outliers. This returned an acceptable rate of identifying both outliers outside of the data domain and outliers "hidden in plain sight".

Designing a more sophisticated threshold model would require extensive research outside of the scope of this chapter.

Online Model Update

After estimating whether a datapoint deviates from the dataset substantially, it could be labelled as an outlier. Depending on the purpose of the data, the outlier can then be extracted and removed or set aside for further analysis. On the other hand, if a datapoint lays within the allowed range of deviation, it is sent through a different pipeline of the algorithm. Because of its nature, the datapoint allows the model to learn from its success, therefore as seen in Algorithm 3, all variance and covariance values are updated using the new estimations.

Algorithm 3. Updating the Outlier Detection Model

Set $meanVector \leftarrow newMeanVector$ ▷ Updating the mean vector
Set $standardDeviationVector \leftarrow newStandardDeviationVector$
▷ Updating the standard deviation vector
Set $covarianceMatrix \leftarrow newCovarianceMatrix$
▷ Updating the covariance matrix
Set $eigenvector \leftarrow newEigenvector$ ▷ Updating the eigenvector

First, the previous mean vector is replaced with the data estimation given the new datapoint as part of the whole dataset. Then it is followed by updating the standard deviation vector and the covariance matrix with their up-to-date values including the datapoint. The final step is replacing the old eigenvector, which represented the dataset excluding the newly received non-anomalous data, with the new eigenvector. This is done in order to track the flow of the data in real time. An example of such a flow is the temperature data from the Intel Berkeley Lab, Figure 2, where the data change in time t. In case the eigenvector is not updated, then after a given time $t+s$, the rise in the temperature due to multiple natural factors will be inaccurately detected as an anomalous behaviour of the sensor or the environment.

GROUP VOTING APPROACH & POLICIES

In order to explore as much as possible from the proposed solution of this chapter given the amount of time for this experiment, three different voting policies are considered driven by the three perspectives: confidence, independence and opinion. Each policy achieves a different level of decision certainty affecting the trade-off between error rate and communication cost.

Given the uncertainty around the most accurate angle between eigenvectors, which divides all normal data from outliers, it was decided to perform an analysis for each policy given all possible 90 angles from 0° to 90°. All policies were analysed so that each one was used on a local neighbourhood 90 times and each time the angle was set from the range of 0° to 90°. Before the execution 5% or 33% of the data were artificially modified to track the performance of each policy.

In order to analyse each policy in different cases, two types of artificial outliers were generated and injected into the dataset for each analysis – "obvious" and "less-visible". The first kind would be simpler to identify as a best-case scenario for the algorithm. Such outliers would lie outside the values range of the data. The second type would have a higher difficulty when being identified – "hidden in plain sight", helping measure the performance in the worst-case scenario. Outliers in this case are hidden in the normal distribution of the data, close to the lower or upper quartile of the data.

$$\begin{aligned}&temperature + (\pm 1) \times 0.5 \times \sigma_{temperature}\\&humidity + (\pm 1) \times 0.5 \times \sigma_{humidity}\\&light + (\pm 1) \times 0.5 \times \sigma_{light}\\&voltage + (\pm 1) \times 0.5 \times \sigma_{voltage}\\&\triangleright\ less\ visible\end{aligned} \tag{1}$$

$$\begin{aligned}&temperature + 4 \times \sigma_{temperature} \times \bar{x}_{temperature}\\&humidity + 7 \times \sigma^2_{humidity} \times 3\\&light + 3 \times \bar{x}^{2}_{light} \times 3\\&voltage + 3 \times \bar{x}_{voltage}\\&\triangleright\ obvious\end{aligned} \tag{2}$$

Policy 1: Opinion-Driven Policy

Description

The opinion-driven policy represents the concept of allowing a group of neighbourhood sensors to decide on the nature (anomalous/normal) of an incoming datapoint without considering the source sensor's opinion. The essence of this technique is its ability to eliminate bias created by outliers in the training data of the relevant sensor but at the cost of constant data exchange between all devices creating a significant communication overhead.

Algorithm 4, for processing a data stream of a single sensor, has the following pattern. Firstly, a model for each sensor in the neighbourhood is trained using the first 1000 datapoints of the already gathered data producing a single eigenvector. The next step is training a threshold model using the most appropriate heuristic for the data, in this case the used heuristic is the one discussed in section Eigenvector Angle Threshold Estimation. Thirdly, given a new datapoint from the sensor's datastream, the authors are able to compute the new mean and variance estimations and output the new eigenvector. Afterwards, the new datapoint is passed on to the other sensing devices in the neighbourhood. After each device has completed the processing of the shared datapoint, the neighbourhood sends each opinion back to the source. The final step is the accumulation of all received messages. Based on the majority of the neighbourhood opinions, the source either labels the datapoint as an outlier or updates the model presuming the normality of the datapoint.

Algorithm 4. Policy 1

$trainNeighbourhoodModelForTheFirst1000datapoints()$
$trainThresholdModel()$
While $streaming$ do
$datapoint \leftarrow fromStream$
$newMeanVector \leftarrow calculateOnlineMeanVector(datapoint)$
$newAdjustedMatrix \leftarrow datapoint - newMeanVector$
$newCovarianceMatrix \leftarrow calculateOnlineCovarianceMatrix()$
$newEigenVector \leftarrow powerMethod(newCovarianceMatrix, \epsilon)$
$normalize(newEigenVector)$
$angle \leftarrow getAngleBetween(oldEigenVector, newEigenVector)$
For $i = 0$ to $neighbours$ do
$sensors[i] \leftarrow getVote(datapoint)$
End For
If $groupVote == NONOUTLIER$ then
$updateModel()$
Else
$SIGNAL$ $OUTLIER$
End If
End While

Algorithm Analysis

Trying to analyse the accuracy of the algorithm on outlier identification, a confusion matrix test was performed. The model for Policy 1 was trained using 1000 datapoints gathered from the sensor and the next 100 datapoints were modified with 5% outliers. Meanwhile, the corresponding neighbours of the source sensor formed an eigenvector representation of their first 1000 datapoints. Next, the altered data from the source sensor were ran 89 times against Policy 1 using a different threshold angle each time for the range from 1° to 90° assigned to each sensor in the neighbourhood. The majority of sensors had confusion metrics as the ones shown in Figure 6, where first row A is analysis performed using only 5% "less visible" outliers and row B - using 5% "obvious" outliers.

The analysis showed poor and inexhaustive results due to two factors. First, despite that a neighbourhood is expected to have similar data due to their similar location to the source of their data, the confusion matrices showed that not all sensors have the same eigenvector threshold. This is shown by the pattern of the confusion matrices above. All datapoints are considered outliers up until the threshold angle of 65°. The angle value 65° appears to be a value around which all neighbours agree on as a threshold, however, having only falsely predicted any outliers appears to be an inaccurate angle for the source sensor's data. For angles above 65° sensors identify all datapoints as normal due to all the deviation of the datapoints residing within the larger angle values.

The second factor taken into consideration was that the confusion matrix limits the possibility for analysing the effect of the different opinions on the predictions with respect to the ground truth. This is the reason for exploring a more detailed version of the confusion matrix, which concerns only the outlier decisions. Therefore, the true and false outlier predictions were divided into 4 subcategories: "consistent", "helpful", "inherent" and "misleading".

- The "consistent" type of predictions represents the agreement of the whole neighbourhood on the anomalous nature of the datapoint and their correctness on the topic. For Policy 1 it is labelled as "consistent" when the neighbourhood considers a datapoint an outlier, the source agrees with the decision and the ground truth states that the datapoint is in fact an outlier.
- The "helpful" type is acknowledged whenever a sensor makes a wrong decision on the nature of a datapoint, however, the vote of the neighbourhood correctly identifies the datapoint and positively influences the output of the pipeline. A prediction is labelled as "helpful" for the current policy

Figure 6. Policy 1: Confusion matrices for sensor 45 with group $[46,44,43]$ *with angles 1° to 90°*

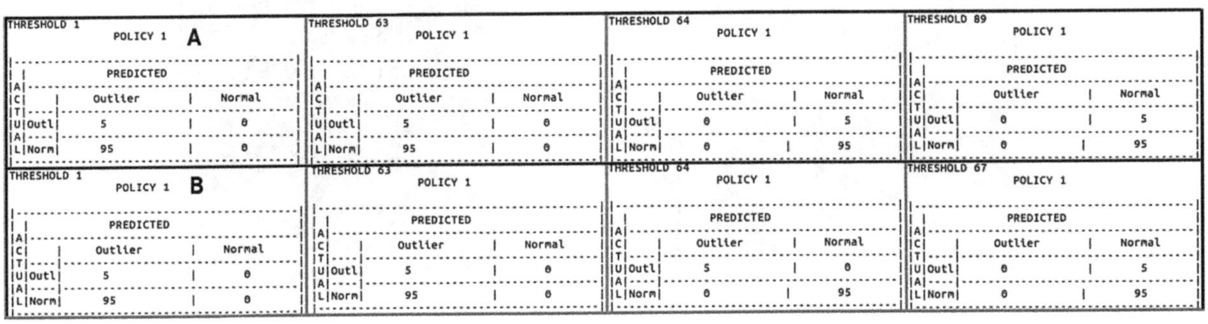

Figure 7. Policy 1: Further analysis on outlier detection accuracy for sensor 45 with group [46,44,43]

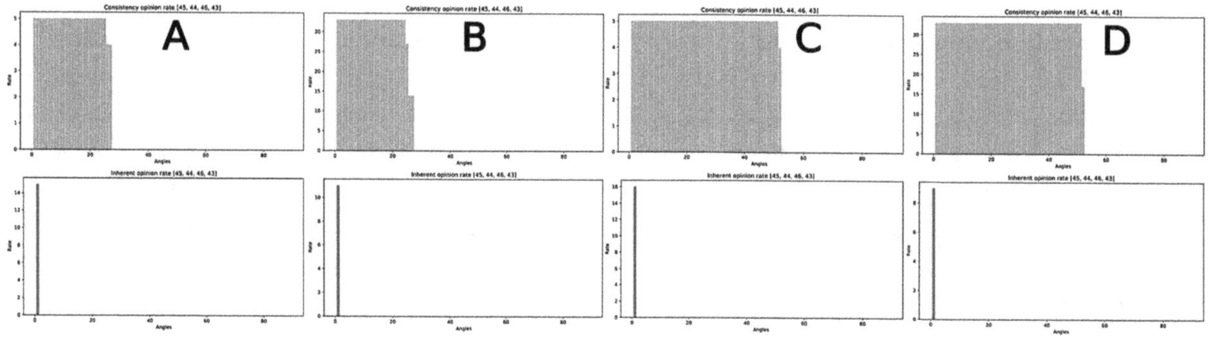

when the neighbourhood identifies an outlier and the source would have had the opposite opinion but the ground truth states that the datapoint is in fact an outlier.

- The third type of prediction – "inherent", represents all datapoints which the whole neighbourhood has incorrectly identified as opposed to the ground truth. A prediction is labelled as "inherent" for this policy when the neighbourhood has labelled a datapoint as an outlier and the source has the same opinion, however, the ground truth states that the datapoint lies in the range of normality.
- The last type is the "misleading" prediction. For this type the source has correctly identified the datapoint, however, the neighbourhood has inaccurately altered the final decision which later has incorrectly updated or not the model with regards to the processed datapoint. For Policy 1 a prediction is labelled as "misleading" when the neighbourhood has made the decision that a datapoint is a normal datapoint, but the neighbourhood has decided that the datapoint is an anomaly, which has resulted in labelling the datapoint as an outlier - wrong compared to the ground truth.

Another set of analysis performed on Policy 1 was visualising the difference received between the angle deviation of a neighbour and the source sensor given the same processed datapoint. The results were plotted in a heatmap, where the darker the colours, the less the difference between the opinions is. More thoroughly explained, whenever a source sensor senses a datapoint and processes it, as already

Figure 8. Policy 1: Heatmaps for the angle differences between source and each neighbour ("less visible" outliers in A, "obvious" outliers in B)

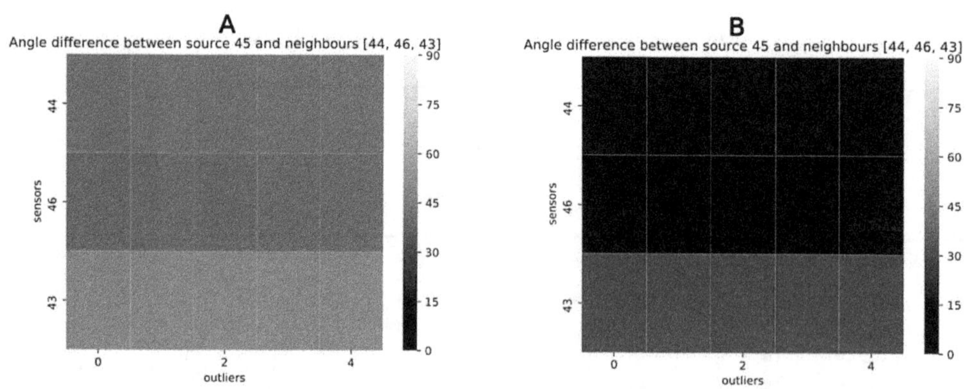

described, a deviation angle is calculated. Then whenever a neighbour receives the given datapoint at that point in time, an angle representing the deviation from their data is produced.

The difference of the two angles is expressed in a heatmap plot for each detected outlier, where each row corresponds to the relevant neighbour and each column - to the processed datapoint labelled as an outlier. The outlier decisions represented by the plot are only for the ones output by the algorithm after applying the policy.

Results & Discussion

The confusion matrix analysis showed that according to the neighbourhood the whole dataset with mild outliers deviate from the neighbourhood's eigenspace from 63° to 64°, which creates difficulties identifying the nature for any of the datapoints. As for the extreme outliers, the analysis in the second row showed that the exact threshold for the neighbours is between 64° and 67°. Such small result domains and unvaried tables did not provide a basis for meaningful conclusions; therefore, it was taken a decision to further investigate only outlier decisions on their own, split into four subgroups.

The analysis of the four subgroups of decisions in Figure 7 for sensor 45 with local neighbours $\left[43,44,46\right]$ were conducted in four different scenarios. The given output is only for the "consistent" decision rate - first row and the "inherent" decision rate - second row. Column A represents the "consistency" and "inherency" output of the algorithm ran with a 100 datapoints stream size in which were artificially injected 5% "less visible" outliers. On the other side, column B is for a datastream of size 100 with 33% injected "less visible" outliers. The "consistency" shows that the surrounding sensors agree with the decision of the source until they reach a decision angle of about 26°.

The "inherency" rate for the "less visible" outliers show that the minimum confusion threshold angle for the whole neighbourhood is between 0° and 2°, where all local sensors predict wrong 15% of the data and label them as an outlier and the ground truth shows that it is not. Both plots lean towards the fact that the deviation of the normal datapoints from the source's model lays around 1°. It also shows that the datapoints deviation with respect to the neighbours' data is up until 26°.

The two subtypes "helping" and "misleading" are empty as no relevant datapoints, which positively or negatively affect the final decision for a datapoint, were recorded. This leans to the conclusion that Policy 1 does not affect the outcome for the outliers of the sensor being processed.

The results of that analysis raised the question whether the data in a neighbourhood are too similar or too dissimilar. If the data are too similar then sensors are capable of assigning outlier detection to only two sensors from the neighbourhood, where both monitor the environment and report faults in any of the two sensors interchangeably and the rest of the sensors can be assigned other tasks reducing energy cost. However, if the data are too dissimilar then the neighbourhood cannot contribute to any local outlier decisions and each sensor then needs to perform the outlier detection job on its own. The "inherency" analysis shows that there is disagreement between the neighbours and the source sensor, which leans towards the statement that the data between different devices are more dissimilar.

An appropriate analysis was to compare what difference a datapoint made in the sensor's data and in the neighbour's data. This was done using the heatmap comparing the differences between the angle deviations for source and neighbour. The heatmaps given in Figure 8 for sensor 45 and group $[46,44,43]$ show a general tendency for all sensors where the angle distance is about 45° for "less visible" outliers and about 15° angle distance for "obvious" outliers. This indicates that sensors in the chosen neighbourhoods cannot decide on the nature of a sensor's data using Policy 1.

Policy 2: Confidence-Driven Policy

Description

The confidence-driven policy represents the concept of allowing a group of "local neighbourhood" sensors to decide on the nature (anomalous/normal) of an incoming datapoint only when the source sensor's opinion leans towards "outlier". The essence of this technique is its attempt to solve the communication overhead between the devices in the "neighbourhood" and preserve the accuracy of the results.

The processing algorithm in Algorithm 5 using this policy has the pattern of firstly producing an eigenvector model for each local sensing device based on the 1000 training datapoints of the already collected data from the incoming stream. The next step is to train a threshold model using the most appropriate heuristic for the data, discussed in section Eigenvector Angle Threshold Estimation. Afterwards, a new datapoint is taken from the stream and then processed against the already trained model. If the results are showing a strong deviation from the already known data, the sensing device sends the corresponding datapoint to all other devices in the group, otherwise updates the model and continues with the next datapoint from the stream. If a sensor has received a message containing the information for a datapoint, it calculates the correlation score against its own data and sends back the output of the inspection.

After receiving all scores, the source sensor makes a decision based on the majority of received votes and either labels the datapoint as an outlier or updates the model.

Algorithm 5. Policy 2

$trainNeighbourhoodModelForTheFirst1000datapoints()$
$trainThresholdModel()$
While $straming$ do
$datapoint \leftarrow fromStream$
$newMeanVector \leftarrow calculateOnlineMeanVector(datapoint)$
$newAdjustedMatrix \leftarrow datapoint - newMeanVector$
$newCovarianceMatrix \leftarrow calculateOnlineCovarianceMatrix()$
$newEigenVector \leftarrow powerMethod(newCovarianceMatrix, \epsilon)$
$normalize(newEigenVector)$
$angle \leftarrow getAngleBetween(oldEigenVector, newEigenVector)$
$getLocalVote(angle)$
If $localVote == OUTLIER$ then
For $i = 0$ to $neighbours$ do
$sensors[i] \leftarrow getVote(datapoint)$
End For
If $groupVote == NONOUTLIER$ then

```
updateModel()
Else
SIGNAL OUTLIER
End If
Else
updateModel()
End If
End While
```

Algorithm Analysis

Analysing Policy 2 was done following the same pattern as Policy 1, therefore a confusion matrix test was performed in order to analyse how each angle affected the final output of the algorithm. As before, row A of Figure 9, contains all results from applying Policy 2 on sensor 45 with a neighbourhood containing sensors $[46, 44, 43]$ where the stream size is from 100 datapoints and 5% of them are "less visible" outliers and for row B are "obvious" outliers.

As part of the confusion matrix test, the algorithm monitored the communication cost for each detected outlier. The communication cost is measured in number of messages exchanged between the source and each of its neighbours in total per execution with a stream size of 100 datapoints.

Furthermore, the same analysis on the types of outliers as the one for Policy 1 was also performed for Policy 2. All outlier decisions were divided into four subgroups – "consistent", "inherent", "helping", "misleading". For Policy 2 a prediction is "consistent" whenever a source sensor decides that a datapoint is an outlier in unison with the decision from the neighbourhood and the ground truth. An "inherent" prediction is having the source decide that a datapoint is not an outlier and the neighbourhood agrees with the label. This, however, contradicts with the ground truth, which states that the datapoint is an outlier and the model should not be updated. On the other hand, a prediction is labelled as "helpful" for this policy when a source has found an outlier, but the neighbourhood disagrees with the made decision, which positively affects the result since the ground truth states that the datapoint is not an outlier. The final type of prediction for Policy 2 is the "misleading" one, where a source sensor has labelled a data-

Figure 9. Policy 2: Confusion matrices for sensor 45 with group $[46, 44, 43]$ with angles from 1° to 90°

A

THRESHOLD 1 — POLICY 2

OUTPUT	PREDICTION: Outlier	PREDICTION: Normal
Outl	0	5
Norm	0	95

THRESHOLD 33 — POLICY 2

OUTPUT	PREDICTION: Outlier	PREDICTION: Normal
Outl	0	5
Norm	0	95

THRESHOLD 74 — POLICY 2

OUTPUT	PREDICTION: Outlier	PREDICTION: Normal
Outl	0	5
Norm	0	95

THRESHOLD 89 — POLICY 2

OUTPUT	PREDICTION: Outlier	PREDICTION: Normal
Outl	0	5
Norm	0	95

B

THRESHOLD 1 — POLICY 2

OUTPUT	PREDICTION: Outlier	PREDICTION: Normal
Outl	1	4
Norm	0	95

THRESHOLD 23 — POLICY 2

OUTPUT	PREDICTION: Outlier	PREDICTION: Normal
Outl	1	4
Norm	0	95

THRESHOLD 69 — POLICY 2

OUTPUT	PREDICTION: Outlier	PREDICTION: Normal
Outl	1	4
Norm	0	95

THRESHOLD 89 — POLICY 2

OUTPUT	PREDICTION: Outlier	PREDICTION: Normal
Outl	1	4
Norm	0	95

Figure 10. Policy 2: Further analysis on outlier detection accuracy for sensor 45 with group $[46,44,43]$

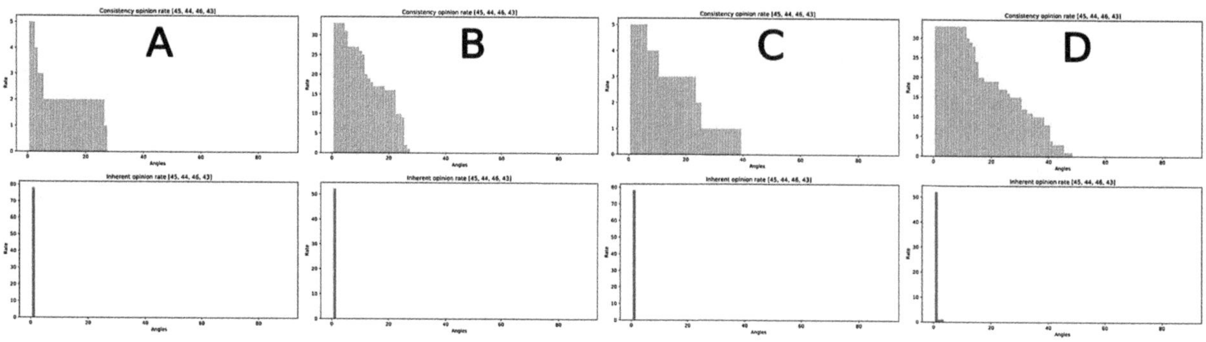

Figure 11. Policy 2: Heatmaps for the angle differences between source and each neighbour ("less visible" outliers in A, "obvious" outliers in B)

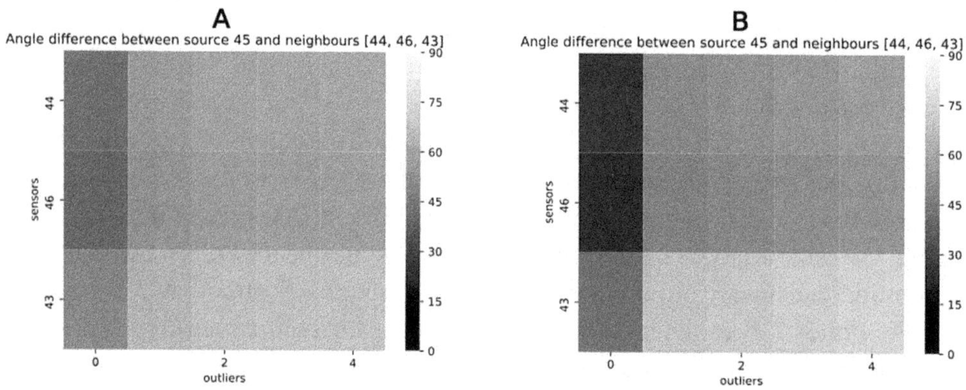

point as an outlier and the neighbourhood has altered that decision. However, according to the ground truth that contradicts with the actual anomalous nature of the datapoint.

The bar chart plot produced for Policy 2 required conducting an angle distance analysis as well. The heatmaps have the same format as the one for Policy 1. Each row relates to the distance of a neighbour's angle to the source's angle regarding the same processed datapoint and each column stands for the datapoint being processed sorted by order of processing time.

Results and Discussion

The results from the confusion matrix analysis, Figure 9, showed that all datapoints are labelled as normal for the mild outliers despite the presence of 5% outliers. This could have been explained by the cumulative voting of the neighbourhood on the normality of the source's data, however, the communication counter for each confusion matrix indicated 0 exchanged messages. The lack of communication leads to the assumption that the deviation angles of the source's data are under 1°. As for the second scenario, where the data are injected with extreme outliers, the analysis for sensor 45 showed that there is one correctly identified and 4 ignored outliers, which in comparison to the analysis of other neighbourhoods stood out as a good representation of most local environments. Such division in the predictions led to the second phase of the built analytical pipeline.

The results from the categorised predictions showed a steep staircase bar chart plot among all four scenarios indicating the consistent agreement of the neighbourhood with the source's prediction of an outlier. This behaviour could be expected based on the analysis of Policy 1, where it was found a big difference between the deviations with the voting neighbours and the source sensor, which leads to all neighbours agreeing with the outlier prediction of the source. There were again neither helping nor misleading predictions on the side of the neighbourhood. However, it is again possible to observe an inherent behaviour when there is an extremely low angle assigned to the whole neighbourhood.

Given such results allows further investigation on the reason for such similarities between the analyses of all four scenarios. The heatmaps visualisation of the difference between the eigenvector distance for the source sensor and each neighbour, shows some enlightenment on the problem. Observing plot A shows that the angle between the eigenvectors received when a neighbour processes a local datapoint is in the range from 40° to 75° for the mild outliers. As for plot B containing only extreme outliers shows an angle difference from 25° to 75°. The presence of such significant differences allows false outliers to be introduced into the model given that such one is initially gathered by the source sensor.

Policy 3: Independence-Driven Policy

Description

The independence-driven policy represents the concept of a sensor independently deciding the nature of an incoming datapoint without the need of locally shared opinion. The essence of this technique is the complete reduction of the communication cost, however, at the possible decrease in accuracy due to an initial bias in the training data. The algorithm in Algorithm 6 under this policy contains three main steps. The first one is training the model based on the first 1000 datapoints gathered from the datastream and a model on the threshold of correlation using the most appropriate algorithm for the type and shape of the data. The next step is accepting a new datapoint from the stream and processing it against the eigenvector model and then computing its correlation score. The final step is making a decision on the nature of the datapoint by observing whether the score is below or above the trained threshold. Given this observation, the sensor either labels the datapoint as an outlier or updates the algorithm's model with respect to a high score or a low score.

Algorithm 6. Policy 3

$trainModelForTheFirst1000datapoints()$
$trainThresholdModel()$
`While` $streaming$ `do`
$datapoint \leftarrow fromStream$
$newMeanVector \leftarrow calculateOnlineMeanVector(datapoint)$
$newAdjustedMatrix \leftarrow datapoint - newMeanVector$
$newCovarianceMatrix \leftarrow calculateOnlineCovarianceMatrix()$
$newEigenVector \leftarrow powerMethod(newCovarianceMatrix, \epsilon)$
$normalize(newEigenVector)$
$angle \leftarrow getAngleBetween(oldEigenVector, newEigenVector)$

$getLocalVote(angle)$
If $localVote == OUTLIER$ then
$updateModel()$
Else
$SIGNAL\ OUTLIER$
End If
End While

Algorithm Analysis

Policy 3 is analysed in a slightly different variation of the previous analysis due to its independent technique. The first part of using a confusion matrix on the policy remains the same as in section Policy 1: Opinion-driven Policy and section Policy 2: Confidence-driven Policy. The preparation for the analysis was done with building a model using 1000 training datapoints. As seen in Figure 12 first row A is once again the result of applying the current policy on a datastream of size 100 with 5% injected mild outliers and second row B with datastream of the same size but with 5% extreme outliers.

Further analysis investigates and indicates how the lack of neighbourhood decision affects the accuracy of the algorithm given that the policy prevents any communication expenses as opposed to the

Figure 12. Policy 3: Confusion matrices for sensor 45 with group $[46, 44, 43]$ *with angles from 1° to 90°*

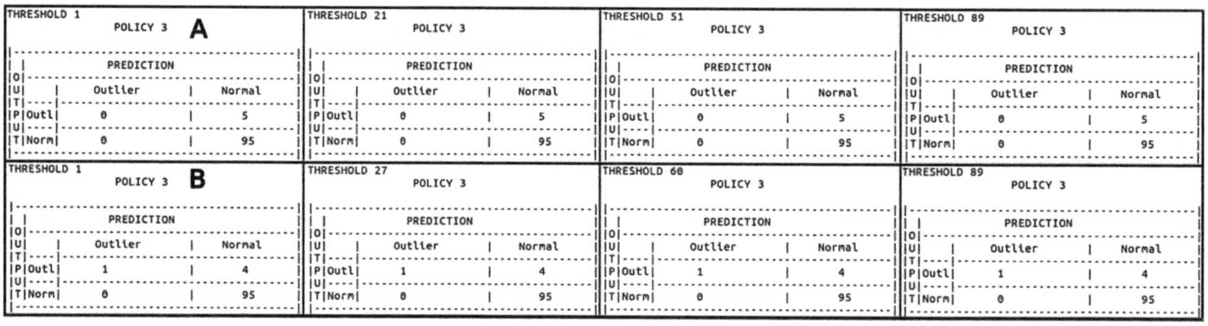

Figure 13. Policy 3: Further analysis on outlier detection accuracy for sensor 45 with group $[46, 44, 43]$

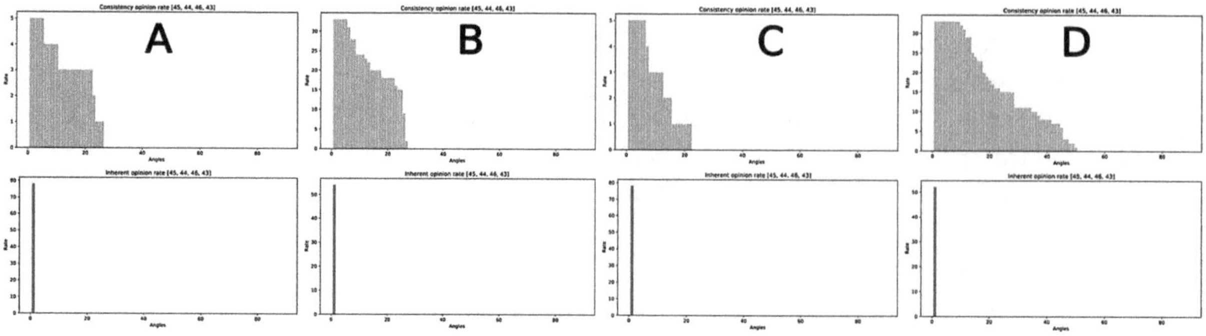

other two policies. Once again, dividing all predictions into four subgroups – “consistent”, “inherent”, “helping”, “misleading”, the analysis visualises all results using a bar chart. Having a consistent or inherent prediction is also a prediction with “no regret” as the result cannot be altered by the neighbourhood decision. Exposing predictions labelled as “helping” indicate that a source has inaccurately labelled a datapoint as an outlier and has missed a learning point in the process. A “misleading” prediction is one which would have been inaccurately altered by the neighbourhood if the algorithm had not applied the independent policy.

Results

The results from the analysis of Policy 3 are reflected by the changes occurring in the confusion matrices of the algorithm’s performance or as in this case the constant values of each field. As seen in row A, the “less visible” outliers are not detected by the source sensor and are considered normal due to the small deviation explaining the data. Row B shows a similar situation for the extreme outliers of the data. There appears to be one outlier correctly identified by the algorithm, however the rest remain hidden for all analysed thresholds.

The continued investigation using the subtype prediction analysis shows only “no regret” decisions, for which the neighbourhood does not affect the outcome neither in a positive nor in a negative way. The lack of “helping” or “misleading” predictions lessens the importance of the neighbourhood due to the lack of an accurate threshold estimator and more accurate representatives of the local environment.

EXPERIMENTAL EVALUATION

Dynamic Threshold Evaluation

The estimated thresholds from the boxplot outlier formulas in section Eigenvector Angle Threshold Estimation were evaluated with regards to the neighbourhood with sensors $[45, 46, 44, 43]$. For it was formed using a minimal grouping distance algorithm and each member gathered data and trained a predicting model using a set of 1000 valid datapoints. Each neighbour gathered an additional set of 100 valid datapoints used only to calculate their deviation from the main dataset, later processed as part of the incoming datastream. From the produced 100 deviation angles an extreme threshold is calculated using the formulas for extreme statistical outliers in the boxplot analysis. The evaluation for each policy is performed over a validation stream of 300 datapoints with 5% and 33% modified values appearing as in two of the scenarios as “less visible” and in the other two as “obvious” outliers.

Given such validation data, the perfect outcome of the evaluation would be to have at least one sensor for which there are exactly either 15 or 99 found “helping” outlier predictions from the neighbourhood and 0 “inherent” and “misleading” ones. Such results would mean that the neighbourhood’s predictions can positively affect at least one sensor’s outlier detection accuracy. However, as seen from the three set of plots Figure 14, Figure 15 and Figure 16 for policies 1, 2 and 3, the algorithm shows poor results.

The evaluation for Policy 1 indicates that the neighbourhood identifies correctly all extreme outliers and less than half of the “less visible” outliers, and they are consistent with the predictions of the source sensor. Moreover, there appears to be mistakenly identified outliers by the whole neighbourhood, and

Figure 14. Policy 1: Evaluation for sensor 45 with group $[46, 44, 43]$

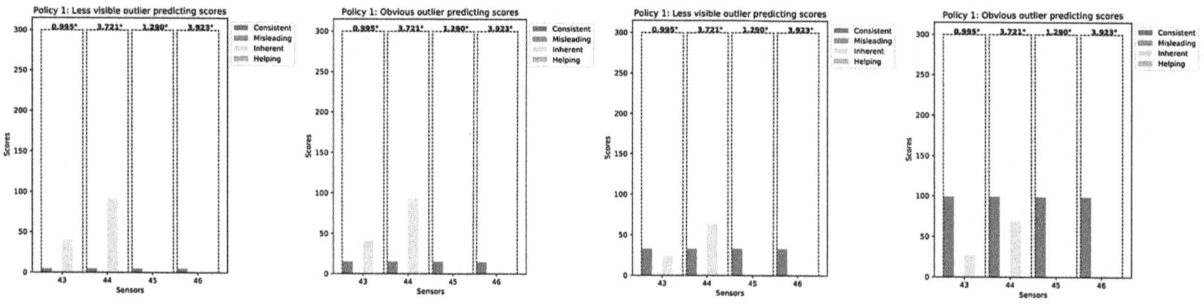

Figure 15. Policy 2: Evaluation for sensor 45 with group $[46, 44, 43]$

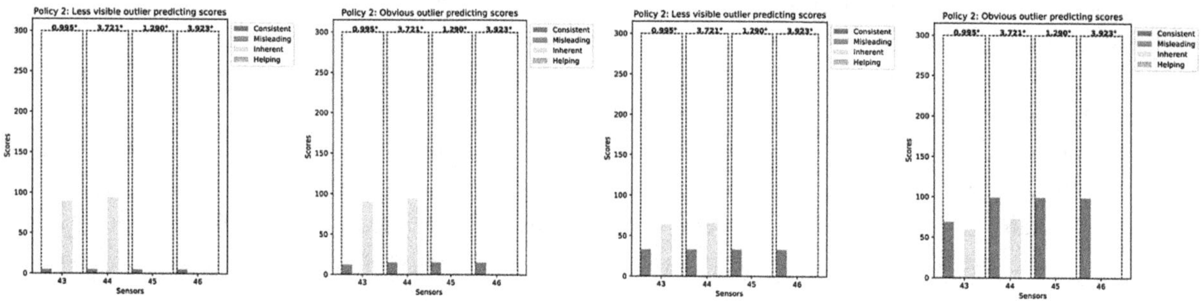

the amount doubles when the number of outliers decreases. That could be due to the difference between the threshold angles of the two sensors 44 and 46 with the sensors 43 and 45. This assumption explains the spike for sensor 44, where a decision for an outlier is outvoted by sensors 43 and 45, which given the calculated threshold could have easily mistaken normal datapoints for outliers from sensor 44. However, that creates inconsistency with the results shown from sensor 43, where there is a significant number of falsely labelled outliers. Moreover, having no misleading or inherent outlier predictions for sensors 45 and 46, leans towards the conclusion that the estimated threshold is imprecise causing inaccurate results.

The evaluation of Policy 2 shows from similar to marginally worse results, as there appears to be more inherently false predictions for sensor 43 while the consistency between source and neighbourhood predictions stay the same. Having less communication between sensors in the current neighbourhood compared to Policy 1 is as follows: with 5% "obvious outliers" - 369 messages, with 33% "obvious outliers" - 748 messages, with 5% "less obvious outliers" - 306 messages and with 33% "less obvious outliers" - 391 messages. Such results show substantially less communication than the previous policy but at the cost of doubling the number of falsely identified outliers for a sensor.

The evaluation for Policy 3 showed better results in identifying the outliers, however, the algorithm inaccurately labelled a significant part of the normal datastream as outliers for the mild outliers' scenario. The policy, however, showed significantly good performance when identifying extreme outliers, especially for sensors 45 and 46 where the false rate was recorded as 0%.

Figure 16. Policy 3: Evaluation for sensor 45 with group $[46, 44, 43]$

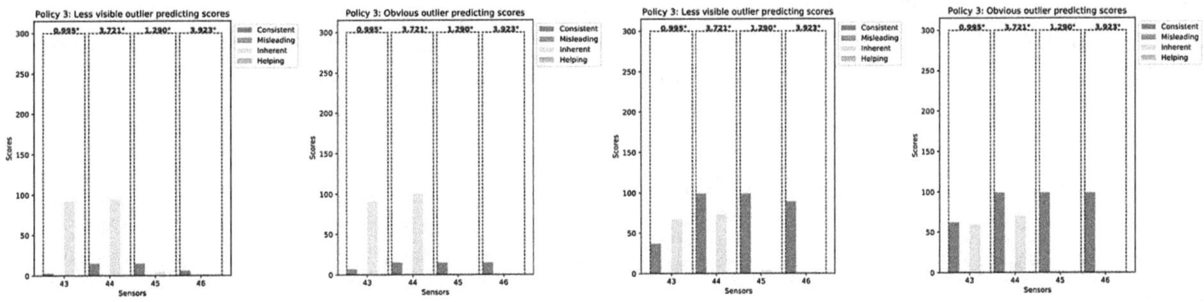

Evaluation Summary

All the aforementioned results show that the current choice of neighbourhood does not affect the final outcome in neither positive nor negative direction. All sensors preserve the ability to individually detect any outliers, however, the current neighbourhood only agrees on any outlier decisions, making all the carried communication a significant disadvantage. Therefore, the initial hypothesis of the project shall be rejected for the current state of the algorithm.

FUTURE RESEARCH DIRECTIONS

Given the analysis of the difference found between the sensors' eigenvector angles when processing the same datapoints in section Policy 1: Opinion-driven Policy, a basis for future research is further inves-

Figure 17. Heatmaps on the distance between angle values for different sensors in the same neighbourhood

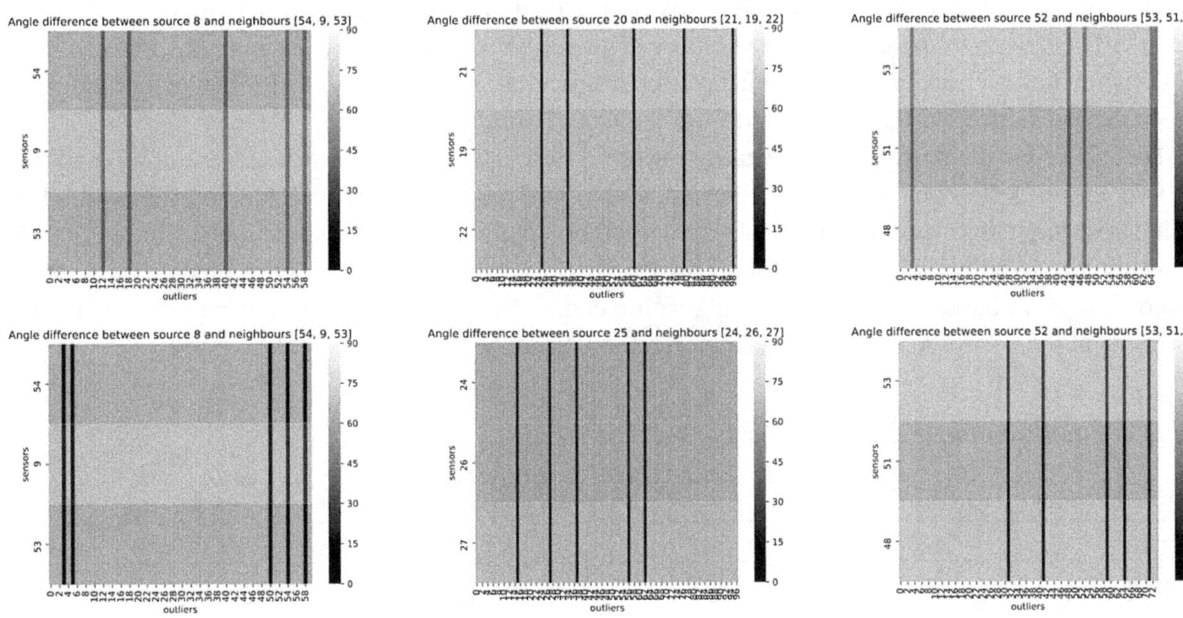

tigating the reasons for such differences. In the heatmaps in Figure 17 is seen sensor 8 with group $[54, 9, 53]$ and their corresponding differences between their eigenvector angles when processing the detected outliers. The plots are from a model trained over 1000 datapoints and where 100 datapoints were further processed with 5% artificially modified "less visible" or "obvious" outliers. It can be easily distinguished the five injected outliers (darker lines in the heatmap) among all detected suspicious datapoints.

This raises three potential questions for further analysis:

Investigation Question 1: Does the neighbour's angle decrease when an outlier is present?

Investigation Question 2: Does the neighbour's angle decrease, and the source sensor's angle **increase** when an outlier is present?

Investigation Question 3: Does the source sensor's angle increase, and the neighbour's angle stay the same due to its inability to understand data other than its own?

Better understanding how outliers affect the angles between the eigenvectors can lead to improving the local neighbourhood definition algorithm in order to form neighbourhoods which can comprehend each other's data and form a universal voting policy. The conclusion of such investigation could lead to either improving each neighbourhood's accuracy or completely rejecting the initial hypothesis of inter-helping sensor communities.

CONCLUSION

Performing the aforementioned analysis in section Policy 1: Opinion-driven Policy, section Policy 2: Confidence-driven Policy, section Policy 3: Independence-driven Policy on the Berkeley dataset and evaluating the performance of the implemented outlier detection algorithm on the same multivariate dataset, showed a lack of any positive effect introduced by a neighbourhood decision.

The results for the majority of the currently defined neighbourhoods showed huge discrepancies in the calculated eigenvector angles between source and neighbours as seen by the heatmap analysis in the sections for Policy 1 and Policy 2, which caused neighbours to signal any datapoints as outliers since the processed source's datapoints varied from the neighbours' values range.

Therefore, for the four distinct scenarios where different amounts of outliers were combined with the two outlier definitions for "less visible" and "obvious" anomalous datapoints, the evaluation of the built algorithm with respect to the initially set hypothesis could be summarised into the following table.

The initial hypothesis is rejected for all reviewed scenarios; therefore, the current state of the introduced local PCA-based outlier detection algorithm cannot be used to infer higher accuracy on anomaly decisions at the cost of higher communication rate in close-range sensor neighbourhoods.

Table 4. Final hypothesis analysis: reject or fail to reject

% Outliers in the Data	5%	33%
Less visible	Reject	Reject
Obvious	Reject	Reject

ACKNOWLEDGMENT

This research is funded by the EU H2020 GNFUV Project Grant #645220.

REFERENCES

Anagnostopoulos, C., & Hadjiefthymiades, S. (2014). Advanced Principal Component-Based Compression Schemes for Wireless Sensor Networks. *ACM Trans. Sen. Netw., 11*, 7:1-7:34. doi:10.1145/2629330

Anagnostopoulos, C., Hadjiefthymiades, S., & Georgas, P. (2012). PC3: Principal component-based context compression. *Journal of Parallel and Distributed Computing*, *72*(2), 155–170. doi:10.1016/j.jpdc.2011.10.001

Anastasi, G., Conti, M., Di Francesco, M., & Passarella, A. (2009). Energy conservation in wireless sensor networks: A survey. *Ad Hoc Networks*, *7*(3), 537–568. doi:10.1016/j.adhoc.2008.06.003

Ashton, K. (2009). That Internet of Things Thing. *RFID Journal*.

Aziz, N. (2014). *Eigenstructure-Based Angle for Detecting Outliers in Multivariate Data*. Academic Press. PMID:26356647

Bodik, P., Hong, W., Guestrin, C., Madden, S., Paskin, M., & Thibaux, R. (2004). *Intel Lab Data*. Intel Berkeley Research Laboratory. Retrieved from http://db.csail.mit.edu/labdata/labdata.html

Cheng, H., Feng, D., Shi, X., & Chen, C. (2018). Data quality analysis and cleaning strategy for wireless sensor networks. *EURASIP Journal on Wireless Communications and Networking*, *2018*(1), 61. doi:10.118613638-018-1069-6

Cheng, H., Zhihuang, S., Xiong, N., & Yang, X. (2016). Energy-efficient nodes scheduling algorithms for wireless sensor networks using Markov random field model. *Inf Sci*, *329*(1), 461–477. doi:10.1016/j.ins.2015.09.039

Dayan, P., & Berridge, K. C. (2014). Model-based and model-free Pavlovian reward learning: Revaluation, revision, and revelation. *Cognitive, Affective & Behavioral Neuroscience*, *14*(2), 473–492. doi:10.375813415-014-0277-8 PMID:24647659

Deng, J. D. (2016). Online Outlier Detection of Energy Data Streams Using Incremental and Kernel PCA Algorithms. In *2016 IEEE 16th International Conference on Data Mining Workshops (ICDMW)*, (pp. 390-397). IEEE. doi:10.1109/ICDMW.2016.0062

Edelman, A. (2018, Spring). The Power Method. *Linear Algebra*.

Harrington, P. (2012). *Machine Learning in Action*. Manning Publications.

Harth, N., & Anagnostopoulos, C. (2017). *Quality-aware aggregation & predictive analytics at the edge, 2017 IEEE International Conference on Big Data* (pp. 17–26). Boston, MA: Big Data; Retrieved from http://eprints.gla.ac.uk/149980/

Harth, N., Anagnostopoulos, C., & Pezaros, D. (2018). Predictive intelligence to the edge: Impact on edge analytics. *Evolving Systems*, *9*(1), 91–91. doi:10.100712530-017-9210-z

Haykin, S. (1999). *Neural networks: a comprehensive foundation*. Pearson Education.

Hill, T., & Lewicki, P. (2006). *Statistics: Methods and Applications: A Comprehensive Reference for Science, Industry, and Data Mining*. StatSoft, Inc.

Holmes, S. (2002). *Example of the Power method with Matlab*. Academic Press.

Intanagonwiwat, C., Estrin, D., Govindan, R., & Heidemann, R. (2002). Impact of network density on data aggregation in wireless sensor networks. *Proceedings of the 22nd International Conference on Distributed Computing Systems*. 10.1109/ICDCS.2002.1022289

Lischinski, D. (2007). *The Power Method*. Academic Press.

Lopez, P. G., Montresor, A., Epema, D., Datta, A., Higashino, T., Iamnitchi, A., ... Riviere, E. (2015). Edge-centric Computing. *Computer Communication Review*, *45*(5), 37–42. doi:10.1145/2831347.2831354

McKinney, W. (2018). *Python for data analysis: data wrangling with Pandas, NumPy, and IPython*. Sebastopol, CA: O'Reilly Media, Inc.

NIST/SEMATECH. (n.d.). *What are outliers in the data?* Retrieved from http://www.itl.nist.gov/div898/handbook/

Pearson, K. (1901). On lines and planes of closest fit to systems of points in space. *Philosophical Magazine*, *2*, 559–572.

Python Graphing Library. PlotlyAPI. (n.d.). Retrieved from https://plot.ly/python/

Sharma, S. K., & Wang, X. (2017). Live Data Analytics With Collaborative Edge and Cloud Processing in Wireless IoT Networks. *IEEE Access: Practical Innovations, Open Solutions*, *5*, 4621–4635. doi:10.1109/ACCESS.2017.2682640

Smith, L. (2002). *A tutorial on principal components analysis*. Tech. rep., Cornell University. Retrieved from http://www.cs.otago.ac.nz/cosc453/student_tutorials/principal_components.pdf

Walt, S., Colbert, S. C., & Varoquaux, G. (2011). The NumPy Array: A Structure for Efficient Numerical Computation. *Computing in Science & Engineering*, *13*(2), 22–30. doi:10.1109/MCSE.2011.37

Wang, L., Zhang, Y., & Feng, J. (2005). On the Euclidean distance of images. *IEEE Transactions on Pattern Analysis and Machine Intelligence*, *27*(8), 1334–1339. doi:10.1109/TPAMI.2005.165 PMID:16119271

Wei, Z., He, Y., He, W., & Liu, X. (2017). Information Dissemination Model Based on Clustering Analysis of Information Network Development. In *4th International Conference on Information Science and Control Engineering (ICISCE)* (pp. 403-407). Changsha: Academic Press. 10.1109/ICISCE.2017.91

Welford, B. P. (1962). Note on a Method for Calculating Corrected Sums of Squares and Products. *Technometrics*, *4*(3), 419–420. doi:10.1080/00401706.1962.10490022

Yao, Y., & Gehrke, J. (2002). The cougar approach to in-network query processing in sensor networks. *SIGMOD Record*, *31*(3), 9–18. doi:10.1145/601858.601861

ADDITIONAL READING

Fahmy, H. M. (2016). *Wireless sensor networks: Concepts, applications, experimentation and analysis.* Singapore: Springer. doi:10.1007/978-981-10-0412-4

Ganguly, A. R., Gama, J., Omitaomu, O. A., Gaber, M., & Vatsavai, R. R. (2008). *Knowledge Discovery from Sensor Data* (1st ed.). CRC Press. doi:10.1201/9781420082333

Hastie, T. J., Tibshirani, R., & Friedman, J. (2009). *The Elements Of Statistical Learning: Data Mining, Inference, and Prediction* (2nd ed.). New York: Springer-Verlag New York. doi:10.1007/978-0-387-84858-7

Jolliffe, I. (2018). *Principal component analysis.* I. T. Jolliffe., Springer-Verlag New York.

Lee, J., & Verleysen, M. (2007). *1). Nonlinear Dimensionality Reduction.* New York: Springer-Verlag New York. doi:10.1007/978-0-387-39351-3

Petersen, P. (2012). *Linear Algebra.* New York: Springer-Verlag New York. doi:10.1007/978-1-4614-3612-6

Rogers, S., & Girolami, M. (2016). *A First Course in Machine Learning* (2nd ed.). Chapman & Hall/CRC.

KEY TERMS AND DEFINITIONS

Datapoint: A single unit of data.

Eigenvector: A non-zero vector representation in the eigenspace of a given set of data.

IoT: Internet of things; a smart connected network of devices.

Model: The input to a machine learning algorithm built upon some training data and used to make a future prediction about some relevant unseen data.

Neighborhood: A group of odd or even number of sensors, which share the same environment and communicate directly; they are not separated by walls or any other substantial obstacles.

Online Algorithm: An algorithm executed on a real-time streaming set of data.

Outlier: An anomaly in some data, which would cause disturbance and unexpected results.

PCA: Principle component analysis; in the context of this chapter, a technique used for lossless data compression and decompression.

Policy: Rules for voting in a neighborhood of sensing devices.

Power Method: An iterative approach for estimating the first eigenvector and eigenvalue for a set of data.

Chapter 12
Intelligent Tracking and Positioning of Targets Using Passive Sensing Systems

Saad Iqbal
NUST School of Electrical Engineering and Computer Science (SEECS), Pakistan

Usman Iqbal
NUST School of Electrical Engineering and Computer Science (SEECS), Pakistan

Syed Ali Hassan
NUST School of Electrical Engineering and Computer Science (SEECS), Pakistan

ABSTRACT

Target localization and tracking has always been a hot topic in all eras of communication studies. Conventional system used radars for the purpose of locating and/or tracking an object using the classical methods of signal processing. Radars are generally classified as active and passive, where the former uses both transmitter and receivers simultaneously to perform the localization task. On the other hand, passive radars use existing illuminators of opportunity such as wi-fi or GSM signals to perform the aforementioned tasks. Although they perform detection using classical correlation methods and CFAR, recently machine learning has been used in various application of passive sensing to elevate the system performance. The latest developed models for intelligent RF passive sensing system for both outdoor and indoor scenarios are discussed in this chapter, which will give insight to the readers about their designing.

INTRODUCTION

Wireless communication systems are extensively used in both indoor and outdoor scenarios. The radio frequency (RF) signals transmitted from majority of these system can be used by passive receivers as non-cooperative opportunistic signals for tracking moving objects around them. RF signals, e.g., frequency modulation (FM), satellite signals, global system for mobile communication (GSM) and digital television (DTV) signals can be used in outdoor environments for passive listening. On the other side, Wi-Fi

DOI: 10.4018/978-1-5225-7458-3.ch012

is a suitable option for the indoor environment because most of the buildings these days possess Wi-Fi systems that are used to connect users with the internet. Different security and surveillance applications can be designed for the buildings' premises by using Wi-Fi-based passive sensing system. As the passive RF system has no dedicated transmitters, they are considered covert and power effective in nature.

The prominent application of passive sensing system is determining the motion profile of target object in the surroundings. By using advance techniques and complex algorithms, such systems are now able to provide more information about the motion of objects rather than just detecting them. However, there is always a big question i.e., "can we bypass the complex conventional techniques in implementing such systems?" This question can be answered by drawing attention to the emerging field of *machine learning and deep learning*.

Over the last decade, many researchers analyzed and developed different kinds of intelligent tracking techniques for moving bodies based on RF passive sensing systems. The classical RF passive sensing system has two or more receiving channels. One is the reference channel which provides unaltered echoes of the transmitted signals and other is the surveillance channel which provides the echoes of the transmitted signals from the area of interest (AOI). As the number of surveillance channels increases, the resolution of tracking gets better. Furthermore, the adaptive filtering techniques such as least mean square (LMS), extensive cancellation algorithm (ECA) etc. are used for obtaining the output which is free from the variation of transmitted signals. Therefore, this resultant time series signal only has information related to the changes happening in the AOI as demonstrated by (Iqbal et al., 2018).

(Chen et al., 2015; Colone et al., 2012) have used complex mathematical techniques to develop the algorithms which help in determining the nature of motions in AOI. However, the recent use of deep learning framework by (Iqbal et al., 2018; Khan et al., 2017b; Wang et al., 2017) in this field proved its success in both accuracy and efficiency. This chapter is about the design and analysis of the RF passive system based on machine learning and deep learning frameworks in outdoor and indoor scenarios. The rest of the chapter is organized as follows. History, origin and evolution of the passive sensing system is discussed in the *next* section, whereas general working of passive RF sensing system is elaborated in *RF passive sensing system* section. Afterword, design of outdoor and indoor passive sensing system is discussed in *passive sensing system for outdoor scenarios* section and *passive sensing system for indoor scenarios* section respectively. At the end, prominent applications of passive sensing system based on machine learning/deep learning are discussed in *application of passive RF sensing system.*

BACKGROUND

The concept of passive radars is not new. Back in 1935, Robert Watson Watt demonstrated the working of the radar by detecting a Handley Page Heyford bomber from a distance of 12 km using the BBC shortwave transmitter at Daventry as mentioned in (Chodos et al., 2006). During World War 2, the project named "Klein Heidelberg" was initiated by the Germans as described in (Griffiths & Willis, 2010) for long-range air surveillance. Six of these bi-static radars were deployed along the Belgian, Dutch and French side of the English Channel and North Sea. These radars didn't have any dedicated transmitters and worked by using the reflections from the Chain Home (British coastal radar system).

In 1955, unique type of passive radars named as "AN/FPS-23 Fluttar Radars" were deployed on the North America Distant Early Warning (DEW) line for detecting the penetration in the territory by low-flying bombers. According to (Skolnik, 2007), they detect the presence of aircraft/ bomber by extracting the Doppler frequency shift from the received bi-static echoes of the signals from the TV transmitter scattered from the aircraft. At the end of 20th century, the system named "Silent Sentry system" was launched by Lockheed-Martin Mission Systems. This system exploits the analogue television and FM radio transmitters for detecting and positioning of flying aircrafts.

After the 20th century, excessive research has been done to improve the reliability and quality of the passive radars in outdoor scenarios. Both signal processing and machine learning techniques were developed to solve the in-hand problems and mature the passive sensing systems. Nowadays, the use of indoor tracking system through passive sensing system has increased because multiple non-cooperative opportunistic RF signals i.e. Wi-Fi, DTV, GSM, etc. are available in almost every building. However, there is no doubt to say that development of passive radar for indoor environment is more challenging than outdoor environment because of inherent properties of RF signals.

RF PASSIVE SENSING SYSTEM

RF passive sensing systems are the types of the bi-static radars that do not need any dedicated transmitter as shown in Figure 1. These systems work by exploiting already existed RF signals in the environment for detecting and positioning of moving target objects.

There are two main types of RF passive sensing system, which are as follow:

1. Classical Passive Sensing System
2. CSI based Passive Sensing System

Classical Passive Sensing System

This system is composed of a reference channel and one or more surveillance channels. Reference channel provides direct echo signals of the transmitter whereas surveillance channels provide the echo signals of

Figure 1. Bi-static passive RF sensing model

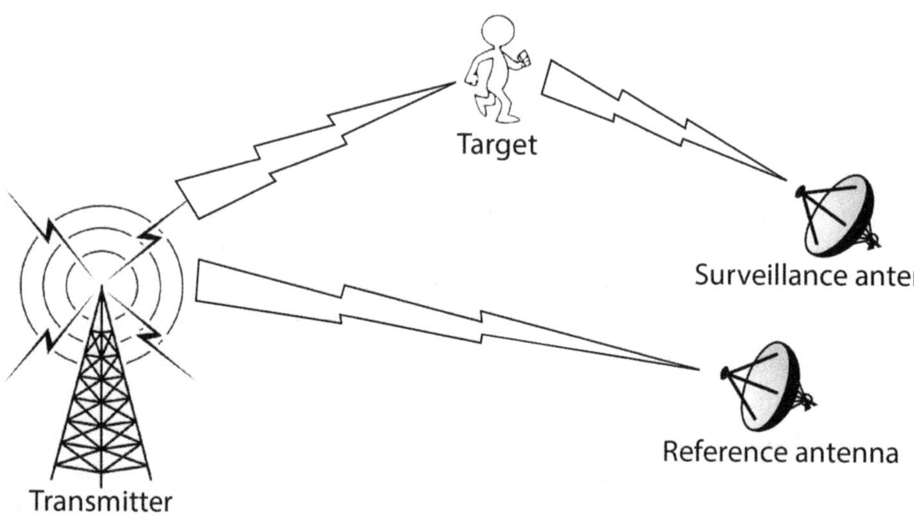

the transmitter which bounce back from the target bodies in the area of interest (AOI). Area of interest is defined as the cross-sectional area where target object can appear.

CSI Based Passive Sensing System

They are the special type of passive sensing systems. They are usually used in buildings with Wi-Fi as non-cooperative opportunistic signal source. They sense the variation in channel state information (CSI) of the signal to predict the motion profile of the moving target bodies. Rather than using signal processing, the deep learning models are used for classification and activity recognition in CSI-based passive sensing system.

PASSIVE SENSING SYSTEM FOR OUTDOOR SCENARIOS

In the outdoor environment, passive sensing systems are used for detecting, positioning and tracking of the target objects. Most likely target objects are aircrafts or helicopters in the sky and cars or trucks on the highways. Therefore, such passive sensing systems have some prominent characteristics, which are as follow:

- Target body is far away from the passive sensing system
- Target objects are more likely to be excellent RF reflector
- Air is acting as a medium in such environment
- Usually, non-cooperative opportunistic signals i.e. GPS, FM, DTV, GSM are available with quite good SNR

The first mentioned characteristic enables the passive radars to detect the target body by measuring delay in the echoes of the transmitter signals. Moreover, Doppler frequency shift in the echoes of the transmitter signals due to moving target body helps in determining the speed. Combining these two operations, it gives rise to the powerful signal processing tool named as "Cross ambiguity function (CAF)" which will be discussed briefly in later topics. The direction and position of target body are generally determined by two methods:

- Triangulation (By receiving multiple transmitters' echoes) as demonstrated in (Malanowski, 2012)
- Array antenna used as Surveillance receiver (increase antenna's directivity)

In recent past, many innovative designs of the RF passive sensing systems for outdoor scenarios are proposed which provide much more information than just tracking and positioning of target bodies. (Malanowski, 2012) developed a complex algorithm for converting bi-static parameters of a target into Cartesian coordinate obtained from FM-based passive sensing system. One of the most critical requirement of the passive sensing system is channel equalization of the reference signal. This problem is addressed by (Krysik et al., 2014) using Constant Magnitude Algorithm on FM-based Passive sensing. (Zhang et al., 2016) developed the practical FM-based passive system, which can exploit up to eight carrier frequencies for detection and tracking of the targets. It is developed from commercial on-the-shelf products and works in real-time.

The weak target echoes in the reference signal heavily degrade direction of arrival (DOA) estimate. To cope with this problem, the robust (DOA) estimate method has been proposed by (Xiao-Yong et al., 2015) for the FM-based passive sensing system. By the help of Germany Ministry of Defense, the Stationary Cassidian FM Passive Radar has been developed as mentioned in (Schroeder et al., 2013) which is operational for 24/7 and provides very good detection over the range up to 100km. Digital Audio/Video Broadcast (DAB/DVB) are the excellent candidates for the outdoor passive sensing system and they both use OFDM modulation scheme for transmission. In order to use these broadcast as opportunistic signals for passive sensing system, (Berger et al., 2010) discussed the signal processing scheme for Passive sensing systems working on OFDM waveform. Also, (Zeng et al., 2016) discussed the method to suppress the strong direct path interference in DAB/DVB-based passive sensing system. This enables system to identify the slow moving targets. (Shi et al., 2017) proposed an algorithm sequential processing-joint integrated probabilistic data association (SP-JIPDA) for multi-static passive Radar based on DAB/DVB. Every proposed methodology and algorithm refined the working of classical Passive Radar but also increased the computation as well as hardware cost.

It can be noticed that the above-mentioned works use classical signal processing algorithms for target localization and/or detection. However, recently a machine learning-based solution was developed by (Del-Rey-Maestre et al., 2017) for RF passive sensing and it outperformed the conventional passive sensing techniques. The model is briefly discussed in the next section.

Machine Learning Based Passive Sensing System for Outdoor Scenarios

As shown in Figure 2, every passive sensing system is consisted of three main stages, which are as follow:

- Reception stage
- Processing stage
- Detection stage

Figure 2. Basic architecture of ML-based passive RF sensing system for outdoor scenarios

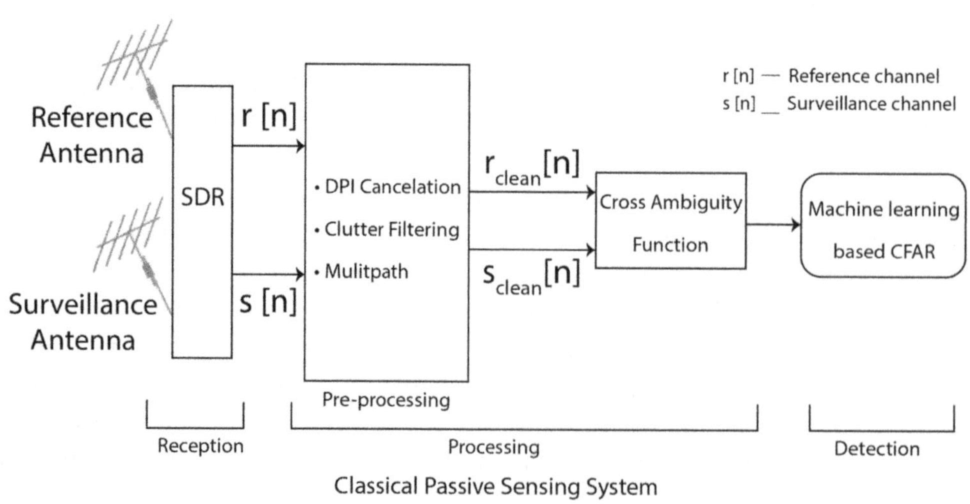

Reception Stage

The main function of this stage is to provide synchronized reference and surveillance channel. The reference channel receives the direct echoes of the transmitter signals and surveillance channel receives echoes of the transmitter signals bounced back from the target body. There can be more than one surveillance receivers to increase the resolution and accuracy of the tracking and positioning of the target objects.

In order to design such system on hardware, there are some points to be noted:

- For better results, the reference antenna should be directed towards the transmitter tower and surveillance antenna should be focused toward the area of interest (AOI). In some cases, however the surveillance antenna is omni in nature and capture signals from all directions. In such situation, the impact of Direct Path interference (DPI), clutter and multipath on the results gets highly increased.
- Software defined radios (SDRs) are good options for prototyping passive sensing system. The sampling frequency of the SDR should be chosen on the basis of the minimum path delay specification of the passive radar. Mathematically:

$$f_s \geq \frac{1}{\tau_{\min}},$$

where, $\tau_{\min}$ is the first excess delay of the multipath channel.

- The synchronization of the whole reception system is critical. Therefore, there should be a dedicated block that can be used for multi-channel synchronization in the entire system.

Processing Stage

After the reception, the pre-processing is required before applying cross ambiguity function (CAF). The pre-processing includes following operations:

- DPI Cancellation
- Clutter and Multipath Cancelation

DPI Cancellation

The DPI or direct echoes of the signals from the transmitter received by surveillance receiver as shown in Figure 3 causes the broad line in the zero Doppler region of the CAF. This makes impossible for extracting the information about slow moving target objects. Therefore, its effect should be cancelled in the pre-processing stage.

Figure 3. Modeling DPI and multipath effect

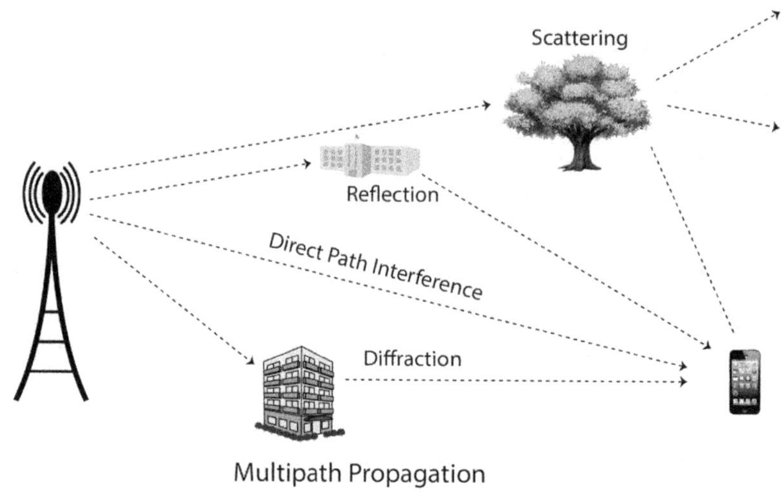

Clutter and Multipath Cancellation

Clutter and multipath effects arise from the echoes of the signals bounced from the random static/moving objects i.e. sea, earth, buildings, etc., as shown in Figure 3. This causes false alarms in the passive sensing system.

Both DPI and Clutter/Multipath effect is overcome by using adaptive filtering or modified adaptive filtering techniques. Following steps are used in this filtering process:

- Firstly, adaptive filtering (functions as echo cancellation) is applied on the reference channel $r\left[n\right]$ to obtain single direct echo signal $r_{clean}\left[n\right]$. Some researcher also used the technique of regeneration of transmitted signal in which they demodulate and decode the reference signal and then regenerate the signal that was actually used in transmission.
- Secondly, filter out $r_{clean}\left[n\right]$ and its copies from the surveillance channel $s\left[n\right]$ which will eventually result in further attenuation of DPI and Clutter.

Thus, refined reference signal $r_{clean}\left[n\right]$ and filtered surveillance signal $s_{clean}\left[n\right]$ are available after pre-processing stage. Now, it is time to apply cross ambiguity function (CAF) on $r_{clean}\left[n\right]$ and $s_{clean}\left[n\right]$ to obtain ambiguity surface.

Cross Ambiguity Function

The CAF is used to extract the range-Doppler. Mathematically, it is expressed as,

$$CAF\left(\tau, f\right) = \sum_{n=0}^{N} s_{clean}\left[n\right] \cdot r_{clean}^{*}\left[n-\tau\right] e^{-\frac{j2\pi nf}{N}},$$

where, τ and f are representing path delay and frequency shift respectively. The maximum range of $\tau_{\max}$ and $f_{\max}$ are calculated using pre-defined limits of the range and speed of the target body. Mathematically,

$$\tau_{\max} \cong f_s * T_{\max path},$$

$$f_{\max} \cong \frac{v_{\max}}{c} f_c f_s N,$$

where, $T_{\max path}$ is the path delay occurring at the maximum range, f_c is the center frequency of transmitted signal, f_s is sampling frequency of the system and N is the length of the $CAF(\tau, f)$. Moreover, Doppler resolution is given as:

$$f_d = (N \cdot f_s)^{-1}.$$

Most of the time, CAF computation causes the bottle neck issue in the real time performance of the Radars. However, there are different proposed techniques to compute the ambiguity surface using as minimum computation as possible. The most effective method for finding the CAF consists of following steps:

Step 1: Cross-correlation between $r_{clean}[n]$ and $s_{clean}[n]$
Step 2: Low-pass filtering (optional)
Step 3: Discrete Fourier Transform (DFT) or Fast Fourier Transform (FFT)

Some authors used ZFFT (zoom-FFT) instead of FFT as in (Zhihai & Tao, 2008) to enhance the overall performance. After this process, the detection techniques are used to detect and localize the target bodies.

Detection Stage

Mostly, every radar system uses the CFAR (Constant False Alarm Rate) for the target detection over the ambiguity surface. (Del-Rey-Maestre et al., 2017) successfully designed the approximation of Neyman Pearson or CFAR detector based on the machine learning techniques. The (Del-Rey-Maestre et al., 2017) used Gaussian model for high frequency Doppler and likelihood ratio detectors for Doppler closed to zero. For the false alarm detection, neural network (N-N) based adaptive threshold technique was used. Interested readers are encouraged to refer to (Del-Rey-Maestre et al., 2017) for better understanding.

PASSIVE SENSING SYSTEM FOR INDOOR SCENARIOS

The application of passive sensing system in indoor scenarios are quite different from outdoor ones. In the buildings, the passive sensing system are used for human gesture recognition, through-the-wall movement detection, gesture-based communication and controlling systems, etc. Moreover, the indoor passive radars are also used in developing non-invasive health monitoring systems, i.e., tremor detection, breathing rate estimation, fall-detection, etc.

As, it is known fact that mostly Wi-Fi signals are available in the indoor environments with good SNR than any other type of RF signal, therefore, the majority of research is carried out using the Wi-Fi signals as the source of illumination for the passive sensing system. Before moving towards the designing of the passive radars for indoor use, following are the some facts to remember:

- Indoor surroundings are composed of small area, i.e., rooms, galleries etc. compared to the outdoor.
- The medium for RF signals varies from walls to air. This causes nondeterministic power variation in received signal, multipath effect, clutter, etc.
- Target bodies are usually humans and their limbs which are not good RF reflector for low radio frequencies.

From these facts, it is concluded that positioning of target bodies measuring through path delay is quite un-convenient because of small distances. As, target bodies are humans themselves and their average indoor speed is quite low. So, Doppler frequency shift would be not more than 10-15 Hz. However, Received Signal Strength Index (RSSI) is good candidate in positioning and tracking of moving body but this need survey of place beforehand. Other than RSSI, channel state information (CSI) also used for tracking target objects in buildings. Moreover, this approach also reduce the hardware of passive sensing system to just single receiver.

Recently, a lot of work has been done in indoor passive sensing system based on classical approach as well as CSI approach. Using the classical passive sensing system, (Colone et al., 2012) developed Inverse Synthetic Aperture Radar (ISAR) based model for high resolution cross-range profiling of moving target objects using Wi-Fi signals. To take this to the next step, (Chetty et al., 2011) presented the Wi-Fi based bi-static passive radar model for through the wall personnel surveillance. The echoes from the moving bodies contain Doppler frequency shift. Using this concept, (Chen et al., 2015) developed Doppler only indoor passive Wi-Fi tracking methods based on high Doppler resolution passive radar. They worked on two filters i.e. extended Kalman filter and the sequential importance resampling (SIR) particle filter for target tracking.

Other than just target tracking, the majority of indoor applications required activity recognition and classification. Therefore, machine learning as well as deep learning is required in the modeling of such applications. In the same context, (Tan et al., 2014) created practical system reliable for extracting the micro-Doppler pattern from the indoor motion profile. These profiles include waving of hand, falling from chair, typing, etc. (Chen et al., 2016a) extended this system for classification of six daily routine motion profiles. They used two surveillance receiver instead of one for obtaining 2-D motion profile. The Sparse Representation Classifier (SRC) opted for the classification of activities based on micro-Doppler patterns of activities. Rather than using micro-Doppler patterns, (Chen et al., 2016b) used phase variation patterns for classification of medical activities i.e. tremor, breathing, etc. using the machine learning techniques. Moreover, (Khan et al., 2017a) used Convolutional Neural Network (CNN) model on RSSI patterns for classification of breathing activity as well as used random forest estimator for breathing rate estimation using passive Wi-Fi sensing system. Similarly, (Iqbal et al., 2018) classified basic human movements i.e. forward motion, backward motion and no motion from their RSSI pattern using CNN model.

Almost, every CSI-based indoor passive sensing system is based on machine learning or deep learning. (Wang et al., 2017) purposed CSI-based fingerprint for indoor localization incorporating deep learning framework. Their system has two stage i.e. training stage and test stage. This paper also emphasized over the fact that CSI pattern contain more motion profile information than the RSS pattern. They also developed the methodology named as CARM (CSI based human Activity Recognition and Monitoring system), which can recognize and classify the human activities by just using Laptop receiver. The main problem in activity classification is that at different orientations same activity dies not produce same CSI- pattern. To address this issue, (Virmani & Shahzad, 2017) developed system WiAG (a Wi-Fi based Agnostic Gesture recognition system) which need training data in just one orientation rather than in all orientations. This intelligent system itself produce the CSI-pattern for all other orientation and save the time in training of model.

Machine Learning-Based Passive Sensing System for Indoor Scenarios

Now, the both classical and CSI-based design for the indoor passive sensing system will be discussed.

Classical Design

This system is quite different from the classical design of outdoor passive sensing system. As, the attributes of phases i.e. reception, processing and detection are totally specific to the indoor environment as shown in Figure 4.

Figure 4. ML-based classical passive sensing system

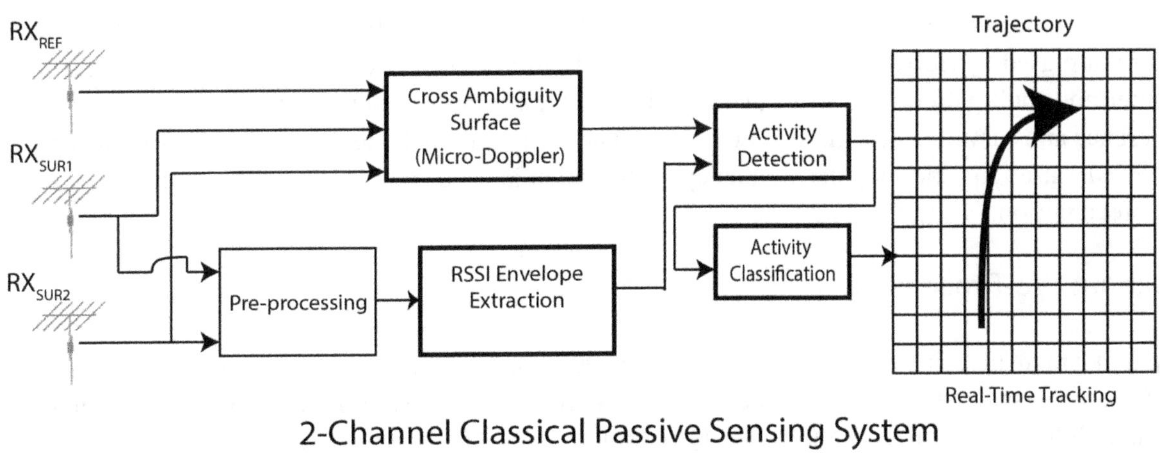

Reception Phase

This stage of passive sensing system receives the direct echoes and echoes bounced from the target objects of the transmitter signals and provides the reference and surveillance channel to the processing stage. As, the indoor scenarios are composed of small place and target bodies are not far away from the transmitter (access point) and passive sensing system. Thus, even maximum path delay are in nano-seconds. Therefore, if system is designed to also receive the delayed copies of the transmitter signal then required sampling frequency will be around ~20 MHz or more for path delay of 50 nsec. Mathematically,

$$f_s = \frac{1}{\tau_{\min}} = \frac{1}{50n\,\mathrm{sec}} \cong 20\,Mhz\,.$$

This concludes that indoor passive sensing system will not give best results to track or identify target object using path delayed echoes of the transmitter. Thus, only micro-Doppler and RSSI variation will be able to work for indoor system. Now, the reception stage can have low sampling frequency (f_s MHz up to 1 MHz) to work with indoor sensing system as we are only considering micro-Doppler and RSSI variation.

Processing Phase

Two prominent processing methods are proposed by the research community, which are as follow:

- With pre-processing
- Without pre-processing

The accuracy is expected to increase by implementing both processing methods on the same passive RF sensing system as shown in figure 4. However, there is no considerable work/analysis has been done on this topic.

With Pre-Processing

In this module, adaptive filtering tools like LMS adaptive filter and ECA (extensive cancellation Algorithm) are used by (Iqbal et al., 2018) as shown in Figure 5 to filter out direct path interference (DPI), clutter and multipath effect. After the filtration stage, complex conjugate multiplication of the clean reference and surveillance signal is carried out. The resultant would be the time series signal composed of RSSI variations and Doppler frequency shift, which is used in next stage for activity classification and recognition.

Without Pre-Processing

System which are using deep learning or machine learning model for detection are usually works without pre-processing stage. Because, according to (Chen et al., 2016a), DPI variation also contains the reliable information regarding the activity/motion profile. Therefore, pre-processing will do more bad

Figure 5. Passive RF sensing system with pre-processing block

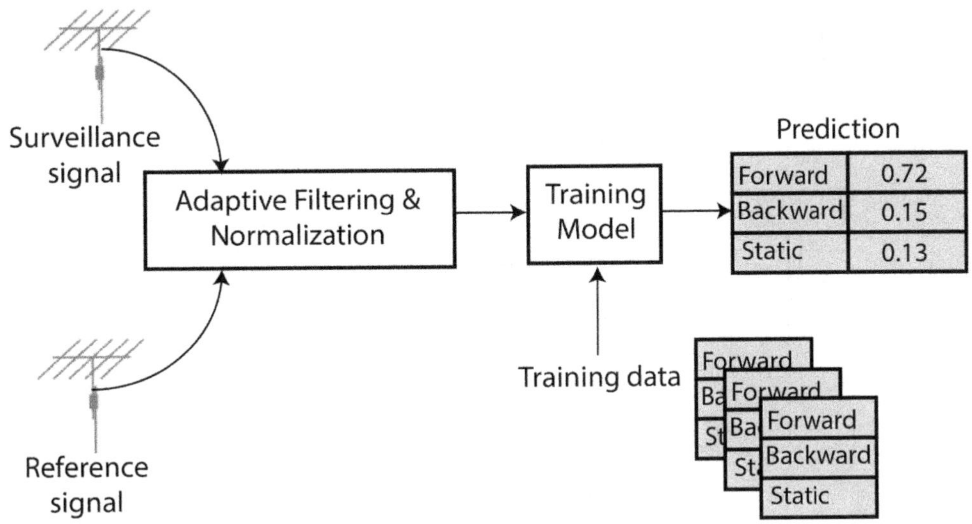

rather than good. Afterward, the batch processing technique of CAF is used to extract micro-Doppler and phase variation due to disturbance/activity in the surroundings. Single or few range bins are used in the modified CAF technique.

Batch processing technique of CAF or modified CAF is formulated to decrease the computation in computing with increasing the frequency resolution of CAF. This help to magnify the little details hidden in Doppler region close to zero line in CAF, which is the foremost design requirement of indoor passive sensing system. Afterward, the resultant time series signal is used in detection stage for activity classification and recognition.

Detection Phase

In the detection phase, the time series signal is first sliced and then activity detection algorithm provides the information regarding the initial point and end point of the motion profile. Afterward, the motion profile is adjusted to fixed sample size (Depend on the requirement of trained model) by decimation or interpolation block. Then, these unknown motion profiles are tagged by trained model.

The trained model can be based on machine learning algorithm i.e. Sparse Representation Classifier (SRC), Support Vector Machine (SVM), etc. Moreover as stated in this (Iqbal et al., 2018; Khan et al., 2017b), Convolutional Neural Network (CNN) model with 1-D input argument can also be used for the identification of motion profile.

CSI-Based Design

The Channel State information (CSI) is refer as the channel properties of communication link. It represent the combined effect of multipath, power decay with distance, clutter, etc., present in the channel.

Let, $\overrightarrow{X_i}$ and $\overrightarrow{Y_i}$ denote the transmitted and received signal vectors respectively. Then, we have

Figure 6. Basic architecture of CSI-based passive sensing system

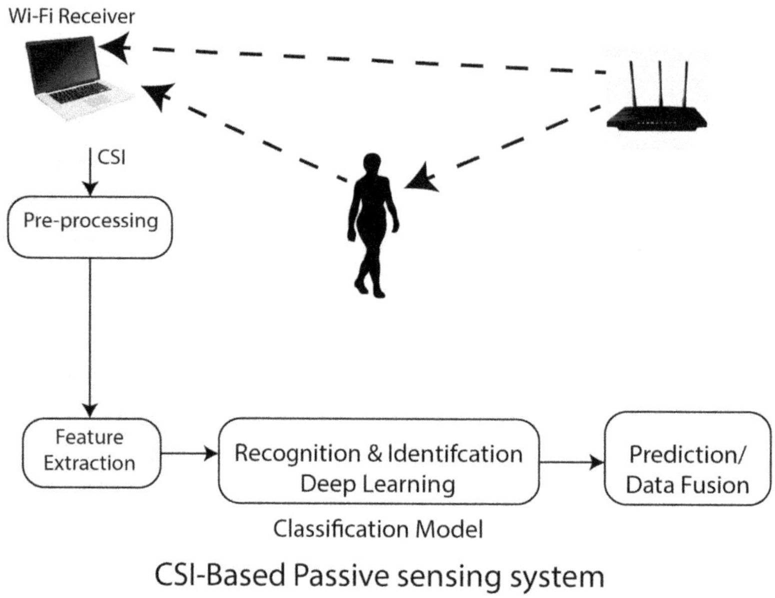

$$\overrightarrow{Y_i} = CSI.\overrightarrow{X_i} + \vec{N}$$

where, $\vec{N}$ is the additive white Gaussian noise and CSI represents the channel's frequency response.

In order to extract CSI of any communication channel, the dedicated hardware is required to record physical layer (PHY) information. However, in case of Wi-Fi, almost every laptop is equipped with NIC (Network Interface Controller) card which can provide CSI information of wired channel as well as of wireless channel by accessing the device driver. Now, it make things easy for obtaining CSI for passive Wi-Fi sensing system.

The basic architecture of CSI-based passive sensing system is shown in Figure 6. The reception stage for CSI-based Wi-Fi sensing systems is straight-forward and carried out by the integrated mobile/laptop NIC cards. Now, remaining stages of CSI-based passive sensing system are given below:

Processing

The main function of the processing is to attenuate noise present in the raw CSI data. Mostly, the average is computed over the CSI values from multiple antennas to reduce noise margin. However, (Wang et al., 2017) skipped the processing part and used all the CSI values for indoor fingerprint to exploit the diversity of the channel.

The power of deep learning can be estimate from the fact that there is no more processing is required. Rest of things are taken care by deep learning model.

Deep Learning Model

There will be two phases:

1. Offline phase (Training phase)
2. Online phase (Test phase)

Offline phase or training phase is composed of gathering tagged CSI raw data and using it for training of the model. After successful training, the CSI raw data other than used in training is used to determine the accuracy of the model. This phase is called test phase.

The (Wang et al., 2017) purposed the DeepFi (deep learning model) for the CSI-based Fingerprinting for indoor localization. The model consist four layers of deep network and the number of neurons are gradually increasing in upper layers to reduce the dimensions of CSI data. As, the CSI characteristics varies from location to location. Thus for each location, the model weights between connected layers are also unique. These weight are exploited as fingerprints to facilitate the positioning and localization. The whole implementation is not discussed here. Interested readers are encouraged to read this paper for better understanding.

APPLICATION OF PASSIVE RF SENSING SYSTEM

There are some prominent applications of passive RF sensing system, which are designed by incorporating deep learning frame work in passive sensing system. Applications are as follows:

1. Respiration/Breathing estimation
2. Activity recognition
3. Motion profile classification

Respiration/Breathing Rate Estimation

With the passage of time, the requirement of non-invasive medical diagnosis drastically increasing. People want to know about their health condition without wearing any medical instruments. The one of the health indicator is respiration/breathing rate.

Figure 7. CNN model for detection of breathing activity in passive sensing system

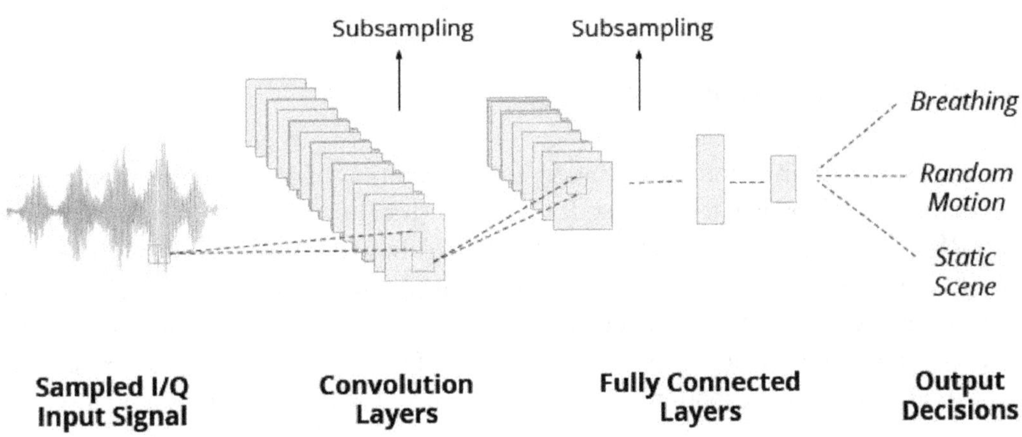

Figure 8. Accuracy of CNN models with variations in kernel depth

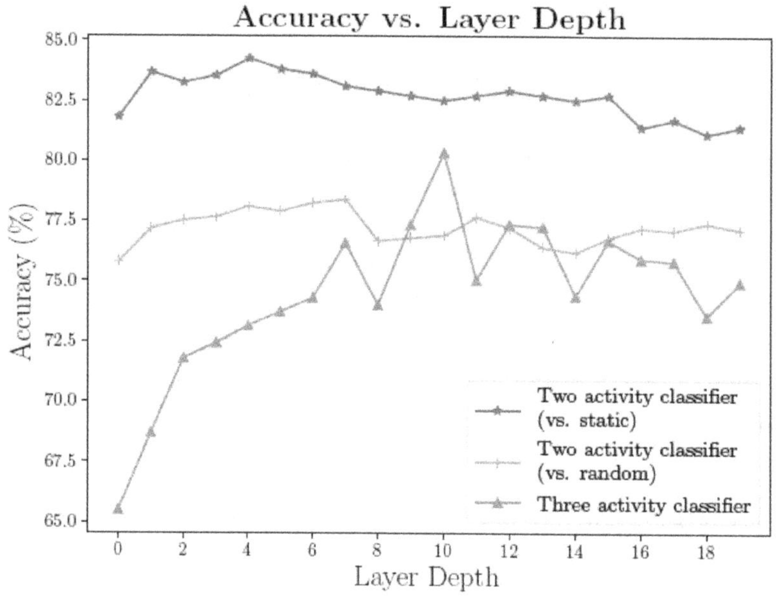

The novel solution for this application is stated in (Khan et al., 2017a), which use the Wi-Fi passive sensing system incorporating deep learning framework. Specifically, CNN model as shown in Figure 7 is used for detection of breathing activity and random forest estimator for determining the breathing rate. The Figure 8 shows the relation of model accuracy vs kernel layer depth stated in (Khan et al., 2017a). This helps in determining the optimal number of layers required for better accuracy with least computation.

Activity Recognition in Health Monitoring System

There is huge research going on in designing health monitoring system for patient rooms. The passive sensing based solution is purposed by (Khan et al., 2017b), which can classify the activities of the patient i.e. breathing, Tremor, fall, etc. They utilized the pattern of phase variation for classification of activity, whose results are shown in figure 9. It shows that fall detection has the highest accuracy, while breathing has the least. Moreover, accuracy drastically decreased by shifting from the LOS to NLOS environment in breathing measurement.

Classification of Basic Motion Profiles

This passive sensing system is for classification and recognition of basic indoor motion i.e. forward, backward and no motion by using CNN model stated in (Iqbal et al., 2018). The system exploits the variation in RSSI and frequency Doppler shift for classification of motion profiles.

The results from online testing of model are shown in figure 10. In real-time testing of this model, the setup receives signals for two seconds and then predicts the movement profile. After excessive evaluation, accuracy of the model in real-time is around 70%. The model fails to correctly predict the transitions (deceleration) of the motions. The signals received at the transitions are closely resembled

Figure 9. Accuracy comparison of health activities. Breathing rate accuracy is determined by comparing actual breaths per minute to monitored breaths. Tremor classification accuracy is determined the ability to classify slow (4-7Hz) and fast (7-11Hz) tremor

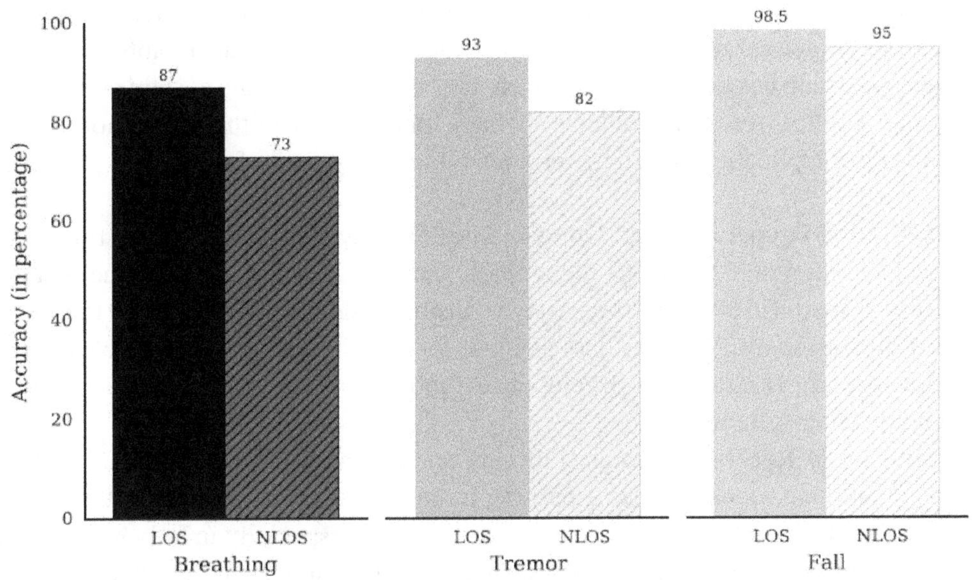

Figure 10. Results from online evaluation of the purposed deep learning model

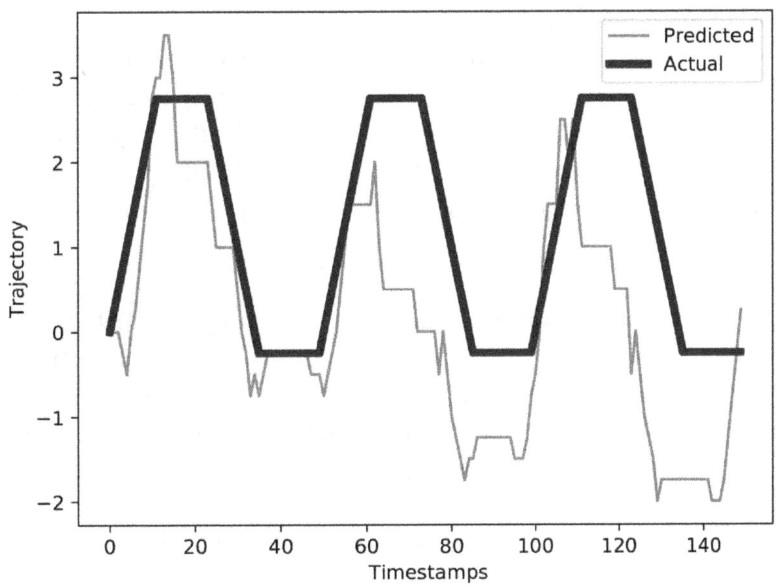

to the motion profile of opposite movement. In order to mitigate this issue, (Iqbal et al., 2018) stated that the previous predictions from the model should be considered along with the current predictions.

FUTURE RESEARCH DIRECTIONS

Deep learning based passive sensing is a broad research area and has potential to offer countless new and innovative ideas to ease the daily life problems such as home surveillance, patient monitoring, gesture communication, target localization, activity recognition, through-the-wall monitoring and movement tracking. As deep learning has been started to be applied in passive sensing system only few years ago, a lot of engineering as well as research problems are there for designing of the applications stated as above.

Some of the key complications faced by research community are as follows:

- **Optimization of Deep Learning Models:** The majority of pre-existing Artificial Intelligence (AI) models are designed for image recognition and classification in which the input data is 2 or 3 dimensional matrix. However, the passive sensing system requires the AI models which will take the n-dimensional time series data as input for activity recognition and classification. Thus, whole new research is needed for developing AI models which will be specifically optimized for the passive sensing system.
- **Pre-Processing of Received Signals:** Filtering and slicing of received signals, before it is used by the deep learning model, is also emerging issue. As received signals contain DPI, scattering and multipath which pretty much arise in random manner especially in indoor scenarios therefore these effects should be eliminated. The current pre-processing systems are computationally expensive which cause bottleneck in the performance. However, CSI-based passive sensing systems are one of their kind, as they don't have any pre-processing block and still work like charm in activity recognition and classification. But, not every transceiver allows the extraction of CSI which limit its use for research purpose only.
- **Reduction in Hardware Devices:** Nowadays, the research regarding passive sensing system is mostly carried out for the indoor scenarios. Thus, the portability and compactness of the hardware is really a big issue. The research is now carried out to make these system as compact as possible to install in regular homes or commercial areas. Some of the mobile companies are also indulged to integrate such system in smart phones, which will bring revolution in domain of passive sensing systems.

Furthermore, 5G will be expected to be launched successfully in 2019-2020. The localization of mobile users is one of the feature of 5G technology. Now, question arises that can passive sensing system incorporating ML and/or deep learning models be able to address this issue? Also, big mobile companies are doing research in developing modules which can estimate the surrounding activities other than by using camera. ML based solutions of passive sensing systems are compact in nature and required less computation. Therefore, they are expected to be the best candidates for implementing such smart modules. But, still extensive research is required to make the ML-based passive sensing system reliable for such applications.

CONCLUSION

This chapter provided detailed history, description, design metrics, design scheme and applications of passive sensing systems. The upcoming passive sensing system are designed by using Deep learning/

Machine learning models instead of matched filters based on signal processing. They are efficient in computation and better in accuracy than the systems based on signal processing techniques. Both the indoor and outdoor applications of passive sensing system i.e. target positioning, activity recognition, through-the-wall monitoring, etc., are adopting the deep learning models for their detection phase. The novel approaches discussed in the chapter for designing of both indoor and outdoor passive sensing systems based on machine learning provided enough knowledge to develop and analyze any intelligent RF sensing system. This chapter also discussed latest trends and methods for boosting performance of ML-based passive sensing system, which give diverse choices to research community for choosing their future research directions.

REFERENCES

Berger, C. R., Demissie, B., Heckenbach, J., Willett, P., & Zhou, S. (2010). Signal processing for passive radar using OFDM waveforms. *Signal processing for passive radar using OFDM waveforms, 4*(1), 226-238.

Chen, Q., Chetty, K., Woodbridge, K., & Tan, B. (2016b). Signs of life detection using wireless passive radar. *International Radar Conference (RadarConf)*. Philadelphia: Academic Press. 10.1109/RADAR.2016.7485313

Chen, Q., Tan, B., Chetty, K., & Woodbridge, K. (2016a). Activity recognition based on micro-doppler signature with in-home Wi-Fi. *IEEE International Conference on e-Health Networking, Applications and Services (Healthcom)*. Munich, Germany: IEEE. 10.1109/HealthCom.2016.7749457

Chen, Q., Tan, B., Woodbridge, K., & Chetty, K. (2015). Indoor target tracking using high Doppler resolution passive Wi-Fi radar. *IEEE International Conference on Acoustics, Speech and Signal Processing (ICASSP)*. Brisbane, Australia: IEEE. 10.1109/ICASSP.2015.7179036

Chetty, L., Smith, G. E., & Woodbridge, K. (2011). hrough-the-wall sensing of personnel using passive bi-static WiFi radar at standoff distances. *IEEE Transactions on Geoscience and Remote Sensing, 50*(4), 1218–1226. doi:10.1109/TGRS.2011.2164411

Chodos, A., Ouellette, J., & Tretkoff, E. (2006, April). April, 1935: British Patent for Radar System for Air Defense Granted to Robert Watson-Watt. *APS NEWS: This Month in Physics History, 15*. Retrieved from https://www.aps.org/publications/apsnews/200604/history.cfm

Colone, F., Pastina, D., Falcone, P., & Lombardo, P. (2012). Wi-Fi-based passive ISAR for high-resolution cross range profiling of moving targets. *European Conference on Synthetic Aperture Radar (EUSAR)*. Nuremberg, Germany: Academic Press.

Del-Rey-Maestre, N., Jarabo-Amores, M., Mata-Moya, D., Barcena-Humanes, J., & Ael-Hoyo, P. D. (2017). Machine learning techniques for coherent CFAR detection based on statistical modeling of UHF passive ground clutter. *IEEE Journal of Selected Topics in Signal Processing, 12*(1), 104–118. doi:10.1109/JSTSP.2017.2780798

Griffiths, H., & Willis, N. (2010). Klein Heidelberg – the world's first modern bi-static radar system. *IEEE Transactions on Aerospace and Electronic Systems, 44*(4), 1571–1588. doi:10.1109/TAES.2010.5595580

Iqbal, S., Iqbal, U., Hassan, S. A., & Saleem, S. (2018). Indoor motion classification using passive RF sensing incorporating deep learning. In *IEEE 87th Vehicular Technology Conference (VTC)*. Porto, Portugal: IEEE.

Khan, U. M., Kabir, Z., & Hassan, S. A. (2017a). A deep learning framework using passive WiFi sensing for respiration monitoring. *IEEE Global Communication Conference (Globecom)*. Singapore: IEEE. 10.1109/GLOCOM.2017.8255027

Khan, U. M., Kabir, Z., & Hassan, S. A. (2017b). Wireless health monitoring using passive Wi-Fi sensing. In *International Wireless Communications and Mobile Computing Conference (IWCMC)*. Valencia, Spain: Academic Press.

Krysik, P., Gajo, Z., Kulpa, J. S., & Malanowski, M. (2014). Reference channel equalization in FM passive radar using the constant magnitude algorithm. In *International Radar Symposium (IRS)*. Gdansk, Poland: Academic Press. 10.1109/IRS.2014.6869284

Malanowski, M. (2012). Algorithm for target tracking using passive radar. *International Journal of Electronics and Telecommunications*, *58*(4), 345–350. doi:10.2478/v10177-012-0047-x

Schroeder, A., Edrich, M., & Winkler, V. (2013). Stationary cassidian FM passive radar demonstrator for 24/7 operation and sensor cluster measurements. In *International Radar Symposium (IRS)*. Dresden, Germany: Academic Press.

Shi, Y. F., Park, S. H., & Song, T. L. (2017). Multi-target tracking in cluttered environment for a multistatic passive radar system under the DAB/DVB network. *EURASIP Journal on Advances in Signal Processing*, *2017*(1), 11. doi:10.118613634-017-0445-4

Skolnik, M. L. (2007). Fluttar DEW-Line Gap-Filler. In H. Griffiths & N. Willis (Eds.), *Advances in bi-static radar* (pp. 35–47). Raleigh, NC: Scitech Publishing, Inc.

Tan, B., Woodbridge, K., & Chetty, K. (2014). A real-time high resolution passive WiFi doppler-radar and its applications. In *International Radar Conference*. Lille, France: Academic Press. 10.1109/RADAR.2014.7060359

Virmani, A., & Shahzad, M. (2017). Position and orientation agnostic gesture recognition using WiFi. In *International Conference on Mobile Systems, Applications, and Services (MobiSys)*. Niagara Falls, NY: Academic Press. 10.1145/3081333.3081340

Wang, X., Gao, L., Mao, S., & Pandey, S. (2017). CSI-based fingerprinting for indoor localization: A deep learning approach. *IEEE Transactions on Vehicular Technology*, *66*(1), 763–776.

Xiao-Yong, L., Jun, W., & Jue, W. (2015). Robust direction of arrival estimate method in FM-based passive bi-static radar with a four-element Adcock antenna array. *IET Radar, Sonar & Navigation*, *9*(4), 392–400. doi:10.1049/iet-rsn.2014.0270

Zeng, Y., Ai, X., Wang, L., Wang, X., & Zheng, G. (2016). Experimental research of dual-polarization passive radar based on DTTB signal. *Journal of Computer and Communications*, *4*(3), 101–107. doi:10.4236/jcc.2016.43016

Zhang, C., Wu, Y., Wang, J., & Luo, Z. (2016). FM-based multi-frequency passive radar system. In *IEEE International Conference on Signal Processing, Communications and Computing (ICSPCC)*. Hong Kong, China: IEEE.

Zhihai, Z., & Tao, S. (2008). *Research on fast computation of ambiguity function*. Sanya, Hainan, China: Congress on Image and Signal Processing. doi:10.1109/CISP.2008.458

ADDITIONAL READING

Del-Rey-Maestre, N., Jarabo-Amores, M., Mata-Moya, D., Barcena-Humanes, J., & Ael-Hoyo, P. D. (2017). Machine learning techniques for coherent CFAR detection based on statistical modeling of UHF passive ground clutter. *IEEE Journal of Selected Topics in Signal Processing*, *12*(1), 104–118. doi:10.1109/JSTSP.2017.2780798

Wang, X., Gao, L., Mao, S., & Pandey, S. (2017). CSI-based fingerprinting for indoor localization: A deep learning approach. *IEEE Transactions on Vehicular Technology*, *66*(1), 763–776.

KEY TERMS AND DEFINITIONS

Ambiguity Function: An ambiguity function is a two-dimensional function of time delay and Doppler frequency $X\left(\tau, f\right)$ showing the variation in returned pulse due to the receiver matched filter.

CSI: In wireless communications, channel state information (CSI) refers to known channel properties of a communication link. This information describes how a signal propagates from the transmitter to the receiver and represents the combined effect of, for example, scattering, fading, and power decay with distance.

Deep Learning: Deep learning is part of a broader family of machine learning methods based on learning data representations, as opposed to task-specific algorithms. Learning can be supervised, semi-supervised, or unsupervised.

DPI: Direct path interference are composed of direct echoes of the transmitter signals received by the surveillance receiver in passive sensing system.

Multi-Path: In wireless telecommunications, multipath is the propagation phenomenon that results in radio signals reaching the receiving antenna by two or more paths.

Chapter 13
CHEERBOT:
A Step Ahead of Conventional ChatBot

Chintan Bimal Maniyar
Charotar University of Science and Technology, India

Chintan Bhatt
Charotar University of Science and Technology, India

Tejas Nimeshkumar Pandit
Charotar University of Science and Technology, India

Dewanshi Harishankar Yadav
Charotar University of Science and Technology, India

ABSTRACT

The objective of this chapter is to discuss the authors' interaction and involvement with technology and bots, opening a whole new and wide scope of possibilities while letting bots comfort us. The prevalence of bots and automation is increasing by every passing day – Cortana, Siri have been here for a long time and have now been overtaken by Alexa and other home automation systems that provide a two-way dialogue conversations. This chapter explores the possibilities of creating bots that can cheer us up when we are sad. Analyzing the semantics of our sentences and analyzing the pitch of our voice to identify our emotional state, and then providing an n-way dialogue conversation, relevant to the then existing context, instead of the mundane two-way dialogue conversation is the lucid content of this chapter. Summing it up, this chapter examines the possibility of creating bots that can serve as an emotional support to us humans.

INTRODUCTION

Our society is now being reshaped by rapid advances in information technology, leading towards the age of auto-intelligence. Over the past couple of years, we have been swarmed by various new kinds of bots. Bots are automated with software or hardware that are powered by the advances in Artificial Intelligence technologies. The two most appealing disciplines of AI are Machine Learning and Deep Learning. These

DOI: 10.4018/978-1-5225-7458-3.ch013

have a great wave of change coming in for our society, reshaping the existing norms and trends –some of which are making things easier to be done - automation and maybe percipient emotional support up to a certain level. Advances in deep learning and other machine learning algorithms are currently causing a tectonic shift in the technology landscape. Technology behemoths like Google, Microsoft, Amazon, Facebook and Salesforce are engaged in an artificial intelligence (AI) arms race, gobbling up machine learning talent and startups at an alarming pace. They are building AI technology war chests in an effort to develop an insurmountable competitive advantage.

Historically, AI has been defined by the ability of a computer to pass the Turing test, which meant the public wasn't going to be happy with AI until they had a walking, talking robot. Anything less was considered a failure. We're still far away from creating this kind of general AI, but we're already solving some types of advanced problems with machine learning, a subset of AI proper. Rather than focus on general intelligence, machine learning algorithms work by improving their ability to perform specific tasks using data. Problems that used to be the exclusive domain of humans – computer vision, speech recognition, autonomous movement – are being solved today by machine learning algorithms.

In fact, machine learning has become such a huge area of focus and, for all practical purposes, the term machine learning has become synonymous with AI. Ultimately this is a good thing. The more consumers and companies start associating the term AI with real-world applications of machine learning like self-driving cars, the more they realize that AI is a real thing. It's here to stay, and it holds the promise of reshaping the technology landscape over the next several years.

The best instance to epitomize the facilitation of Machine Learning to our daily lives is a chatbot, or, better known as a *personal assistant*. With the help of this advanced technology in bots, we are able to interact with devices and gadgets at our home and offices using speech. These bots give ease to our lives, but they cannot make the human like conversation. These bots just follow the commands, which are instructed by us, they are not intelligent enough to understand the sentiments. Due to this reason, they cannot emotionally attach to us when we are unhappy and wanted to share our feelings.

To make it happen, we are introducing the CHEERBOT - "A bot which tries to cheer you up when you are sad". CHEERBOT will try to make the conversation more like humans, interacting with her is easy for any person. The main aim of CHEERBOT is to make people happy or make them feel good when they are not happy by talking with them.

MOTIVATION AND INSPIRATION

This project initially started as a vague embryo of idea – to create something that would break the stereotype of being 'addicted' to the technology and its catering devices. In today's age, we are so driven by technology extensively - which is a good thing. We are embracing the new changes and developments to facilitate our lives. But at the same time, we apparently end up blaming technology for out casting us and barring ourselves from social interactions. We end up saying we don't have anyone to talk to just because of being so into technology. Then why not use the technology itself to solve this problem? This social stigma of barring oneself from social interactions allegedly because of technology is the exact problem that our project has addressed, and solved using the technology itself.

This was just an ignitor towards a wider perspective that would then serve as the main purpose of pursuing CHEERBOT. The impetus of it is right as in its namesake – to cheer everyone up, but delving a little deep, when does anyone need cheering up? There are so many reasons why someone would be sad, especially in today's age – loneliness, depression, anxiety etc. But the worst outcome of these sentiments is outrageously alarming.

In this age of youth and energy, advancement and opportunities, even in an environment of such vigor, energy and excitement, a survey by the WHO states that 1 out of 3 people, globally, suffer from anxiety and depression. This is a global societal issue – stress and unpleasantness getting the better of people and keeping them from being happy. Prolonged depression and anxiety, loneliness, recklessness can lead to a very unfortunate turn of incidents – suicide. All over the world, according to a global survey, there are 123 suicides a day. Although there is single reason as to why someone would try to kill themselves, several surveys put depression as the top risk factor (about 90% of the people have a mental illness when they die).

The best way anyone can feel even a bit better is by talking about what they feel – bullying, negligence from loved ones, social isolation are some of the major factors contributing to depression. Talking about these things can definitely help the victims – for that is what suicide prevention helplines do – have good Samaritans talk to people in need, to help them. But there are certain things that everyone might not be comfortable with sharing with anyone – any other person in fact apart from their own selves, what then? We always need a listener!

This did it. We were determined to give the world such a listener and at the same time, punch the social stigma of technology being a leech, right in its face. Evidently speaking, the society itself was the motivation as well as the inspiration behind CHEERBOT.

WHAT ARE CHATBOTS?

Chatbots are a software or a computer program that you interact through in conversational way via a chat or voice interface. Alternatively, chatbots are an intelligent interface that can make conversations in a coherent manner (Behera, 2016). It is a personal assistant integrated into messengers, applications or websites. The task performed by them is to follow the command or to respond to the command. Generally, chatbots are used for a dialog system for purposes like customer services or information acquisition. Unlike goal oriented systems, chatbots do not have any specific goal that guides the conversation - rendering some proclaimed evaluation metrics such as task completion rate quite inappropriate(Yu et al., 2016). A chatbot works on the basis of a set of rules by matching the patterns or keywords in an input string, then pull a reply from the database or uses artificial intelligence to converse with humans. Current chatbots use methods varying from database retrieval to machine translation(Ritter et al., 2011) and from retrieval based response selection(Banchs and Li, 2012) to recurrent neural network sequence generation(Vinyals and Le, 2015). But despite this diversity, even after a thorough cross referencing of responses from more than one database, the variability of the resulting database responsible for providing responses is surprisingly low(Yu et al., 2016). Most of the time, conventional chatbots are not able to identify the sentiments of a person to whom they are chatting, they are unable to understand the emotions and feelings of a person. Chatbots are made for a dialog system between the user and the chatbot, although it is using natural language processing and artificial intelligence, but they are not made to understand their feelings.

Speaking in a typical technological lingo, a chatbot is another way of humans interacting with software. While there are overlaps with functionality offered by websites and apps, interacting with a chatbot is different to interacting with a website or with an app. The best examples include Google Voice Assistant,

Microsoft Cortana, Apple Siri and thus. These are voice enabled chatbots – capable of performing an auditory input output conversation. But these do not *usually* offer an 'n-dialogue conversation'. They offer a bi-dialogue conversation. You say a command, they acknowledge it, interact with the underlying OS, achieve your task and give you the acknowledgement of the same. Every process between you feeding in the input (speaking) and the bot giving you the acknowledgement of its completion happens in the background in an incredibly fast fashion. This is what makes it seem as a conversation. For instance, *'Okay Google, wake me up after 10 minutes!'* → *'Sure, I've set an alarm of 21:10!'* Now, this seeming conversation which actually takes no longer than 10-15 seconds, serves evidence that the Google Voice Assistant did just set an alarm in less than 5 seconds.

Another simple example of chatbots would be Natasha (Hike), Ruhh (Microsoft India) – these are text based chatbots. They give you the feel and illusion of chatting with an actual person on the other side. There are platforms such as Streebo Chatbot Builder, DialogFlow etc. which provide an excellent interface – extremely user-friendly – to build a chatbot.

A chatbot once built can then be integrated with a system to provide a continuous, simplified service to the user. The voice enabled chatbots mentioned earlier - Google Voice Assistant, Microsoft Cortana, Apple Siri - provide the service of a personal assistant to their users on their machines. The text enabled chatbots are integrated with working systems in real time to balance the user load and to direct the users towards the services that they need. For e.g. a chatbot window pops up on a travel website asking you about your trip and other details such as time and date of travel, budget, etc. and shows you relevant suggestions – all of this in a chat based text conversation maybe in a chat window. This spares the user the hustle of manually booking the tickets on the website as, at the end of the conversation the chatbot also shows a travel plan and offers to book the same. All of these things give the user the feel of interacting with a human while it is actually a smart snippet of code running quietly in the back. But this 'feel of interacting with other human' plays a very significant role here – the user feels comforted, as if talking to a person. Hence the presence of chatbots somehow comforts the user, which makes the user prefer the application with a chatbot instead of one without a chatbot. This shows the lucrative upside of chatbots.

WHAT IS CHEERBOT AND HOW IS IT DIFFERENT FROM EXISTING CHATBOTS?

'A good companion shortens the longest road' this quote is penned very rightly, saying that loneliness is always overcome by companionship. Suppose, if chatbots had the functionality of sentiment analysis, then they could have easily identified the mood of the user and would have further continued the conversation with the user, just like a companion. The CHEERBOT does exactly that in a more refined way.

Conventional chatbots are more like personal digital assistants, expecting and accepting commands from the user and eager to execute them in on their respective devices of residence. There are other chatbots such as chatbots for cultural heritage (Pilao et al., 2005), security training(Kowalski et al., 2013), blind high school students(Bigham et al., 2008) which are not just conventional chatbots - they operate within a specific scope and domain with a more concentrated objective. Similarly, CHEERBOT here is not just a conventional chatbot, it rooms the scope of sentiment analysis, emotion detection and emotional confrontation.

As mentioned in the previous section, chatbots as such offer a bi-dialogue conversation – accepting your command and acknowledging to you its execution. CHEERBOT offers an 'n-dialogue conversation' in text as well as auditory mode, one which goes on until you say a specific syllable - in this case, *'bye'*. Also, it serves as a listener. CHEERBOT is not a conventional chatbot accepting your commands and assisting you use your device. It is instead a chatbot that senses your emotional state based on how you converse and then tries to cheer you up – hence the name. Based on what the user types in (text mode), or even says (auditory mode), CHEERBOT tries to analyze the user's emotional state by keyword based sentiment analysis and voice pitch detection and accordingly builds up a conversation. If the user is sad and it seems so from the textual input or from his/her voice pitch, the reply is framed so that it could please the user – something funny or a gentle confrontation of the user not being alone.

These features are what make the CHEERBOT stand out from the crowd of conventional chatbots. Not only does it offer an 'n-dialogue conversation', bit it is a chatbot that provides emotional confrontation as well as moral support to the user based on his/her mood. If the user says that he/she is sad, the bot will come up with something funny or confronting. If the expresses that he/she is leaving in a sad state, the bot tries to make the user want to stay.

CONSTITUENTS OF THE CHEERBOT

Before getting into how the CHEERBOT functions, it is important to understand its constituents:

1. Google Text to Speech implementer
2. Google Speech Recognizer
3. NLTK implementer, tokenizer and sentence builder
4. Sentiment predictor
5. Driver program

GTTS: Google Text to Speech

gTTS (*Google Text-to-Speech*), a Python library and CLI tool to interface with Google Translate's text-to-speech API. Writes spoken mp3 data to a file, a file-like object (bytestring) for further audio manipulation, or stdout.

Features

- Customizable speech-specific sentence tokenizer that allows for unlimited lengths of text to be read, all while keeping proper intonation, abbreviations, decimals and more;
- Customizable text pre-processors which can, for example, provide pronunciation corrections;
- Automatic retrieval of supported languages.

GTTS is used by many people to develop products which should be independent of a single language i.e. it should not be dependent on only one language like English. For example,

GTTS System for the Spoken Web Search Task at MediaEval 2012(Varona et al., 2012).The project requirement was that it should be supported by all the languages which was made possible by GTTS. Another instance is of Albayzin 2010 Speaker diarization evaluation (Diez et al., 2010). Speaker Diarization partition the speech of a speaker and then merge it for same speaker to identify the speaker by their audio signals

Google Speech Recognition

Google Speech-to-Text enables the developers to transform audio to text by using advance deep-learning neural network algorithm on audio for conversion. Accuracy of Google Speech-to-Text ameliorates over the time. Other than converting Speech-to-Text, it provides the facility of auto-correcting the words, recognizing proper nouns and accurately punctuates the transcriptions. Furthermore, it has the ability to recognize 120 other different languages, to support multilingualism globally - user has to specify the language codes and it will automatically identify the correct language. It supports real-time streaming or pre-recorded audio, using Google's machine learning technology. All of these features are made available in an easy to use API - The Google Cloud Speech-to-Text.

Google Speech Recognition has been effectively and successfully used in many products like Google 411 which was voice-based calling service (Google Inc. 2007), Google Mobile App that includes voice search (Google Inc. 2008), Voice Command System using Raspberry PI (Kaur et al., 2016). As google has an uncountable word dictionary, it helps machine to understand the meaning of each word and ultimately the meaning of command that is given to a machine so that machine can react or reply back.

Given the wide expanse of features and the ease of usage and implementation, Google Speech Recognition is the best approach for CHEERBOT to identify the user's input in auditory mode.

NLTK: Natural Language Toolkit

The Natural Language Toolkit (NLTK) is a platform used for building Python programs that work with human language data for applying in statistical natural language processing (NLP). It contains text processing libraries for tokenization, parsing, classification, stemming, tagging and semantic reasoning. It also includes graphical demonstrations and sample data sets as well as accompanied by a cook book and a book which explains the principles behind the underlying language processing tasks that NLTK supports.

The Natural Language Toolkit is an open source library for the Python programming language originally written by Steven Bird, Edward Loper and Ewan Klein for use in development and education. It comes with a hands-on guide that introduces topics in computational linguistics as well as programming fundamentals for Python which makes it suitable for linguists who have no deep knowledge in programming, engineers and researchers that need to delve into computational linguistics, students and educators.

The library contains

- Lexical analysis: Word and text tokenizer
- n-gram and collocations
- Part-of-speech tagger
- Tree model and Text chunker for capturing
- Named-entity recognition

Is it possible for computer program to understand the natural languages like English? The answer is Yes. The technique from which a natural language can be processed is called NLP. NLP helps to process large amount of texts and to do analysis on that text. NLP has many applications. It is used in sentiment

Figure 1. The NLP Process

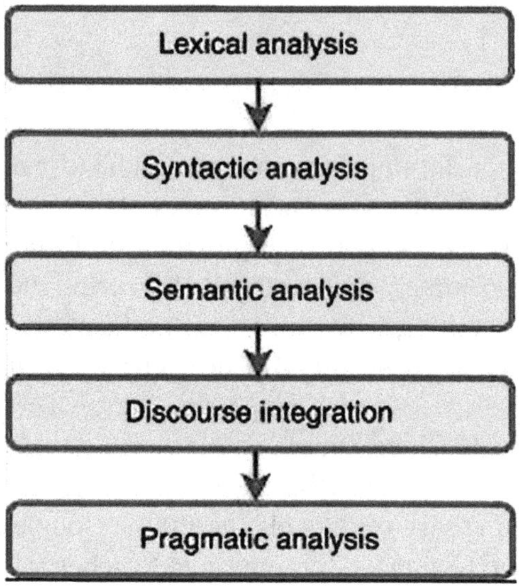

analysis which helps data scientists to find the response from comments on social media that is the response is positive, negative or neutral. Some applications require converting the natural language in some target representation which would help an algorithm to proceed further. One such application is our CHEERBOT. It requires to understand the meaning of a sentences given by the user with the help of which algorithm can generate a proper response. Therefor we need to understand the steps of NLP.

NLP consists of the following five steps:

1. Morphological Analysis
2. Syntactic Analysis
3. Semantic Analysis
4. Discourse Integration
5. Pragmatic Analysis

Morphological Analysis

Morphological means structural analysis. It is the analysis of morphemes of a word. It analyze the core part and affixes of a word. For example a word is "Unfortunately". In this word "un" is a prefix and "ly" is a suffix and "fortunate" is the core word. These distinctions have made in this phase. Another Example we can take as the word "cats". Morphological Parser parses this word and gives output as (cat + noun + plural).

Syntactic Analysis

This phase use the output of morphological analysis as an input to build a structural description of a sentence. This process is called "parsing". Parsing means extracting meaning from a sequence of words.

In this phase the sentence is converted into a structural form like a tree and all the words are given its meaning, or better say, a flat sentence is converted into a hierarchical structure.

Semantic Analysis

The tree like structure serves as an input to the semantic analysis. This phase logically processes this structure and identifies the most relevant elements from the sentence. In NLP, the most crucial part is understanding what is being said in a sentence and semantic analysis is therefore the heart of the NLP process. To interpret the meaning, this phase starts by reading all the words – nodes of the tree like structure, identifies the text elements and assigns them to their logical and grammatical role. Also, surrounding context is analyzed to accurately disambiguate the proper meaning of words which might have more than one meaning.

Discourse Integration

The meaning of a sentence may depend upon the sentence preceding it and may affect the meaning of the sentences that follow it. For example in a sentence consisting of the words like "he/she/it" and the respective name for "He" is given in the preceding sentences. Once the reference has been made we can also know exactly who is referred in the sentence.

For example two sentences are "James has two bikes. He also has a car." In the second sentence a reference is made from "He" to "James". Therefore the computer can understand that "John also has a car".

Pragmatic Analysis

This phase decides that what to do with the final result? Either save it or process it.

For an instance if a sentence is "Save xyz.docx file" then this sentence is converted into a command and the translation is done in this phase. In the same way if the sentence is a fact then it will be stored in the knowledge base or database and for some sentences it will generate a corresponding response.

SENTIMENT PREDICTOR

This constituent is the heart of the CHEERBOT. It contains logic to perform keyword based sentiment analysis based on user's textual input as well as the logic to impute the auditory input of the user to detect the voice pitch in order to refine the sentiment analysis of the user. It is implemented in Python 3.5.4.

DRIVER PROGRAM

Finally, this is the main program of the CHEERBOT, responsible for calling all the functions, seeking input, switching between the textual and auditory input modes. Essentially, it is the driver program which can be turned into an executable file for further distribution of the CHEERBOT.

HOW IT WORKS

CHEERBOT works in an ample flow of rhythmic steps. The topics preceding this one are fit prerequisites for the CHEERBOT. It is a step ahead of conventional chatbots, being able to perform sentiment analysis on its knowledge based out of user interaction. The input to CHEERBOT is auditory as well as text based.

In case of the text based input, it performs sentiment analysis based on the input provided by the user, purely a keyword based sentiment analysis – cross validating the user's input string and then determining whether the user is sad *or* not sad. This assertion here is very essential – since it is a bot that is cheering up, it needs to know whether the user is sad or not. So the analysis of the user input either results in a *sad* state determination or a *not sad* state determination. All other states are irrelevant for the CHEERBOT. There is a functionality of auditory input/output too – the user can directly speak to the bot and the reply is in an auditory form. But the analysis and determination is based purely on the text of the user's input. This write up describes the CHEERBOT v1.0.

The succeeding version shall carry an added functionality of voice pitch detection, hence modularizing and at the same time, bettering the sentiment analysis – on top of the keyword based sentiment analysis, determination would be further based upon the pitch of the user's auditory input. Naturally, this cherry works only with the auditory input/output mode because a voice clip is needed to recognize its pitch and strengthen the keyword based sentiment analysis.

Here is what happens in the CHEERBOT v2.0, auditory input/output mode – the keyword based sentiment analysis plays its part right in place, but apart from the text based analysis, the pitch of the voice of the user is analyzed, using voice intelligence. How does this help? Well, the pitch of our voice directly points towards the emotional state that we are in, at that moment. Hence, analyzing the pitch of user's auditory input proves helpful in making an inclination as to what his emotional state could be. Of course, here also, the emotional states include *sad state* and *not sad* state only, hence concluding whether the user needs cheering up. The analysis here is based on both the parametric outcomes – it is a cumulative result of what the keyword based sentiment analysis indicated as well as what the voice pitch analysis indicated. Hence, the second version does a better job at identifying the emotional state of the user and accordingly carrying on the conversation.

Before getting into how the CHEERBOT functions, it is important to understand its constituents:

1. Google Text to Speech implementer
2. Google Speech Recognizer
3. NLTK implementer, tokenizer and sentence builder
4. Sentiment predictor
5. Driver program

The technical prerequisites for CHEERBOT include a computing device embedded with a microphone, speaker, viewport, keypad/keyboard and a RAM of at least 2 GB running on a workable operating system. The outcome of this project can be integrated with any working system as a third party application.

Figure 2 concisely sums up the working of CHEERBOT.

To better understand its working, following is the flow of operation of the CHEERBOT:

1. Fetching Input
2. Semantically analyzing the input

Figure 2. Workflow

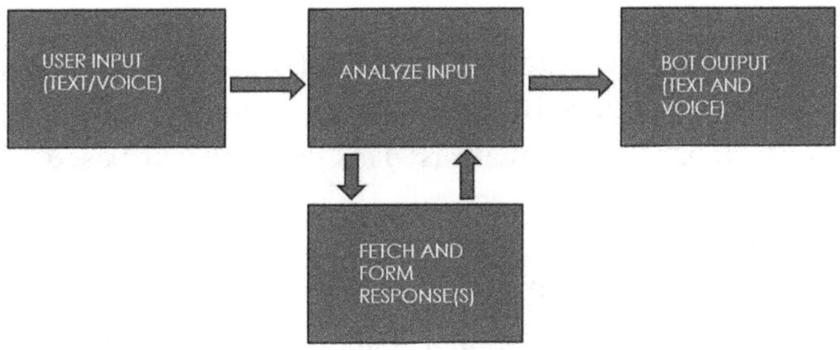

3. Imputing the input for analysis
4. Determining the nature of reply
5. Constructing a valid and a relevant reply based on the input
6. Presenting the reply

Fetching Input

The very first step to start any conversation is to be able to listen to the conversationee. This involves listening to what the user has to say. Fetching input is the very first step for starting an 'n-dialogue conversation' between the user and the bot.

There are two modes of input – a text based chat and a seemingly real auditory conversation. For the first method, the input is fetched simply from the console in the form of the string entered by the user in lieu of the chat conversation. The input is accepted in the form of string and stored in a temporary variable for interpretational processing.

For the second method which includes an auditory front, the input is via the microphone which records the audio clip of the user's assertion. This audio clip is then fed to the Google Speech Recognition module which identifies the syllables present in the audio clip and outputs a string of the same – hence we ultimately get a string from the auditory input. Again, this string is stored in a temporary variable for interpretational processing.

Semantically Analyzing the Input

The next step after fetching the input is making sense out of it. For this, we have a string rooming the user's input. This string is tokenized so that it can be semantically analyzed. The sentences in the string are structurally broken down after being tokenized. From these unit components, semantic analysis is performed in an attempt to understand the user's input.

For e.g. if the input string is 'How are you?', then the sentence is tokenized as follows: How-*adverb*, are-*helping verb*, you-*pronoun*. After tokenization, the helping verbs are dropped. So 'How are you?' becomes 'How you?' Now, 'How' has been tokenized as an adverb, so it is definite that the reply to this token has to be an adjective (say 'fine'). The token 'you' helps in deciding the grammatical narrative (first person/ second person/ third person). This process is achieved via NLTK. Thus, semantic analysis helps in understanding the intended meaning of the sentence.

Imputing the Input for Analysis

After the semantic analysis, we need to determine the emotional state of the user and hence the nature of reply that the bot must construct. But before analyzing the input for determination of user's emotional state, we need to prepare the input for that analysis. This goes for text-based input as well as the auditory input.

In case of text based input, for keyword based sentiment analysis, the string containing the input sentences needs to be split in individual syllables, so that each syllable can be cross validated with a set of predefined keywords indicating a *sad* emotional state. If the cross validation of each syllable results into a negative outcome, it indicates a *not sad* emotional state.

In case of auditory input, the audio clip is operated upon. It is divided into small chunks and the pitch frequency of each chunk is detected. Based on this frequency, the emotional state of the user is determined.

Determining the Nature of Reply

After the semantic analysis, the next step is to analyze the input based on its form – text based or auditory, to determine the emotional state of the user (*sad* or *not sad*) and hence the course of further conversation.

In case of text based input, keyword based sentiment analysis is performed and a determination of the emotional state of the user is made while in case of auditory input, apart from the keyword based sentiment analysis, voice pitch is also analyzed for a finer determination of the emotional state of the user.

In both the scenarios, there are two possible outcomes/ determinations of the user's emotional state – *sad* or *not sad*. Based on this determination, the type of reply is selected. Say if the user is *sad*, then a funny reply generation might be opted for and otherwise a straightforward reply generation to the user's input is opted for.

Constructing a Valid and a Relevant Reply Based on the Input

The next step after the determination of user's emotional state and the nature of reply is to construct the reply. This step is aided a lot by the second step which included tokenizing the sentences in the input string. Based on the nature of tokens and of course the emotional state of the user, a reply is generated in the form of string.

Say for example, the input string contains 'How are you?' and after tokenization, it turns into – How-*adverb*, are-*helping verb*, you-*pronoun*. According to the semantic analysis the reply should contain an *adjective* (in reply to the *adverb*), so an adjective is selected, say 'fine – *adjective*'. Now the *pronoun* ('you') helps in identifying the correct grammatical narrative – here it should be first person – 'I' (in reply to the third person narrative). And finally based on the helping verb and the grammatical narrative in the input, the helping verb for the reply is determined.

So, 'How are you?' becomes 'How you?' From 'How', 'fine' is determined. From 'you', 'I' is determined. So, 'How you?' generates 'I fine'. And eventually the fitting helping verb is amalgamated into the reply, forming 'I am fine'. This reply is returned as a string.

Presenting the Reply

The reply constructed in the previous step is in the form of a string.

In case of text based conversation mode, this string is printed out in on the chat console as a reply from the bot.

In case of auditory mode, this string is fed to the Google Text to Speech Implementer which converts this string into an auditory output which in turn is outputted from the device speaker.

ADDENDUM

This idea of CHEERBOT emphasizes the concept of *embracing technology to achieve emotional confrontation*. It proves to be a listener and an emotional confronter during hard times. A person who is on the verge of being adversely affected from loneliness or targeted isolation shouldn't keep the musings inside of oneself. But many a times, a feeling of violation keeps certain individuals from talking about what they are feeling. And not being able to talk is the worst of things that anyone should ever got through.

Following are some screenshots of the CHEERBOT console in execution.

Figure 3 illustrates the keyword based sentiment analysis and its result. As the user expresses his sadness the CHEERBOT tries to confront the user.

Figure 4 shows the pitch detection and analysis. Here a single audio clip of the user saying 'hey' is imputed and analyzed for the pitch frequency. Based on the frequency, the emotional state of the user is estimated – different emotional states reflect different pitches in a person's voice. Depending on the same an appropriate reply is constructed.

Figure 5 demonstrates the voice pitch detection and the following voice pitch based sentiment analysis. The user says 'hey' in a very low pitch, which prompts the CHEERBOT to check on the user's cheeriness.

Figure 6 demonstrates the confronting reply from the CHEERBOT when the user is in a *sad* state. The mode of conversation is auditory.

Figure 7 demonstrates how the CHEERBOT wants to make the user stay when he/she seems to be leaving in a *sad* state. The mode of conversation is auditory.

CHEERBOT is born out of technology and right now operates only on console. Figure 8 shows a proposed UI design.

CHEERBOT v2.0 supports text and auditory input methods. Apart from the proposed UI, enhancements in the future versions include support of regional, local and national languages other than English. In the text input method, future enhancements include the integration of smileys and emoticons in the replies generated by the bot as a whole.

CONCLUSION

All in all, CHEERBOT is an attempt to make everyone realize that they are never truly alone and a humble manifestation of technology to make this world a better place. It is a refinement of and from the conventional chatbots. CHEERBOT is a result of integrating some intelligent techniques and a purpose with the existing idea of chatbots. Sentiment analysis via mediums of text and voice pitch and generating confronting responses accordingly is what gives this bot a purpose and scope of operation. A real life conversation in which the user can share his/her endurances or talk his/her heart out or even rant out his/her badgers while the bot calmly listens and gives reassuring and confronting responses, can be made with the CHEERBOT and that itself is its novelty.

Figure 3. Keyword-based sentiment analysis evaluation

```
BOT : Hi! You are?
YOU : Chintan
BOT : Hello Chintan! How are you doing?
YOU : I am alone
INFO:root:Broback: respond to I am alone
INFO:root:Pronoun=You, noun=None, adjective=None, verb=('am', 'VBP')
BOT : I'm here for you
YOU : Please help me
INFO:root:Broback: respond to Please help me
INFO:root:Pronoun=None, noun=None, adjective=None, verb=('Please', 'VB')
BOT : I understand
YOU : Dont go
INFO:root:Broback: respond to Dont go
INFO:root:Pronoun=None, noun=None, adjective=None, verb=('go', 'VBP')
BOT : I understand
YOU : Dont go
INFO:root:Broback: respond to Dont go
INFO:root:Pronoun=None, noun=None, adjective=None, verb=('go', 'VBP')
BOT : You're never alone
YOU : Thank you
INFO:root:Broback: respond to Thank you
INFO:root:Found noun: Thank
INFO:root:Pronoun=I, noun=Thank, adjective=None, verb=(None, None)
INFO:root:Returning phrase 'Yeah but I know a lot about Thank'
BOT : Yeah but I know a lot about Thank
YOU : :)
INFO:root:Broback: respond to :)
INFO:root:Pronoun=None, noun=None, adjective=None, verb=(None, None)
INFO:root:Returning phrase 'meet me at our lair, bro?'
BOT : meet me at our lair, bro?
```

Figure 4. Voice Pitch Detection

```
 RESTART: F:\Cl
on.py
262.110115 Hz.
262.110166 Hz.
262.110113 Hz.
262.110093 Hz.
262.110156 Hz.
262.110087 Hz.
262.110105 Hz.
262.110150 Hz.
```

Figure 5. Voice pitch-based sentiment analysis evaluation

```
YOU :  chat
INFO:root:Broback: respond to chat
INFO:root:Found noun: chat
INFO:root:Pronoun=None, noun=chat, adjective=None, ver
INFO:root:Returning phrase 'meet me at our lair, bro?'
39
DEBUG:urllib3.connectionpool:Starting new HTTPS connec
.com
DEBUG:urllib3.connectionpool:https://translate.google.
00 None
DEBUG:urllib3.connectionpool:Starting new HTTPS connec
.com
DEBUG:urllib3.connectionpool:https://translate.google.
s?client=tw-ob&ttsspeed=1&tl=en&textlen=25&tk=669347.8
r%2C+bro%3F&idx=0&total=1&ie=UTF-8 HTTP/1.1" 200 7776
BOT : meet me at our lair, bro?
I'm listening...
YOU :  hey
40
DEBUG:urllib3.connectionpool:Starting new HTTPS connec
.com
DEBUG:urllib3.connectionpool:https://translate.google.
00 None
DEBUG:urllib3.connectionpool:Starting new HTTPS connec
.com
DEBUG:urllib3.connectionpool:https://translate.google.
s?client=tw-ob&ttsspeed=1&tl=en&textlen=27&tk=5976.421
seem+sad.&idx=0&total=1&ie=UTF-8 HTTP/1.1" 200 8832
BOT : Are you okay? You seem sad.
I'm listening...
```

Figure 6. Conversation demonstration 1

```
BOT : meet me at our lair, bro?
I'm listening...
YOU :  no not now
INFO:root:Broback: respond to no not now
INFO:root:Pronoun=None, noun=None, adjective=None, ver
52
DEBUG:urllib3.connectionpool:Starting new HTTPS connec
.com
DEBUG:urllib3.connectionpool:https://translate.google.
00 None
DEBUG:urllib3.connectionpool:Starting new HTTPS connec
.com
DEBUG:urllib3.connectionpool:https://translate.google.
s?ie=UTF-8&client=tw-ob&q=I%27m+here+for+you&idx=0&tex
&tk=187356.303177&tl=en HTTP/1.1" 200 4608
BOT : I'm here for you
I'm listening...
YOU :  I am going away
INFO:root:Broback: respond to I am going away
INFO:root:Pronoun=You, noun=None, adjective=None, verb
53
DEBUG:urllib3.connectionpool:Starting new HTTPS connec
.com
DEBUG:urllib3.connectionpool:https://translate.google.
00 None
DEBUG:urllib3.connectionpool:Starting new HTTPS connec
```

Figure 7. Conversation demonstration 2

```
I'm listening...
YOU :  I am going away
INFO:root:Broback: respond to I am going away
INFO:root:Pronoun=You, noun=None, adjective=None, verb
53
DEBUG:urllib3.connectionpool:Starting new HTTPS connec
.com
DEBUG:urllib3.connectionpool:https://translate.google.
00 None
DEBUG:urllib3.connectionpool:Starting new HTTPS connec
.com
DEBUG:urllib3.connectionpool:https://translate.google.
s?ie=UTF-8&client=tw-ob&q=Don%27t+leave+just+yet&idx=0
al=1&tk=114746.506799&tl=en HTTP/1.1" 200 6432
BOT : Don't leave just yet
I'm listening...
YOU :  thank you
INFO:root:Broback: respond to thank you
INFO:root:Found noun: thank
INFO:root:Pronoun=I, noun=thank, adjective=None, verb=
INFO:root:Returning phrase 'Actually I am a very Thank
54
```

Figure 8. Proposed UI design

FUTURE RESEARCH AND ENHANCEMENTS

As of now, CHEERBOT exists through two versions - CHEERBOT v1.0 and CHEERBOT v2.0 which function on keyword based sentiment analysis and voice pitch based emotion detection. The upcoming version of CHEERBOT is scheduled to have an integration of Long short-term memory (LSTM) model with the response generation phase in order to incorporate smileys and other emoticons relevant to the ongoing conversation. This will help in humanizing the bot's response more.

Apart from the LSTM, the upcoming version shall also have an added parameter to refine the process of analysing and detecting the user's emotional state - Facial Gesture and Contour Analysis. This would cascade the requirement of having a video interface in order to capture the user's face while the user is conversing. In this method, the movement and states of facial expression are to be converted into gestures, which then will be calibrated against the facial contours of the user and would eventually help in determining the emotional state of the user based on his facial expressions, while he/she is conversing with the CHEERBOT.

REFERENCES

Banchs, R. E., & Li, H. (2012). Iris: a chat-oriented dialogue system based on the vector space model. *Proceedings of the ACL 2012 System Demonstrations, Association for Computational Linguistics*, 37-42.

Behera, B. (2016). Chappie - A Semi-Automatic Intelligent Chatbot. *LCPST*. Retrieved from https://www.chatbots.org/paper/chappie_-_a_semi-automatic_intelligent_chatbot/

Bigham, J. P., Aller, M. B., Brudvik, J. T., Leung, J. O., Yazzolino, L. A., & Ladner, R. E. (n.d.). Inspiring blind high school students to pursue computer science with instant messaging chatbots. *ACM SIGCSE Bulletin, 40*, 449-453.

Daly, L. (2011). *An interactive guide to writing bots in Python*. Retrieved from https://apps.worldwritable.com/tutorials/chatbot/

Dasgupta, P. B. (2017). Detection and Analysis of Human Emotions through Voice and Speech Pattern Processing. *International Journal of Computer Trends and Technology*, *52*(1), 1–3. doi:10.14445/22312803/IJCTT-V52P101

Diez, M., Penagarikano, M., Varona, A., Rodriguez-Fuentes, L. J., & Bordel, G. (2010). GTTS System for the Albayzin 2010 Speaker Diarization Evaluation. *VI Jornadas EnTecnología Del Habla and II Iberian SLTech Workshop*. Retrieved August 22, 2018, from https://www.researchgate.net/publication/229048611_GTTS_System_for_the_Albayzin_2010_Speaker_Diarization_Evaluation

Kaur, S., Sharma, S., Jain, U., & Raj, A. (2016, July). Voice command system using raspberry pi. *Advanced Computational Intelligence: An International Journal, 3*. doi:10.5121/acii.2016.3306

Kowalski, S., Pavlovska, K., & Goldstein, M. (2013). Two case studies in using chatbots for security training. In *Information Assurance and Security Education and Training* (pp. 265–272). Springer. doi:10.1007/978-3-642-39377-8_31

Pilato, G., Vassallo, G., Augello, A., Vasile, M., & Gaglio, S. (2005). Expert chat-bots for cultural heritage. *IntelligenzaArtificiale*, *2*(2), 25–31.

Rich, E., Knight, K., & Nair, S. B. (2009). *Artificial Intelligence* (3rd ed.). Retrieved August 22, 2018, from https://i4iam.files.wordpress.com/2013/08/artificial-intelligence-by-rich-and-knight.pdf

Ritter, A., Cherry, C., & Dolan, W. B. (2011). Datadriven response generation in social media. *Proceedings of the conference on empirical methods in natural language processing, Association for Computational Linguistics*, 583–593.

Schalkwyk, J., Beeferman, D., Beaufays, F., Byrne, B., Chelba, C., Cohen, M., . . . Strope, B. (n.d.). *Google Search by Voice: A Case Study.* Retrieved August 22, 2018, from https://static.googleusercontent.com/media/research.google.com/en//pubs/archive/36340.pdf

Udantha, M. (2015, May 15). *Natural Language Toolkit (NLTK) Sample and Tutorial: Part 1*. Retrieved from https://dzone.com/articles/natural-language-toolkit-nltk

Varona, Penagarikano, Rodriguez-Fuentes, Bordel, & Diez. (2013). GTTS system for the spoken web search task at MediaEval2012. Proc. *MediaEval, 35.*

Vinyals, O., & Le, Q. (2015). A neural conversational model. *ICML Deep Learning Workshop 2015.*

Yu, Z., Yu, Z., Black, A. W., & Rudnicky, A. I. (2016). Chatbot Evaluation and Database Expansion via Crowdsourcing. *Allen Institute for Artificial Intelligence*. Retrieved from https://www.semanticscholar.org/paper/Chatbot-Evaluation-and-Database-Expansion-via-Yu-Xu/a950b42d3e46f4e1fb0e7e6959b-c85a66246adb8

KEY TERMS AND DEFINITIONS

CHEERBOT: A chatbot that operating in the scope of user's emotional wellbeing by performing sentiment analysis and determining user's emotional state.

Emotional Confrontation: An assurance, and a form of emotional support provided in order to give hope, when a person is sad and/or in a depressed state.

Facial Gesture and Contour Analysis: The process of analyzing facial expressions in the form of gestures by mapping and calibrating every expression against the facial contours to eventually obtain a facial gesture.

LSTM: Long short-term memory is the building block of a specific kind of recurrent neural network.

NLP: Natural language processing is the facilitative process enabling a computer to understand natural languages such as English.

Voice Pitch: In a stream of audio input of the user, voice pitch defines the most distinct frequency of from the chunk of the voice clip, for all the chunks of the voice clip.

Compilation of References

3GPP. (2012). *Technical Specification Group Services and System Aspects; Policy and Charging Control Architecture Release 12*. Retrieved from http://www.qtc.jp/3GPP/Specs/23203-a60.pdf

Abdolkarimi, E., Abaei, G., & Mosavi, M. (2017). A wavelet-extreme learning machine for low-cost INS/GPS navigation system in high-speed applications. *GPS Solutions*, *22*(1), 15. doi:10.100710291-017-0682-x

Abhipriya, R., Aysha, S., Gayathri, K., & Kathiravan, K. (2017). 3S: A Radio Identification based Continuous Spectrum Sensing Protocol for Safety of Women in Cognitive Radio Networks. In *2017 International Conference on Communication and Signal Processing (ICCSP)* (pp. 2042 - 2046). Chennai: IEEE. 10.1109/ICCSP.2017.8286762

Abu-Mostafa, Y. S., Magdon-Ismail, M., & Lin, H. T. (2012). Learning from data (Vol. 4). New York, NY: AMLBook.

Ademoye, O. A., & Ghinea, G. (2009). Synchronization of olfaction-enhanced multimedia. *IEEE Transactions on Multimedia*, *11*(3), 561–565. doi:10.1109/TMM.2009.2012927

Ademoye, O. A., Murray, N., Muntean, G. M., & Ghinea, G. (2016). Audio masking effect on inter-component skews in olfaction-enhanced multimedia presentations. *ACM Transactions on Multimedia Computing Communications and Applications*, *12*(4), 51. doi:10.1145/2957753

Agrawal, R., & Srikant, R. (2000). Privacy-preserving data mining: Vol. 29. *No. 2*. ACM.

Akyildiz, I. F., & Vuran, M. C. (2010). *Wireless sensor networks* (Vol. 4). John Wiley & Sons. doi:10.1002/9780470515181

Alcatel, L. (2009). *The LTE network Architecture—A comprehensive tutorial." Strategic Whitepaper*. Lucent Alcatel.

Al-Karaki, J. N., & Kamal, A. E. (2004). Routing techniques in wireless sensor networks: A survey. *IEEE Wireless Communications*, *11*(6), 6–28. doi:10.1109/MWC.2004.1368893

Alpaydın, E. (2014). *Introduction to Machine Learning*. MIT Pres.

Alsheikh, M. A., Lin, S., Niyato, D., & Tan, H.-P. (2014). Machine Learning in Wireless Sensor Networks: Algorithms, Strategies, and Applications. *IEEE Communications Surveys and Tutorials*, *16*(4), 1996–2018. doi:10.1109/COMST.2014.2320099

Amos, B., Turner, H., & White, J. (2013). Applying machine learning classifiers to dynamic android malware detection at scale. *2013 9th International Wireless Communications and Mobile Computing Conference, IWCMC 2013*, 1666–1671. 10.1109/IWCMC.2013.6583806

Anagnostopoulos, C., & Hadjiefthymiades, S. (2014). Advanced Principal Component-Based Compression Schemes for Wireless Sensor Networks. *ACM Trans. Sen. Netw.*, *11*, 7:1-7:34. doi:10.1145/2629330

Anagnostopoulos, C., Hadjiefthymiades, S., & Georgas, P. (2012). PC3: Principal component-based context compression. *Journal of Parallel and Distributed Computing*, *72*(2), 155–170. doi:10.1016/j.jpdc.2011.10.001

Anandakumar, H., & Umamaheswari, K. (2017). Supervised machine learning techniques in cognitive radio networks during cooperative spectrum handovers. *Cluster Computing*, *20*(2), 1505–1515. doi:10.100710586-017-0798-3

Anastasi, G., Conti, M., Di Francesco, M., & Passarella, A. (2009). Energy conservation in wireless sensor networks: A survey. *Ad Hoc Networks*, *7*(3), 537–568. doi:10.1016/j.adhoc.2008.06.003

Andrews, J. G., Buzzi, S., Choi, W., Hanly, S. V., Lozano, A., Soong, A. C. K., & Zhang, J. C. (2014). What Will 5G Be? *IEEE Journal on Selected Areas in Communications*, *32*(6), 1065–1082. doi:10.1109/JSAC.2014.2328098

Andrews, M., Qian, L., & Stolyar, A. (2005). Optimal Utility based Multi-User Throughput Allocation Subject to Throughput Constraints. In *Proceedings IEEE 24th Annual Joint Conference of the IEEE Computer and Communications Societies*, (pp. 2415- 2424), Miami, FL: IEEE. 10.1109/INFCOM.2005.1498527

Arroyo-Valles, R., Alaiz-Rodriguez, R., Guerrero-Curieses, A., & Cid-Sueiro, J. (2007, December). Q-probabilistic routing in wireless sensor networks. In *Intelligent Sensors, Sensor Networks and Information, 2007. ISSNIP 2007. 3rd International Conference on* (pp. 1-6). IEEE. 10.1109/ISSNIP.2007.4496810

Ashton, K. (2009). That Internet of Things Thing. *RFID Journal.*

Avocanh, F. T. S., Abdennebi, M., & Ben-Othman, J. (2014). An Enhanced Two Level Scheduler to Increase Multimedia Services Performance in LTE Networks. In *Proceedings of IEEE International Conference on Communications (ICC)* (pp. 2351-2356). Sydney, NSW, Australia: IEEE. 10.1109/ICC.2014.6883674

Ayodele, T. O. (2010). Types of machine learning algorithms. In *New advances in machine learning*. InTech.

Aziz, N. (2014). *Eigenstructure-Based Angle for Detecting Outliers in Multivariate Data*. Academic Press. PMID:26356647

Bae, S. J., Choi, B.-G., & Chung, M. (2011). Delay-Aware Packet Scheduling Algorithm for Multiple Traffic Classes in 3GPP LTE System. In *Proceedings of 17th Asia-Pacific Conference on Communications (APCC)* (pp. 33 – 37). Sabah, Malaysia: APCC. 10.1109/APCC.2011.6152755

Bahrepour, M., Meratnia, N., Poel, M., Taghikhaki, Z., & Havinga, P. J. (2010, November). Distributed event detection in wireless sensor networks for disaster management. In *Intelligent Networking and Collaborative Systems (INCOS), 2010 2nd International Conference on* (pp. 507-512). IEEE.

Bampis, C. G., Li, Z., Katsavounidis, I., & Bovik, A. C. (2018). Recurrent and Dynamic Models for Predicting Streaming Video Quality of Experience. *IEEE Transactions on Image Processing*, *27*(7), 3316–3331. doi:10.1109/TIP.2018.2815842 PMID:29641409

Banchs, R. E., & Li, H. (2012). Iris: a chat-oriented dialogue system based on the vector space model. *Proceedings of the ACL 2012 System Demonstrations, Association for Computational Linguistics*, 37-42.

Bandyopadhyay, S., & Coyle, E. J. (2003, April). An energy efficient hierarchical clustering algorithm for wireless sensor networks. In *INFOCOM 2003. Twenty-Second Annual Joint Conference of the IEEE Computer and Communications* (Vol. 3, pp. 1713-1723). IEEE.

Behera, B. (2016). Chappie - A Semi-Automatic Intelligent Chatbot. *LCPST*. Retrieved from https://www.chatbots.org/paper/chappie_-_a_semi-automatic_intelligent_chatbot/

Behrooz & Moh. (2016). Intelligent mobile messaging for urban networks: Adaptive intelligent mobile messaging based on reinforcement learning. *Proceedings of 12th IEEE International Conference on Wireless and Mobile Computing, Networking and Communications (WiMob)*, 1–8.

Bengio, Y. (2009). Learning deep architectures for AI. *Foundations and trends® in Machine Learning, 2*(1), 1-127.

Bengio, Y., Lamblin, P., Popovici, D., & Larochelle, H. (2007). Greedy layerwise training of deep networks. Proc. Adv. Neural Information Processing System, 153–160.

Bengio, Y., & Delalleau, O. (2011). On the expressive power of deep architectures. In *Algorithmic Learning Theory* (pp. 18–36). Berlin, Germany: Springer. doi:10.1007/978-3-642-24412-4_3

Bennaceur, J., Souihi, S., Idoudi, H., Saidane, L. A., & Mellouk, A. (2017). Game-based Secure Sensing for the Mobile Cognitive Radio Network. In *2017 IEEE 28th Annual International Symposium on Personal, Indoor, and Mobile Radio Communications (PIMRC)* (pp. 1 - 7). Montreal: IEEE.

Berger, C. R., Demissie, B., Heckenbach, J., Willett, P., & Zhou, S. (2010). Signal processing for passive radar using OFDM waveforms. *Signal processing for passive radar using OFDM waveforms, 4*(1), 226-238.

Beyer, K., Goldstein, J., Ramakrishnan, R., & Shaft, U. (1999, January). When is "nearest neighbor" meaningful? In *International conference on database theory* (pp. 217-235). Springer.

Bharathi, S., Kumar, D., & Ram, D. (2018). Defence against responsive and non-responsive jamming attack in cognitive radio networks: an evolutionary game theoretical approach. *The Journal of Engineering, 2018*(2), 68 - 75.

Bigham, J. P., Aller, M. B., Brudvik, J. T., Leung, J. O., Yazzolino, L. A., & Ladner, R. E. (n.d.). Inspiring blind high school students to pursue computer science with instant messaging chatbots. *ACM SIGCSE Bulletin, 40*, 449-453.

Bi, S., Zhang, R., Ding, Z., & Cui, S. (2015). Wireless communications in the era of big data. *IEEE Communications Magazine, 53*(10), 190–199. doi:10.1109/MCOM.2015.7295483

Bkassiny, M., Li, Y., & Jayaweera, S. K. (2013). A survey on machine-learning techniques in cognitive radios. *IEEE Communications Surveys and Tutorials, 15*(3), 1136–1159. doi:10.1109/SURV.2012.100412.00017

Bodik, P., Hong, W., Guestrin, C., Madden, S., Paskin, M., & Thibaux, R. (2004). *Intel Lab Data.* Intel Berkeley Research Laboratory. Retrieved from http://db.csail.mit.edu/labdata/labdata.html

Borisova, N. (2014). An Approach for Ontology Based Information Extraction. *Information Technology and Control, 12*(1). doi:10.1515/itc-2015-0007

Boyan, J. A., & Littman, M. L. (1994). Packet routing in dynamically changing networks: A reinforcement learning approach. In Advances in neural information processing systems (pp. 671-678). Academic Press.

Branch, J. W., Giannella, C., Szymanski, B., Wolff, R., & Kargupta, H. (2013). In-network outlier detection in wireless sensor networks. *Knowledge and Information Systems, 34*(1), 23–54. doi:10.100710115-011-0474-5

Bräysy, T., Couturier, S., Smit, N., Le Nir, V., Tuukkanen, T., Verheul, E., ... Krygier, J. (2017). Network management issues in military cognitive radio networks. *International Conference on Military Communications and Information Systems (ICMCIS)*. 10.1109/ICMCIS.2017.7956487

Brookman, J., Rouge, P., Alva, A., & Yeung, C. (2017). Cross-Device Tracking: Measurement and Disclosures. *Proceedings On Privacy Enhancing Technologies, 2017*(2). doi:10.1515/popets-2017-0020

Brunato, M., & Battiti, R. (2005). Statistical learning theory for location fingerprinting in wireless LANs. *Computer Networks, 47*(6), 825–845. doi:10.1016/j.comnet.2004.09.004

Calheiros, R. N., Ranjan, R., Beloglazov, A., De Rose, C. A. F., & Buyya, R. (2011). CloudSim: A toolkit for modeling and simulation of cloud computing environments and evaluation of resource provisioning algorithms. *Software, Practice & Experience*, *41*(1), 23–50. doi:10.1002pe.995

Callado, A., Kamienski, C., Szabo, G., Peter-Gero, B., Kelner, J., Fernandes, S., & Sadok, D. (2009). A survey on internet traffic identification. *IEEE Communications Surveys and Tutorials*, *11*(3), 1–16. doi:10.1109/SURV.2009.090304

Capozzi, F., Piro, G., Grieco, L. A., Boggia, G., & Camarda, P. (2013). Downlink Packet Scheduling in LTE Cellular Networks: Key Design Issues and a Survey. *IEEE Communications Surveys and Tutorials*, *15*(2), 678–700. doi:10.1109/SURV.2012.060912.00100

Casana, M., & Redondi, A. E. C. (2017). IoT communicatin technologies for smart city. In V. Angelakis & ... (Eds.), *Designing, Developing, and Facilitating Smart Cities*. Springer Publishing. doi:10.1007/978-3-319-44924-1_8

Casa, P., & Wassermann, S. (2017). Improving QoE prediction in mobile video through machine learning. *8th International Conference on the Network of the Future (NOF)*, 1-7. 10.1109/NOF.2017.8251212

Casey, K., Lim, A., & Dozier, G. (2006, July). Evolving general regression neural networks for tsunami detection and response. In *Evolutionary Computation. CEC 2006. IEEE Congress on* (pp. 2451-2458). IEEE.

Checko, A., Christiansen, H. L., Yan, Y., Scolari, L., Kandaras, G., Berger, M., & Dittmann, L. (2015). Cloud RAN for mobile networks – A technology overview. *IEEE Communications Surveys and Tutorials*, *17*(1), 405–426. doi:10.1109/COMST.2014.2355255

Chen, B., Hu, F., & Zhu, K. (2018). Full Spectrum Sharing in Cognitive Radio Networks toward 5G: A Survey. *IEEE Access: Practical Innovations, Open Solutions*, (99), 1–1.

Chen, D., Liu, Z., Wang, L., Dou, M., Chen, J., & Li, H. (2013). Natural disaster monitoring with wireless sensor networks: A case study of data-intensive applications upon low-cost scalable systems. *Mobile Networks and Applications*, *18*(5), 651–663. doi:10.100711036-013-0456-9

Cheng, H., Feng, D., Shi, X., & Chen, C. (2018). Data quality analysis and cleaning strategy for wireless sensor networks. *EURASIP Journal on Wireless Communications and Networking*, *2018*(1), 61. doi:10.118613638-018-1069-6

Cheng, H., Zhihuang, S., Xiong, N., & Yang, X. (2016). Energy-efficient nodes scheduling algorithms for wireless sensor networks using Markov random field model. *Inf Sci*, *329*(1), 461–477. doi:10.1016/j.ins.2015.09.039

Chen, Q., Chetty, K., Woodbridge, K., & Tan, B. (2016b). Signs of life detection using wireless passive radar. *International Radar Conference (RadarConf)*. Philadelphia: Academic Press. 10.1109/RADAR.2016.7485313

Chen, Q., Tan, B., Chetty, K., & Woodbridge, K. (2016a). Activity recognition based on micro-doppler signature with in-home Wi-Fi. *IEEE International Conference on e-Health Networking, Applications and Services (Healthcom)*. Munich, Germany: IEEE. 10.1109/HealthCom.2016.7749457

Chen, Q., Tan, B., Woodbridge, K., & Chetty, K. (2015). Indoor target tracking using high Doppler resolution passive Wi-Fi radar. *IEEE International Conference on Acoustics, Speech and Signal Processing (ICASSP)*. Brisbane, Australia: IEEE. 10.1109/ICASSP.2015.7179036

Chen, W., Liu, T., & Yang, X. (2016). Reinforcement learning behaviors in sponsored search. *Applied Stochastic Models in Business and Industry*, *32*(3), 358–367. doi:10.1002/asmb.2157

Chen, Z., & Liu, B. (2016). Lifelong Machine Learning. *Synthesis Lectures On Artificial Intelligence And Machine Learning*, *10*(3), 1–145. doi:10.2200/S00737ED1V01Y201610AIM033

Chetty, L., Smith, G. E., & Woodbridge, K. (2011). hrough-the-wall sensing of personnel using passive bi-static WiFi radar at standoff distances. *IEEE Transactions on Geoscience and Remote Sensing*, *50*(4), 1218–1226. doi:10.1109/TGRS.2011.2164411

Chien, C., Elgorriaga, I., & McConaghy, C. (2001, August). Low-power direct-sequence spread-spectrum modem architecture for distributed wireless sensor networks. In *Proceedings of the 2001 international symposium on Low power electronics and design* (pp. 251-254). ACM. 10.1145/383082.383151

Chodos, A., Ouellette, J., & Tretkoff, E. (2006, April). April, 1935: British Patent for Radar System for Air Defense Granted to Robert Watson-Watt. *APS NEWS: This Month in Physics History, 15*. Retrieved from https://www.aps.org/publications/apsnews/200604/history.cfm

Choudhari, T., & Moh, M. (2018). Prioritized Task Scheduling in Fog Computing. In *Proc. of 2018 Annual ACM Southeast Conference*. Richmond, KY. ACM Press. 10.1145/3190645.3190699

Chung, W. C., Chang, C. J., & Wang, L. C. (2012). An Intelligent Priority Resource Allocation Scheme for LTE-A Downlink Systems. *IEEE Wireless Communications Letters*, *1*(3), 241–244. doi:10.1109/WCL.2012.033012.120146

Cisco. (2017). *Cisco Visual Networking Index: Global Mobile Data Traffic Fore-cast Update, 2016-2021 White Paper* Retrieved from http://www.cisco.com/c/en/us/solutions/collateral/service-provider/ visual-networking-index-vni/mobile-white-paper-c11-520862.html

Clancy, C., Hecker, J., Stuntebeck, E., & O'Shea, T. (2007). Applications of Machine Learning to Cognitive Radio Networks. *IEEE Wireless Communications*, *14*(4), 47–52. doi:10.1109/MWC.2007.4300983

Colone, F., Pastina, D., Falcone, P., & Lombardo, P. (2012). Wi-Fi-based passive ISAR for high-resolution cross range profiling of moving targets. *European Conference on Synthetic Aperture Radar (EUSAR)*. Nuremberg, Germany: Academic Press.

Comşa, I. S., Aydin, M., Zhang, S., Kuonen, P., & Wagen, J.-F. (2011). Reinforcement learning based radio resource scheduling in LTE-advanced. In *Proceedings of 17th International Conference on Automation and Computing* (pp. 219 - 224). Huddersfield, UK: Academic Press.

Comşa, I. S., Aydin, M., Zhang, S., Kuonen, P., & Wagen, J.-F. (2012). Multi Objective Resource Scheduling in LTE Networks Using Reinforcement Learning. *International Journal of Distributed Systems and Technologies*, *3*(2), 39–57. doi:10.4018/jdst.2012040103

Comşa, I.-S. (2014a). *Sustainable Scheduling Policies for Radio Access Networks Based on LTE Technology*. Bedfordshire, UK: University of Bedfordshire.

Comşa, I.-S., Aydin, M., Zhang, S., Kuonen, P., Wagen, J.-F., & Lu, Y. (2014b). Scheduling Policies Based on Dynamic Throughput and Fairness Tradeoff Control in LTE-A Networks. In *Proceedings of IEEE Conference on Local Computer Networks (LCN)* (pp. 418-421). Edmonton, AB, Canada: IEEE. 10.1109/LCN.2014.6925806

Comşa, I.-S., De Domenico, A., & Ktenas, D. (2017). QoS-Driven Scheduling in 5G Radio Access Networks - A Reinforcement Learning Approach. In *Proceedings of IEEE Global Communications Conference (GLOBECOM)* (pp. 1-7). Singapore: IEEE. 10.1109/GLOCOM.2017.8254926

Comşa, I.-S., Trestian, R., & Ghinea, G. (2018). 360° Mulsemedia Experience over Next Generation Wireless Networks - A Reinforcement Learning Approach. In *Proceedings of Tenth International Conference on Quality of Multimedia Experience (QoMEX)* (pp. 1-6). Cagliari, Italy: Academic Press. 10.1109/QoMEX.2018.8463409

Comşa, I.-S., Zhang, S., Aydin, M., Chen, J., Kuonen, P., & Wagen, J.-F. (2014c). Adaptive Proportional Fair Parameterization Based LTE Scheduling Using Continuous Actor-Critic Reinforcement Learning. In *Proceedings of IEEE Global Communication Conference (GLOBECOM)* (pp. 4387 - 4393). Austin, TX: IEEE. 10.1109/GLOCOM.2014.7037498

Comşa, I.-S., Zhang, S., Aydin, M., Kuonen, P., Lu, Y., Trestian, R., & Ghinea, G. (2018). Towards 5G: A Reinforcement Learning-based Scheduling Solution for Data Traffic Management. *IEEE eTransactions on Network and Service Management*, 1–15.

Comşa, I.-S., Zhang, S., Aydin, M., Kuonen, P., & Wagen, J. (2012). A Novel Dynamic Q-Learning-Based Scheduler Technique for LTE-advanced Technologies Using Neural Networks. In *Proceedings of IEEE Conference on Local Computer Networks (LCN)* (pp. 332 – 335). Clearwater, FL: IEEE.

Coufal, D. (2016). On convergence of kernel density estimates in particle filtering. *Kybernetika*, 735–756. doi:10.14736/kyb-2016-5-0735

Couturier, S., Krygier, J., Bentstuen, O. I., & Nir, V. L. (2015). Challenges for Network Aspects of Cognitive Radio. *2015 International Conference on Military Communications and Information Systems (ICMCIS)*. 10.1109/ICMCIS.2015.7158680

Daly, L. (2011). *An interactive guide to writing bots in Python*. Retrieved from https://apps.worldwritable.com/tutorials/chatbot/

Dasgupta, P. B. (2017). Detection and Analysis of Human Emotions through Voice and Speech Pattern Processing. *International Journal of Computer Trends and Technology*, *52*(1), 1–3. doi:10.14445/22312803/IJCTT-V52P101

Dayan, P., & Berridge, K. C. (2014). Model-based and model-free Pavlovian reward learning: Revaluation, revision, and revelation. *Cognitive, Affective & Behavioral Neuroscience*, *14*(2), 473–492. doi:10.375813415-014-0277-8 PMID:24647659

Dehestani, D., Eftekhari, F., Guo, Y., Ling, S., Su, S., & Nguyen, H. (2011). Online Support Vector Machine Applicationfor Model Based Fault Detection and Isolationof HVAC System. *International Journal Of Machine Learning And Computing*, 66-72. doi:10.7763/ijmlc.2011.v1.10

Del-Rey-Maestre, N., Jarabo-Amores, M., Mata-Moya, D., Barcena-Humanes, J., & Ael-Hoyo, P. D. (2017). Machine learning techniques for coherent CFAR detection based on statistical modeling of UHF passive ground clutter. *IEEE Journal of Selected Topics in Signal Processing*, *12*(1), 104–118. doi:10.1109/JSTSP.2017.2780798

Demestichas, P., Dimitrakopoulos, G., Strassner, J., & Bourse, D. (2006). Introducing Reconfigurability and Cognitive Networks Concepts in the Wireless World. *IEEE Vehicular Technology Magazine*, *1*(June), 32–39. doi:10.1109/MVT.2006.283572

Deng, J. D. (2016). Online Outlier Detection of Energy Data Streams Using Incremental and Kernel PCA Algorithms. In *2016 IEEE 16th International Conference on Data Mining Workshops (ICDMW)*, (pp. 390-397). IEEE. doi:10.1109/ICDMW.2016.0062

Dharne, A. G., Lee, J., & Jayasuriya, S. (2006, June). Using fuzzy logic for localization in mobile sensor networks: simulations and experiments. In *American Control Conference*, 2006 (pp. 6-pp). IEEE. 10.1109/ACC.2006.1656524

Di, M., & Joo, E. M. (2007, December). A survey of machine learning in wireless sensor netoworks from networking and application perspectives. In *Information, Communications & Signal Processing, 2007 6th International Conference on* (pp. 1-5). IEEE.

Diez, M., Penagarikano, M., Varona, A., Rodriguez-Fuentes, L. J., & Bordel, G. (2010). GTTS System for the Albayzin 2010 Speaker Diarization Evaluation. *VI Jornadas EnTecnología Del Habla and II Iberian SLTech Workshop*. Retrieved August 22, 2018, from https://www.researchgate.net/publication/229048611_GTTS_System_for_the_Albayzin_2010_Speaker_Diarization_Evaluation

Di, M., & Joo, E. M. (2007). A Survey of Machine Learning in Wireless Sensor Networks. *Proceedings of the 6th International Conference on Information, Communications and Signal Processing (ICICS)*, 1–5. 10.1109/ICICS.2007.4449882

Ding, G., Wu, Q., Yao, Y., Wang, J., & Chen, Y. (2013). Kernel-Based Learning for Statistical Signal Processing in Cognitive Radio Networks. *IEEE Signal Processing Magazine*, *30*(4), 126–136. doi:10.1109/MSP.2013.2251071

Dong, S., Agrawal, P., & Sivalingam, K. (2007, November). Reinforcement learning based geographic routing protocol for UWB wireless sensor network. In *Global Telecommunications Conference, 2007. GLOBECOM'07. IEEE* (pp. 652-656). IEEE. 10.1109/GLOCOM.2007.127

Dzbor, M., Stutt, A., Motta, E., & Collins, T. (2007). Representations for semantic learning webs: Semantic Web technology in learning support. *Journal of Computer Assisted Learning*, *23*(1), 69–82. doi:10.1111/j.1365-2729.2007.00202.x

Edelman, A. (2018, Spring). The Power Method. *Linear Algebra*.

Ehrlich, M., Wisniewski, L., & Jasperneite, J. (2018). State of the art and future applications of industrial wireless sensor networks. In Kommunikation und Bildverarbeitung in der Automation (pp. 28-39). Springer.

Elbamby, M. S., Perfecto, C., Bennis, M., & Doppler, K. (2018). Toward Low-Latency and Ultra-Reliable Virtual Reality. *IEEE Network*, *32*(2), 78–84. doi:10.1109/MNET.2018.1700268

Elnahas, O., Elsabrouty, M., Muta, O., & Furukawa, H. (2018). Game Theoretic Approaches for Cooperative Spectrum Sensing in Energy-Harvesting Cognitive Radio Networks. *IEEE Access: Practical Innovations, Open Solutions*, *6*, 11086–11100. doi:10.1109/ACCESS.2018.2810107

Elshaer, H., Boccardi, F., Dohler, M., & Irmer, R. (2015). Load & Backhaul Aware Decoupled Downlink/Uplink Access in 5G Systems. In *Proceedings of IEEE International Conference on Communications (ICC)*, pp. 5380–5385. 10.1109/ICC.2015.7249179

Endo, E. T., & Murray, T. (1991). Real-time seismic amplitude measurement (RSAM): A volcano monitoring and prediction tool. *Bulletin of Volcanology*, *53*(7), 533–545. doi:10.1007/BF00298154

Erdelj, M., Natalizio, E., Chowdhury, K. R., & Akyildiz, I. F. (2017). Help from the sky: Leveraging UAVs for disaster management. *IEEE Pervasive Computing*, (1), 24–32.

Erdelj, M., Król, M., & Natalizio, E. (2017). Wireless sensor networks and multi-UAV systems for natural disaster management. *Computer Networks*, *124*, 72–86. doi:10.1016/j.comnet.2017.05.021

Err, M.-J., & Deng, C. (2004). *Online tuning of fuzzy interference systems using dynamic fuzzy Q-learning. IEEE Trans. on Systems, Man, and Cybernetics, Part B*, 34, 1478–1489.

Eze, J., Zhang, S., Liu, E., & Eze, E. (2017). Cognitive Radio Technology assisted Vehicular Ad-Hoc Networks (VANETs): Current Status,Challenges, and Research Trends. In *23rd International Conference on Automation and Computing (ICAC)* (pp. 1 - 6). Huddersfield: IEEE. 10.23919/IConAC.2017.8082035

Fan, M., Zhang, H., Hu, H., Li, J., & Wang, Q. (2017). Cooperative Spectrum Sensing in RF-powered Full-Duplex Cognitive Radio Networks. In *2017 First International Conference on Electronics Instrumentation & Information Systems (EIIS)* (pp. 1 - 6). Harbin: IEEE. 10.1109/EIIS.2017.8298712

FANTASTIC-5G. (2016). *D4.1: Technical Results for Service Specific Multi-Node/MultiAntenna Solutions.* Retrieved from http://fantastic5g.com/wp-content/uploads/2016/06/FANTASTIC-5G_D4.1_ Final.pdf

Farabet, C., Couprie, C., Najman, L., & LeCun, Y. (2013). Learning hierarchical features for scene labeling. *IEEE Transactions on Pattern Analysis and Machine Intelligence, 35*(8), 1915–1929. doi:10.1109/TPAMI.2012.231 PMID:23787344

Farrugia, R. A., & Debono, C. J. (2008). A Robust Error Detection Mechanism for H.264/AVC Coded Video Sequences Based on Support Vector Machines. *IEEE Transactions on Circuits and Systems for Video Technology, 18*(12), 1766–1770. doi:10.1109/TCSVT.2008.2004919

Farrugia, R. A., & Debono, C. J. (2009). A Support Vector Machine Approach for Detection and Localization of Transmission Errors Within Standard H.263++ Decoders. *IEEE Transactions on Multimedia, 11*(7), 1323–1330. doi:10.1109/TMM.2009.2030651

Fathy, M., Tammam, A., & Saafan, A. (2017). Mitigating the Impact of Malicious Behavior via Utilizing Multiple Routes in a Cooperative Sensing Cognitive Radio Network. *IEEE 15th Student Conference on Research and Development (SCOReD).*

Feldman, D., Schmidt, M., & Sohler, C. (2013, January). Turning big data into tiny data: Constant-size coresets for k-means, pca and projective clustering. In *Proceedings of the twenty-fourth annual ACM-SIAM symposium on Discrete algorithms* (pp. 1434-1453). Society for Industrial and Applied Mathematics. 10.1137/1.9781611973105.103

Foo, M. C., Lim, H. B., Zeng, Y., Lam, V. T., Teo, R., & Ng, G. W. (2005, December). Impact of distributed resource allocation in sensor networks. In *Intelligent Sensors, Sensor Networks and Information Processing Conference, 2005. Proceedings of the 2005 International Conference on* (pp. 69-74). IEEE.

Forster, A. (2007). Machine Learning Techniques Applied to Wireless Ad-Hoc Networks: Guide and Survey. *2007 3rd International Conference on Intelligent Sensors, Sensor Networks and Information*, 365–370. 10.1109/ISSNIP.2007.4496871

Forster, A. (2007, December). Machine learning techniques applied to wireless ad-hoc networks: Guide and survey. In *Intelligent Sensors, Sensor Networks and Information, 2007. ISSNIP 2007. 3rd International Conference on* (pp. 365-370). IEEE.

Forster, A., & Murphy, A. L. (2007, December). FROMS: Feedback routing for optimizing multiple sinks in WSN with reinforcement learning. In *Intelligent Sensors, Sensor Networks and Information, 2007. ISSNIP 2007. 3rd International Conference on* (pp. 371-376). IEEE.

Förster, A., & Murphy, A. L. (2009). CLIQUE: Role-free clustering with Q-learning for wireless sensor networks. *Proceedings - International Conference on Distributed Computing Systems*, 441–449. 10.1109/ICDCS.2009.43

Gao, L., & Moh, M. (2018). Joint Computation Offloading and Prioritized Scheduling in Mobile Edge Computing. In *Proceedings of IEEE International Conference on High Performance Computing and Simulation.* Orleans, France: IEEE.

Gao, W., Kwong, S., & Jia, Y. (2017). Joint Machine Learning and Game Theory for Rate Control in High Efficiency Video Coding. *IEEE Transactions on Image Processing, 26*(12), 6074–6089. doi:10.1109/TIP.2017.2745099 PMID:28858792

Gharavi, H., & Ban, K. (2003). Multihop sensor network design for wide-band communications. *Proceedings of the IEEE, 91*(8), 1221–1234. doi:10.1109/JPROC.2003.814919

Ghinea, G., & Ademoye, O. (2012). The sweet smell of success: Enhancing multimedia applications with olfaction. *ACM Transactions on Multimedia Computing Communications and Applications, 8*(1), 2. doi:10.1145/2071396.2071398

Ghinea, G., Timmerer, C., Lin, W., & Gulliver, S. R. (2014). Mulsemedia: State of the Art Perspectives and Challenges. *ACM Transactions on Multimedia Computing Communications and Applications, 11*(17), 1–23.

Ghotekar, N. (2016). Analysis and Data Mining of Call Detail Records using Big Data Technology. *IJARCCE, 5*(12), 280–283. doi:10.17148/IJARCCE.2016.51264

Gohil, A., Modi, H., & Patel, S. K. (2013). 5G technology of mobile communication: A survey. *2013 International Conference on Intelligent Systems and Signal Processing, ISSP 2013*, 288–292. 10.1109/ISSP.2013.6526920

Goswami, S., Chakraborty, S., Ghosh, S., Chakrabarti, A., & Chakraborty, B. (2016). *A review on application of data mining techniques to combat natural disasters. Ain Shams Engineering Journal.*

Gottapu, S. K., & Vallabhaneni, P. (2018). Coordination between Multiple Collocated Cognitive Radio Networks. In *2018 Conference on Signal Processing And Communication Engineering Systems (SPACES)* (pp. 33 - 36). Vijayawada: IEEE. 10.1109/SPACES.2018.8316310

Grazia, M. D. F. D., Zucchetto, D., Testolin, A., Zanella, A., Zorzi, M., & Zorzi, M. (2018). QoE Multi-Stage Machine Learning for Dynamic Video Streaming. *IEEE Transactions on Cognitive Communications and Networking*, *4*(1), 146–161. doi:10.1109/TCCN.2017.2784449

Griffiths, H., & Willis, N. (2010). Klein Heidelberg – the world's first modern bi-static radar system. *IEEE Transactions on Aerospace and Electronic Systems*, *44*(4), 1571–1588. doi:10.1109/TAES.2010.5595580

Gu, C., Zhang, S., Xue, X., & Huang, H. (2011). Online Wireless Mesh Network Traffic Classification using. *Machine Learning*, *5*, 1524–1532.

Guzella, T. S., & Caminhas, W. M. (2009). A review of machine learning approaches to spam filtering. *Expert Systems with Applications*, *36*(7), 10206–10222. doi:10.1016/j.eswa.2009.02.037

H.264/AVC Software Coordination. (2007). Retrieved from http://iphome.hhi.de/suehring/tml/

H.264/MPEG-4AVC Reference Software Manual. (2007). Retrieved from http://iphome.hhi.de/suehring/tml/JM(JVT-X072).pdf

Harrington, P. (2012). *Machine Learning in Action.* Manning Publications.

Harth, N., & Anagnostopoulos, C. (2017). *Quality-aware aggregation & predictive analytics at the edge, 2017 IEEE International Conference on Big Data* (pp. 17–26). Boston, MA: Big Data; Retrieved from http://eprints.gla.ac.uk/149980/

Harth, N., Anagnostopoulos, C., & Pezaros, D. (2018). Predictive intelligence to the edge: Impact on edge analytics. *Evolving Systems*, *9*(1), 91–91. doi:10.100712530-017-9210-z

Hasan, M., & Mishra, P. (2012). Robust Gesture Recognition Using Gaussian Distribution for Features Fitting. *International Journal Of Machine Learning And Computing*, 266-273. doi:10.7763/ijmlc.2012.v2.128

Hasselt, H. P. (2011). *Insights in Reinforcement Learning Formal Analysis and Empirical Evaluation of Temporal-Difference Learning Algorithms*. University of Utrecht.

Haykin, S. (1999). *Neural networks: a comprehensive foundation.* Pearson Education.

Haykin, S. (2005). Cognitive radio: Brain-empowered wireless communications. *IEEE Journal on Selected Areas in Communications*, *23*(2), 201–220. doi:10.1109/JSAC.2004.839380

Hazar, M. A., Odabasioglu, N., Ensari, T., Kavurucu, Y., & Sayan, O. F. (2017). Performance analysis and improvement of machine learning algorithms for automatic modulation recognition over Rayleigh fading channels. *Neural Computing & Applications*, *29*(9), 351–360. doi:10.100700521-017-3040-6

Heinzelman, W. R., Chandrakasan, A., & Balakrishnan, H. (2000, January). Energy-efficient communication protocol for wireless microsensor networks. In *System sciences, Proceedings of the 33rd annual Hawaii international conference on* (pp. 10-pp). IEEE. 10.1109/HICSS.2000.926982

Hill, T., & Lewicki, P. (2006). *Statistics: Methods and Applications: A Comprehensive Reference for Science, Industry, and Data Mining.* StatSoft, Inc.

Hinton, G. E., Osindero, S., & Teh, Y. W. (2006). A fast learning algorithm for deep belief nets. *Neural Computation, 18*(7), 1527–1554. doi:10.1162/neco.2006.18.7.1527 PMID:16764513

Hinton, G., & Salakhutdinov, R. (2006). Reducing the dimensionality of data with neural networks. *Science, 313*(5786), 504–507. doi:10.1126cience.1127647 PMID:16873662

Holmes, S. (2002). *Example of the Power method with Matlab.* Academic Press.

Houeland, T., & Aamodt, A. (2017). A learning system based on lazy metareasoning. *Progress In Artificial Intelligence, 7*(2), 129–146. doi:10.100713748-017-0138-0

Huang, J., Wu, K., & Moh, M. (2014). Dynamic Virtual Machine Migration Algorithms Using Enhanced Energy Consumption Model for Green Cloud Data Centers. In *Proceedings of the International Conference on High Performance Computing and Simulation (HPCS).* Bologna, Italy. IEEE Press. 10.1109/HPCSim.2014.6903785

Huby, G., Cockram, J., & Fleming, M. (2013). Through-life Data Exploitation to Reduce Downtime and Costs. *Procedia CIRP, 11*, 50–55. doi:10.1016/j.procir.2013.07.070

Hussain, A., & Cambria, E. (2018). Semi-supervised learning for big social data analysis. *Neurocomputing, 275*, 1662–1673. doi:10.1016/j.neucom.2017.10.010

Hussain, S., Schaffner, S., & Moseychuck, D. (2009). Applications of Wireless Sensor Networks and RFID in a Smart Home Environment. *2009 Seventh Annual Communication Networks and Services Research Conference*, 153–157. 10.1109/CNSR.2009.32

Hu, Y., Wang, Y., & Liu, B. (2017). Latency reduction and load balancing in coded storage systems. In *Proceedings of ACM Symposium on Cloud Computing.* Santa Clara, CA. ACM Press. 10.1145/3127479.3131623

Hwang, K. (2017). *Cloud Computing for Machine Learning and Cognitive Applications.* The MIT Press.

Ibnkahla, M. (2000). Applications of neural networks to digital communications–a survey. *Signal Processing, 80*(7), 1185–1215. doi:10.1016/S0165-1684(00)00030-X

Int. Telecommun. Union. (2000). *Methodology for the Subjective Assessment of the Quality of Television Pictures.* ITU-R Recommendation BT.500-11, Tech. Rep.

Intanagonwiwat, C., Estrin, D., Govindan, R., & Heidemann, R. (2002). Impact of network density on data aggregation in wireless sensor networks. *Proceedings of the 22nd International Conference on Distributed Computing Systems.* 10.1109/ICDCS.2002.1022289

Intanagonwiwat, C., Govindan, R., & Estrin, D. (2000, August). Directed diffusion: A scalable and robust communication paradigm for sensor networks. In *Proceedings of the 6th annual international conference on Mobile computing and networking* (pp. 56-67). ACM. 10.1145/345910.345920

Internet of Things & Creation of the Fifth V of Big Data. (2017). *International Journal Of Science And Research, 6*(1), 1363–1366. doi:10.21275/art20164394

Iqbal, S., Iqbal, U., Hassan, S. A., & Saleem, S. (2018). Indoor motion classification using passive RF sensing incorporating deep learning. In *IEEE 87th Vehicular Technology Conference (VTC)*. Porto, Portugal: IEEE.

Islam, N. (2018). *Business Intelligence and Analytics for Operational Efficiency*. SSRN Electronic Journal. doi:10.2139srn.3163429

Jain, R., Chiu, D. M., & Hawe, W. A. (1984). Quantitative Measure of Fairness and Discrimination for Resource Allocation in Shared Computer System. In Research Report, DEC-TR-301, 1984.

Janakiram, D., Reddy, V. A., & Kumar, A. P. (2006, January). Outlier detection in wireless sensor networks using Bayesian belief networks. In *Communication System Software and Middleware, 2006. Comsware 2006. First International Conference on* (pp. 1-6). IEEE.

Jan, M. A., Nanda, P., He, X., & Liu, R. P. (2018). A Sybil attack detection scheme for a forest wildfire monitoring application. *Future Generation Computer Systems*, *80*, 613–626. doi:10.1016/j.future.2016.05.034

Jara, A., Olivieri, A., Bocchi, Y., Jung, M., Kastner, W., & Skarmeta, A. (2014). Semantic Web of Things: An analysis of the application semantics for the IoT moving towards the IoT convergence. *International Journal of Web and Grid Services*, *10*(2/3), 244. doi:10.1504/IJWGS.2014.060260

Jayaraman, P. P., Zaslavsky, A., & Delsing, J. (2010, June). Intelligent processing of k-nearest neighbors queries using mobile data collectors in a location aware 3D wireless sensor network. In *International Conference on Industrial, Engineering and Other Applications of Applied Intelligent Systems* (pp. 260-270). Springer. 10.1007/978-3-642-13033-5_27

Jayasri, C., & Elanagai, G. (2017). Implementation of network security based data hauling by collaborative spectrum sensing in cognitive radio network. In *2017 International Conference on Innovations in Information, Embedded and Communication Systems (ICIIECS)* (pp. 1 - 5). Coimbatore: IEEE.

Jaya, T., Raaja, V. K., Sharanya, C., & Rajendran, V. (2017). An Efficient Power Allocation and Joint Optimal Sensing of Spectrum in CDMA-based Cognitive Radio Networks. In *2017 International Conference on Communication and Signal Processing* (pp. 1769 - 1773). Chennai: IEEE. 10.1109/ICCSP.2017.8286697

Jiang, C., Zhang, H., Ren, Y., Han, Z., Chen, K. C., & Hanzo, L. (2017). Machine Learning Paradigms for Next-Generation Wireless Networks. *IEEE Wireless Communications*, *24*(2), 98–105. doi:10.1109/MWC.2016.1500356WC

Jolliffe, I. (2011). Principal component analysis. In *International encyclopedia of statistical science* (pp. 1094–1096). Berlin: Springer. doi:10.1007/978-3-642-04898-2_455

Jouffe, L. (1998). *Fuzzy Inference System Learning by Reinforcement Methods. IEEE Trans. on Systems, Man, and Cybernetics, Part C* , 28, 338–355.

Kanungo, T., Mount, D. M., Netanyahu, N. S., Piatko, C. D., Silverman, R., & Wu, A. Y. (2002). An efficient k-means clustering algorithm: Analysis and implementation. *IEEE Transactions on Pattern Analysis and Machine Intelligence*, *24*(7), 881–892. doi:10.1109/TPAMI.2002.1017616

Kaplantzis, S., Shilton, A., Mani, N., & Sekercioglu, Y. A. (2007, December). Detecting selective forwarding attacks in wireless sensor networks using support vector machines. In *Intelligent Sensors, Sensor Networks and Information, 2007. ISSNIP 2007. 3rd International Conference on* (pp. 335-340). IEEE.

Karlsson, A., Al-Saadeh, O., Gusarov, A., Renuka, V. R. C., Tombaz, S., & Sung, K. W. (2016). Energy-efficient 5G deployment in rural areas. *Proceedings of IEEE 12th International Conference on Wireless and Mobile Computing, Networking and Communications (WiMob)*, 1–7.

Karneyenka, V., Mohta, K., & Moh, M. (2017). Location and Mobility Aware Resource Management for 5G Cloud Radio Access Networks. In *Proc. of the Int. Conf. on High Performance Computing and Simulation (HPCS)*. Genoa, Italy: IEEE Press. 10.1109/HPCS.2017.35

Kashyap, R., & Piersson, A. (2018). Big Data Challenges and Solutions in the Medical Industries. In Handbook of Research on Pattern Engineering System Development for Big Data Analytics. IGI Global. doi:10.4018/978-1-5225-3870-7.ch001

Kashyap, R., Gautam, P., & Tiwari, V. (2018). Management and Monitoring Patterns and Future Scope. In Handbook of Research on Pattern Engineering System Development for Big Data Analytics. IGI Global. doi:10.4018/978-1-5225-3870-7.ch014

Kashyap, R. (2018). Object boundary detection through robust active contour based method with global information. *International Journal Of Image Mining*, *3*(1), 22. doi:10.1504/IJIM.2018.093008

Kashyap, R., & Gautam, P. (2017). 'Fast Medical Image Segmentation Using Energy-Based Method', Biometrics. *Concepts, Methodologies, Tools, and Applications*, *3*(1), 1017–1042. doi:10.4018/978-1-5225-0983-7.ch040

Kashyap, R., & Gautam, P. (2017). *Fast Medical Image Segmentation Using Energy-Based Method.* Pattern and Data Analysis in Healthcare Settings, Medical Information Science Reference.

Kashyap, R., & Piersson, A. (2018). *Impact of Big Data on Security. In Handbook of Research on Network Forensics and Analysis Techniques* (pp. 283–299). IGI Global. doi:10.4018/978-1-5225-4100-4.ch015

Kashyap, R., & Tiwari, V. (2017). Energy-based active contour method for image segmentation. *International Journal of Electronic Healthcare*, *9*(2–3), 210–225. doi:10.1504/IJEH.2017.083165

Kassim, M. R. M., & Harun, A. N. (2016, October). Applications of WSN in agricultural environment monitoring systems. In *Information and Communication Technology Convergence (ICTC), 2016 International Conference on* (pp. 344-349). IEEE.

Kaur, H., Sawhney, R. S., & Komal, N. (2012). Wireless sensor networks for disaster management. *International Journal of Advanced Research in Computer Engineering & Technology (IJARCET)*, *1*(5).

Kaur, G., & Moh, M. (2018). Cloud Computing Meets 5G Networks: Efficient Cache Management in Cloud Radio Access Networks. In *Proc. of 2018 Annual ACM Southeast Conference*. Richmond, KY: ACM Press. 10.1145/3190645.3190674

Kaur, S., Sharma, S., Jain, U., & Raj, A. (2016, July). Voice command system using raspberry pi. *Advanced Computational Intelligence: An International Journal*, *3*. doi:10.5121/acii.2016.3306

Kelly, F. (1997). Charging and Rate Control for Elastic Traffic. *European Transactions of Telecommunications*, *8*(1), 33–37. doi:10.1002/ett.4460080106

Khan, N., Martini, M. G., Bharucha, Z., & Auer, G. (2012). Opportunistic Packet Loss Fair Scheduling for Delay-Sensitive Applications over LTE Systems. In *Proceedings of IEEE Wireless Communications and Networking Conference* (pp. 1456 – 1461). Shanghai, China: IEEE. 10.1109/WCNC.2012.6214010

Khan, U. M., Kabir, Z., & Hassan, S. A. (2017a). A deep learning framework using passive WiFi sensing for respiration monitoring. *IEEE Global Communication Conference (Globecom)*. Singapore: IEEE. 10.1109/GLOCOM.2017.8255027

Khan, U. M., Kabir, Z., & Hassan, S. A. (2017b). Wireless health monitoring using passive Wi-Fi sensing. In *International Wireless Communications and Mobile Computing Conference (IWCMC)*. Valencia, Spain: Academic Press.

Kim, J., Ko, S.-W., Cha, H., & Kim, S.-L. (2017). Testbed Verification of Spectrum Access Opportunity Detection in Cognitive Radio Networks. *23rd Asia-Pacific Conference on Communications (APCC)*. 10.23919/APCC.2017.8304086

Kim, L. (2018). DeepX: Deep Learning Accelerator for Restricted Boltzmann Machine Artificial Neural Networks. *IEEE Transactions on Neural Networks and Learning Systems*, *29*(5), 1441–1453. doi:10.1109/TNNLS.2017.2665555 PMID:28287986

Kim, M. H., & Park, M. G. (2009). Bayesian statistical modeling of system energy saving effectiveness for MAC protocols of wireless sensor networks. In *Software Engineering, Artificial Intelligence, Networking and Parallel/Distributed Computing* (pp. 233–245). Berlin: Springer. doi:10.1007/978-3-642-01203-7_20

Kolding, T. E. (2006). QoS-Aware Proportional Fair Packet Scheduling with Required Activity Detection. In *Proceedings of IEEE Vehicular Technology Conference*, (pp. 1-5). Montreal, Quebec, Canada: IEEE. 10.1109/VTCF.2006.277

Kotobi, K., & Bilén, S. G. (2018). Secure Blockchains for DYNAMIC Spectrum Access. *IEEE Vehicular Technology Magazine*, *13*(1), 32–39. doi:10.1109/MVT.2017.2740458

Koushika, A. M., & Thamarai Selvi, S. (2014). Load Balancing Using Software Defined Networking in cloud environment. In *Proceedings of 2014 International Conference on Recent Trends in Information Technology (ICRTIT)*. IEEE Press. 10.1109/ICRTIT.2014.6996164

Kowalski, S., Pavlovska, K., & Goldstein, M. (2013). Two case studies in using chatbots for security training. In *Information Assurance and Security Education and Training* (pp. 265–272). Springer. doi:10.1007/978-3-642-39377-8_31

Krause, A., Guestrin, C., Gupta, A., & Kleinberg, J. (2006). *Near-optimal sensor placements*. Nashville, TN: IPSN. doi:10.1145/1127777.1127782

Krishnamachari, L., Estrin, D., & Wicker, S. (2002). The impact of data aggregation in wireless sensor networks. In *Distributed Computing Systems Workshops, 2002. Proceedings. 22nd International Conference on* (pp. 575-578). IEEE.

Krizhevsky, A., Sutskever, I., & Hinton, G. E. (2012). Imagenet classification with deep convolutional neural networks. Proc. Advance Neural Information Processing System, 25, 1097–1105.

Krysik, P., Gajo, Z., Kulpa, J. S., & Malanowski, M. (2014). Reference channel equalization in FM passive radar using the constant magnitude algorithm. In *International Radar Symposium (IRS)*. Gdansk, Poland: Academic Press. 10.1109/IRS.2014.6869284

Kulkarni, R. V., & Venayagamoorthy, G. K. (2009, June). Neural network based secure media access control protocol for wireless sensor networks. In *Neural Networks, 2009. IJCNN 2009. International Joint Conference on* (pp. 1680-1687). IEEE.

Kulkarni, R. V., Member, S., Förster, A., & Venayagamoorthy, G. K. (2011). Computational Intelligence in Wireless Sensor Networks : A Survey. *IEEE Communications Surveys and Tutorials*, *13*(1), 68–96. doi:10.1109/SURV.2011.040310.00002

Kulkarni, S. R., Lugosi, G., & Venkatesh, S. S. (1998). Learning pattern classification-a survey. *IEEE Transactions on Information Theory*, *44*(6), 2178–2206. doi:10.1109/18.720536

Kwan, R., Leung, C., & Zhang, J. (2009). Resource Allocation in an LTE Cellular Communication System. In *Proceedings of IEEE International Conference on Communications*, (pp. 1-5). Dresden, Germany: IEEE. 10.1109/ICC.2009.5199412

Langley, P., & Simon, H. A. (1995). Applications of machine learning and rule induction. *Communications of the ACM*, *38*(11), 54–64. doi:10.1145/219717.219768

Lan, P., Zhai, C., Chen, L., Gao, B., & Sun, F. (2018). Optimal power allocation for bi-directional full duplex underlay cognitive radio networks. *IET Communications*, *12*(2), 220–227. doi:10.1049/iet-com.2016.1384

LeCun, Y., Bengi, Y., & Hilton, G. (2015). Deep Learning. *Nature, 521*(7553), 436–444. doi:10.1038/nature14539 PMID:26017442

Lee, M., Marconett, D., Ye, X., & Yoo, S. (2007). Cognitive network management with reinforcement learning for wireless mesh networks. *IP Operations and Management*, 168–179. doi:10.1007/978-3-540-75853-2_15

Liang, W., Ng, S. X., & Hanzo, L. (2017). Cooperative Overlay Spectrum Access in Cognitive Radio Networks. *IEEE Communications Surveys and Tutorials, 19*(3), 1924–1944. doi:10.1109/COMST.2017.2690866

Li, G., Jiang, S., Zhang, W., Pang, J., & Huang, Q. (2014). Online web video topic detection and tracking with semi-supervised learning. *Multimedia Systems, 22*(1), 115–125. doi:10.100700530-014-0402-0

Lim, H. B., Foo, M. C., & Zeng, Y. (2006, December). An adaptive distributed resource allocation scheme for sensor networks. In *International Conference on Mobile Ad-Hoc and Sensor Networks* (pp. 770-781). Springer. 10.1007/11943952_65

Lippmann, R. (1987). An introduction to computing with neural nets. *IEEE ASSP Magazine, 4*(2), 4–22. doi:10.1109/MASSP.1987.1165576

Lischinski, D. (2007). *The Power Method.* Academic Press.

Little, B. (2010). Concerns with Learning-Management Systems and Virtual Learning Environments. *Elearn, 2010*(7), 2. doi:10.1145/1833513.1837142

Liu, X. (2012). A survey on clustering routing protocols in wireless sensor networks. *Sensors, 12*(8), 11113-11153.

Liu, B., Tian, H., & Xu, L. (2013). An Efficient Downlink Packet Scheduling Algorithm for Real Time Traffics in LTE Systems. In *Proceedings of IEEE Consumer Communications and Networking Conference (CCNC)* (vol. 1, pp. 364-369). Las Vegas, NV: IEEE.

Liu, L., Yang, F., Wang, R., Shi, Z., Stidwell, A., & Gu, D. (2012). Analysis of handover performance improvement in cloud-RAN architecture. In *Proc. 7th International ICST Conference on Communications and Networking in China.* IEEE Press.

Liu, X., Chong, E., & Shroff, N. (2001). Opportunistic Transmission Scheduling with Resource- Sharing Constraints in Wireless Networks. *IEEE Journal on Selected Areas in Communications, 19*(10), 2053–2064. doi:10.1109/49.957318

LIVE -Labaratory for Image and Video Quality Engineering, an Image Quality Assessment Database. (2003). Retrieved from http://live.ece.utexas.edu/research/quality/subjective.htm

Li, Y., Pateromichelakis, E., Vucic, N., Luo, J., Xu, W., & Caire, G. (2017). Radio Resource Management Considerations for 5G Millimeter Wave Backhaul and Access Networks. *IEEE Communications Magazine, 55*(6), 86–92. doi:10.1109/MCOM.2017.1601118

Lopez, P. G., Montresor, A., Epema, D., Datta, A., Higashino, T., Iamnitchi, A., ... Riviere, E. (2015). Edge-centric Computing. *Computer Communication Review, 45*(5), 37–42. doi:10.1145/2831347.2831354

Lu, C. H., & Fu, L. C. (2009). Robust location-aware activity recognition using wireless sensor network in an attentive home. *IEEE Transactions on Automation Science and Engineering, 6*(4), 598–609. doi:10.1109/TASE.2009.2021981

Lundevall, M., Olin, B., Olsson, J., Wiberg, N., Wanstedt, S., Eriksson, J., & Eng, F. (2004). Streaming Applications over HSDPA in Mixed Service Scenarios. In *Proceedings of IEEE Vehicular Technology Conference (VTC-Fall)* (vol. 2, pp. 841 - 845). Los Angeles, CA: IEEE. 10.1109/VETECF.2004.1400139

Luo, X., & Chang, X. (2015). A novel data fusion scheme using grey model and extreme learning machine in wireless sensor networks. *International Journal of Control, Automation, and Systems, 13*(3), 539–546. doi:10.100712555-014-0309-8

Malanowski, M. (2012). Algorithm for target tracking using passive radar. *International Journal of Electronics and Telecommunications*, *58*(4), 345–350. doi:10.2478/v10177-012-0047-x

Mao, G. (2009). *Localization Algorithms and Strategies for Wireless Sensor Networks.* Science And Technology; doi:10.4018/978-1-60566-396-8

Mawdsley, G., Tyson, A., Peressotti, C., Jong, R., & Yaffe, M. (2009). Accurate estimation of compressed breast thickness in mammography. *Medical Physics*, *36*(2), 577–586. doi:10.1118/1.3065068 PMID:19291997

McKinney, W. (2018). *Python for data analysis: data wrangling with Pandas, NumPy, and IPython*. Sebastopol, CA: O'Reilly Media, Inc.

Meng, M., Xu, H., Wu, X., d'Auriol, B. J., Jeong, B. S., Lee, S., & Fan, X. (2007, June). PBR: Priority Based Routing in Multi-Sink Sensor Networks. In ICWN (pp. 380-386). Academic Press.

Misra, S., Krishna, P., & Abraham, K. (2010). A stochastic learning automata-based solution for intrusion detection in vehicular ad hoc networks. *Security and Communication Networks*, *4*(6), 666–677. doi:10.1002ec.200

Mnih, V., Kavukcuoglu, K., Silver, D., Graves, A., Antonoglou, I., Wierstra, D., & Riedmiller, M. (2013). Playing Atari with Deep Reinforcement Learning. *Proceedings of NIPS Deep Learning Workshop*. arXiv:1312.5602

Mnih, V., Badia, A. P., Mirza, M., Graves, A., Harley, T., Silver, D., & Kavukcuoglu, K. (2016). Asynchronous Methods for Deep Reinforcement Learning. In *Proceedings of the 33rd International Conference on Machine Learning*. New York, NY: W&CP.

Monghal, G., Laselva, D., Michaelsen, P.-H., & Wigard, J. (2010). Dynamic Packet Scheduling for Traffic Mixes of Best Effort and VoIP Users in E-UTRAN Downlink. In *Proceedings of IEEE Vehicular Technology Conference (VTC-Spring)* (pp. 1-5). Taipei, Taiwan: IEEE. 10.1109/VETECS.2010.5493737

Morelande, M. R., Moran, B., & Brazil, M. (2008, March). Bayesian node localisation in wireless sensor networks. In *Acoustics, Speech and Signal Processing, 2008. ICASSP 2008. IEEE International Conference on* (pp. 2545-2548). IEEE.

Muñoz, P., Barco, R., Ruiz-Avilés, J. M., de la Bandera, I., & Aguilar, A. (2013). *Fuzzy Rule-Based Reinforcement Learning for Load Balancing Techniques in Enterprise LTE Femtocells* (Vol. 62). IEEE Transactions on Vehicular Technology.

Neath, A., & Cavanaugh, J. (2011). The Bayesian information criterion: Background, derivation, and applications. *Wiley Interdisciplinary Reviews: Computational Statistics*, *4*(2), 199–203. doi:10.1002/wics.199

Nguyen, T. T. T., & Armitage, G. (2008). A survey of techniques for internet traffic classification using machine learning. *IEEE Communications Surveys and Tutorials*, *10*(4), 56–76. doi:10.1109/SURV.2008.080406

Ning, X., Ting, Z., Ying, W., & Ping, Z. (2006). A MC-GMR Scheduler for Shared Data Channel in 3GPP LTE System. In *Proceedings of IEEE Vehicular Technology Conference (VTC-Fall)* (pp. 1-5). Montreal, Quebec, Canada: IEEE. 10.1109/VTCF.2006.279

NIST/SEMATECH. (n.d.). *What are outliers in the data?* Retrieved from http://www.itl.nist.gov/div898/handbook/

Nocedal, J., & Wright, S. J. (2006). Penalty, Barrier, and Augmented Lagrangian Methods. In *Numerical Optimization* (pp. 488–524). Springer Series in Operations Research. doi:10.1007/978-0-387-40065-5_17

Noel, K. (2016). *Application of Machine Learning to Systematic Allocation Strategies.* SSRN Electronic Journal; doi:10.2139srn.2837664

O'Shea, T., & Hoydis, J. (2017). An introduction to deep learning for the physical layer. *IEEE Transactions on Cognitive Communications and Networking*, *3*(4), 563–575. doi:10.1109/TCCN.2017.2758370

Olwal, T. O., Djouani, K., & Kurien, A. M. (2016). A Survey of Resource Management Toward 5G Radio Access Networks. *IEEE Communications Surveys and Tutorials*, *18*(3), 1656–1686. doi:10.1109/COMST.2016.2550765

Ott, D., & Houdek, F. (2014). Automatic Requirement Classification: Tackling Inconsistencies Between Requirements and Regulations. *International Journal of Semantic Computing*, *08*(01), 47–65. doi:10.1142/S1793351X14500020

Palattella, M., Dohler, M., Grieco, A., Rizzo, G., Torsner, J., Engel, T., & Ladid, L. (2017). Internet of Things in the 5G era: Enables, architecture, and business model. *IEEE Journal of Selected Advanced Topics in Communications.*

Pandana, C., & Liu, K. R. (2005). Near-optimal reinforcement learning framework for energy-aware sensor communications. *IEEE Journal on Selected Areas in Communications*, *23*(4), 788–797. doi:10.1109/JSAC.2005.843547

Paradis, L., & Han, Q. (2007). A survey of fault management in wireless sensor networks. *Journal of Network and Systems Management*, *15*(2), 171–190. doi:10.100710922-007-9062-0

Pearson, K. (1901). On lines and planes of closest fit to systems of points in space. *Philosophical Magazine*, *2*, 559–572.

Pilato, G., Vassallo, G., Augello, A., Vasile, M., & Gaglio, S. (2005). Expert chat-bots for cultural heritage. *IntelligenzaArtificiale*, *2*(2), 25–31.

Pinto, A. R., & Montez, C. (2010). Autonomic approaches for enhancing communication QoS in dense Wireless Sensor Networks with real time requirements. *Proceedings - International Test Conference*, 1–10. 10.1109/TEST.2010.5699288

Pinto, A. R., Montez, C., Araújo, G., Vasques, F., & Portugal, P. (2014). An approach to implement data fusion techniques in wireless sensor networks using genetic machine learning algorithms. *Information Fusion*, *15*(1), 90–101. doi:10.1016/j.inffus.2013.05.003

Proebster, M., Mueller, C. M., & Bakker, H. (2010). Adaptive Fairness Control for a Proportional Fair LTE Scheduler. *Proceedings of IEEE International Symposium on Personal Indoor and Mobile Radio.* 10.1109/PIMRC.2010.5671970

Punyavachira, T. (2013). *Forecasting Stock Indices Movement Using Hybrid Model: A Comparison of Traditional and Machine Learning Approaches*. SSRN Electronic Journal. doi:10.2139srn.2416494

Puthal, D. (2018). Lattice-modelled Information Flow Control of Big Sensing Data Streams for Smart Health Application. *IEEE Internet of Things Journal*. doi:10.1109/jiot.2018.2805896

Python Graphing Library. PlotlyAPI. (n.d.). Retrieved from https://plot.ly/python/

Rajasegarar, S., Leckie, C., & Palaniswami, M. (2008). Anomaly detection in wireless sensor networks. *IEEE Wireless Communications*, *15*(4), 34–40. doi:10.1109/MWC.2008.4599219

Rajkumari, R., & Marchang, N. (2018). Secure Non-consensus Based Spectrum Sensing in Non-Centralized Cognitive Radio Networks. *IEEE Sensors Journal*, *18*(9), 3883–3890. doi:10.1109/JSEN.2018.2816595

Raju, R., & Moh, M. (2018). Machine learning techniques for security of Internet of Things (IoT) and fog computing systems. In *Proceedings of IEEE International Conference on High Performance Computing and Simulation*. Orleans, France: IEEE.

Rappaport, T. S. (1996). *Wireless communications: Principles and practice* (vol. 2). Prentice Hall PTR.

Rastegardoost, N., & Etemadyrad, N. (2015). *Machine Learning in Wireless Sensor Networks*. Academic Press.

Rathore, M., Paul, A., Ahmad, A., & Jeon, G. (2017). IoT-Based Big Data. *International Journal on Semantic Web and Information Systems*, *13*(1), 28–47. doi:10.4018/IJSWIS.2017010103

Reguri, V. R., Kogatam, S., & Moh, M. (2016). Energy Efficient Traffic-Aware Virtual Machine Migration in Green Cloud Data Centers. In *Proc. the 2nd IEEE Int. Conf. on High Performance and Smart Computing (HPSC)*. IEEE Press. 10.1109/BigDataSecurity-HPSC-IDS.2016.55

Rhee, J. H., Holtzman, J. M., & Kim, D.-K. (2003). Scheduling of Real/Non-Real Time Services: Adaptive EXP/PF Algorithm. In *Proceedings of IEEE Semiannual Vehicular Technology Conference (VTC-Spring)* (vol. 1, pp. 462 – 466). Jeju, South Korea: IEEE.

Rich, E., Knight, K., & Nair, S. B. (2009). *Artificial Intelligence* (3rd ed.). Retrieved August 22, 2018, from https://i4iam.files.wordpress.com/2013/08/artificial-intelligence-by-rich-and-knight.pdf

Ritter, A., Cherry, C., & Dolan, W. B. (2011). Datadriven response generation in social media. *Proceedings of the conference on empirical methods in natural language processing, Association for Computational Linguistics*, 583–593.

Rodriguez, J. D., Perez, A., & Lozano, J. A. (2010). Sensitivity Analysis of k-Fold Cross Validation in Prediction Error Estimation. *IEEE Transactions on Pattern Analysis and Machine Intelligence*, *32*(3), 569–575. doi:10.1109/TPAMI.2009.187 PMID:20075479

Roos, T., Myllymaki, P., Tirri, H., Misikangas, P., & Sievanen, J. (2002). A Probabilistic Approach to WLAN User Location Estimation. *International Journal of Wireless Information Networks*, *9*(3), 155–164. doi:10.1023/A:1016003126882

S, J. P., & Sudha, T. (2017). Energy Efficient Wireless Sensor Network Assisted Spectrum Sensing for Cognitive Radio Network. *IEEE International Conference on Intelligent Techniques in Control, Optimization and Signal Processing (INCOS)*.

Sabbah, A., & Ibnkahla, M. (2017). Optimizing Power Allocation in Mission Critical Cognitive Radio Networks. In *2017 IEEE 28th Annual International Symposium on Personal, Indoor, and Mobile Radio Communications (PIMRC)* (pp. 1-7). Montreal: IEEE.

Sądel, B., & Śnieżyński, B. (2017). Online Supervised Learning Approach for Machine Scheduling. *Schedae Informaticae*, *1/2016*. doi:10.4467/20838476SI.16.013.6194

Sadiq, B., Madan, R., & Sampath, A. (2009). Downlink Scheduling for Multiclass Traffic in LTE. *EURASIP Journal on Wireless Communications and Networking*, *2009*(1), 1–18. doi:10.1155/2009/510617

Salakhutdinov, R., & Hinton, G. E. (2009). Deep Boltzmann machines. *Proc. Int. Conf. AI Statististics.*, 448–455.

Salameh, H. B., Almajali, S., Ayyash, M., & Elgala, H. (2018). Spectrum Assignment in Cognitive Radio Networks for Internet-of-Things Delay-sensitive Applications under Jamming Attacks. *IEEE Internet of Things Journal*.

Sandra, P. R. (2013). A Survey on Recommendation Techniques in E-Commerce. *IEEE Communications Surveys and Tutorials*, *2*(12), 1586–1590.

Sathyanarayana, S., & Moh, M. (2016). Joint Route-Server Load Balancing in Software Defined Networks using Ant Colony Optimization. In *Proceedings of the International Conference on High Performance Computing and Simulation (HPCS)*. Innsbruck, Austria. IEEE Press. 10.1109/HPCSim.2016.7568330

Saxena, N., Roy, A., Sahu, B., & Kim, H. (2017, February). Efficient IoT gateway over 5G wireless: New design with prototype and implementation results. *IEEE Communications Magazine*.

Schalkwyk, J., Beeferman, D., Beaufays, F., Byrne, B., Chelba, C., Cohen, M., . . . Strope, B. (n.d.). *Google Search by Voice: A Case Study*. Retrieved August 22, 2018, from https://static.googleusercontent.com/media/research.google.com/en//pubs/archive/36340.pdf

Schroeder, A., Edrich, M., & Winkler, V. (2013). Stationary cassidian FM passive radar demonstrator for 24/7 operation and sensor cluster measurements. In *International Radar Symposium (IRS)*. Dresden, Germany: Academic Press.

Schumaker, R. (2013). Machine learning the harness track: Crowdsourcing and varying race history. *Decision Support Systems*, *54*(3), 1370–1379. doi:10.1016/j.dss.2012.12.013

Schwab, D., & Ray, S. (2017). Offline reinforcement learning with task hierarchies. *Machine Learning*, *106*(9-10), 1569–1598. doi:10.100710994-017-5650-8

Schwarz, S., Mehlfuhrer, C., & Rupp, M. (2011). Throughput Maximizing Multiuser Scheduling with Adjustable Fairness. In *Proceedings of IEEE International Conference on Communications (ICC)* (pp.1-5). Kyoto, Japan: IEEE. 10.1109/icc.2011.5963489

Seshadrinathan, K., & Bovik, A. C. (2010). Motion tuned spatio-temporal quality assessment of natural videos. *IEEE Transactions on Image Processing*, *19*(2), 335–350. doi:10.1109/TIP.2009.2034992 PMID:19846374

Seshadrinathan, K., Soundararajan, R., Bovik, A. C., & Cormack, L. K. (2010). Study of Subjective and Objective Quality Assessment of Video. *IEEE Transactions on Image Processing*, *19*(6), 1427–1441. doi:10.1109/TIP.2010.2042111 PMID:20129861

Shahriari, B., & Moh, M. (2017). Generic Online Learning for Partial Visible Dynamic Environment with Delayed Feedback - Online Learning for 5G C-RAN Load-Balancer. In *Proc. of the International Conference on High Performance Computing and Simulation (HPCS)*. Genoa, Italy. IEEE Press. 10.1109/HPCS.2017.36

Shareef, A., Zhu, Y., & Musavi, M. (2008, February). Localization using neural networks in wireless sensor networks. In *Proceedings of the 1st international conference on MOBILe Wireless MiddleWARE, Operating Systems, and Applications* (p. 4). ICST (Institute for Computer Sciences, Social-Informatics and Telecommunications Engineering). 10.4108/ICST.MOBILWARE2008.2901

Sharma, V. (2015, May 21). *Cognitive radio networks: Technology Survey & Research Challenges.* Retrieved April 15, 2018, from https://www.slideshare.net/vatsalasharma180/cognitive-radio-networks-48446892

Sharma, D. K., Singh, A., Khanna, A., & Jain, A. (2017). Evaluation of Parameters and Techniques for Genetic Algorithm based Channel Allocation in Cognitive Radio Networks. In *2017 Tenth International Conference on Contemporary Computing (IC3)* (pp. 1 - 6). Noida: IEEE. 10.1109/IC3.2017.8284341

Sharma, S. K., & Wang, X. (2017). Live Data Analytics With Collaborative Edge and Cloud Processing in Wireless IoT Networks. *IEEE Access: Practical Innovations, Open Solutions*, *5*, 4621–4635. doi:10.1109/ACCESS.2017.2682640

Shen, Y. J., & Wang, M. S. (2008). Broadcast scheduling in wireless sensor networks using fuzzy Hopfield neural network. *Expert Systems with Applications*, *34*(2), 900–907. doi:10.1016/j.eswa.2006.10.024

Shi, Y. F., Park, S. H., & Song, T. L. (2017). Multi-target tracking in cluttered environment for a multi-static passive radar system under the DAB/DVB network. *EURASIP Journal on Advances in Signal Processing*, *2017*(1), 11. doi:10.118613634-017-0445-4

Shnayderman, A., Gusev, A., & Eskicioglu, A. M. (2006). An SVD-based grayscale image quality measure for local and global assessment. *IEEE Transactions on Image Processing*, *15*(2), 422–429. doi:10.1109/TIP.2005.860605 PMID:16479812

Shu, H., & Liang, Q. (2005, March). Fuzzy optimization for distributed sensor deployment. In *Wireless Communications and Networking Conference*, 2005 *IEEE* (Vol. 3, pp. 1903-1908). IEEE.

Sinha, A., & Chandrakasan, A. (2001). Dynamic power management in wireless sensor networks. *IEEE Design & Test of Computers*, *18*(2), 62–74. doi:10.1109/54.914626

Skolnik, M. L. (2007). Fluttar DEW-Line Gap-Filler. In H. Griffiths & N. Willis (Eds.), *Advances in bi-static radar* (pp. 35–47). Raleigh, NC: Scitech Publishing, Inc.

Sleeman, R., & Van Eck, T. (1999). Robust automatic P-phase picking: An on-line implementation in the analysis of broadband seismogram recordings. *Physics of the Earth and Planetary Interiors, 113*(1-4), 265–275. doi:10.1016/S0031-9201(99)00007-2

Smith, L. (2002). *A tutorial on principal components analysis.* Tech. rep., Cornell University. Retrieved from http://www.cs.otago.ac.nz/cosc453/student_tutorials/principal_components.pdf

Sohrabi, K., Gao, J., Ailawadhi, V., & Pottie, G. J. (2000). Protocols for self-organization of a wireless sensor network. *IEEE Personal Communications, 7*(5), 16-27.

Song, G. (2005). *Cross-Layer Resource Allocation and Scheduling in Wireless Multicarier Networks.* Atlanta, GA: Georgia Institute of Technology.

Song, J., Yoo, T., & Song, P. J. (2016). Mobility level management for 5G network. *Proceedings of International Conference on Information and Communication Technology Convergence (ICTC)*, 940–943.

Specht, D. F. (1991). A general regression neural network. *IEEE Transactions on Neural Networks, 2*(6), 568–576. doi:10.1109/72.97934 PMID:18282872

Sriharipriya, K. C., & Sanju, R. (2017). Artifical Neural Network Based Multi Dimensional Spectrum Sensing In Full Duplex Cognitive Radio Networks. In *2017 International Conference on Computing Methodologies and Communication (ICCMC)* (pp. 307 - 312). Erode: IEEE. 10.1109/ICCMC.2017.8282697

Srinivasan, V., Lapuschkin, S., Hellge, C., Müller, K. R., & Samek, W. (2017, March). Interpretable human action recognition in compressed domain. In *Acoustics, Speech and Signal Processing (ICASSP), 2017 IEEE International Conference on* (pp. 1692-1696). IEEE.

Statistica. (2018). *Global IP Data Traffic from 2016 to 2021.* Retrieved from https://www.statista.com/statistics/499431/global-ip-data-traffic-forecast

Steinwart, I., & Christmann, A. (2008). *Support vector machines.* Springer Science & Business Media.

Stone, P. (2000, June). *TPOT-RL applied to network routing.* ICML.

Stone, P., & Veloso, M. (1999, April). Team-partitioned, opaque-transition reinforcement learning. In *Proceedings of the third annual conference on Autonomous Agents* (pp. 206-212). ACM. 10.1145/301136.301195

Su, G., & Moh, M. (2018). Improving Energy Efficiency and Scalability for IoT Communications in 5G Networks. In *Proc. of 12th ACM Int. Conf. on Ubiquitous Information Management and Communication (IMCOM).* Langkawi, Malaysia: ACM. 10.1145/3164541.3164547

Sumi, M. S., & Ganesh, R. S. (2017). Performance Enhancing Techniques in Cognitive Radio Networks. In *IEEE International Conference on Circuits and Systems (ICCS 2017)* (pp. 172 - 178). Kerala: IEEE. 10.1109/ICCS1.2017.8325984

Šumrada, R. (2011). Free OGC standardized web services for spatial data. *Geodetski Vestnik, 55*(1), 046-056. doi:10.15292/geodetski-vestnik.2011.01.046-056

Sun, R., Tatsumi, S., & Zhao, G. (2002, October). Q-map: A novel multicast routing method in wireless ad hoc networks with multiagent reinforcement learning. In *TENCON'02. Proceedings. 2002 IEEE Region 10 Conference on Computers, Communications, Control and Power Engineering* (Vol. 1, pp. 667-670). IEEE.

Sun, H., & Betti, R. (2015). A Hybrid Optimization Algorithm with Bayesian Inference for Probabilistic Model Updating. *Computer-Aided Civil and Infrastructure Engineering*, *30*(8), 602–619. doi:10.1111/mice.12142

Sussner, P., Esmi, E., & Barros, L. C. (2016). Controlling the width of the sum of interactive fuzzy numbers with applications to fuzzy initial value problems. *Proceedings of IEEE International Conference on Fuzzy Systems*, 1453–1460.

Sutton, R. S., & Barto, A. G. (2017). *Reinforcement Learning: An Introduction*. London: IMT Press. doi:10.1007/978-1-4899-7687-1_720

Tam, V., Cheng, K. Y., & Lui, K. S. (2006, November). A descend-based evolutionary approach to enhance position estimation in wireless sensor networks. In Null (pp. 568-574). IEEE.

Tan, B., Woodbridge, K., & Chetty, K. (2014). A real-time high resolution passive WiFi doppler-radar and its applications. In *International Radar Conference*. Lille, France: Academic Press. 10.1109/RADAR.2014.7060359

Tango, F., Minin, L., Tesauri, F., & Montanari, R. (2010). Field tests and machine learning approaches for refining algorithms and correlations of driver's model parameters. *Applied Ergonomics*, *41*(2), 211–224. doi:10.1016/j.apergo.2009.01.010 PMID:19286165

Telecommunication Standardization Sector of, ITU. (2008). Subjective video quality assessment methods for multimedia applications. [Tech. Rep.]. *ITU-T P*, *910*, 4–5.

Thakur, P., Kumar, A., Pandit, S., Singh, G., & Satashia, S. N. (2017). Effect of Imperfect Spectrum Monitoring on Cognitive Radio Network Performance. *Fourth International Conference on Image Information Processing (ICIIP)*. 10.1109/ICIIP.2017.8313716

Thomas, R., & Poonguzhali, P. K. (2014). An energy efficient multi-sink clustering algorithm for wireless sensors networks. *International Journal of Advanced Engineering and Global Technology*, *2*(4).

Tibshirani, R., James, G., Witten, D., & Hastie, T. (2013). Linear Regression, Support Vector Machines, Tree-Based Methods. In An Introduction to Statistical Learning with Applications in R (pp. 59-102, 344-354, 316-321). New York, NY: Springer.

Tofan, C. A. (2014). Optimization Techniques of Decision Making - Decision Tree. *Advances In Social Sciences Research Journal*, *1*(5), 142–148. doi:10.14738/assrj.15.437

Toivonen, H., & Gross, O. (2015). Data mining and machine learning in computational creativity. *Wiley Interdisciplinary Reviews. Data Mining and Knowledge Discovery*, *5*(6), 265–275. doi:10.1002/widm.1170

Toufik, I., & Knopp, R. (2011). Multi-User Scheduling and Interference Coordination. In LTE. The UMTS Long Term Evolution. From Theory To Practice (pp. 279-292). John Wiley & Sons Ltd. doi:10.1002/9780470978504.ch12

Trestian, R., Muntean, G., & Ormond, O. (2009). Signal Strength-based Adaptive Multimedia Delivery Mechanism. In *IEEE 34th Conference on Local Computer Networks, 2009. LCN 2009* (pp. 297–300). IEEE. 10.1109/LCN.2009.5355092

Trestian, R., Comşa, I.-S., & Tuysuz, M. F. (2018). Seamless Multimedia Delivery within a Heterogeneous Wireless Networks Environment: Are We There Yet? *IEEE Communications Surveys and Tutorials*, *20*(2), 945–977. doi:10.1109/COMST.2018.2789722

Trestian, R., Ormond, O., & Muntean, G.-M. (2012). On the Impact of Wireless Network Traffic Location and Access Technology on Mobile Device Energy Consumption. In *Proceedings of the 2012 IEEE 37th Conference on Local Computer Networks (LCN 2012)* (pp. 200–203). Washington, DC: IEEE. 10.1109/LCN.2012.6423608

Tropp, J. A., Gilbert, A. C., & Strauss, M. J. (2006). Algorithms for simultaneous sparse approximation. Part I: Greedy pursuit. *Signal Processing*, *86*(3), 572–588. doi:10.1016/j.sigpro.2005.05.030

Tsagkaris, K., Katidiotis, A., & Demestichas, P. (2008). Neural network-based learning schemes for cognitive radio systems. *Computer Communications*, *31*(14), 3394–3404. doi:10.1016/j.comcom.2008.05.040

Tsai, C. & Moh, M. (2018). Cache Management for 5G Cloud Radio Access Networks. In *Proc. of 12th ACM Int. Conf. on Ubiquitous Information Management and Communication (IMCOM)*. Langkawi, Malaysia: ACM Press. 10.1145/3164541.3164559

Tsai, C. W., Lai, C. F., Chiang, M. C., & Yang, L. T. (2014). Data mining for Internet of Things: A survey. *IEEE Communications Surveys and Tutorials*, *16*(1), 77–97. doi:10.1109/SURV.2013.103013.00206

Tsai, C., & Moh, M. (2017a). Abstract: Cache Management and Load Balancing for 5G Cloud Radio Access Networks. In *Proc. ACM Symposium on Cloud Computing*. Santa Clara, CA: ACM Press. 10.1145/3127479.3132690

Tsai, C., & Moh, M. (2017b). Load Balancing in 5G Cloud Radio Access Networks Supporting IoT Communications for Smart Communities. In *Proc. of 2017 IEEE International Symposium on Signal Processing and Information Technology (ISSPIT)*. Bilbao, Spain: IEEE. 10.1109/ISSPIT.2017.8388652

Udantha, M. (2015, May 15). *Natural Language Toolkit (NLTK) Sample and Tutorial: Part 1*. Retrieved from https://dzone.com/articles/natural-language-toolkit-nltk

United Nations. (2012). United Nations Secretary-General's high-level panel on global sustainability. In *Resilient People, Resilient Planet: A future worth choosing*. Author.

Urquhart, L., & Rodden, T. (2016). *A Legal Turn in Human Computer Interaction? Towards Regulation by Designn for the Internet of Things*. SSRN Electronic Journal. doi:10.2139srn.2746467

Van Hasselt, H., & Wiering, M. A. (2009). Using Continuous Action Spaces to Solve Discrete Problems. In *Proceedings of International Joint Conference on Neural Networks* (pp. 1149–1156). Atlanta, GA: Academic Press. 10.1109/IJCNN.2009.5178745

van Norden, W., de Jong, J., Bolderheij, F., & Rothkrantz, L. (2005, July). Intelligent task scheduling in sensor networks. In *Information Fusion, 2005 8th International Conference on* (Vol. 2, pp. 8-pp). IEEE. 10.1109/ICIF.2005.1592013

Varona, Penagarikano, Rodriguez-Fuentes, Bordel, & Diez. (2013). GTTS system for the spoken web search task at MediaEval2012. Proc. *MediaEval, 35.*

Vien, Q. T., Akinbote, T., Nguyen, H. X., Trestian, R., & Gemikonakli, O. (2015). On the coverage and power allocation for downlink in heterogeneous wireless cellular networks. In *2015 IEEE International Conference on Communications (ICC)* (pp. 4641–4646). IEEE. 10.1109/ICC.2015.7249055

Vien, Q. T., Nguyen, H. X., Trestian, R., Shah, P., & Gemikonakli, O. (2016). A Hybrid Double-Threshold Based Cooperative Spectrum Sensing over Fading Channels. *IEEE Transactions on Wireless Communications*, *15*(3), 1821–1834. doi:10.1109/TWC.2015.2496598

Vinyals, O., & Le, Q. (2015). A neural conversational model. *ICML Deep Learning Workshop 2015.*

Virmani, A., & Shahzad, M. (2017). Position and orientation agnostic gesture recognition using WiFi. In *International Conference on Mobile Systems, Applications, and Services (MobiSys)*. Niagara Falls, NY: Academic Press. 10.1145/3081333.3081340

Walt, S., Colbert, S. C., & Varoquaux, G. (2011). The NumPy Array: A Structure for Efficient Numerical Computation. *Computing in Science & Engineering*, *13*(2), 22–30. doi:10.1109/MCSE.2011.37

Wang, D., McMahan, C., & Gallagher, C. (2015). A general regression framework for group testing data, which incorporates pool dilution effects. *Statistics in Medicine*, *34*(27), 3606–3621. doi:10.1002im.6578 PMID:26173957

Wang, F., & Liu, J. (2011). Networked wireless sensor data collection: Issues, challenges, and approaches. *IEEE Communications Surveys and Tutorials*, *13*(4), 673–687. doi:10.1109/SURV.2011.060710.00066

Wang, H., Yao, Y.-D., & Peng, S. (2018). Prioritized Secondary User Access Control in Cognitive Radio Networks. *IEEE Access: Practical Innovations, Open Solutions*, *6*, 11007–11016. doi:10.1109/ACCESS.2018.2811341

Wang, K., Li, X., Ji, H., & Zhang, X. (2013). Heterogeneous Traffic Scheduling in Downlink High Speed Railway LTE Systems. In *Proceedings of IEEE Global Communications Conference (GLOBECOM)* (pp. 1452-1457). Atlanta, GA: IEEE. 10.1109/GLOCOM.2013.6831278

Wang, L., Zhang, Y., & Feng, J. (2005). On the Euclidean distance of images. *IEEE Transactions on Pattern Analysis and Machine Intelligence*, *27*(8), 1334–1339. doi:10.1109/TPAMI.2005.165 PMID:16119271

Wang, T., Wen, C.-K., Wang, H., Gao, F., Jiang, T., & Jin, S. (2017, November). Deep learning for wireless physical layer: Opportunities and challenges. *China Communications*, *14*(11), 92–111. doi:10.1109/CC.2017.8233654

Wang, X., Ekin, S., & Serpedin, E. (2018). Joint Spectrum Sensing and Resource Allocation in Multi-Band-Multi-User Cognitive Radio Networks. *IEEE Transactions on Communications*, 1–1.

Wang, X., Gao, L., Mao, S., & Pandey, S. (2017). CSI-based fingerprinting for indoor localization: A deep learning approach. *IEEE Transactions on Vehicular Technology*, *66*(1), 763–776.

Wang, Y., Martonosi, M., & Peh, L. S. (2006, May). A supervised learning approach for routing optimizations in wireless sensor networks. In *Proceedings of the 2nd international workshop on Multi-hop ad hoc networks: from theory to reality* (pp. 79-86). ACM. 10.1145/1132983.1132997

Wang, Y., Martonosi, M., & Peh, L. S. (2007). Predicting link quality using supervised learning in wireless sensor networks. *ACM SIGMOBILE Mobile Computing*, *11*(3), 71–83. doi:10.1145/1317425.1317434

Wang, Z., Bovik, A. C., Sheikh, H. R., & Simoncelli, E. P. (2004). Image quality assessment: From error visibility to structural similarity. *IEEE Transactions on Image Processing*, *13*(4), 600–612. doi:10.1109/TIP.2003.819861 PMID:15376593

Wang, Z., Chen, C., & Li, H.-X. (2016). A novel incremental learning scheme for reinforcement learning in dynamic environments. *Proceedings of 12th World Congress on Intelligent Control and Automation (WCICA)*, 2426–2431. 10.1109/WCICA.2016.7578530

Wang, Z., Lu, L., & Bovik, A. C. (2004). Video quality assessment using structural distortion measurement. *Signal Processing Image Communication*, *19*(2), 121–132. doi:10.1016/S0923-5965(03)00076-6

Watanabe, K., & Machida, M. (2012). Outdoor LTE Infrastructure Equipment (eNodeB). *Fujitsu Scientific and Technical Journal*, *48*(1), 27–32.

Watkins, C. J., & Dayan, P. (1992). Q-learning. *Machine Learning*, *8*(3-4), 279–292. doi:10.1007/BF00992698

Wei, Z.-H., & Hu, B.-J. (2018). A Fair Multi-Channel Assignment Algorithm With Practical Implementation in Distributed Cognitive Radio Networks. *IEEE Access: Practical Innovations, Open Solutions*, *6*, 14255–14267. doi:10.1109/ACCESS.2018.2808479

Wei, Z., He, Y., He, W., & Liu, X. (2017). Information Dissemination Model Based on Clustering Analysis of Information Network Development. In *4th International Conference on Information Science and Control Engineering (ICISCE)* (pp. 403-407). Changsha: Academic Press. 10.1109/ICISCE.2017.91

Welford, B. P. (1962). Note on a Method for Calculating Corrected Sums of Squares and Products. *Technometrics*, *4*(3), 419–420. doi:10.1080/00401706.1962.10490022

Werbos, P. J. (2011). Computational intelligence for the smart grid-history, challenges, and opportunities. *IEEE Computational Intelligence Magazine*, *6*(3), 14–21. doi:10.1109/MCI.2011.941587

Wiering, M. A., & van Hasselt, H. (2009). The QV Family Compared to Other Reinforcement Learning Algorithms. In *Proceedings of IEEE International Symposium on Adaptive Dynamic Programming and Reinforcement Learning* (pp. 101–108). Nashville, TN: IEEE. 10.1109/ADPRL.2009.4927532

Winter, J., Xu, Y., & Lee, W. C. (2005, July). Energy efficient processing of k nearest neighbor queries in location-aware sensor networks. In *Mobile and Ubiquitous Systems: Networking and Services, 2005. MobiQuitous 2005. The Second Annual International Conference on* (pp. 281-292). IEEE.

Wu, J. W., Liu, J., Huang, Z., & Zheng, S. (2015). Dynamic Fuzzy *Q*-Learning for Handover Parameters Optimization in 5G multi-tier networks. *Proceedings of International Conference on Wireless Communications & Signal Processing (WCSP)*, 1–5. 10.1109/WCSP.2015.7341220

Xiao-Yong, L., Jun, W., & Jue, W. (2015). Robust direction of arrival estimate method in FM-based passive bi-static radar with a four-element Adcock antenna array. *IET Radar, Sonar & Navigation*, *9*(4), 392–400. doi:10.1049/iet-rsn.2014.0270

Xie, P., Zhang, M., Zhang, G., Zheng, R., Xing, L., & Wu, Q. (2018). On physical-layer security for primary system in underlay cognitive radio networks. *IET Networks*, *7*(2), 68–73. doi:10.1049/iet-net.2017.0138

Yacoab, M. M., & Sundaram, V. (2011). Multiple sink based compressive data aggregation technique for wireless sensor network. *International Journal of Wireless & Mobile Networks*, *3*(2). doi:10.5121/ijwmn.2011.3215

Yao, M. (2014). Research on Learning Evidence Improvement for kNN Based Classification Algorithm. *International Journal Of Database Theory And Application*, *7*(1), 103–110. doi:10.14257/ijdta.2014.7.1.10

Yao, Y., & Gehrke, J. (2002). The cougar approach to in-network query processing in sensor networks. *SIGMOD Record*, *31*(3), 9–18. doi:10.1145/601858.601861

Ye, F., & Zhang, X. (2017). Evidence-theory-based Collaborative Spectrum Sensing with Efficient Creditability Evaluation in Cognitive Radio Networks. In *Progress in Electromagnetics Research Symposium - Fall (PIERS - FALL), 2017* (pp. 46 - 54). Singapore: IEEE.

Yildirim, P., Birant, D., & Alpyildiz, T. (2017). Data mining and machine learning in textile industry. *Wiley Interdisciplinary Reviews. Data Mining and Knowledge Discovery*, *8*(1), e1228. doi:10.1002/widm.1228

Younis, O., & Fahmy, S. (2004). HEED: A hybrid, energy-efficient, distributed clustering approach for ad hoc sensor networks. *IEEE Transactions on Mobile Computing*, *3*(4), 366–379. doi:10.1109/TMC.2004.41

Youssef, K., Messiha, N., & Abd-Elnaby, M. (2017). Throughput and fairness enhancement based resource allocation scheme for underlay cognitive radio networks. In *2017 Japan-Africa Conference on Electronics, Communications and Computers (JAC-ECC)* (pp. 86 - 90). Alexandria: IEEE. 10.1109/JEC-ECC.2017.8305780

Yu, L., Wang, N., & Meng, X. (2005, September). Real-time forest fire detection with wireless sensor networks. In *Wireless Communications, Networking and Mobile Computing, 2005. Proceedings. 2005 International Conference on* (Vol. 2, pp. 1214-1217). IEEE.

Yu, Z., Yu, Z., Black, A. W., & Rudnicky, A. I. (2016). Chatbot Evaluation and Database Expansion via Crowdsourcing. *Allen Institute for Artificial Intelligence.* Retrieved from https://www.semanticscholar.org/paper/Chatbot-Evaluation-and-Database-Expansion-via-Yu-Xu/a950b42d3e46f4e1fb0e7e6959bc85a66246adb8

Yuan, Z., Chen, S., Ghinea, G., & Muntean, G. M. (2014). User quality of experience of mulsemedia applications. *ACM Transactions on Multimedia Computing Communications and Applications*, *11*(1s), 15. doi:10.1145/2661329

Yuan, Z., Ghinea, G., & Muntean, G. M. (2015). Beyond multimedia adaptation: Quality of experience-aware multi-sensorial media delivery. *IEEE Transactions on Multimedia*, *17*(1), 104–117. doi:10.1109/TMM.2014.2371240

Yu, Z., & Tsai, J. J. P. (2008). A Framework of Machine Learning Based Intrusion Detection for Wireless Sensor Networks. *2008 IEEE International Conference on Sensor Networks, Ubiquitous, and Trustworthy Computing (Sutc 2008)*, 272–279. 10.1109/SUTC.2008.39

Zeng, Y., Ai, X., Wang, L., Wang, X., & Zheng, G. (2016). Experimental research of dual-polarization passive radar based on DTTB signal. *Journal of Computer and Communications*, *4*(3), 101–107. doi:10.4236/jcc.2016.43016

Zhang, B., Xiang, Y., & Wang, J. (2010). Information Filtering Algorithm Based on Semantic Understanding. *Dianzi Yu Xinxi Xuebao*, *32*(10), 2324–2330. doi:10.3724/SP.J.1146.2009.01393

Zhang, C., Wu, Y., Wang, J., & Luo, Z. (2016). FM-based multi-frequency passive radar system. In *IEEE International Conference on Signal Processing, Communications and Computing (ICSPCC)*. Hong Kong, China: IEEE.

Zhang, H., & Guo, X. (2014). SDN-based load balancing strategy for server cluster. In *2014 3rd Int. Conference on Cloud Computing and Intelligence Systems (CCIS)*. IEEE Press.

Zhang, J., Yuan, D., & Zhang, H. (2011). Joint Radio Resource Allocation and Scheduling in a Backhaul Constrained Multicell OFDMA Networks. In *Proceedings of 13th International Conference on Communication Technology (ICCT)* (pp. 47 – 51). Jinan, China: Academic Press.

Zhang, N. (2016). Semi-supervised extreme learning machine with wavelet kernel. *International Journal Of Collaborative Intelligence*, *1*(4), 298. doi:10.1504/IJCI.2016.10004854

Zhang, X., Zhang, X., Han, L., & Xing, R. (2018). Utilization-Oriented Spectrum Allocation in an Underlay Cognitive Radio Network. *IEEE Access: Practical Innovations, Open Solutions*, *6*, 12905–12912. doi:10.1109/ACCESS.2018.2808473

Zhang, Y., Kwong, S., Wang, X., Yuan, H., Pan, Z., & Xu, L. (2015). Machine learning-based coding unit depth decisions for flexible complexity allocation in high efficiency video coding. *IEEE Transactions on Image Processing*, *24*(7), 2225–2238. doi:10.1109/TIP.2015.2417498 PMID:25826804

Zhang, Y., Meratnia, N., & Havinga, P. (2009). Adaptive and Online One-Class Support Vector Machine-Based Outlier Detection Techniques for Wireless Sensor Networks. *2009 International Conference on Advanced Information Networking and Applications Workshops*, 990–995. 10.1109/WAINA.2009.200

Zhang, Y., Meratnia, N., & Havinga, P. J. M. (2013). Distributed online outlier detection in wireless sensor networks using ellipsoidal support vector machine. *Ad Hoc Networks*, *11*(3), 1062–1074. doi:10.1016/j.adhoc.2012.11.001

Zhao, L., & Liang, Q. (2005, March). Fuzzy deployment for wireless sensor networks. In *Computational Intelligence for Homeland Security and Personal Safety, 2005. CIHSPS 2005. Proceedings of the 2005 IEEE International Conference on* (pp. 79-83). IEEE.

Zhao, W., & Du, S. (2016). *Spectral–Spatial Feature Extraction for Hyperspectral Image Classification: A Dimension Reduction and Deep Learning Approach* (Vol. 54). IEEE Transactions on Geoscience and Remote Sensing.

Zhihai, Z., & Tao, S. (2008). *Research on fast computation of ambiguity function*. Sanya, Hainan, China: Congress on Image and Signal Processing. doi:10.1109/CISP.2008.458

Zhou, Y., & Wei, Y. (2016). *Learning Hierarchical Spectral–Spatial Features for Hyperspectral Image Classification* (Vol. 46). IEEE Transactions on Cybernetics.

Zou, L., Trestian, R., & Muntean, G. M. (2017). E3-DOAS: Balancing QoE and Energy-Saving for Multi-Device Adaptation in Future Mobile Wireless Video Delivery. *IEEE Transactions on Broadcasting*, 1–15.

Zou, L., Trestian, R., & Muntean, G.-M. (2013). DOAS: Device-Oriented Adaptive Multimedia Scheme for 3GPP LTE systems. In *2013 IEEE 24th International Symposium on Personal Indoor and Mobile Radio Communications (PIMRC)* (pp. 2180–2184). IEEE.

Zou, L., Trestian, R., & Muntean, G.-M. (2014). eDOAS: Energy-aware device-oriented adaptive multimedia scheme for Wi-Fi offload. In *2014 IEEE Wireless Communications and Networking Conference (WCNC)* (pp. 2916–2921). IEEE.

Zou, L., Trestian, R., & Muntean, G.-M. (2013). A utility-based priority scheduling scheme for multimedia delivery over LTE networks. In *2013 IEEE International Symposium on Broadband Multimedia Systems and Broadcasting (BMSB)* (pp. 1–7). IEEE. 10.1109/BMSB.2013.6621745

About the Contributors

Ioan-Sorin Comşa is a Research Scientist in 5G radio resource scheduling at Brunel University, London, UK. He was awarded the B.Sc. and M.Sc. degrees in Telecommunications from the Technical University of Cluj-Napoca, Romania in 2008 and 2010, respectively. He received his Ph.D. degree from the Institute for Research in Applicable Computing, University of Bedfordshire, UK in June 2015. He was also a PhD Researcher with the Institute of Complex Systems, University of Applied Sciences of Western Switzerland, Switzerland. Since 2015, he worked as a Research Engineer at CEA-LETI in Grenoble, France. He is the main author of a patent in machine learning domain with direct application in 5G radio resource scheduling. His research interests include intelligent radio resource and QoS management, reinforcement learning, distributed and parallel computing, adaptive multimedia/mulsemedia delivery.

Ramona Trestian received the B.Eng. degree in Telecommunications from the Electronics, Telecommunications and the Technology of Information Department, Technical University of Cluj-Napoca, Romania in 2007, and the Ph.D. degree from the School of Electronic Engineering, Dublin City University (DCU), Dublin, Ireland in 2012 for her research in adaptive multimedia systems and network selection mechanisms. She was with Dublin City University as an IBM/Irish Research Council Exascale Postdoctoral Researcher, from December 2011 to August 2013. She is currently a Senior Lecturer with the Design Engineering and Mathematics Department, Faculty of Science and Technology, Middlesex University, London, UK. She has published in prestigious international conferences and journals and has four edited books. Her research interests include mobile and wireless communications, multimedia streaming, user quality of experience, handover and network selection strategies, and software-defined networks. She is a reviewer for international journals and conferences and a member of the IEEE Young Professionals, IEEE Communications Society and IEEE Broadcast Technology Society.

* * *

Ekaterina Aleksandrova is an Undergraduate student at the University of Glasgow currently pursuing a degree in Computing Science. Her recent projects include online machine learning algorithms in the field of Internet of Things and Natural Language Processing.

Christos Anagnostopoulos is an Assistant Professor in the School of Computing Science at the University of Glasgow. His expertise is in the areas of network-centric intelligent & adaptive systems and in-network statistical predictive modelling in large-scale sensor/UxV/Edge networks. He has received funding for his research by the EC/H2020 and the industry. Dr Anagnostopoulos is coordinating (Principal Investigator) the projects: EU H2020/GNFUV and EU H2020 Marie Skłodowska-Curie (MSCA)/INNOVATE, and is a co-PI of the EU PRIMES and UK EPSRC CLDS. Dr Anagnostopoulos is an author of over 110 publications, with over 1700 citations in refereed scientific journals/conferences. He is an academic member of the Information, Data, and Analysis (IDA) and Networked Systems Research (NETLAB), and associate member of the Glasgow Systems Section (GLASS) at Glasgow. Dr Anagnostopoulos before joining Glasgow was appointed as an Assistant Professor at Ionian University and served as an Adjunct Assistant Professor at the University of Athens (Dept Informatics and Telecommunications) and University of Thessaly in the area of network-centric information systems. He has held postdoctoral research positions at University of Glasgow (UK/EPSRC) and University of Athens (EC-funded projects) in the areas of large-scale statistical learning & predictive analytics in distributed environments. He holds a BSc (Hons. and Valedictorian) in Informatics & Telecommunications, MSc (distinction) in Advanced Information Systems, and PhD in Computing Science, University of Athens. He is an associate fellow of the HEA, member of ACM and IEEE.

Mehmet Emin Aydin is Senior Lecturer in the Department of Computer Science and Creative Technology of the University of the West of England. He has received BSc, MA and PhD degrees from Istanbul Technical University, Istanbul University and Sakarya University, respectively. His research interests include parallel and distributed metaheuristics, resource planning, scheduling and optimization, combinatorial optimization, evolutionary computation and intelligent agents and multi agent systems. He has conducted guest editorial for a number of special issues of peer-reviewed international journals. In addition to being the member of advisory committees of many international conferences, he is editorial board member of various peer-reviewed international journals. He is currently a fellow of Higher Education Academy, UK, a member of EPSRC College, member of The OR Society UK, ACM and IEEE.

Chintan Bhatt (Ph.D.) is currently working as an Assistant Professor in Computer Engineering department, Chandubhai S. Patel Institute of Technology, CHARUSAT. He is a member of IEEE, EAI, ACM, CSI, AIRCC and IAENG (International Association of Engineers). His areas of interest include Internet of Things, Data Mining, Web Mining, Networking, Security Mobile Computing, Big Data and Software Engineering. He has more than 5 years of teaching experience and research experience, having good teaching and research interests. He has chaired a track in CSNT 2015 and ICTCS 2014. He has been working as Reviewer in Wireless Communications, IEEE (Impact Factor-6.524) and Internet of Things Journal, IEEE, Knowledge-Based Systems, Elsevier (Impact Factor-2.9) Applied Computing and Informatics, Elsevier and Mobile Networks and Applications, Springer. He has delivered an expert talk on Internet of Things at Broadcast Engineering Society Doordarshan, Ahmedabad on 30/09/2015. He has been awarded Faculty with Maximum Publication in CSIC Award and Paper Presenter Award at International Conference in CSI-2015, held at New Delhi.

Richard Demo Souza was born in Florianópolis-SC, Brazil. He received the B.Sc. and D.Sc. degrees in electrical engineering from the Federal University of Santa Catarina (UFSC), Brazil, in 1999 and 2003, respectively. In 2003, he was a Visiting Researcher with the Department of Electrical and Computer Engineering, University of Delaware, USA. From 2004 to 2016, he was with the Federal University of Technology–Paraná, Brazil. Since 2017, he has been with UFSC, Brazil, where he is an Associate

Professor. His research interests are in the areas of wireless communications and signal processing. He has served as an Associate Editor for the IEEE Communications Letters, EURASIP Journal on Wireless Communications and Networking, and the IEEE Transactions on Vehicular Technology. He was co-recipient of the 2014 IEEE/IFIP Wireless Days Conference Best Paper Award and the 2016 Research Award from the Cuban Academy of Sciences, the supervisor of the awarded Best Ph.D. Thesis in Electrical Engineering in Brazil in 2014. He is a Senior Member of the Brazilian Telecommunications Society.

Tolga Ensari is an assistant professor at Computer Engineering, Istanbul University, Turkey. His research area are machine learning, deep learning and pattern recognition. He has been visiting scholar between 2011-2012 in the Department of Electrical and Computer Engineering at University of Louisville, USA. He is member of IEEE. He was awarded as the international successful researcher by Istanbul University.

Gheorghiţă Ghinea is a Professor in the Computer Science Department at Brunel University, United Kingdom. He received the B.Sc. and B.Sc. (Hons) degrees in Computer Science and Mathematics, in 1993 and 1994, respectively, and the M.Sc. degree in Computer Science, in 1996, from the University of the Witwatersrand, Johannesburg, South Africa; he then received the Ph.D. degree in Computer Science from the University of Reading, United Kingdom, in 2000. His work focuses on building adaptable cross-layer end-to-end communication systems incorporating user perceptual requirements. He is a member of the IEEE and the British Computer Society.

Monalisa Ghosh is a research scholar at Indian Institute of Technology, India, Department of Electrical and Electronics Communication Engineering.

Melike Günay is a researcher and teaching assistant at Computer Engineering Department, Istanbul Kultur University, Turkey since 2012. She completed her M.Sc. degree at İstanbul University. Her research areas are artificial intelligence, machine learning, deep learning and bioinformatics.

Malka N. Halgamuge is a Research Fellow in the Department of Electrical and Electronic engineering, University of Melbourne. She received the PhD degree from the same department in 2007. Since then she has published in areas including wireless communication, life sciences and Data Science/ Big Data. She has published more than 90 peer-reviewed technical articles attracting over 872 Google Scholar Citations with h-index = 14 and her Research Gate RG Score is 32.42. She is an experienced researcher and educator with a demonstrated history of working with highly reputed research institutes all over the world on life sciences.

Syed Ali Hassan received his PhD in Electrical Engineering from Georgia Institute of Technology, Atlanta USA in 2011. He received his MS Mathematics from Georgia Tech in 2011 and MS Electrical Engineering from University of Stuttgart, Germany in 2007. He was awarded BE degree in Electrical Engineering from National University of Sciences and Technology (NUST), Pakistan, in 2004. His broader area of research is signal processing for communications with a focus on cooperative communications for wireless networks, stochastic modeling, estimation and detection theory, and smart grid communications.

Currently, he is working as an Assistant Professor at the School of Electrical Engineering and Computer Science (SEECS), NUST, where he is the director of Information Processing and Transmission (IPT) Lab, which focuses on various aspects of theoretical communications. He was a visiting professor at Georgia Tech in Fall 2017 and also holds senior membership of IEEE. He also held industry positions, in Cisco Systems Inc. CA, USA, and Center for Advanced Research in Engineering, Islamabad, Pakistan.

Muhammad Ali Imran received the M.Sc. (Distinction) and Ph.D. degrees from Imperial College London, U.K., in 2002 and 2007, respectively. He is the Vice Dean Glasgow College UESTC and a Professor of Communication Systems with the School of Engineering, University of Glasgow. He is an Affiliate Professor with the University of Oklahoma, USA, and a Visiting Professor with 5G Innovation Centre, University of Surrey, U.K. He has over 18 years of combined academic and industry experience, working primarily in the research areas of cellular communication systems. He has been awarded 15 patents, has authored/co-authored over 300 journal and conference publications, and has been a Principal/Co-Principal Investigator on over six million in sponsored research grants and contracts. He has supervised over 30 successful Ph.D. graduates. He was a recipient of the Award of Excellence in Recognition of his academic achievements, conferred by the President of Pakistan, the IEEE Comsocs Fred Ellersick Award 2014, the FEPS Learning and Teaching Award 2014, and the Sentinel of Science Award 2016. He was twice nominated for Tony Jeans Inspirational Teaching Award. He is a shortlisted finalist for the Wharton-QS Stars Awards 2014, QS Stars Reimagine Education Award 2016 for innovative teaching and VCs learning, and Teaching Award in University of Surrey. He is a Senior Fellow of Higher Education Academy, U.K.

Saad Iqbal is an Electrical Engineering graduate from National University of Sciences and Technology (NUST), Islamabad. Currently, he is a Research Assistant at Information Processing and Transmission (IPT) Lab SEECS-NUST.

Ramgopal Kashyap's areas of interest are image processing, pattern recognition, and machine learning. He has published many research papers, and book chapters in international journals and conferences like Springer, Inderscience, Elsevier, ACM, and IGI-Global indexed by Science Citation Index (SCI) and Scopus (Elsevier). He has Reviewed Research Papers in the Science Citation Index Expanded, Springer Journals and editorial board member and conferences programme committee member of the IEEE, Springer international conferences and journals held in countries: Czech Republic, Switzerland, UAE, Australia, Hungary, Poland, Taiwan, Denmark, India, USA, UK, Austria, Vietnam Dubai and Turkey. He has written many book chapters published by IGI Global, USA.

Pierre Kuonen obtained the Master degree in electrical engineering from the Ecole Polytechnique Federale de Lausanne (EPFL) in 1982. After six years of experience in the industry (oil exploration in Africa, then development of computer-assisted manufacturing tools in Geneva), he joined laboratory of computer science theory at EPFL in 1988 and began working in the field of parallel and distributed systems for high-performance computing. He obtained the PhD in computer science in 1993. Since then he has been working in this field. First at EPFL, where he founded and led the GRIP (Parallel Computing Research Group), then at the University of Applied Sciences of Western Switzerland (HES-SO) he joined in 2000. Since 2003, he has been a full professor at the University of Applied Sciences and Architecture in Fribourg, where he leads the GRID and Cloud Computing Group and co-leads the Institute of Complex Systems (iCoSys). In 2014, in collaboration with Prof. Jean Hennebert (HES-SO, Fribourg) and Prof. Philippe Cudre-Mauroux (University of Fribourg) he founded the DAPLAB (Data Analysis and Processing Lab) that aims at facilitating the access for enterprises and universities to emerging technologies in the area of big data and intelligent data analysis.

Melody Moh obtained her MS and Ph.D., both in computer science, from Univ. of California - Davis. She joined San Jose State University in 1993, and has been a Professor since Aug 2003. Her research interests include cloud computing, mobile wireless networking, security/privacy for cloud and network systems, and machine learning applications. She has received over 500K dollars of research grants from both NSF and industry, has published over 140 refereed papers in international journals, conferences and as book chapters, and has consulted for various companies.

Yağız Nalçakan is a research assistant at Computer Engineering, Altinbas University, Istanbul, Turkey. He is at M.Sc. program at Istanbul University. His work interests are artificial intelligence, deep learning, machine learning, abductive logic programming (ALP) and data visualization.

Oluwakayode Onireti received the B.Eng. (Hons.) degree in electrical engineering from the University of Ilorin, Ilorin, Nigeria, in 2005, the M.Sc. (Hons.) degree in mobile and satellite communications, and the Ph.D. degree in electronics engineering from the University of Surrey, Guildford, U.K., in 2009 and 2012, respectively. From 2013 to 2016, he was a Research Fellow with ICS/5GIC, University of Surrey. He is currently a Research Associate with the School of Engineering, University of Glasgow. His main research interests include self-organizing cellular networks, energy efficiency, multiple-input multiple-output, and cooperative communications. He has been actively involved in projects such as ROCKET, EARTH, Greencom, QSON, and Energy proportional EnodeB for LTE-Advanced and Beyond. He is currently involved in the DARE project, a ESPRC funded project on distributed autonomous and resilient emergency management systems.

Madhushi P. Ranasinghe received her BEng (Hons) degree in Electronic Engineering at Sri Lanka Institute of Information Technology affiliated with Sheffield Hallam University, United Kingdom in 2014, Graduate Diploma in Electronics, Communications and Computer Engineering at Institute of Engineers Sri Lanka College of Engineering, in 2016 and Master of Information Technology degree from Charles Sturt University, Melbourne, in 2018. Her main research interests include evaluation of challenges in cognitive radio networks based on past research work and utilization of artificial neural networks for pattern recognition in commercial applications.

Neha Sharma is working as Assistant Professor in Manipal University Jaipur. Currently pursuing PhD in Wireless Communications.

Paulo Valente Klaine received the B.Eng. (with Distinction) degree in electrical and electronic engineering from the Federal University of Technology–Paraná, Brazil, in 2014 and the M.Sc. (with Distinction) degree in mobile communications systems from the University of Surrey, Guildford, U.K., in 2015. He is currently pursuing the Ph.D. degree with the School of Engineering, University of Glasgow. He was with the 5G Innovation Centre, University of Surrey in 2016. His main interests include self organizing cellular networks and the application of machine learning algorithms in wireless networks.

Narendra Singh Yadav is an associate professor in the department of Information Technology, Manipal University Jaipur. His research interests include Clustering, Routing and Security in ad hoc wireless networks, wireless sensor networks, digital forensics, malware analysis and reverse engineering.

Eyyüp Yıldız is a research assistant at Computer Engineering, Erzincan Binali Yıldırım University, Turkey. He is at M.Sc. program at Istanbul University. His work interests are machine learning, flow cytometry, mixture models, clustering and bioinfomatics.

Sijing Zhang obtained his B.Sc. and M.Sc. degrees, both in Computer Science, from Jilin University, Changchun, China in 1982 and 1988, respectively. He earned a PhD degree in Computer Science from the University of York, UK in 1996. He then joined the Network Technology Research Centre (NTRC) of Nanyang Technological University, Singapore as a post-doctoral fellow. In 1998, he returned to the UK to work as a research fellow with the Centre for Communications Systems Research (CCSR) of the University of Cambridge. He joined the School of Computing and Technology, University of Derby, UK, as a senior lecturer in 2000. Since October 2004, he has been working as a senior lecturer in the Department of Computer Science and Technology, University of Bedfordshire, UK. His research interests include wireless networking, schedulability tests for hard real-time traffic, performance analysis and evaluation of real-time communication protocols, QoS provision, vehicular ad-hoc networks, and wireless networks for real-time industrial applications.

Index

L

M

N

O

P

Q

R

S

U

V

W

www.ingramcontent.com/pod-product-compliance
Lightning Source LLC
LaVergne TN
LVHW062142200926
848484LV00007B/1843